Transportation Engineering: Theory and Practice

Transportation Engineering: Theory and Practice

Editor: Jay Rhodes

MURPHY & MOORE
www.murphy-moorepublishing.com

Published by Murphy & Moore Publishing
1 Rockefeller Plaza,
New York City, NY 10020, USA
www.murphy-moorepublishing.com

Transportation Engineering: Theory and Practice
Edited by Jay Rhodes

International Standard Book Number: 978-1-63987-540-5 (Hardback)

Cataloging-in-Publication Data

Transportation engineering : theory and practice / edited by Jay Rhodes.
p. cm.
Includes bibliographical references and index.
ISBN 978-1-63987-540-5
1. Transportation engineering. 2. Highway engineering. I. Rhodes, Jay.
TA1145 .T73 2022
629.04--dc23

Contents

Preface

Transportation engineering is a field of engineering which applies the principles of science and technology to the planning, design, operation and management of the different modes of transportation. This includes safe, rapid, efficient, economical and environmentally compatible transit of goods and passengers. It encompasses air, water, railroad and highway transportation. Although planning and design are central to transportation engineering, the areas of network analysis, operations planning, logistics, financing and policy analysis are also relevant to highway and urban transportation. Modern technologies such as advanced traveler information systems, intelligent transportation systems and vehicle infrastructure integration systems are integrated for the smooth movement of vehicles on roads or tracks. This book discusses the fundamentals as well as modern approaches of transportation engineering. It traces the progress of this field and highlights some of its key concepts and applications. It is a collective contribution of a renowned group of international experts.

The information shared in this book is based on empirical researches made by veterans in this field of study. The elaborative information provided in this book will help the readers further their scope of knowledge leading to advancements in this field.

Finally, I would like to thank my fellow researchers who gave constructive feedback and my family members who supported me at every step of my research.

Editor

1

Highway Design and Safety Consequences: A Case Study of Interstate Highway Vertical Grades

Zongxin Tang,[1] Sikai Chen,[2] Jianchuan Cheng,[1] Seyed Ali Ghahari,[2] and Samuel Labi[2]

[1]*School of Transportation, Southeast University, 2 Sipailou, Nanjing, Jiangsu 210096, China*
[2]*Lyles School of Civil Engineering, Purdue University, 550 Stadium Mall Dr., West Lafayette, IN 47907, USA*

Correspondence should be addressed to Jianchuan Cheng; jccheng@seu.edu.cn

Academic Editor: Zhi-Chun Li

Vertical alignment, which includes vertical grades and lengths, is a critical aspect of highway design policy that influences safety. A full understanding of the effect of vertical grade and segment length on highway safety can help agencies to evaluate or adjust their design policies regarding vertical alignment design features (grade and length). For this reason, it is useful to assess the current relationships between design policy and safety performance. To address this task, this paper uses data from interstate segments to first establish the relationship between these design features and safety. Safety is expressed in terms of the three different levels of crash severity (fatal, injury, and property damage only). In its analysis, the paper departs from the traditional univariate models (where each crash severity is modeled separately) and instead uses a seemingly unrelated negative binomial (SUNB) technique, a multivariate model that duly accounts for the unobserved shared effects between the different levels of crash severity. In addition, the paper's models duly recognize and account for the holistic nature of the grade and tangent length effects: the effect of the sum (interaction) of the vertical grade and length is different from the sum of their individual effects. The paper investigates the relationships for rural and urban interstate highway segments. Against the background of the developed models, the paper evaluates current design policies (specifications on vertical alignment grade and length) for similar classes of highways at a number of countries and presents a set of nomograms that feature lines representing points of equal safety performance. These charts can be used by the highway agencies to evaluate and compare their current or possible future highway design policies.

1. Introduction

The vertical alignment of a highway segment, which is guided by geometric design policy, is the end result of an evaluation of the benefits and costs associated with alternative route locations with due consideration given to existing terrain, safety, and construction haulage and cost. Grades or straight vertical segments are designed to be steep enough to allow for longitudinal drainage, but not so steep as to pose a danger to vehicles through inadvertent excessive speed at downhill segments (and, conversely, difficulty of climbing steep uphill segments that pose safety risks where there are inadequate passing opportunities).

Therefore, highway vertical alignment design involves a specification of the vertical grade and the length over which such grade occurs. Evidence from the literature suggests that vehicles with relatively high weight-to-power ratio tend to lose speed as they travel uphill at such tangents. Recent studies have assessed the extent to which the operating speed of trucks on highways is sensitive to vertical grade: speeds significantly decrease at upgrade segments and increase at downgrade segments [1–4]. Due to such speed reduction, there can be significant heterogeneity of operating speeds on the vertical grade segment, with resulting adverse effects on safety. Recognizing that the highway vertical alignment plays a pivotal role in highway safety, the design manuals of national highway agencies and organizations, including the AASHTO Green Book-A Policy on Geometric Design of Highways and Streets [5] in the United States, specify that vertical tangent grade and length must be kept within certain limits to promote safety.

Empirical studies that have investigated the relationship between highway vertical grade and safety have been rather sparse compared to the general body of highway safety literature. St. John and Kobett [6] investigated the safety effects of long, steep grades on rural two-lane highways using traffic speed distributions generated with computer simulation and found that where trucks constitute a fifth of the traffic stream, the crash involvement rate increases from 175% of passenger car rates for 4 percent grades, to 250% of passenger cars for 8% grades. They also found that at steep downgrades (greater than 4%) trucks tend to use crawl speeds to maintain control which tends to increase crash rates. At crest curves, steep grades cause inadequate sight distance that could impair safety. Over the past two decades, a number of researchers have established explicit relationships between highway vertical grade and safety, including Easa [7, 8], Hassan and Easa [9], and Labi [10, 11]. Other noteworthy contributors include Hassan et al. [12–14] who discussed the importance of sight distance and operating speed along with combined horizontal and vertical alignments at highway cut and fill sections. Hassan [15] showed that segment grade and length play key roles in the stopping distance, which in turn affects safety. It is not surprising, therefore, that posting a speed indicator at the crest of vertical curves of rural highway segments has been found to have safety benefits [16]. Miaou [17] and Daniel and Chien [18] determined that vertical grades exceeding 2% could cause increased crash rates and recommended that grades be kept as low as 0.005 per 100ft. In addition, a horizontal curve appears to be flatter where it overlaps with a sag vertical curve and sharper where it overlaps with a crest vertical curve; such distortion could impair the ability for drivers to maintain appropriate speeds at such sections, thereby possibly causing unsafe driving conditions [19]. It has been suggested that, in order to minimize the erroneous perception of drivers, the sight distance and the horizontal curve radius should be decreased, and the length of vertical curve per 1% change in grade should be increased [20]. Crash rates on curves with similar tangents are reported to be 1.5 to 4 times higher, compared to other types of curves [21]. This suggests that one way of reducing crash fatalities and injuries is to improve the road alignment consistency.

With regard to modeling technique for highway crash analyses, the literature is replete with a variety of model types and forms. Abbas [22] evaluated traffic safety conditions using a wide variety of functional forms. Hanley and Forkenbrock [23] applied Monte Carlo techniques to analyze two-lane highway safety regarding passing maneuvers. Elvik et al. [24] used power functions to investigate the relationship between operating speed and road safety. Shively et al. [25] used a semiparametric Poisson-gamma model with a Bayesian nonparametric estimation procedure to investigate the number of crashes based on geometric features and operating conditions. Hosseinpour et al. [26], using zero-inflated negative binomial (NB) and Poisson, hurdle NB and Poisson, standard and random-effect NB, and Poisson models, identified terrain type (e.g., flat and rolling terrain) as an influential factor of head-on severe crashes. Mohammadi et al. [27] used a longitudinal NB model that includes autoregressive correlation arrangement to model crash frequency. Bauer and Harwood [28] investigated the effect of combined longitudinal grade and horizontal curve on road safety and generated crash modification factors (CMFs) associated with safety performance at tangent sections. Yan et al. [29] developed a highway alignment comprehensive index (ACI), which includes the grade, to predict highway operating speed which can be used to evaluate highway safety performance. The length of vertical grade, which has been found by Wang et al. [4] to be influential to safety, was not considered in the Yan et al. [29] research. Zeng et al. [30] and Zeng et al. [31] proved that higher grade increased the crash risk on highway segments. Zou and Yue [32] analyzed causation of road accidents by a Bayesian Network approach. Factors, including road geometry and road surface condition, were proved to be influential on road accident frequency.

Crash counts across severity levels (e.g., property damage only, injury, and fatal) are correlated in nature. Such correlations emerge from a variety of sources including data on traffic collisions that involve multiple occupant injuries from the same crash. In addition, unobserved factors (e.g., vehicle speed and weather condition at the time of crash) may influence multiple crash counts of different severity levels simultaneously for each highway segment. Therefore, modeling crashes of each severity level separately and not accounting for these correlations will result in less precise estimations of crash factors associated with different crash severity levels ([33]; Anastasopoulos and Mannering, 2016; [34–36]). Winkelmann [37] introduced a seemingly unrelated negative binomial (SUNB) framework to account for these correlations among count data. It also overcomes drawbacks of other regression models, such as seemingly unrelated Poisson (SUP) model, which are unable to account for overdispersion (a widespread phenomenon in count data analysis). AASHTO's policy on geometric design of highways and streets (also referred to as the "Green Book" [5]) suggests that vertical grade and segment length are influential factors of crashes. The Federal Highway Administration's Highway Safety Manual (HSM) and Porter et al. [38] presented methods for predicting the mean crash frequency under specific conditions of highway operations, geometry, and terrain. The literature lacks studies that specifically investigated the simultaneous effect of vertical grade and segment length on crash frequency and the variation of the nature of this relationship across rural and urban highways. In addition, past work on this topic stopped short of evaluating current geometric design practices vis-à-vis the model outcomes or making any recommendations to guide the former.

In attempting to throw light on this important aspect of highway design policy, this paper assesses the safety impacts of highway vertical alignment by first establishing the relationship between vertical alignment parameters (grade and length) on the safety experience at these highway classes. Another objective is to incorporate the holistic nature of the grade and tangent length effects: the effect of the sum (interaction) of the vertical grade and length is different from the sum of their individual effects. Finally, the paper sets out to evaluate current design policies (regarding vertical alignment grade and length specifications) against the background of the developed models and presents a set of nomograms that feature lines representing points of equal safety performance. These charts can be used by the highway agencies to evaluate and compare their current or future highway design policies. The paper's scope covers highway types of interstate quality or similar class. The analysis was carried out separately for rural and urban segments.

2. Methodology

2.1. Estimate SUNB Models. The paper expresses the safety experience in terms of the three different levels of crash severity (fatal, injury, and property damage only). The paper departs from the traditional univariate models (where each crash severity is modeled separately) and instead uses multivariate models that estimates the models for the different severities simultaneously. In doing so, the paper duly accounts for unobserved yet shared influences across the factors that influence the different levels of crash severity. SUNB can be considered a specific case of multivariate models. The general SUNB framework, which was first introduced by Winkelmann ([37]; 2008), can be tailored to the current problem context as follows.

Let z_i represent a J-dimensional vector which counts each element which is independently Poisson distributed with parameter λ_{ij}, where z_i is a vector of counts (the number) of accidents in the ith level of accident severity, on j^{th} highway segment.

$$z_i = (z_{i1}, z_{i2} \dots z_{iJ})'. \quad (1)$$

i is fatal, injury, or PDO; and $j = 1, 2, \dots, J$ is an index referring to a highway segment.

The expected number of accidents of the ith severity level in j^{th} road segment is given by

$$\lambda_{ij} = e^{x_{ij}'\beta_j}. \quad (2)$$

x_{ij} is a vector of independent variables associated with the geometric design (including vertical grade and segment length), traffic volume, and pavement condition.

$\beta_j = [\beta_0, \beta_{vertical\ grade}, \beta_{length}, \beta_{interaction}, \dots]'$ is a vector of coefficients for the corresponding independent variables.

This paper focuses on the effects of vertical grade ($\beta_{\text{vertical grade}}$), segment length β_{length}, and the interaction $\beta_{interaction}$ between these two factors. Therefore, these coefficients are mentioned specifically in the coefficient vector. To account for the presence of an error term in the model, let $z_{ij} \mid \upsilon_{ij}$ a be Poisson distributed random variable with parameter $\lambda_{ij}\upsilon_{ij}$. Then (2) becomes

$$\lambda_{ij}\upsilon_{ij} = e^{x_{ij}'\beta_j + \varepsilon_{ij}}. \quad (3)$$

υ_{ij} is gamma distributed with mean 1 and variance α^{-1}.

The marginal distribution of z_{ij} after integration over υ_{ij} is negative binomial with mean $\text{E}(z_{ij}) = \lambda_{ij}$ and variance $\text{Var}(z_{ij}) = \lambda_{ij} + \alpha^{-1}\lambda_{ij}^{2}$.

In order to develop the probability generating function for the negative binomial distribution, the parameter α is parameterized as follows:

$$\alpha = \frac{\lambda_{ij}}{\sigma}, \quad (4)$$

where symbols have meanings as explained above.

The parameterization will not affect the mean; however, the variance $\text{Var}(z_{ij}) = \lambda_{ij}(1+\sigma)$ becomes a linear function of the mean. After the parameterization, the negative binomial distribution has the following probability generating function:

$$\begin{aligned} P_z(s) &= [1 + \sigma(1 - s)]^{-\lambda_{ij}/\sigma} \\ P_u(s) &= [1 + \sigma(1 - s)]^{-\gamma/\sigma} \\ P_y(s) &= [1 + \sigma(1 - s)]^{-(\lambda_{ij}+\gamma)/\sigma} \end{aligned} \quad (5)$$

where u_i represents a scalar random variable with negative binomial distribution and mean γ. $y_{ij} = z_{ij} + u_i$, $s_i = \min(y_{i1}, y_{i2} \cdots y_{iJ})$.

The covariances between y_{ij} and y_{ik} are given by $\text{Cov}(y_{ij}, y_{ik}) = \gamma(1 + \sigma)$.

The framework allows for overdispersion on the condition that $\sigma > 0$. The joint probability function of the multivariate negative binomial framework is expressed as follows:

$$f_{NB}(z_{ij}) = \frac{\Gamma(\lambda_{ij}/\sigma + z_{ij})}{\Gamma(\lambda_{ij}/\sigma)\Gamma(z_{ij}+1)}\left(\frac{1}{1+\sigma}\right)^{\lambda_{ij}/\sigma}\left(\frac{\sigma}{1+\sigma}\right)^{z_{ij}} \quad (6)$$

where f_{NB} is the negative binomial probability function. Winkelmann ([37]; 2008) provides further details on the framework.

2.2. Nomogram Development. A nomogram is a diagram representing the relationship between at least three variable quantities. This displays two independent variables and shows how they interact to affect a third, dependent variable or outcome. Typical forms include a three-dimensional plot and a contour map. When the dependent variable is placed into distinct levels that have finite boundaries, nomogram can be used to easily compare the effect (on the dependent variable) of different combinations of the independent variables. An example of a contour nomogram is the isopleth (a line on a map connecting points having equal incidence of a specified meteorological feature). Similarly, in this paper, we introduce the term "isodehn" to represent a line drawn on a map through all points having the same safety performance as a function of two variables.

In this paper, after estimating the SUNB models for a certain crash severity level and a highway functional class, isodehns were plotted on a 2-dimensional figure (the dimensions being the segment length and vertical alignment grade) with isodehns along with the current design standards of different countries. In order to provide meaningful and representative information to the practice, we defined three levels of traffic volume based on the quartile values from our data: low traffic volume (corresponds to 25th percentile value), medium traffic volume (corresponds to 50th percentile value), and high traffic volume (corresponds to 75th percentile value). Similarly, the safety level of each isodehn is determined based on the percentiles (e.g., 10%, 20%, 30%) of

TABLE 1: Descriptive statistics of variables.

	Mean	Std Dev	Minimum	Maximum
Rural Interstate				
Total number of fatal crashes on each segment	0.241	0.501	0	3
Total number of injury crashes on each segment	4.265	3.785	0	19
Total number of PDO crashes on each segment	18.030	15.206	0	79
Segment length (miles)	5.417	2.824	0.560	6.94
Average annual daily traffic (10,000 vehicles)	3.271	1.600	1.181	12.865
Lane width (ft.)	12.030	0.784	8.750	15
Right shoulder width (ft.)	10.335	0.981	2.800	12
Pavement roughness (IRI in in/mile)	80.553	29.482	53	219
Vertical curve grade (%)	1.639	1.950	0	4.084
Road segments with left shoulder or median	100%		0	1
Left shoulder width (ft.)	4.562	1.765	3	11
Median width (ft.)	61.236	14.308	2	60
Urban Interstate				
Total number of fatal crashes on each segment	0.650	1.336	0	7
Total number of injury crashes on each segment	30	60.224	0	228
Total number of PDO crashes on each segment	121.688	261.035	0	953
Segment length (miles)	5.372	2.868	0.400	6.44
Average annual daily traffic (10,000 vehicles)	5.275	3.298	1.140	12.385
Lane width (ft.)	11.986	0.561	10	14.460
Right shoulder width (ft.)	9.953	0.723	6	11
Pavement roughness (IRI in in/mile)	77.329	21.613	48	147
Vertical curve grade (%)	1.540	1.883	0	3.682
Road segments with left shoulder or median	97.50%		0	1
Left shoulder width (ft.)	5.965	4.182	0	15
Median width (ft.)	45.241	21.637	0	60

the crash count for a certain crash severity level on specific road class. By this definition, the area between any two consecutive isodehns is statistically considered to be equally safe (the same level of expected crashes). That is, any combinations (points) of the segment length and vertical curve grade, of which the points locate in the same area, statistically have the same impact in terms of the number of crashes.

In this paper, we develop the isodehn plots for injury and PDO crashes only but not for fatal crashes, because the number of expected fatal crashes is too small to show meaningful comparisons. Nevertheless, in estimating the models, fatal crashes are included in the presented SUNB framework because their occurrences share some of the unobserved effects (including weather condition, speed limit, driver age, and conditions) with other crash severity levels. Therefore, inclusion of the fatal crash model in the system of models helps to improve the estimation performance of the injury and PDO models.

After developing the isodehns, the current design standards/recommendations of different countries are plotted on the same figures. By comparing the combination of segment length and vertical curve grade from the existing design standards/recommendations with the isodehns, the existing practice can be evaluated.

3. Description of the Data

A database of 418 highway segments was used to investigate the safety impacts of highway vertical alignment in this paper. This study database was developed by merging two independent datasets: a road safety dataset and a road inventory dataset. The combined database includes information regarding historical data on vehicle crashes at highways in the state of Indiana over a three-year period. In order to investigate the differences of safety impacts across rural and urban locations, the dataset was further divided into two subdatasets: rural interstate (317 highway segments) and urban interstate (101 highway segments). The vehicle crashes were placed into three levels of crash severity: fatal, injury, and property damage only (PDO). Only segments that did not contain any sag or crest curves were used for the analysis, and the grade along the segment was recorded. In most cases, the grade was uniform but, in a few cases, a segment had one or two grades and the average was recorded. Table 1 presents the summary statistics of the key variables.

4. Model Estimation Results and Interpretation

The estimation results of the SUNB models are presented in Tables 2(a) and 2(b) for rural interstate and urban interstate, respectively. The presented variables were found to be statistically significant at a level of confidence of 10%. The signs of the estimated coefficients were found to be intuitive from an engineering standpoint. Based on the mathematical form of the SUNB model, the coefficients are shown as exponents.

TABLE 2

(a) Model estimation results for Rural Interstates

Variables	Fatal		Injury		PDO	
	coefficient	*t* value	coefficient	*t* value	coefficient	*t* value
Constant	−12.8386	−4.30	−22.7048	−14.24	−19.8081	−14.11
Segment length (miles)	0.2146	3.98	0.1548	6.24	0.1540	7.05
Natural log of annual average daily traffic	1.4462	5.13	0.8843	6.48	0.8234	6.85
Average vertical grade (%)			0.0963	3.27	0.0636	2.46
Left shoulder width (ft.)	−1.1704	−10.81	−0.6364	−18.50	−0.7372	−24.38
Median width (ft.)			−0.0123	−2.91	−0.0131	−3.52
Number of lanes			3.4596	15.66	3.4045	17.53
Number of observations = 317; *McFadden pseudo* ρ^2 = 0.203						

(b) Model estimation results for Urban Interstates

Variables	Fatal		Injury		PDO	
	coefficient	*t* value	coefficient	*t* value	coefficient	*t* value
Constant	−5.1638	−8.25	−28.7188	−6.34	−24.2710	−5.77
Segment length (miles)	0.2327	9.16				
Natural logarithm of annual average daily traffic			2.8430	8.19	2.2157	8.28
Lane width (ft.)			−0.9093	−3.65	−0.6286	−3.06
Right shoulder width (ft.)			1.2243	5.01	1.1272	6.00
IRI (in/mile)	0.02317	3.25				
Average vertical grade (%)	0.5034	5.12	0.2086	1.96	0.1852	1.86
Left shoulder width (ft.)	−0.2087	−5.01	−0.08187	−1.99	−0.09313	−2.38
Number of lanes	−0.2480	−2.45				
Average vertical curve grade ∗ Lane width (Interaction term)			0.02367	3.19	0.02615	3.72
Number of observations = 101; *McFadden pseudo* ρ^2 = 0.167						

Figures 1–6 present the expected crash frequency of a given crash type across the rural and urban interstate highways and the sensitivity (using marginal effects) of the expected crash frequency to the average grade of the vertical alignment. The marginal effect reflects the impact of a one-unit change in an independent variable on the dependent variable (i.e., the expected frequency of fatal, injury, or PDO crashes). For crash severity level k, the marginal effect of the mth independent variable is calculated as

$$\mathrm{ME}_{X_{mk}}^{\lambda_{mk}} = \frac{\partial \lambda_k}{\partial X_{mk}} = \beta_{mk} EXP\left(X_k' \beta_k\right) \tag{7}$$

where λ_k is the estimated model for crash severity level k; X_{mk} is the mth independent variable; β_{mk} is the corresponding estimated coefficient; X_k' and β_k are the vectors of independent variables and the corresponding estimated coefficients, respectively.

The expected crash frequencies (the vertical axis in 3-dimensional space) are presented in Figures 1, 2, 3, 4, 5, and 6. The interaction between vertical grade and length is presented in Figures 2, 3, 4, 5, and 6. If there is a significant interaction term between segment length and vertical curve grade in the estimated model, the expected crashes will be further elevated as observed in the spike in Figures 5 and 6 (in Figure 4, the "spike-like shape" is due to the large coefficients of the average vertical grade and the segment length variables). Figures 1–3 do not exhibit any spike. The marginal effects (y-axis) and the vertical curve grade (x-axis) are shown in the figures across different road segment lengths. Where there is no interaction between the vertical curve grade and segment length, the lines that represent marginal effects are horizontal. On the contrary, where such interaction exists, the marginal effects are not linear but curves.

The figures suggest that, compared to their rural counterparts, urban interstate highways generally have higher numbers of expected crashes, and the crashes are more sensitive to changes in the vertical alignment grade (in the marginal effect plots, the change in crashes increases at a faster rate as the grade increases). At the urban interstate highways with relatively steep grades, PDO and injury crashes are very sensitive to the grade length. In other words, at such segments, a small increase in length will lead to a significant increase in the expected PDO and injury crashes. In the case of rural interstate highways, the interaction between the segment length and vertical curve grade is not significant for fatal crashes. However, the interaction is considerable for the injury and PDO crashes (albeit much lower compared to urban interstate highways). Overall, the figures suggest that where there is interaction between the vertical alignment grade and length, the crashes are more sensitive to longer segments at steep segments compared to those at relatively flat segment. In addition, with regard to urban interstate highways with small grades and short lengths, fatal crash

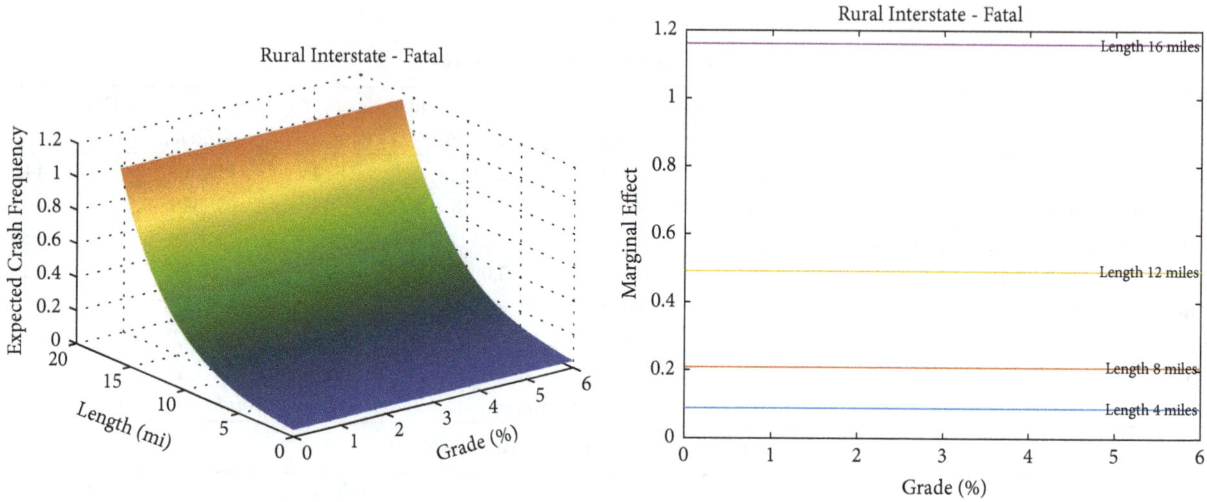

(a) Expected frequency of crashes

(b) Marginal effect (crash sensitivity to grade)

FIGURE 1: Rural interstate fatal crashes.

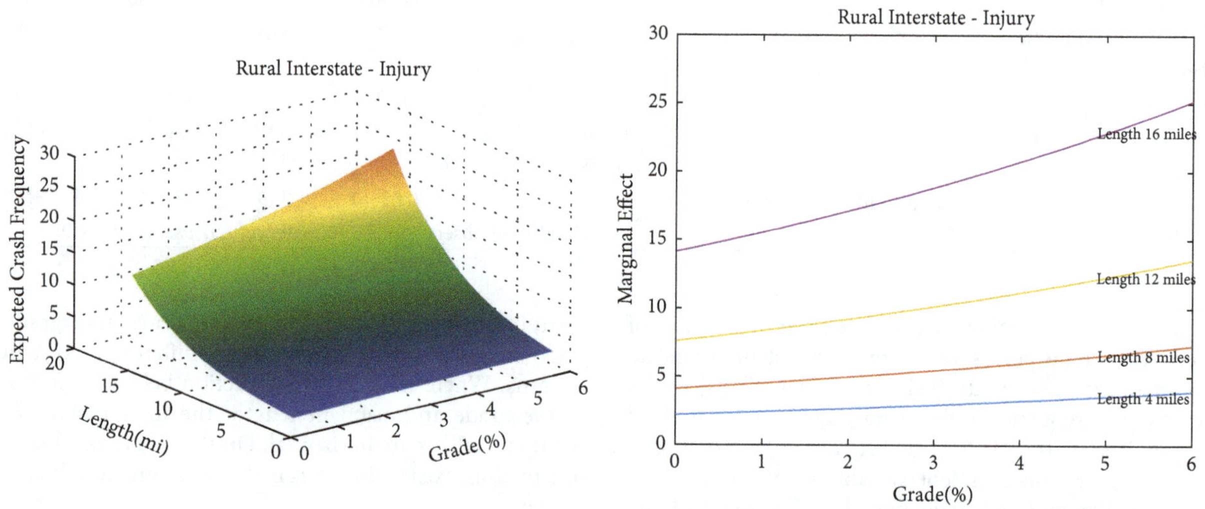

(a) Expected frequency of crashes

(b) Marginal effect (crash sensitivity to grade)

FIGURE 2: Rural interstate injury crashes.

frequency is found to be less sensitive to changes in the grade and segment length (as shown in the light blue area); however, it is considerably sensitive where the grade and length are relatively high.

5. Evaluation of Current Design Policy Across Different Countries

The traffic operating conditions at highways differ considerably from country to country, and therefore any comparison of design standards across countries must be done with circumspection. In this paper, we evaluate the current practice based on design standards, using a set of figures. Figures 7–10 present nomograms that indicate the relationship between the average vertical alignment grade and length, on one hand, and the expected number of crashes on the other hand. These relationships are represented by isodehns on the nomogram. In these figures, areas with lighter color represent a higher safety level of design (with respect to vertical alignment grade and length) compared to those with darker color.

For purposes of illustration and to facilitate a representative and realistic evaluation, we define three levels of traffic volume based on the quartile values from the dataset for rural interstate and urban interstate highways: low traffic volume (corresponds to 25th percentile value, AADT = 29,000), medium traffic volume (corresponds to 50th percentile value, AADT = 37,000), and high traffic volume (corresponds to 75th percentile value, AADT = 74,000). Nomograms for the expected injury and PDO crashes were plotted separately but were not plotted for fatal crashes because the number of

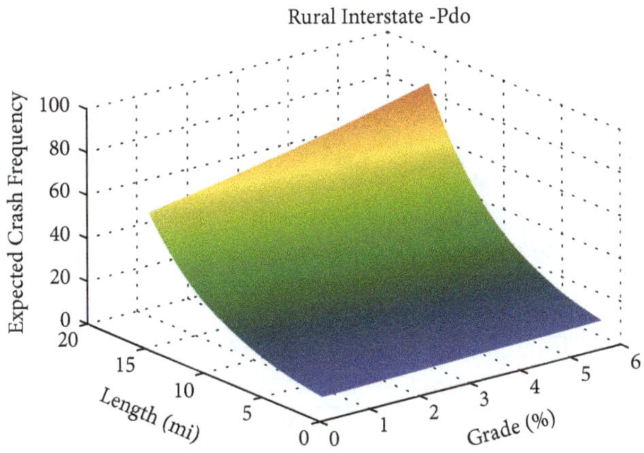

(a) Expected frequency of crashes

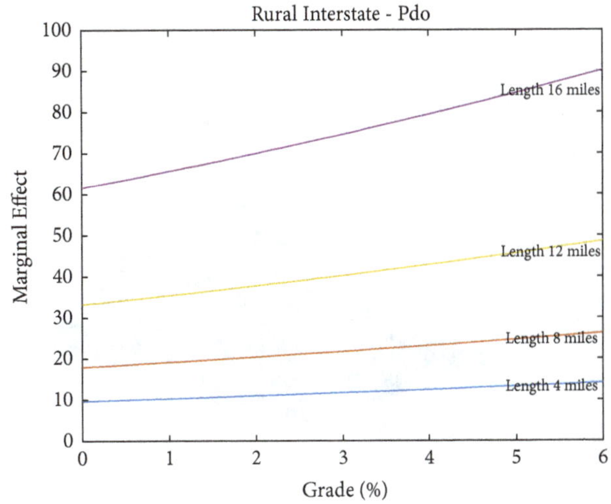

(b) Marginal effect (crash sensitivity to grade)

FIGURE 3: Rural interstate PDO crashes.

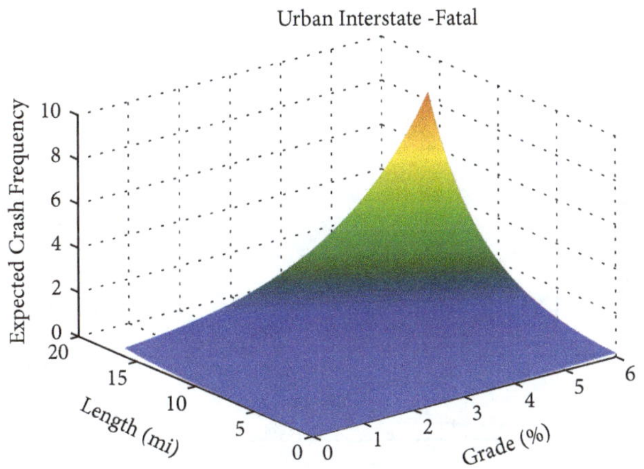

(a) Expected frequency of crashes

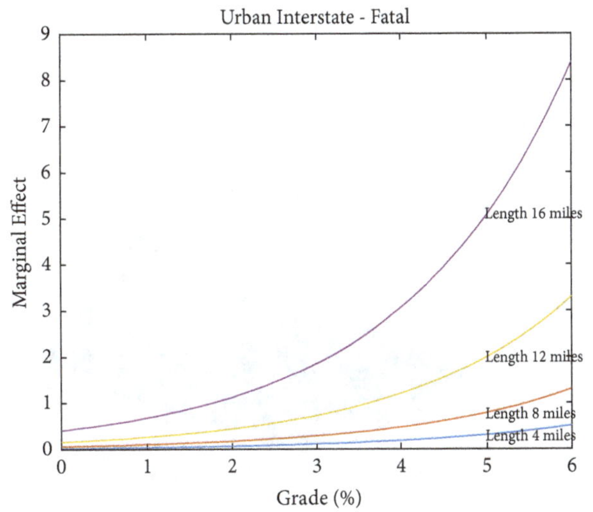

(b) Marginal effect (crash sensitivity to grade)

FIGURE 4: Urban interstate fatal crashes.

fatal crashes is low. Nevertheless, in the modeling process, fatal crashes are included in the presented SUNB framework during the model estimation because unobserved effects (of fatal crashes) associated with the explanatory variables, such as weather condition and speed limit, are shared with those of injury and PDO crashes.

For illustration purposes, the isodehns were plotted and compared with design standards/recommendations at USA, China, Japan, and Iran [5, 39–42]. Again, it is worth noting that the models presented in this paper were based on data collected in the USA, and, therefore, the model results (i.e., the relationship found between the number of crashes, segment length, and grade) are not necessarily transferable to other countries. This is because drivers in different countries have significant different driving behavior or culture, which was not captured in this paper. Nevertheless, it is expected that other countries may obtain useful insights from the results presented. In addition, the study methodology can be replicated by applying the proposed framework in this paper using data from the other countries, to evaluate their highway design. To make the comparisons across different countries more meaningful, the standards shown from other countries pertain to the highest class of roads in those countries: freeways, expressways, and national highways or motorways. Sinha et al. [43] showed that the operating conditions and geometric design policies (such as the design speed, lane and shoulder widths, horizontal curves, access control, and sight distances) for the highest class of highways in many countries share some similarity.

In order to compare design values of vertical curve grade and segment length across these countries, the recommended design values were chosen for a given design

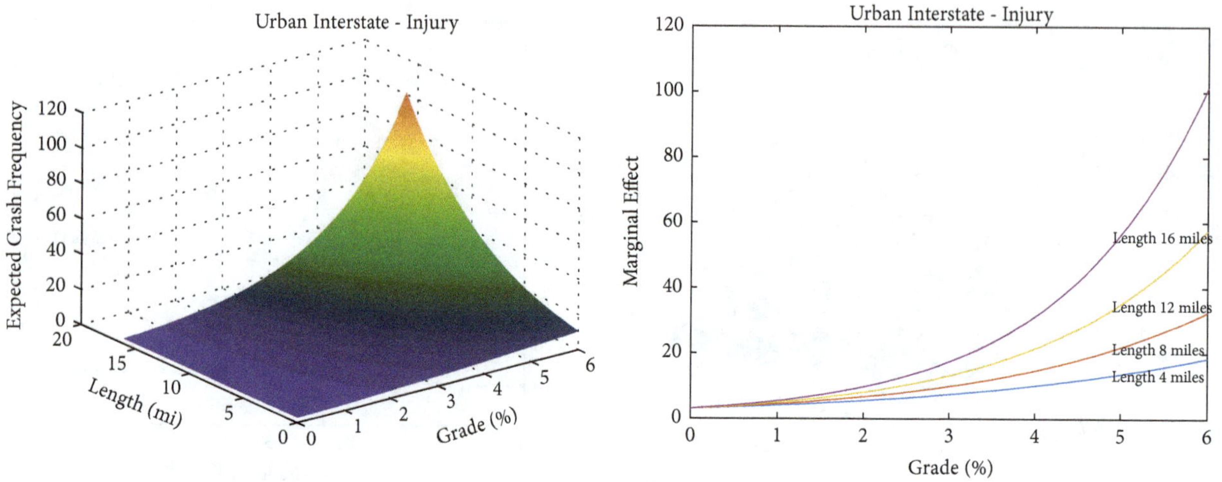

(a) Expected frequency of crashes

(b) Marginal effect (crash sensitivity to grade)

FIGURE 5: Urban interstate injury crashes.

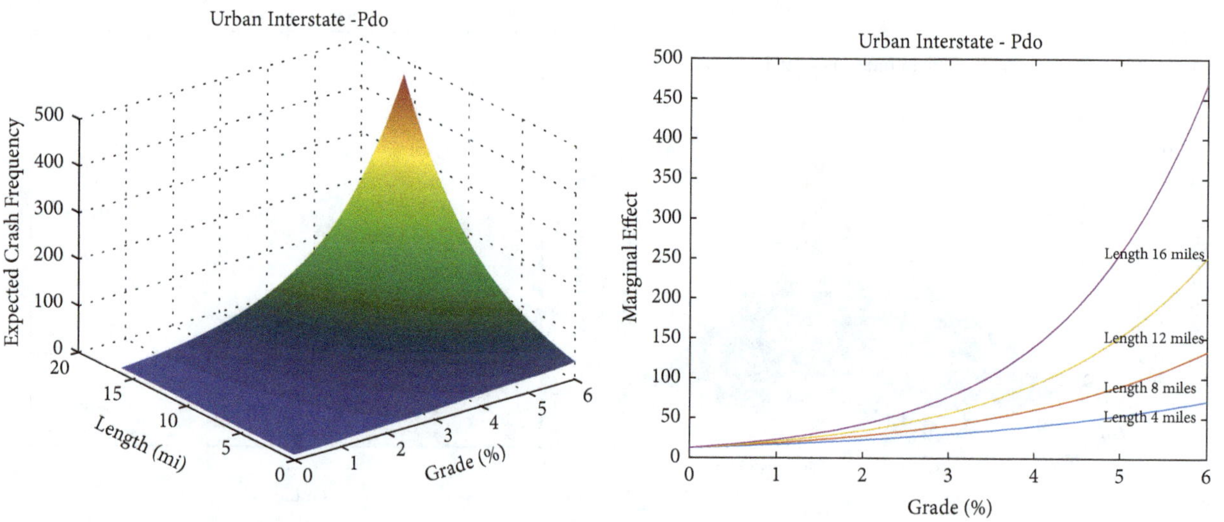

(b) Marginal effect (crash sensitivity to grade)

(a) Expected frequency of crashes

FIGURE 6: Urban interstate PDO crashes.

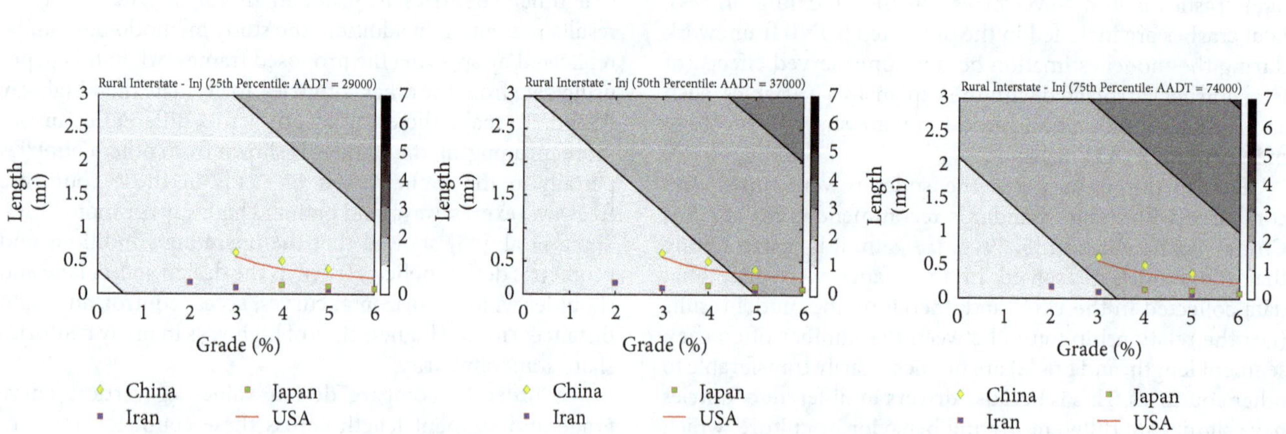

FIGURE 7: Prediction of rural interstate injury crash frequency.

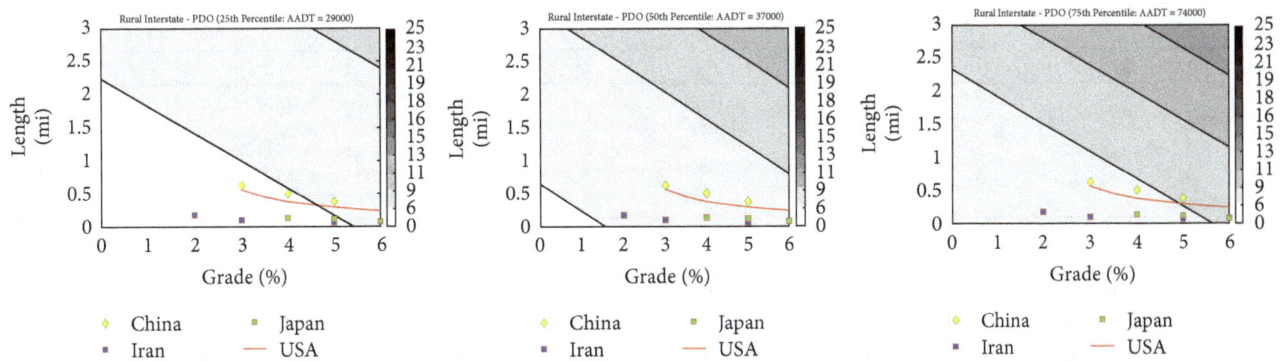

FIGURE 8: Prediction of rural interstate PDO crash frequency.

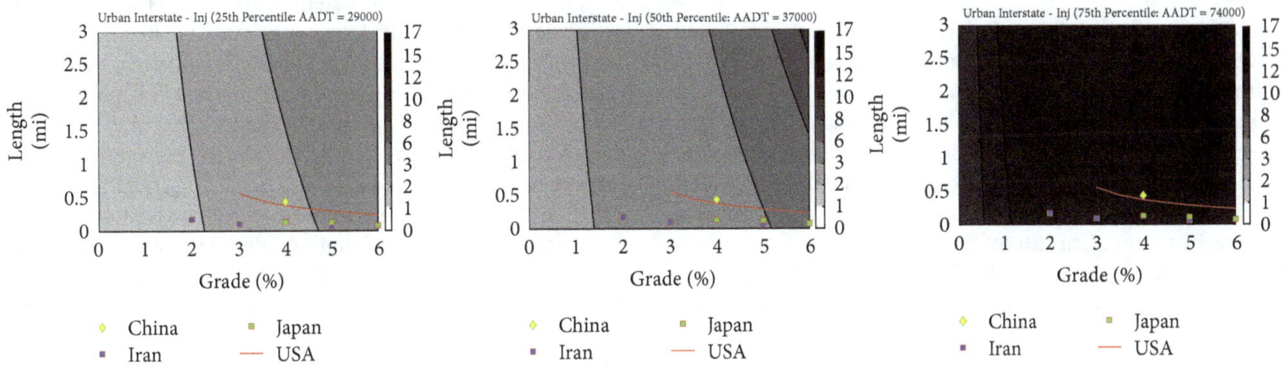

FIGURE 9: Prediction of urban interstate injury crash frequency.

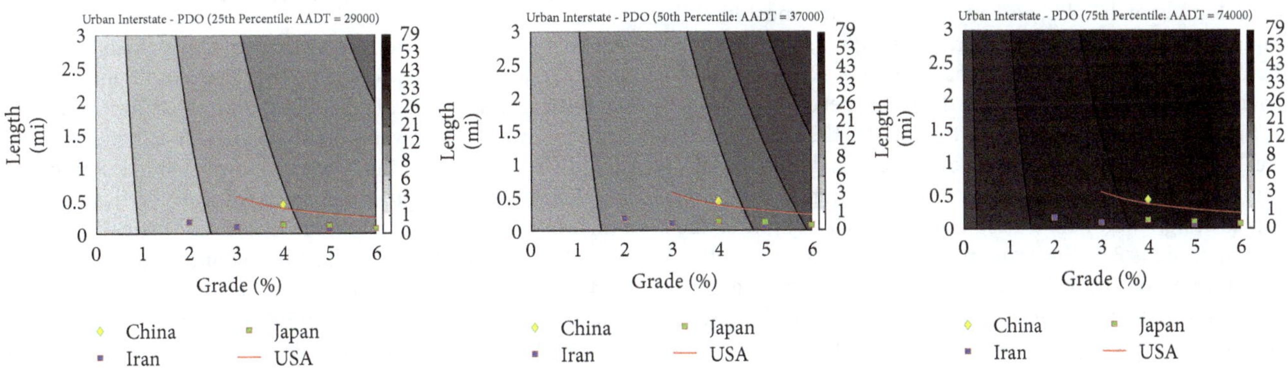

FIGURE 10: Prediction of urban interstate PDO crash frequency.

speed of 100 km/h for all four countries. In the case of China, the design codes for urban roads and nonurban roads in China are different. "Design Specification for Highway Alignments" (Specification) [39] is developed and published by the Ministry of Transport of China, whereas "Code for Design of Urban Road Engineering" (Code) [40] is developed and published by the Ministry of Housing and Urban-Rural Development of China. The Code and Specification specify that the maximum grades are required to be 3% and 4% for urban and nonurban roads with design speed of 100km/h, respectively. Moreover, grade can be 1% higher than regular value where topographical limits are excessive. For urban roads, the maximum length on 4% grade is 700m, according to the Code. For nonurban roads, however, the maximum length is 1000m, 800m, and 600m, on the grade of 3%, 4%, and 5%, respectively.

In the USA, highway design regarding the segment length and vertical curve grade is recommended by *A Policy on Geometric Design of Highways and Streets* (The Green Book) [5], which is developed by the American Association of State Highway and Transportation Officials (AASHTO). A maximum grade of 5% is considered appropriate for a design speed of 100km/h (approximately, 70 mph). The Green Book also suggests different speed reductions to multiple length curves. A speed reduction of 15mph can be used to analyze the relationship between the vertical curve grade and corresponding segment length (see the red lines in Figures 7–10). For Iran's high-class highways, similar values can be

TABLE 3: Maximum length (mi) for a certain grade with design speed of 70mph.

Vertical Grade (%)	2%	3%	4%	5%	6%
China	–	0.6214	0.4971	0.3728	–
Iran	0.1705	0.0947	–	0.0568	–
Japan	–	–	0.1326	0.0947	0.0758
USA	–	0.5592	0.3728	0.2920	–

found in Code 415 developed and published by the "Iran Vice Presidency for Strategic Planning and Supervision" for highway routes [42]. Even though continuous lines similar to the US cases can be drawn, discrete points were plotted on the figures to highlight the difference of the Iran policy from that of the USA.

With regard to highways in Japan, when the design speed is 100km/h, the maximum length of the grade is 700m, 500m, and 400m on grade of 4%, 5%, and 6%, respectively. Moreover, when design speed is 120km/h, maximum length for grade is 800m, 500m, and 400m on grade of 3%, 4%, and 5%, respectively. Unlike the continuous values provided by USA (AASHTO) and Iran, those of China and Japan are discrete in nature. The use of continuous lines rather than discrete points as design guidelines can be considered more suitable to the practice. The maximum length of a straight segment, for a given grade with a design speed of 70mph, is presented in Table 3.

In Table 3, it can be observed that critical length for a given grade varies extensively across the countries. For vertical grade of 3%, 4%, and 5%, China has the most liberal length limitation (longest maximum lengths), much more liberal compared to Iran or Japan. Design policy in USA provides grade-length curves for different speed reductions.

The "dots" for a specific country denotes the same "safety level" (corresponds to designs of 100km/h) based on the design standards/recommendations. For example, regarding China's design policy, for a design speed of 100km/h, when the vertical grade is 3%, the segment length should be less than 1000m. In addition, if the vertical grade is 4%, the segment length should be less than 800m. When the vertical grade is 5%, the segment length should be less than 600m.

It is worth noting where the traffic volume is low that the dots tend to be located in lighter colored regions of the nomogram (i.e., areas representing relatively higher safety). On the other hand, where the traffic volume is high, the dots tend to be located in darker colored regions of the nomogram (i.e., areas representing relatively lower safety). In addition, the nomograms generally suggest that, in the case of urban interstates, safety is more sensitive to changes in traffic volume compared to rural interstates: where the traffic volume is high, the expected number of crashes is far more sensitive to changes in vertical alignment grade and length, compared to their rural counterparts. There are general observations, and, for any specific highway segment, additional factors must be considered to reach a more definitive assessment of the overall safety effects. The linear nature of the isodehns for rural highway nomograms suggests that, with regard to this class of highways, there is relatively little or no interaction between the vertical grade and length, at least within the ranges of lengths and grades in the available dataset. On the other hand, the isodehns for urban highways are not linear, indicating strong interaction between these two features of highway vertical alignment at urban locations.

The nomograms can be used to evaluate if the current design standards are consistent with what is found in the observed data: if the dots for a country are located in the same area (not located on different side of an isodehn), then there is consistency. Regarding the data from the USA, particularly for urban interstates, the figures suggest that the AASHTO design could be adjusted to become more conservative with respect to the grade (i.e., the maximum vertical alignment grade could be decreased) and less conservative with respect to the length (maximum length could be increased). In other words, an increase in length is less likely to make the existing AASHTO design policy cross the isodehn to darker (area of lower safety) territory; however, an increase in vertical alignment grade is likely to have such an effect.

6. Anecdotal Evidence from China

An interesting but extreme case that currently exists in China was used as a basis to provide further discussion on the issue of vertical alignment deign policy and safety implications. This is an expressway segment located at a rural area in Jiangsu Province of China. The expressway has a design speed of 120km/h. The segment consists of two vertical grades and one vertical (crest) curve. The first vertical grade is +2.90% and length is 1,775m. The second grade is −2.90% and length is 1,980m. The curve has radius of 20,000m. The segment is known for its high crash rate. According to specification for highway design in China, the maximum grade is 3% when design speed is 120km/h. In addition, maximum length for this grade is 900m. If the grade is less than 3%, there is no length restriction. Therefore, certain designers may be seeking to skirt the policy narrowly by designing very long segments that do not violate the existing design policy. These segments turn out to be dangerous for driving. For example, in the case study, the lengths of 1,775m and 1,980m are very close to the grade threshold of 3% but far exceed the maximum length of 900m specified for 3% grades according to design code in China. To avoid such borderline designs in future, the vertical alignment design policy needs to be tightened (i.e. made less liberal). This can be done only after investigating the safety consequences of various combinations of length and grade. Using the methodology in this paper, nomograms can be developed to show isodehns within the agency desires to have for their highways. Then appropriate locations on the nomogram can be selected as the new design policy to avoid hazardous designs that involve borderline slopes of excessive length or borderline lengths of excessive slope.

7. Summary, Conclusions, Limitations, and Future Research

At many agencies, design policies were established several decades ago. However, with changing operating characteristics due to changing vehicle designs and capabilities and driver capabilities and perceptions, it is often useful to update design policies by examining them based on the performance measures with which they are closely associated. This paper addressed this issue from the perspective of highway vertical alignment. Vertical grades and lengths are key components of highway vertical alignment and therefore constitute critical aspects of highway design policy that influences safety.

In this respect, a full understanding of the effect of vertical grade and segment length on highway safety can help agencies to evaluate or adjust their design policies regarding vertical alignment design features (grade and length). In throwing some light on this issue, this paper first established the statistical relationship between vertical grade and length on one hand and highway safety on the other hand. The paper expressed safety in terms of the three different levels of crash severity—fatal, injury, and property damage only. In developing the statistical relationship, the paper eschewed the traditional univariate models (where each crash severity is modeled separately) and instead used a multivariate model (specifically, a seemingly unrelated negative binomial (SUNB) technique), which duly accounts for the unobserved shared effects between the different levels of crash severity.

In addition, the paper's models duly recognized and attempted to account for the holistic nature of the grade and tangent length effects particularly where interaction exists between the two variables: the effect of the sum (interaction) of the vertical grade and length is not necessarily the same as the sum of their individual effects. The research findings showed that the isodehns for rural interstate highways were linear since there was little or no statistically significant interaction between the vertical grade and length for this road class or their coefficients of these two variables were rather small. However, the isodehns for urban highway were not linear because there was statistically significant interaction between the vertical grade and length coefficients, because these two variables had rather large coefficients, or both. The paper investigated the problem statement separately for rural and urban interstate highway segments. Against the background of the developed models, the paper evaluated current design policies (vertical alignment grade and length specifications) for similar classes of highways at a number of countries (USA, China, Iran, and Japan) and presented a set of nomograms that feature safety isodehns vis-à-vis existing practice. These charts can be used by the highway agencies to evaluate and compare their current or possible future highway design policies. In addition, the charts are indicative of the sensitivity of highway safety performance to changes in vertical alignment, which can be used to quantify the expected trade-off between geometry-improvement investments and safety benefits (an important aspect of highway asset management, as pointed out by Sinha et al., [44]).

For the analysis in this paper, the average vertical curve grade in a segment is used. This can be considered appropriate in the case of short segments with uniform grade throughout their length or segments with no crest or sag but with slight grade changes along their length. For instance, a segment with half of its length at +2% and half at −2% will have average grade of 0%, which can be misleading. As such, future studies should consider uniform-grade segments, that is, segments that have the same grade over their entire length. Further, a limitation in the data collection is that the Indiana Crash Dataset used in this paper reports the number of crashes at an aggregate level (i.e., for each road segment). The lack of disaggregate crash data may restrict models from calibrating the impact of vertical curve grade more precisely. In addition, the horizontal curve was shown to be not significant and therefore excluded from this paper. In future research where disaggregate crash data are available, the horizontal curve and its interaction with vertical curve should be further tested. In addition, the framework was demonstrated using crash data from a specific state (Indiana) that may not reflect situations at other states or countries. Nonetheless, the framework presented in this study can be considered general and flexible for use by agencies to evaluate the safety consequences of the highway vertical grade thresholds specified their design policies.

Conflicts of Interest

The authors declare that they have no conflicts of interest.

References

[1] R. Archilla and J. Morrall, "Traffic characteristics on two-lane highway downgrades," *Transportation Research Part A: Policy and Practice*, vol. 30, no. 2, pp. 119–133, 1996.

[2] C. M. Morris and E. T. Donnell, "Passenger car and truck operating speed models on multilane highways with combinations of horizontal curves and steep grades," *Journal of Transportation Engineering*, vol. 140, no. 11, Article ID 04014058, 2014.

[3] *Highway Capacity Manual 6th Edition: A Guide for Multimodal Mobility Analysis, Transportation Research Board,* National Research Council, Washington, DC, USA, 2016.

[4] L. Wang, J. Cheng, and Y. Zhang, "Reliability-Based Specification on Critical Length of Highway Sections with Near-Maximum Grade," *KSCE Journal of Civil Engineering*, vol. 22, no. 4, pp. 1406–1417, 2018.

[5] American Association of State Highway & Transportation Officials, *A Policy on Geometric Design of Highways and Streets*, AASHTO, Washington, DC, USA, 2011.

[6] A. D. St. John and D. R. Kobett, *Grade Effects on Traffic Flow Stability and Capacity, NCHRP Report 185*, Transportation Research Board, Washington, DC, USA.

[6] A. D. St. John and D. R. Kobett, *Grade Effects on Traffic Flow Stability and Capacity, NCHRP Report 185*, Transportation Research Board, Washington, DC, USA.

[7] S. M. Easa, "Design considerations for highway sight-hidden dips," *Transportation Research Part A: Policy and Practice*, vol. 28, no. 1, pp. 17–29, 1994.

[8] S. M. Easa, "Reliability-based design of sight distance at railroad grade crossings," *Transportation Research Part A: Policy and Practice*, vol. 28, no. 1, pp. 1–15, 1994.

[9] Y. Hassan, S. M. Easa, and A. O. Abd El Halim, "Passing sight distance on two-lane highways: Review and revision," *Transportation Research Part A: Policy and Practice*, vol. 30, no. 6, pp. 453–467, 1996.

[10] S. Labi, *Effects Of Geometric Characteristics of Rural Two-Lane Roads on Safety, Technical Report FHWA/IN/JTRP-2005/2*, Joint Transportation Research Program, W. Lafayette, IN, USA, 2006.

[11] S. Labi, "Efficacies of roadway safety improvements across functional subclasses of rural two-lane highways," *Journal of Safety Research*, vol. 42, no. 4, pp. 231–239, 2011.

[12] Y. Hassan, S. M. Easa, and A. O. Abd El Halim, "Design considerations for combined highway alignments," *Journal of Transportation Engineering*, vol. 123, no. 1, pp. 60–68, 1997.

[13] Y. Hassan, G. Gibreel, and S. M. Easa, "Evaluation of highway consistency and safety: practical application," *Journal of Transportation Engineering*, vol. 126, no. 3, pp. 193–201, 2000.

[14] Y. Hassan and S. M. Easa, "Effect of vertical alignment on driver perception of horizontal curves," *Journal of Transportation Engineering*, vol. 129, no. 4, pp. 399–407, 2003.

[15] Y. Hassan, "Improved design of vertical curves with sight distance profiles," *Transportation Research Record*, no. 1851, pp. 13–24, 2003.

[16] D. R. Jessen, K. S. Schurr, P. T. McCoy, G. Pesti, and R. R. Huff, "Operating speed prediction on crest vertical curves of rural two-lane highways in Nebraska," *Transportation Research Record*, no. 1751, pp. 67–75, 2001.

[17] S. P. Miaou, "The relationship between truck accidents and geometric design of road sections: Poisson versus negative binomial regressions," *Accident Analysis & Prevention*, vol. 26, no. 4, pp. 471–482, 1994.

[18] J. Daniel and S. I.-J. Chien, "Truck safety factors on urban arterials," *Journal of Transportation Engineering*, vol. 130, no. 6, pp. 742–752, 2004.

[19] B. L. Smith and R. Lamm, "Coordination of horizontal and vertical alinement with regard to highway esthetics," *Transportation Research Record*, no. 1445, pp. 73–85, 1994.

[20] S. Bidulka, T. Sayed, and Y. Hassan, "Influence of vertical alignment on horizontal curve perception. Phase I: Examining the hypothesis," *Transportation Research Record*, no. 1796, pp. 12–23, 2002.

[21] C. Zegeer, J. R. Stewart, F. M. Council, D. W. Reinfurt, and E. Hamilton, "Safety effects of geometric improvements on horizontal curves," *Transportation Research Record*, vol. 1356, 1992.

[22] K. A. Abbas, "Traffic safety assessment and development of predictive models for accidents on rural roads in Egypt," *Accident Analysis & Prevention*, vol. 36, no. 2, pp. 149–163, 2004.

[23] P. F. Hanley and D. J. Forkenbrock, "Safety of passing longer combination vehicles on two-lane highways," *Transportation Research Part A: Policy and Practice*, vol. 39, no. 1, pp. 1–15, 2005.

[24] R. Elvik, P. Christensen, and A. Amundsen, "Speed and road accidents," in *An evaluation of the Power Model. Institute of Transport Economics*, Institute of Transport Economics, Oslo, Norway, 2004.

[25] T. S. Shively, K. Kockelman, and P. Damien, "A Bayesian semi-parametric model to estimate relationships between crash counts and roadway characteristics," *Transportation Research Part B: Methodological*, vol. 44, no. 5, pp. 699–715, 2010.

[26] M. Hosseinpour, A. S. Yahaya, and A. F. Sadullah, "Exploring the effects of roadway characteristics on the frequency and severity of head-on crashes: Case studies from Malaysian Federal Roads," *Accident Analysis & Prevention*, vol. 62, pp. 209–222, 2014.

[27] M. A. Mohammadi, V. A. Samaranayake, and G. H. Bham, "Crash frequency modeling using negative binomial models: An application of generalized estimating equation to longitudinal data," *Analytic Methods in Accident Research*, vol. 2, pp. 52–69, 2014.

[28] K. Bauer and D. Harwood, "Safety effects of horizontal curve and grade combinations on rural two-lane highways," *Transportation Research Record*, vol. 2398, pp. 37–49, 2013.

[29] Y. Yan, G. Li, J. Tang et al., "A novel approach for operating speed continuous predication based on alignment space comprehensive index," *Journal of Advanced Transportation*, vol. 2014, Article ID 9862949, 14 pages, 2017.

[30] Q. Zeng, H. Huang, X. Pei, S. C. Wong, and M. Gao, "Rule extraction from an optimized neural network for traffic crash frequency modeling," *Accident Analysis & Prevention*, vol. 97, pp. 87–95, 2016.

[31] Q. Zeng, J. Sun, and H. Wen, "Bayesian hierarchical modeling monthly crash counts on freeway segments with temporal correlation," in *Journal of Advanced Transportation*, vol. 2017, p. 8, 2017.

[32] X. Zou and W. L. Yue, "A Bayesian network approach to causation analysis of road accidents using netica," *Journal of Advanced Transportation*, vol. 2017, Article ID 2525481, 18 pages, 2017.

[33] J. Lee, M. Abdel-Aty, and X. Jiang, "Multivariate crash modeling for motor vehicle and non-motorized modes at the macroscopic level," *Accident Analysis & Prevention*, vol. 78, pp. 146–154, 2015.

[34] M. T. Sarwar and P. C. Anastasopoulos, "The effect of long term non-invasive pavement deterioration on accident injury-severity rates: A seemingly unrelated and multivariate equations approach," *Analytic Methods in Accident Research*, vol. 13, pp. 1–15, 2017.

[35] S. Chen, T. U. Saeed, and S. Labi, "Impact of road-surface condition on rural highway safety: A multivariate random parameters negative binomial approach," *Analytic Methods in Accident Research*, vol. 16, pp. 75–89, 2017.

[36] S. Chen, T. U. Saeed, S. D. Alqadhi, and S. Labi, "Safety impacts of pavement surface roughness at two-lane and multi-lane highways: accounting for heterogeneity and seemingly unrelated correlation across crash severities," *Transportmetrica A: Transport Science*, vol. 16, pp. 75–89, 2017.

[37] R. Winkelmann, "Seemingly unrelated negative binomial regression," *Oxford Bulletin of Economics and Statistics*, vol. 62, no. 4, pp. 553–560, 2000.

[38] R. Porter, E. Donnell, and J. Mason, "Geometric design, speed, and safety," *Transportation Research Record: Journal of the Transportation Research Board*, no. 2309, pp. 39–47, 2012.

[39] Ministry of Transport of People's Republic of China, *Design Specification for Highway Alignments*, Beijing, China, 2006.

[40] Ministry of Housing and Urban-Rural Development of People's Republic of China, *Code for design of urban road engineering*, Beijing, China, 2012.

[41] Japan Road Association, *Commentary and operation of the road structure Ordinance*, Japan Road Association, Tokyo, Japan, 2015.

[42] V. p. f. S. P. a. Supervision, *Iran Highway Geometric Design Code, No. 415*, Office of Deputy for Strategic Supervision Department of Technical Affairs, Tehran, Iran, 2012.

[43] K. Sinha, S. Labi, and M. Noureldin, *Lessons from international experience cost-effective standards for different types of roads, technical report prepared in conjunction with Harral Winner Thompson Sharp Klein, Inc., for the National Transport Development Policy Committee Planning Commission and the Government of India*, 2011.

[44] K. C. Sinha, S. Labi, and B. R. D. K. Agbelie, "Transportation infrastructure asset management in the new millennium: continuing issues, and emerging challenges and opportunities," *Transportmetrica A: Transport Science*, vol. 13, no. 7, pp. 591–606, 2017.

2

Global and Local Path Planning Study in a ROS-based Research Platform for Autonomous Vehicles

Pablo Marin-Plaza, Ahmed Hussein, David Martin, and Arturo de la Escalera

Universidad Carlos III de Madrid, Leganes, Spain

Correspondence should be addressed to Pablo Marin-Plaza; pamarinp@ing.uc3m.es

Academic Editor: Aboelmaged Noureldin

The aim of this work is to integrate and analyze the performance of a path planning method based on Time Elastic Bands (TEB) in real research platform based on Ackermann model. Moreover, it will be proved that all modules related to the navigation can coexist and work together to achieve the goal point without any collision. The study is done by analyzing the trajectory generated from global and local planners. The software prototyping tool is Robot Operating System (ROS) from Open Source Robotics Foundation and the research platform is the iCab (Intelligent Campus Automobile) from University Carlos III. This work has been validated from a test inside the campus where the iCab has performed the navigation between the starting point and the goal point without any collision. During the experiment, we proved the low sensitivity of the TEB method to variations of the vehicle model configuration and constraints.

1. Introduction

Autonomous vehicles base their decisions on planner modules that create the collision-free waypoints in the path to reach the destination point. These modules are capable of finding the optimal solution minimizing the computational time and distance covered by the vehicle avoiding static and dynamic obstacles. Some research groups test these modules on real vehicles such as PROUD in Parma in 2013 [1] or Karlsruhe Institute of Technology (KIT) driving around the cities with advanced technology in perception, navigation, and decision making [2]. In all cases, the navigation module is divided into a global planner and a local planner, where the first one finds the optimal path with a prior knowledge of the environment and static obstacles, and the second one recalculates the path to avoid dynamic obstacles.

The main contribution of this work is to demonstrate the integration of using TEB local planner in an Ackermann model with real test. The platform used is the iCab (Intelligent Campus Automobile) which works with the software prototyping tool ROS. Additionally, it has been discovered that the capacity of the vehicle to follow the local plan with accuracy is not mandatory. The vehicle is able to reach the goal with modest errors when following the local plan. This characteristic is really important in real life when the vehicle model depends on multiple factors as the pressure of the tires, the number of passengers, and the battery of the vehicle. When this mathematical model varies from the nominal values, the vehicle will not reach point by point the provided local plan, so TEB local planner is more robust to face changes in the configuration of the mathematical model.

In order to prove that the TEB local planner is able to work along the other processes involved in the vehicle, it is necessary to solve several problems related to the localization in the environment (initial localization), the ego-motion of the vehicle (odometry), environment understanding (perception and mapping), which path should the vehicle take (planning), and movement (low level control). Hence, this document gathers all the solutions for each task for navigation and then tests the Dijkstra method for global planning and the Time Elastic Bands method used for local planning.

This document is divided into six sections. Section 2 describes current planning methods used in the autonomous vehicles research field. Section 3 describes the research platform and the modules related to the navigation. Section 4 describes the experimental setup. Section 5 summarizes the results. Finally, the last section contains the conclusions and the future work.

2. Path Planning

The path planning problem is a well-known NP-hardness [3] where the complexity increases with the degrees of freedom of the vehicle. In this work, the ground vehicle is reduced to a 2D space of a single plane X-Y where the Z component is neglected and the vehicle is restricted by some constraints due to the Ackermann configuration [4]. The aim of solving this NP-hardness is not to find one solution that connects the start point and the goal point, but the optimal solution with the minimum distance and the smoothest maneuvers and without hitting any known obstacles. Usually, this module is divided into the global planner, which uses a priori information of the environment to create the best possible path, if any, and the local planner, which recalculates the initial plan to avoid possible dynamic obstacles. The generated plan is a collection of waypoints for global, $P_i = (x_i, y_i)^T$, and local $P_i = (x_i, y_i, \theta_i)^T, \in \mathfrak{R}^2 \times S^1$, where S is the navigation plane where the vehicle is able to move.

2.1. Global Planner. As said before, the global planner requires a map of the environment to calculate the best route. Depending on the analysis of the map, some methods are based on Roadmaps like Silhouette [5] proposed by Canny in 1987 or Voronoi [6] in 2007. Some of them solve the problem by assigning a value to each region of the roadmap in order to find the path with minimum cost. Some examples are the Dijkstra [7] algorithm, Best First [8], and A^* [9]. Another approach is by dividing the map into small regions (cells) called cell decomposition as mentioned in [10]. A similar approach by using potential fields is described in [11], the most extended algorithm used few years ago rapidly exploring random trees [12] or the new approach based on neural networks [13]. Some solutions combine the aforementioned algorithms improving the outcome at the cost of high computational power.

So, for this work and for testing purposes, the selected method is the Dijkstra algorithm because of the simplicity for debugging and its good performance on the experiments. It is necessary to remark that the goal of this paper is not to find the best and fastest solution, but it is to prove that all modules of navigation are working together and the vehicle is able to navigate from one point to another.

Dijkstra uses the roadmap approach to convert the problem into a graphic search method using the information of a grid cell map. This method starts with a set of candidate nodes where the vehicle is able to navigate (free space) assigning a cost value to each of them. From the starting point, this value is increased by the necessary number of nodes to pass through to reach each node. For example, in Figure 1, the free space around the starting point takes the value of one unit and for the second generation of neighbors takes the value of two units and so on. For each cell with value, a checked cell is assigned and the method continues with the next generation of neighbors until the goal point is reached. The minimum value of the sum of all nodes from the starting point and the end point is the shortest path. After the success of finding the global path from the start to the goal, all the selected nodes are translated into positions in the reference axes as the form $P_i = (x_i, y_i)^T$. The global planner divides the map into nodes for each free cell but the outcome is not smooth and some points are not compliant with the vehicle geometry and kinematics.

2.2. Local Planner. In order to transform the global path into suitable waypoints, the local planner creates new waypoints taking into consideration the dynamic obstacles and the vehicle constraints. So, to recalculate the path at a specific rate, the map is reduced to the surroundings of the vehicle and is updated as the vehicle is moving around. It is not possible to use the whole map because the sensors are unable to update the map in all regions and a large number of cells would raise the computational cost. Therefore, with the updated local map and the global waypoints, the local planning generates avoidance strategies for dynamic obstacles and tries to match the trajectory as much as possible to the provided waypoints from the global planner. Different approaches are based on trajectory generation using Clothoids lines [16], Bezier lines [17], arcs and segments [18], or splines [19]. All of them create intermediate waypoints following the generated trajectory. The selected method for local planning is Time Elastic Bands [15, 20] which consists of deforming the initial global plan considering the kinematic model of the vehicle and updating the local path based on dynamic obstacles or the possible deviation from the path. This method has been selected because of the integration of the trajectory planning and obstacle detection with avoidance at low computational cost fulfilling the requirements of the local planner in the research platform iCab.

Time Elastic Bands (TEB) create a sequence of intermediate vehicle poses $p_i = (xi, yi, \theta_i)^T$ modifying the initial global plan. It requires the velocity and acceleration limits of the vehicle, the security distance of the obstacles and the geometric, and kinematic and dynamic constraints of the vehicle. All of this configuration generates a set of commands for speed (v) and steering angle (δ), as $\text{cmd}_i = (v_i, \delta_i)^T$, required to achieve the intermediate waypoints, while the vehicle is moving. Figure 2 shows a robot moving from the starting point to the goal point passing through four waypoints in the smoothest way possible between obstacles. One of the drawbacks of this method is the influence of the dynamic obstacles direction in the generation of the recalculated local path. For example, when a pedestrian is crossing from east to west the trajectory generated which connects a point from the south to the north (Figure 3), the recalculation of the new trajectory will still be on the left of the obstacle but further. The solution is the creation of three elastic bands: one updated each iteration, one is recalculated from the original path, and the last one is created from scratch each time. The selected one is the shortest of the three that is compliant with all requirements.

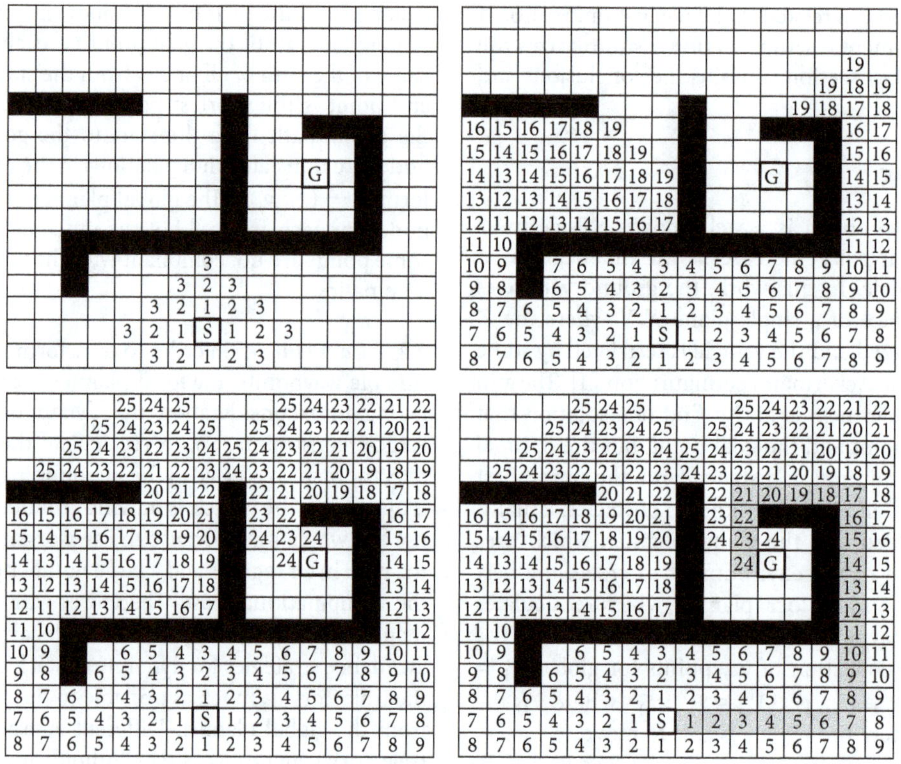

FIGURE 1: Dijkstra method to find the optimal path [14].

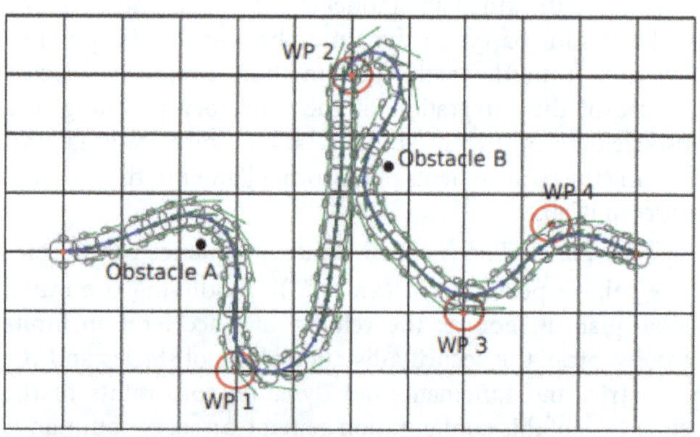

FIGURE 2: Description of Time Elastic Bands from [15].

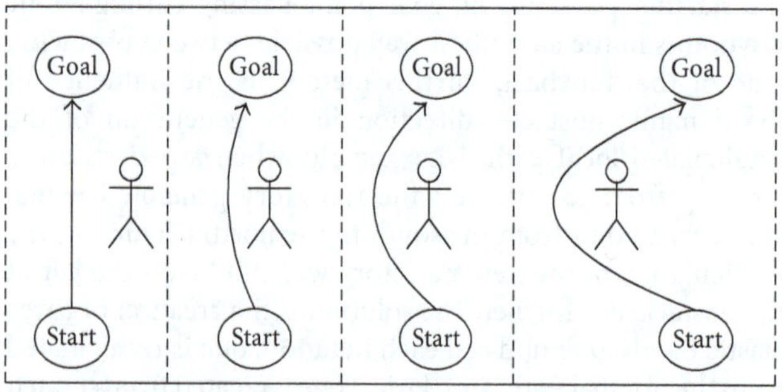

FIGURE 3: Pedestrian crossing from east to west of the trajectory generated.

FIGURE 4: iCab with sensors.

3. Platform Description

The research platform is a modified golf cart equipped with several sensors such as wheel encoders, a Lidar with 16 laser planes on the top, a single plane laser located in the front, a stereo camera located in the front, and a GPS-IMU-Compass located on the top, as shown in Figure 4. The movement of the vehicle is done by a 36 V DC motor for the linear acceleration and another 36 V DC motor for the steering wheel. More information and tools are available in [21–23].

Additionally, the vehicle includes two computers to gather information from the sensors, process it, and perform the navigation decisions. The main computer is in charge of the wheel encoders, images, laser, imu, GPS, and compass

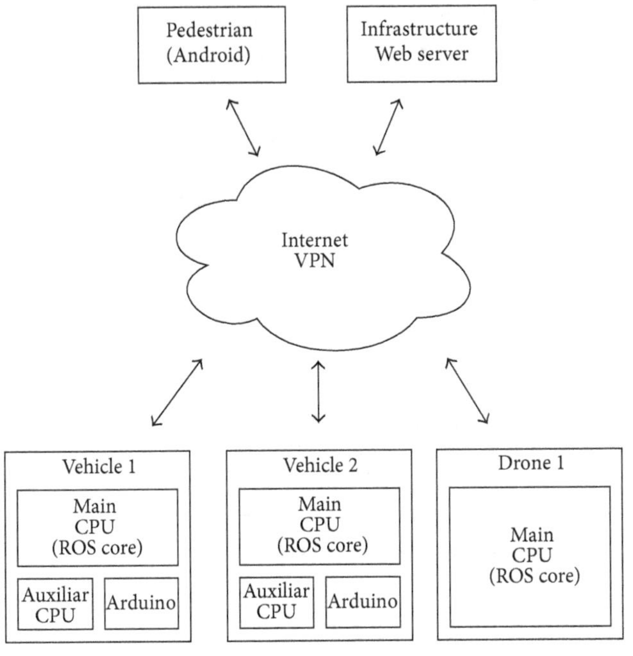

FIGURE 5: System communications.

and generates the maps. Therefore, the system provides wheel odometry, obstacle detection, and maps. The second computer processes the point cloud generated from the Lidar and computes the Lidar odometry. The software used, as aforementioned said, is ROS. This tool allows the vehicle to share and synchronize messages between nodes in the same computer and, additionally, with the computers and microcontrollers in the vehicle by the network using ROS core master. Furthermore, using a master discovery tool for all the ROS core masters in the network (one core master for each vehicle), the systems are able to share messages between vehicles, infrastructures, and pedestrians [24]. The use of one core master for each vehicle allows the inner communications even when there is no Internet connection, as shown in Figure 5.

A brief graphical description of the nodes and connections is shown in Figure 6. From left to right, the nodes are as follows:

(i) *Global map* is responsible for loading the global grid map for the campus as *navigation_msgs/OccupancyGrid.msg*. The selected grid size has a resolution of 0.156 m used in the analysis of the previously recorded point cloud obtained with the Lidar (a priori knowledge).

(ii) *Global planner* uses the Dijkstra algorithm to create the/*global_plan* to reach the goal point using the global map. This path message is a standard type *navigation_msgs/Path.msg* which contains the waypoints without the orientation.

(iii) *Lidar odometry* and *wheel odometry* are in charge of creating the ego-motion measurements from the starting point of the vehicle. The messages used are the type of *navigation_msgs/Odometry.msg*.

(iv) *Odometry manager* is responsible for the fusion and adding the initial localization for the conversion between global and local odometry.

(v) *Costmap 2D* uses the laser sensor information to create a local Costmap in front of the vehicle. The Costmap values represent the proximity of the vehicle to an obstacle using the geometry of the vehicle. The resulting map has the obstacles inflated by a security perimeter where the vehicle should not allow entering. Additionally, this map is used for TEB local planner in order to modify the local path based on the distance from the vehicle to the obstacles.

(vi) *TEB local planner* needs a global plan, Costmap, and vehicle odometry. It publishes the local plan (at a specific rate) with a configurable lookahead distance, and additionally, it publishes the control commands *ackermann_cmd* with the type *ackermannDriveStamped.msg*. The selected lookahead distance is 10 m because it is suitable to avoid big obstacles like other vehicles or groups of students.

(vii) *Obstacle manager* is responsible for sending a warning message to the movement manager for possible threats like pedestrians or obstacles in front of the vehicle. The type of this message is custom and has a list of obstacles with the information of distance and area for each obstacle. This node also returns the distance to the closest obstacle to take into consideration for emergency brake.

(viii) *Task manager* is in charge of selecting the task for the autonomous vehicle. For other projects, it is possible to configure different behaviors like platooning mode, drive between boundaries, or manual mode. This node receives information of the possible threats in the environment in order to stop the vehicle if necessary. This package publishes the reference velocity and reference steering angle.

(ix) *PID velocity* is responsible for creating the PWM necessary to the rear wheels DC motor. It takes as input the current velocity. The type of message is standard as *std_msgs/Float64.msg*.

(x) *Movement manager* is a driver in charge of checking all limits and possible wrong velocity messages. It sends the desired acceleration to the DC motor and the selected angle for the steering angle.

Transformation Trees (tf). One of the most important aspects to consider when working with vehicle odometry is the frame of reference and the coordinates. This node is in charge of transforming the movement of the vehicle from its ego-motion local reference to the global reference adding the initial position and orientation. Therefore, global and local path need the transformation between local coordinates and global coordinates. The *odometry manager* module is in charge of fusing the odometries and publishing the transformation between global and local coordinates using the initial localization in the map of the vehicle. Figure 7 shows all the frames involved in the architecture for one vehicle. The

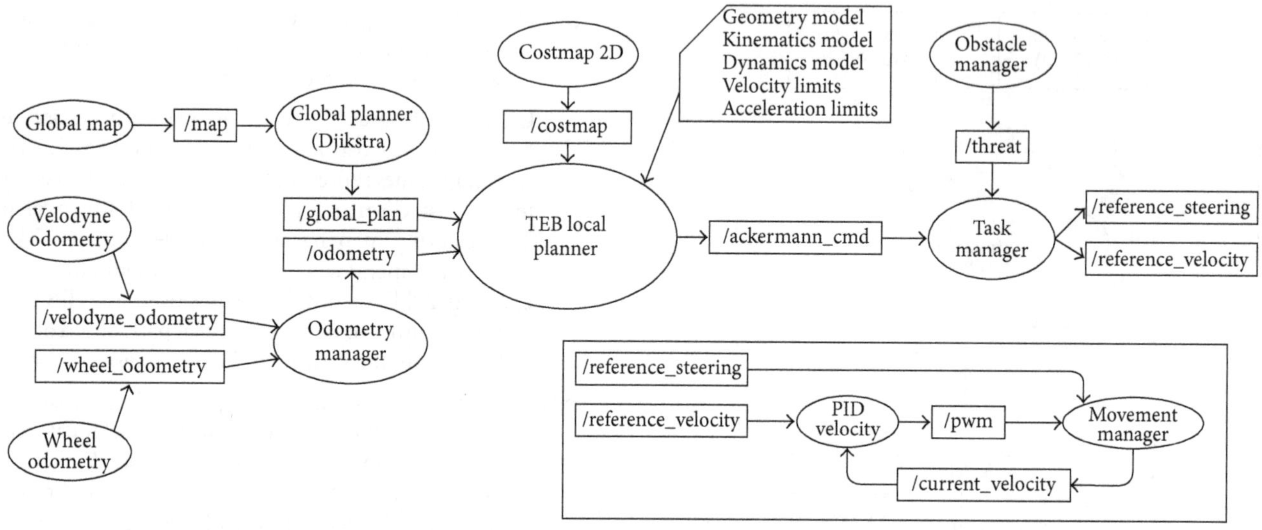

FIGURE 6: Software nodes sharing messages by topics in the partial software architecture.

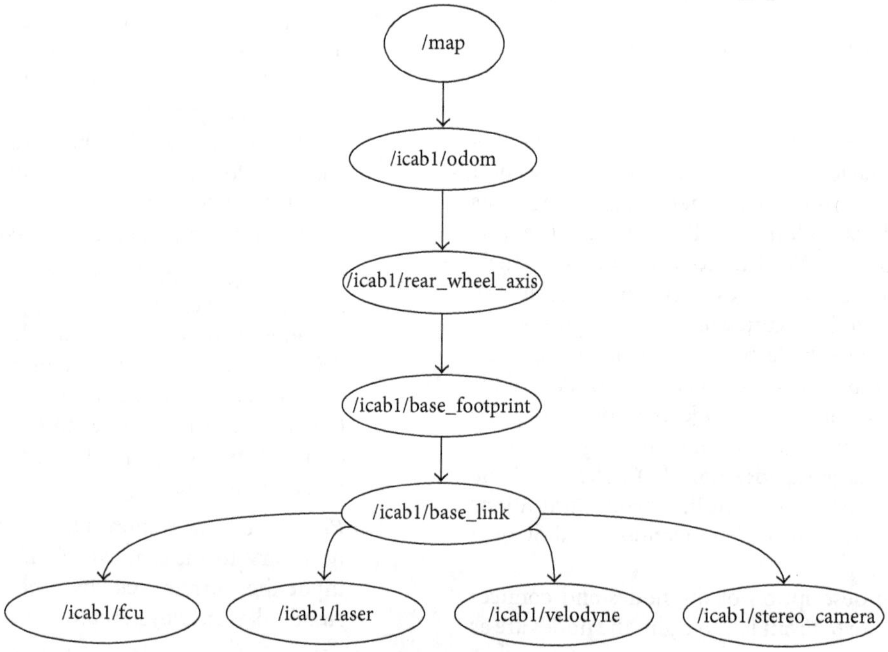

FIGURE 7: Transformation Tree for iCab1 showing all the sensors.

appropriate configuration of these transformations allows the system to collaborate with other coexisting heterogeneous vehicles under a vehicle namespace such as in this case icab1.

4. Experimental Setup

Aiming to test the path planner in the real platform, a simple test has been prepared where the vehicle navigates from a starting point and finishes in a goal point. The selected location takes place in a zone where a circular building is occluding the direct path trajectory. The data is recorded into a bag file, while the vehicle performs the experiment for posterior offline analysis. In Figure 8, the global map used to calculate the global plan is shown and the experiment zone is located inside of the red square.

This map has been generated using the Lidar 3D points with the Lidar odometry and mapping [25] adapted to work with the correct axes reference. Afterwards, the point cloud has been processed to extract the floor at a specific height. The last step is to project this point cloud to the 2D space to have all the static obstacles (notice that the cup of the trees is taken into account as an obstacle due to the possible collision with the leaves and the top of the vehicle).

The representation of the vehicle, the maps, and the movements are easily analyzed using the RVIZ tool. Figure 9 describes the localization of the vehicle in the environment

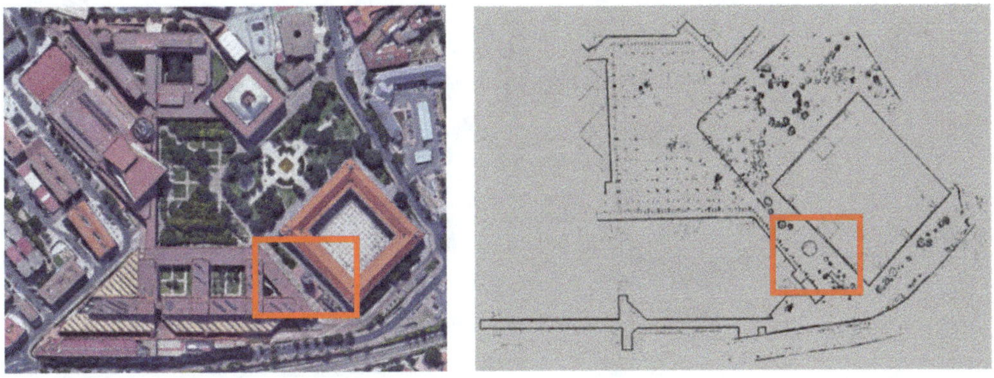

(a) (b)

FIGURE 8: Satellite image of the Campus granted by Google Maps on (a). Campus gridmap on (b), the experiment is located inside of the red square.

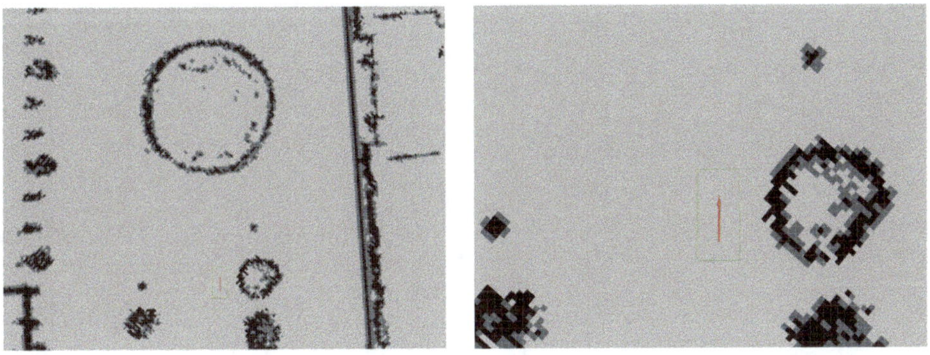

(a) (b)

FIGURE 9: On (a), starting point of the vehicle oriented to the local odometry. On (b), zoom in the details of the vehicle geometry.

where the green rectangular shape represents the geometry and the red arrow represents the orientation.

After introducing the goal point, the representation of the lines for the paths is shown in Figure 10 where the blue line is the global path generated from the global planning using the current localization and the goal point. The input goal point is generated from the RVIZ visualizer using *move_base_msgs/MoveBaseGoal.msg* and clicking on the screen. The orange line is the local path generated from the TEB local planner and each cycle of the node is recalculated.

5. Results

All trajectories generated from the TEB local planner are evaluated offline by the data recorded, while the experiment is running.

5.1. Evaluation Metrics. The evaluation metrics used in this section are the Euclidean distance (DGL) between the Global Plan Waypoints (GPW) and Local Plan Waypoints (LPW) as shown in (1). The algorithm to find the best local plan is constantly updating it in order to avoid obstacles in the route or because the vehicle is not able to follow the previous local path.

$$\mathrm{DGL}_i = \sqrt{(\mathrm{GPW}_i - \mathrm{LPW}_i)^2} \quad \forall i \in Q. \tag{1}$$

5.2. Data Analysis. The experiment has been done by the lookahead distance configured to 10 m for the TEB algorithm because if an obstacle is above the global plan (a dynamic obstacle or a static obstacle which are not on the global map), this lookahead distance should be big enough to find another path even if it is far away from the global plan. Figure 11 represents the global plan generated with Dijkstra method on blue. In Figure 12, each line represents the local plan generated from the position of the vehicle while moving. The representation of multiple lines is due to the recalculation of the local trajectories. It is possible to appreciate the details of some lines that do not end into the goal point because of the lookahead distance configuration. Figure 13 represents the position of the vehicle at each moment in red dots. It is possible to appreciate all the graphs combined in the Figure 14 and with the details of the first three meters of the movement in Figure 15. Each red dot is the current odometry of the vehicle, the big blue line is the Dijkstra global plan, and each

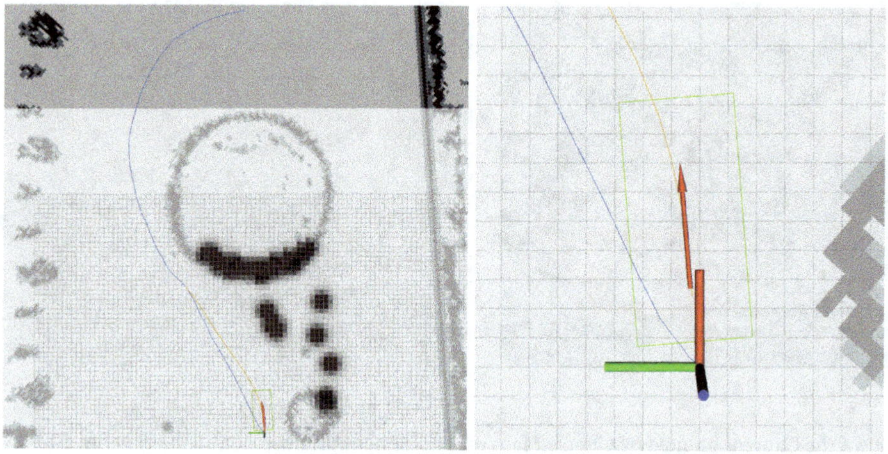

Figure 10: Global plan in blue. Local plan in orange. Overlay of global map and local Costmap 2D.

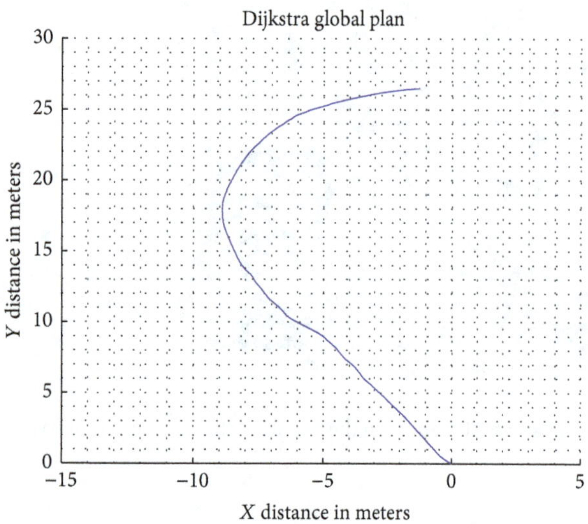

Figure 11: Global path.

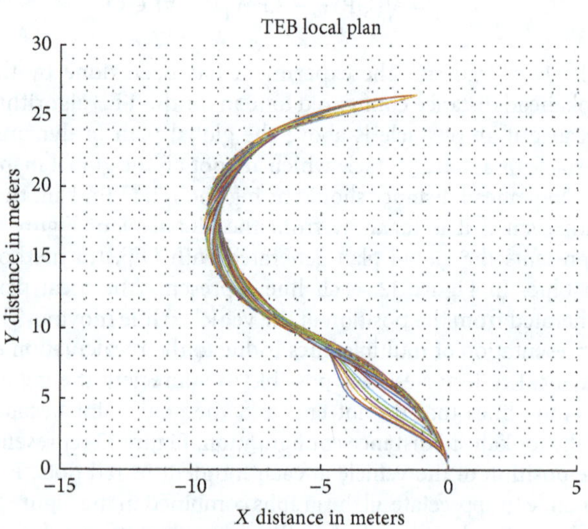

Figure 12: Local path.

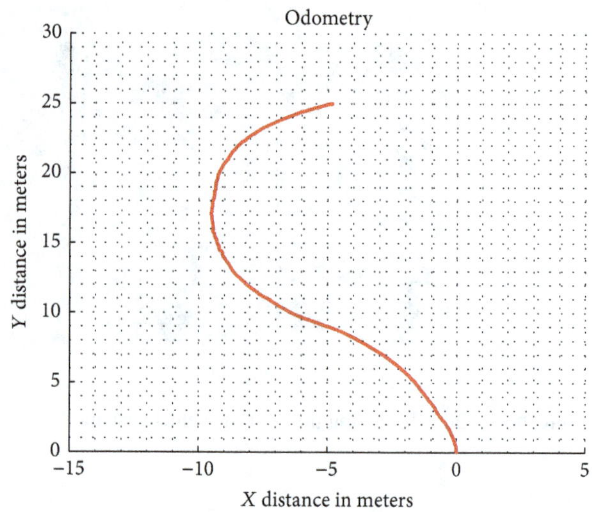

Figure 13: Odometry points.

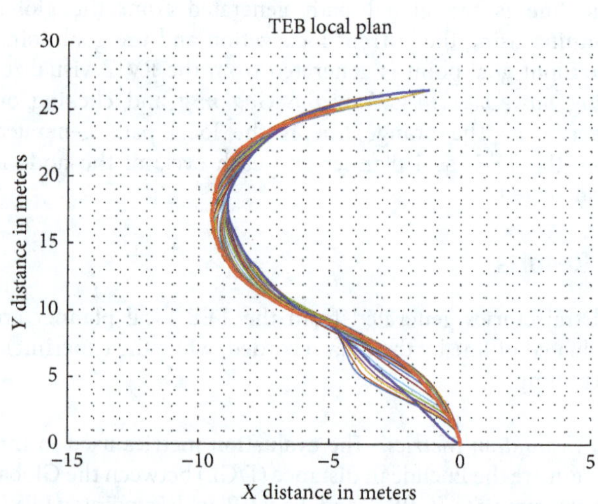

Figure 14: Combined.

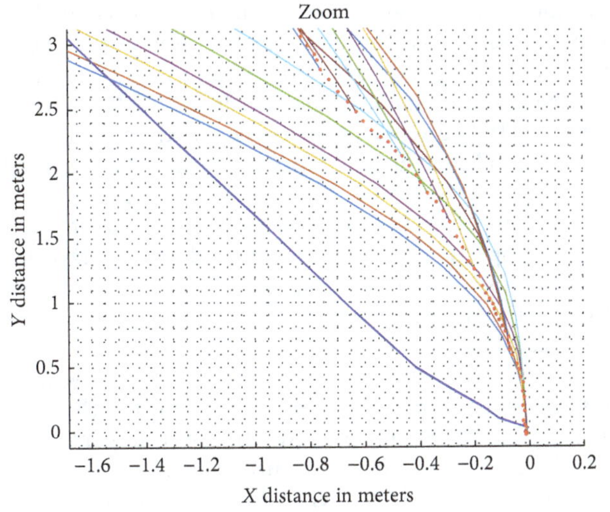

FIGURE 15: Combined zoom.

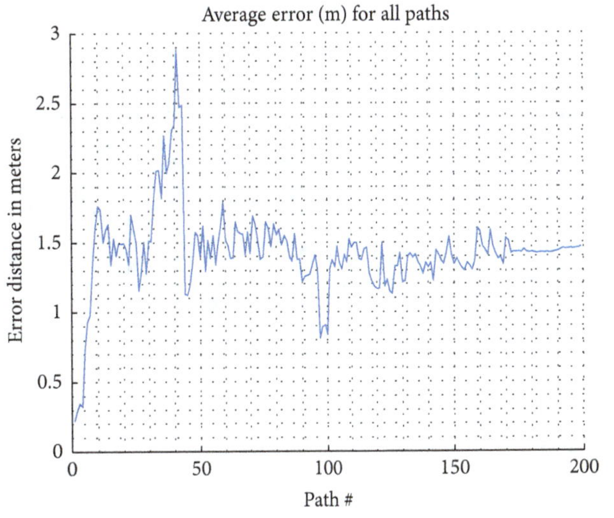

FIGURE 17: Average error.

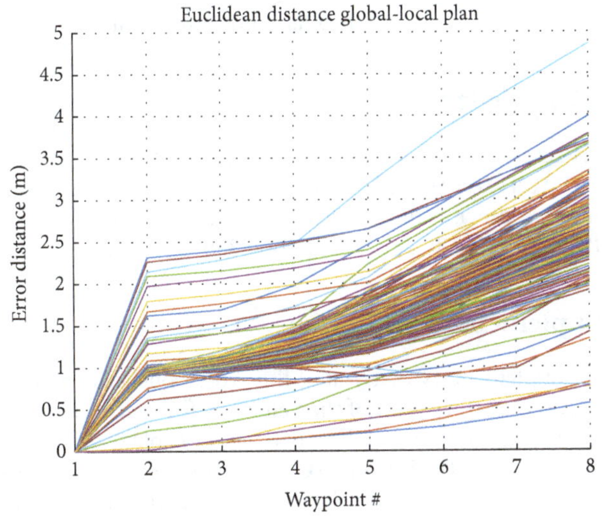

FIGURE 16: Error local plan, global plan.

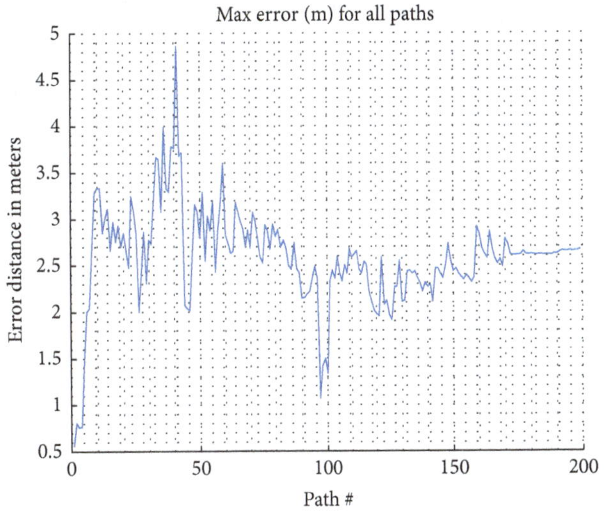

FIGURE 18: Max error.

thin line in multicolor corresponds to each TEB local plan. It is possible to appreciate the recalculations of the local plan in some dots where the vehicle is not able to follow the proposed trajectory (multicolor lines). Even if the vehicle does not follow perfectly the path due to the mathematical model variations configured into the TEB local planner parameters such as steering angle velocity, it seems that the header of the vehicle is different than in the calculated plan which follows into a wrong steering angle velocity configuration.

Euclidean Distance between Local and Global Path. The error between the global path and the local path determines how accurate the recalculation of the trajectory is. In Figure 16, each line represents the distance error for all local trajectories planned, while the vehicle was moving.

In Figure 16, the Euclidean distance between the first points of the local trajectory and the first points of the global trajectory from the position of the vehicle is described for all paths generated. The reason for this first point is because, after the recalculation of the new local path, the old path is not useful anymore. For the analysis, the seven first points have been selected for each local plan. Regarding this graph, it is possible to notice the incremental growth of the Euclidean distance point by point. The reason for this growth is because the vehicle is not able to follow properly the generated local path. Figure 17 is the result of computing the mean of the seven first values for every path. After the initial rise, there is a tendency around 1.42 m. The maximum Euclidean distance error for all paths generated between local path and global path is detailed in Figure 18. Due to the avoidance of an obstacle at the beginning, the local path deviates further from the global path.

Speed and Steering Angle. For the analysis of the control commands of the speed and steering angle granted by the TEB local planner module, Figure 19 shows the values in

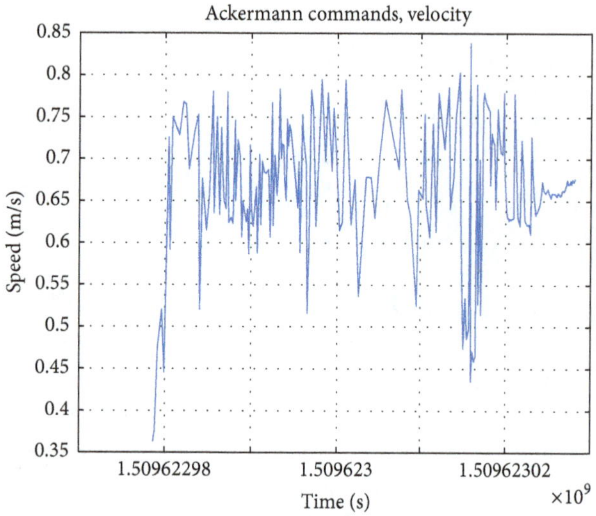

FIGURE 19: Speed.

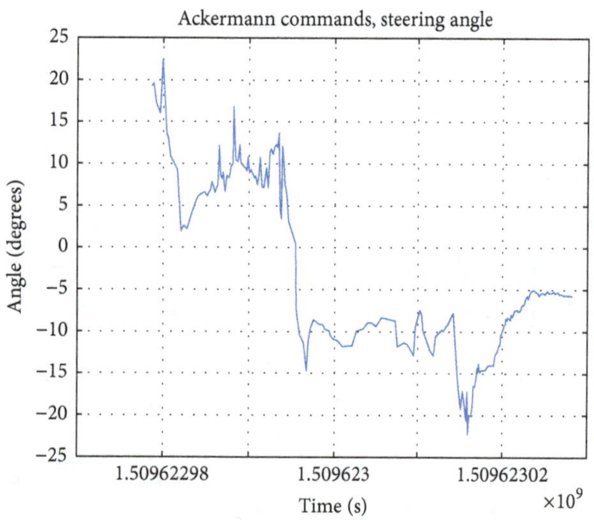

FIGURE 20: Steering.

real time provided to the vehicle. For the steering angle in Figure 20, it is possible to appreciate the variations of the commands each time the local plan is recalculated. At the beginning of the experiment, the steering angle provided is positive (turn to the left) until the vehicle reaches the position in the world where the circular building is left behind and starts to turn to the right with a negative angle.

6. Conclusions and Future Work

The first conclusion of this work determines that the use of Time Elastic Bands is a good choice even if the model for the kinematics of the vehicle is not precisely adjusted or when the vehicle is out of the path. This algorithm is useful in both of these events and when a dynamic obstacle is in front of the vehicle because of the continuous update of the local path. The second conclusion is that this method has been tested in a real vehicle where the proper configuration of the kinematics, dynamics, and geometry model could differ from the real model. This issue generates trajectories slightly different because the expected movement of the vehicle and the real movement of the vehicle are lightly different. Moreover, the TEB local planner generates the commands of velocities and steering angles that allow the vehicle to reach the goal.

For future work, more experiments are planned such as the comparison between this method and the Dynamic Window Approach and changing the control over the line generated with different path followers such as Pure Pursuit, Optimal Path Controller based on LQR, or Follow the Carrot. More complex scenarios will be added where pedestrians intersect with the trajectory generated and the vehicle should act accordingly, avoiding the pedestrian over the correct side in order to test the aforementioned solution for the elastic bands where a pedestrian is crossing in front of the vehicle.

Conflicts of Interest

The authors declare that they have no conflicts of interest.

Acknowledgments

This research is supported by Madrid Community Project SEGVAUTO-TRIES (S2013-MIT-2713) and by the Spanish Government CICYT Projects (TRA2013-48314-C3-1-R, TRA2015-63708-R, and TRA2016-78886-C3-1-R).

Supplementary Materials

The supplementary material of this work consists of a video of the real performance of the vehicle in the place where the experiment was done. In the video, it is possible to appreciate the visualizer RVIZ from ROS with two more windows. The top right window displays the left camera of the stereo vision system mounted on the vehicle. Below this window, the laser readings converted into a local map is placed. The background is composed of the global Costmap 2D and the overlay of the local Costmap 2D. On this map, the black squares correspond to obstacles and the color lines describe the trajectories. The blue line represents the global path generated by the Dijkstra algorithm and the orange represents the local path generated by the Time Elastic Bands algorithm. The green square on the bottom of the video represents the dimensions of the vehicle. As soon as the vehicle moves, the red arrows represent the movement and orientation of the vehicle and all of them are superimposed on each other because of the update of the odometry. The vehicle is able to navigate from one point to another without any collision. *(Supplementary Materials)*

References

[1] A. Broggi, M. Buzzoni, S. Debattisti et al., "Extensive tests of autonomous driving technologies," *IEEE Transactions on Intelligent Transportation Systems*, vol. 14, no. 3, pp. 1403–1415, 2013.

[2] J. Ziegler, P. Bender, M. Schreiber et al., "Making bertha drive-an autonomous journey on a historic route," *IEEE Intelligent Transportation Systems Magazine*, vol. 6, no. 2, pp. 8–20, 2014.

[3] B. Chen and G. Quan, "NP-hard problems of learning from examples," in *Proceedings of the 5th International Conference on Fuzzy Systems and Knowledge Discovery*, pp. 182–186, IEEE, October 2008.

[4] M. Khristamto, A. Praptijanto, and S. Kaleg, "Measuring geometric and kinematic properties to design steering axis to angle turn of the electric golf car," *Energy Procedia*, vol. 68, pp. 463–470, 2015.

[5] J. Canny, "A new algebraic method for robot motion planning and real geometry," in *28th Annual Symposium on Foundations of Computer Science*, pp. 39–48, IEEE, 1987.

[6] P. Bhattacharya and M. L. Gavrilova, "Voronoi diagram in optimal path planning," in *Proceedings of the 4th International Symposium on Voronoi Diagrams in Science and Engineering (ISVD '07)*, pp. 38–47, July 2007.

[7] S. Skiena, "Dijkstra's algorithm," in *Implementing Discrete Mathematics: Combinatorics and Graph Theory with Mathematica*, pp. 225–227, Addison-Wesley, Boston, Mass, USA, 1990.

[8] R. Dechter and J. Pearl, "Generalized best-first search strategies and the optimality of A^*," *Journal of the ACM (JACM)*, vol. 32, no. 3, pp. 505–536, 1985.

[9] D. Gelperin, "On the optimality of A^*," *Artificial Intelligence*, vol. 8, no. 1, pp. 69–76, 1977.

[10] M. Šeda, "Roadmap methods vs. cell decomposition in robot motion planning," in *Proceedings of the 6th WSEAS International Conference on Signal Processing, Robotics and Automation*, pp. 127–132, World Scientific and Engineering Academy and Society (WSEAS), 2007.

[11] P. Vadakkepat, K. C. Tan, and W. Ming-Liang, "Evolutionary artificial potential fields and their application in real time robot path planning," in *Proceedings of the Congress on Evolutionary Computation (CEC 00)*, vol. 1, pp. 256–263, IEEE, July 2000.

[12] S. M. LaValle, *Rapidly-exploring random trees: A new tool for path planning*, 1998.

[13] S. X. Yang and C. Luo, "A neural network approach to complete coverage path planning," *IEEE Transactions on Systems, Man, and Cybernetics, Part B: Cybernetics*, vol. 34, no. 1, pp. 718–725, 2004.

[14] M. Ben-Ari and F. Mondada, "Mapping-based navigation," in *Elements of Robotics*, pp. 165–178, Springer, 2018.

[15] C. Rosmann, W. Feiten, T. Wosch, F. Hoffmann, and T. Bertram, "Efficient trajectory optimization using a sparse model," in *Proceedings of the 2013 6th European Conference on Mobile Robots, ECMR 2013*, pp. 138–143, IEEE, September 2013.

[16] S. Gim, L. Adouane, S. Lee, and J.-P. Dérutin, "Clothoids composition method for smooth path generation of car-like vehicle navigation," *Journal of Intelligent & Robotic Systems*, pp. 1–18, 2017.

[17] J. P. Rastelli, R. Lattarulo, and F. Nashashibi, "Dynamic trajectory generation using continuous-curvature algorithms for door to door assistance vehicles," in *Proceedings of the 25th IEEE Intelligent Vehicles Symposium, IV 2014*, pp. 510–515, IEEE, June 2014.

[18] J. A. Reeds and L. A. Shepp, "Optimal paths for a car that goes both forwards and backwards," *Pacific Journal of Mathematics*, vol. 145, no. 2, pp. 367–393, 1990.

[19] A. Piazzi, C. G. Lo Bianco, M. Bertozzi, A. Fascioli, and A. Broggi, "Quintic G_2-splines for the iterative steering of vision-based autonomous vehicles," *IEEE Transactions on Intelligent Transportation Systems*, vol. 3, no. 1, pp. 27–36, 2002.

[20] C. Rösmann, W. Feiten, T. Wosch, F. Hoffmann, and T. Bertram, "Trajectory modification considering dynamic constraints of autonomous robots," in *Proceedings of the 7th German Conference on Robotics; ROBOTIK 2012*, pp. 1-6, VDE, 2012.

[21] P. Marin-Plaza, J. Beltran, A. Hussein et al., "Stereo vision-based local occupancy grid map for autonomous navigation in ros," in *Joint Conference on Computer Vision, Imaging and Computer Graphics Theory and Applications (VISIGRAPP)*, vol. 3, pp. 703–708, 2016.

[22] R. Moliner and R. Tanda, "Tool for robust tuning of PI/PID controllers with two degree of freedom," *RIAI: Revista Iberoamericana de Automatica e Informatica Industrial*, vol. 13, no. 1, pp. 22–31, 2016.

[23] R. Arnanz, F. J. García, and L. J. Miguel, "Methods for induction motor control: State of art," *RIAI - Revista Iberoamericana de Automatica e Informatica Industrial*, vol. 13, no. 4, pp. 381–392, 2016.

[24] A. Kokuti, A. Hussein, P. Marin-Plaza, A. De La Escalera, and F. Garcia, "V2X communications architecture for off-road autonomous vehicles," in *Proceedings of the 2017 IEEE International Conference on Vehicular Electronics and Safety (ICVES)*, pp. 69–74, IEEE, Vienna, Austria, June 2017.

[25] J. Zhang and S. Singh, "LOAM: Lidar Odometry and Mapping in Real-time," in *Robotics: Science and Systems*, vol. 2, 2014.

3

Structural Analysis of Shipping Fleet Capacity

Lixian Fan,[1] **Sijie Zhang**,[1] **and Jingbo Yin**[2]

[1]*School of Management, Shanghai University, 599 Shangda Rd., Shanghai 200444, China*
[2]*State Key Laboratory of Ocean Engineering, School of Naval Architecture, Ocean and Civil Engineering, Shanghai Jiao Tong University, 800 Dongchuan Rd., Shanghai 200240, China*

Correspondence should be addressed to Jingbo Yin; jingboyin@sjtu.edu.cn

Academic Editor: Stefano de Luca

With the unprecedented growth of the shipping transportation demand, substantial vessels have been built and delivered to the market. This has led to oversupply after the financial crisis in 2008 because of the abrupt decrease in transportation demand. Notwithstanding the importance of shipping market studies in the investment decision-making, there are relatively few empirical studies modelling the impacts on the structural changes of the fleet supply variables. By considering new orders, current fleet size, and demolitions in ship capacity supply, this study develops a systematic model in both bulker and container markets. The three-stage least squares method is employed to estimate the model to avoid endogenous issues. The primary finding suggests the significant impact of market, cost, and operational factors on fleet capacity supply. It also reveals the relatively rational activities in ordering new vessels and cautious in demolition decisions in the container market because of the large capital investment required. These are relevant to investment and demolition decisions in both the bulker and container markets.

1. Introduction

The shipping industry has a long history, with the first cargoes being moved by sea more than 5,000 years ago [1]. It has been becoming indispensable to global trade; it is estimated that more than 80% of the international cargo is transported by ship [2]. According to Clarksons research [3], world trade has grown from 8,491 million tons in 2007 to 11,091 million tons in 2016, a growth rate of 30.62%. Along with the significant increase in world trade, shipping capacity has grown rapidly. According to Clarksons research [3], the global shipping fleet increased by 71.29% in million deadweight tons from 2007 to 2016, despite the 2008 financial crisis. In fact, it has become clear in recent years that, despite the growth in world trade, the shipping capacity supply has surpassed the demand for trade transportation.

Similar to other markets, dynamic changes in supply and demand in the shipping market can pose a variety of problems. For example, when transport capacity exceeds trade needs, the freight rate will decline; stakeholders must respond quickly to reduce losses, and they may reduce the number of fleet in operation by laying up vessels, reducing new orders or scrapping old ships [1, 4]. This asymmetrical movement of supply and demand has attracted many scholars to investigate the dynamics of the shipping market, hoping to identify a relevant model to capture changes in the market.

It is obvious that among the four submarkets in the shipping industry—freight, new-building, second-hand, and demolition—only operations in the new-building and demolition markets can affect the actual fleet numbers in shipping transportation. In a rising market, shipping companies can place new orders to increase capacity, while in a depressed situation, old and obsolete vessels can be sent to scrap dealers for demolition [1]. However, a search of the literature discloses that most previous studies have focused on the characteristics of freight rate or analysed the relationship between various freight variables [5–8]. Although some scholars pointed to factors that affect changes in shipping market capacity [1, 9–11], few of them have investigated these factors using empirical models. Therefore, this study innovatively employs the Simultaneous Equations Model (SEM) to analyse the structural changes in shipping capacity, a model that can help to eliminate endogeneity problems introduced by interdependent dependent variables. This systematic econometric SEM can help in deriving the marginal effect of each factor on endogenous variables. We believe the results of this study could help company managers, investors, and shipowners make strategic decisions to enhance profits.

The study is organized as follows: Section 2 reviews previous research about shipping supply, demand, and demolition markets. Section 3 analyses the dynamics of the shipping market and the factors influencing changes in the market. Section 4 describes the data, introduces the method used in this study, and reports the empirical results. Finally, Section 5 concludes.

2. Literature Review

Many researchers have studied the shipping freight market to gain a clear understanding of the movement of freight. Broadly speaking, two approaches have been developed. First, many researchers have focused on modelling the demand and supply for transportation using the structural model [12–15]. Second, in recent years, inspired by developments in financial economics, the focus has been on modelling the freight rate or freight derivatives directly in a stochastic-statistical model [5, 16–26].

In addition to studying the relationships between various rates, many scholars have studied the factors affecting shipping supply, demand, and investment. Svendsen [9] introduced the determinants of maritime demand and supply. He argued that the volume of cargoes, the distance of transport, political factors affect trade and shipping, total tonnage in operation, the age of ships, and ship operating efficiency determine the supply of capacity. Stopford [1] concluded that there are five key factors influencing the demand for sea transport: the world economy, the level of seaborne commodity trade, average haul, random shocks, and transport costs. He also explained that there are three key aspects influencing the supply of merchant ships: size of the fleet, shipbuilding, and demolition. Therefore, these three variables will be the focus of this study. Volk [10] studied the factors which led to changes in the supply and demand curves related to shipping. He concluded that it is difficult to forecast freight rates in the long term because of the unpredictability of factors influencing these rates. Luo, Fan [15] developed an econometric model to analyse the fluctuation of the container freight rate and fleet capacity. They estimated the Simultaneous Equations Model using three-stage least squares, demonstrating a high degree of accuracy for in-sample and out-sample prediction. Lun and Marlow [27] analysed that shipping capacity is associated with profitability and revenue in the liner shipping industry. It is also claimed that non-mega operators are the most efficient firms in operation. Fan and Luo [11] analysed capacity expansion and ship choice decisions in the liner shipping industry. They developed a theoretical model describing the probability of capacity expansion and tested the model empirically using ship investment data. Their results suggest the market-driven and competitive strategies are relevant to capacity expansion. Notwithstanding of the significant contribution of above studies, a key variable, demolition fleet, which influencing the market supply, has been missing in analysing the shipping market supply.

As pointed out by Stopford [1], ship demolition is a key factor influencing fleet capacity. However, relatively few studies have discussed demolition. Among these, some have focused on the social, environmental, and safety issues, such as Matzluck [28], which noted that the Hong Kong Ship Recycling Convention on ship demolition has not been adopted and implemented by most countries and that this convention does not prohibit the beaching method in ship demolition. In an empirical analysis, Buxton [29] explored the basic characteristics of the shipbreaking market and analysed the trends in the market from the 1960s to the 1990s. He argued that the scrap value of a ship depends on the realisable value of the ship's materials and the cost of demolition. Mikelis [30] argued that shipbreaking prices are affected not only by the supply of steel but also by the needs of the steel market and the fleet supply. Knapp and Kumar [31] analysed the probability of ship demolition using a binary logic regression model. Kagkarakis and Merikas [32] constructed a VAR model to study the causal relationship between international scrap prices and demolition prices and showed that the model can also be used to predict scrapping prices. Yin and Fan [33] discussed that ship demolition is an important strategy in balancing fleet capacity. By employing Cox proportional hazards regression, they analysed the different demolition activities for shipowners from different countries at different market cycles. Different from these studies focusing on the single variable of demolition fleet, this study tries to investigate the impacts on fleet supply of new orders, current fleet size, and ship demolitions. Meanwhile, it tries to investigate the structural connections among them.

Although some of these previous studies have discussed the freight, capacity, and demolition markets, the literature lacks a systematic and quantified analysis of the supply of shipping. To understand how the characteristics determining fleet capacity have evolved, we develop a systematic theoretical model to explain the interdependence among demand and supply variables. We describe the structures behind the three key variables, i.e., new orders, current fleet size, and ship demolitions, by constructing an SEM model. This model is capable of eliminating problems brought about by the existence of interdependent endogenous variables. It can also derive the marginal effect of each of the market-related factors described in Section 3.2 below on the key endogenous variables by systematic estimation methods. This may provide valuable information for various decision- and policy-makers.

3. Structural Movement of the Shipping Fleet Market

It is no doubt that the bulk shipping market is almost a complete competitive market, where thousands of shipping companies are providing transportation service and the freight rate is determined by the interaction of market demand and supply. However, there are some different opinions on the contestability of the container market [34, 35]. Through analysis of three requirements, i.e., same access to technology for entrants and incumbents, low sunk costs (ships can be

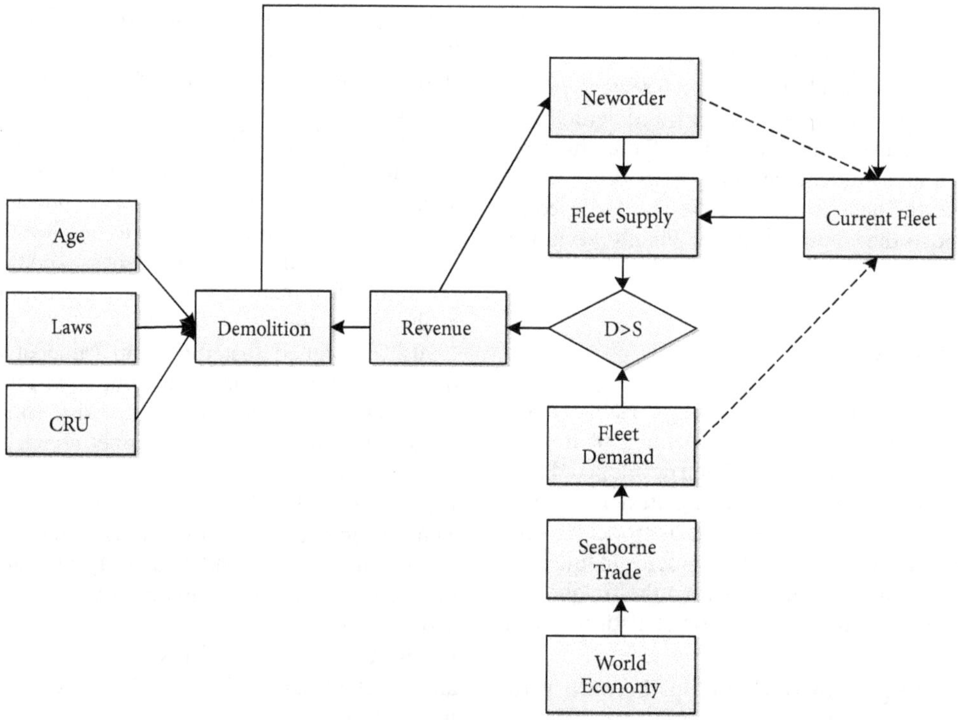

FIGURE 1: *Dynamic analysis of shipping market.* Note: CRU represents the steel price index.

easily diverted from one route to another), and competitions from tramp and bulk carrier, Davies [36] concluded that the container industry is a contestable market. Haralambides [37] also claimed that, irrespective of market structure and capital concentration, the real issue is how accessible (contestable) the market is to new competition. If entry (and eventually exit) to the market by new competitors is easy, the market is said to be contestable. He further analysed that liner shipping is a contestable market. Through explanatory and empirical studies, Notteboom [38], Munim and Schramm [26] and Hirata [39] all concluded that the container market is contestable. As Munim and Schramm [26] claimed "...in times of sudden demand increase, there are many ways to solve the problem, including increasing the cruise speed or increasing loading factor, whereas at the same time, there is a high supply side flexibility present, too...Thus, for the container shipping industry, we can assert that both demand and supply are flexible enough to confirm the assumption of market equilibrium model by Beenstock and Vergottis [14]".

3.1. The Interaction of Supply and Demand. Shipping is a complex industry. On the supply side, there are four different submarkets: freight, second-hand, new-building, and demolition. Each sector has its own unique methods of operation. The main assets (ships) in the shipping industry provide an entire range of services with respect to a variety of goods. The demand for transport is derived from the demand for the goods being transported. International cargo trade is a key factor that affects the demand for shipping transportation services. Stopford [1] observed that the level of world economic output (e.g., GNP or GDP) is by far the most important factor affecting the volume of seaborne trade.

As with any market, the intersection of demand and supply determines the price in the shipping freight market, which in turn affects demand and supply. This dynamic relationship is illustrated in Figure 1. When the demand for seaborne trade exceeds the supply of fleet, the freight rate increases. The higher rate suggests higher profits, so new orders are placed to increase transport capacity. Over time, when the capacity supply exceeds the need for transport, the freight rate decreases. New orders will be decreased and older and inefficient ships are laid up or sent for breaking [10]. It is obvious that activities in the new-building and demolition markets determine changes in the shipping capacity. Because second-hand transactions and lay-ups do not affect the number of ships in the market, they are not depicted in the figure.

3.2. Factors Influencing the Shipping Market. Many factors affect the demand for shipping, including the state of the world economy and international maritime trade, average achieved profit, political events, and transportation costs [40]. The supply of shipping capacity is also affected by various factors, including the size of the world fleet and its productivity, shipbuilding, demolition, and freights (revenue) [41]. Functionally, the fleet capacity supply depends on the size of the existing fleet, plus the delivery of new ships minus the number of ships scrapped.

Shipbreaking has been seen as an essential element for balancing shipping supply and demand because the scrapping of uneconomical vessels can reduce the ship supply [29, 33].

Similar to the shipping supply and demand, ship demolition is also affected by many factors: whether a vessel is technologically or economically outdated, the age of the vessel, shipbreaking industry regulations, scrap prices, and others.

Summarizing the literature, factors influencing the supply of shipping can be classified into market factors, cost, and operational factors. These factors are described in the following subsections and are also illustrated in detail in Table 1.

3.2.1. Market Factors Affecting Shipping Capacity. Shipping companies provide global shipping services to meet the demand for sea transport; i.e., the demand for shipping transportation is derived from the world trade [1]. Shipping companies have to make ship investment decisions according to the transportation demand. When the world economy is booming, the transportation demand is increasing. If the capacity supply cannot satisfy with the demand, the freight rate will increase. This will motivate the shipping companies to place new orders for new ship investment. Because of the construction lags, when the new-buildings are delivered to the market, the market situation may have changed, which determines the new freight level [1, 15, 25]. The uncertain transportation demand plus this shipbuilding lag make the freight market volatile.

On the other hand, when the prosperous economy turns into depression, although new orders will be decreased, the large orders placed in previous increasing situations are kept delivering to the market. If the shipping capacity increases faster than demand, a limited recovery in economic growth may not be effective in bringing the shipping freight rate up. Such as the current situation after the 2008 financial crisis. Some old and obsolete vessels have been sent for demolition. Stopford [1] analysed from the perspective of psychology that the recessionary phase of the shipping cycle will not end until some successive ships are sent for scrapping to decrease the oversupply. In general, ship demolition is a primary element in balancing the supply of and demand for ships [29].

To sum up, the key factors influencing the demand for the shipping market are the world economy and the level of seaborne commodity trade, while the supply of merchant ship capacity is determined by the size of the fleet, shipbuilding (new orders), and demolition. In addition, the relative strength of demand and supply determines the market freight rate level and its profitability.

3.2.2. Cost and Operating Factors Affecting Shipping Capacity. In addition to ordering or demolishing vessels, shipping companies can also control the actual fleet capacity by various supply side flexibility tactics [26], such as laying off vessels, slow steaming, and rescheduling routes. However, these tactics are effective in the short run for individual participants. In the long run, the investment and disinvestment are more strategically important. Since this study is focused on the aggregate market level with yearly data, these tactics are not considered in the empirical model.

The shipping market is different with other markets as in which the main assets (ships) are invested to carry cargoes around the world. When making the investment decision, all buyers pay the same price for the same new ship [10]. That is to say, the new-building price is not only a kind of capital cost; it represents the amount of future investment in the shipping market as well. High freight rates mean high profits, which will attract shipping companies to invest in new-buildings. With increases in demand for new orders, prices in the shipbuilding market increase, leading to an increase in capital costs for shipping companies [42].

As discussed above, there are construction lags in the shipbuilding market. This suggests that there are vessels under construction (it is called 'order book') when shipping companies making investment decisions. This order book actually implies the potential capacity supply in subsequent years. Therefore, it is considered as an influencing factor when shipping companies make investment decisions [43].

In the shipping industry, bunker consumption is a major voyage cost in operating a vessel [1]. Ships face less pressure to operate at slow steam to save fuel when oil prices are low, and higher speeds lead to more capacity. Therefore the bunker price is considered as an influencing factor for fleet deployment.

In addition to above factors, there are some unique variables influencing the demolition market: First, the age of a ship: in general, the average lifespan of a ship is about 30 years. Ships may be scraped because of economic, technical, and regulatory limitations, which could be represented by the age of a ship. As discussed by Evans [44], a ship's cost of repairs, maintenance, surveys, and associated time out of service will increase with the increase of age; since we are focusing on the fleet variables on an aggregated level, a variable, *HighAage*, is designed to be included in the empirical model, which is the ratio of older (20+ years old) vessels to the total fleet (see detail in Table 1). Second, the price of steel: as ship demolition can provide an important source of cash flow when freight rates are low [1], it could be impacted by the global demand of steel resources. Through empirical analysis, Mikelis [45] found a positive correlation between ship demolition and global steel production. So, the steel price is also included in the statistical model in this study. Finally, regulations on ship demolition activities: along with the development of the ship demolition industry, issues of unacceptable worker conditions, pollution, and environmental degradation have aroused much concerns [33]. International conventions and guidelines have attempted to address these problems, such as the Basel Convention, the International Maritime Organization (IMO) guidelines on ship recycling, the International Labor Organization (ILO) guidelines, and Hong Kong International Convention for Safe and Environmentally Sound Recycling of Ships (HKIC). Hence, the impact of demolition laws and regulations on the demolition behaviour is also considered in the statistical model.

4. Data and Methodology

4.1. Data Sources and Description. The data used in this study are collected from two sources: the Shipping Intelligence Network (SIN) by Clarkson Research and the Qianzhan Database (https://qianzhan.com). The SIN database proves various

Table 1: Description of the variables used.

Markets	Variables	Description	Factor	Mean	Maximum	Minimum	Std. Dev.
Bulker	*Neworder*	New order ship (million dwt)	Market	45.378	159.338	8.935	40.627
	Trade	World seaborne iron ore, coking coal, steam coal, phosphate rock, bauxite/alumina, minor bulk trade (Million tonnes)	Market	3102.547	4543.477	1850.778	938.013
	Fleet	Fleet capacity in (million dwt)	Market	422.703	776.653	243.719	184.663
	NBPI	New-building price index	Cost and Operational	133.402	232.09	91.889	36.382
	Freight	Baltic Exchange Dry Index	Market	2265.762	7071.000	673.000	1797.090
	Earings	Clarksons Average Bulker Earnings ($/Day)	Market	14904.048	44267.000	6200.000	10888.390
	Orderbook	Oderbook of Bulkcarrier (million dwt)	Cost and Operational	124.197	326.080	23.973	105.078
	Bunkerprices	380cst bunker prices, Rotterdam	Cost and Operational	289.051	639.635	67.621	191.154
	Demolition	Total Bulkcarrier Demolition (million dwt)	Market	11.516	33.413	0.357	10.351
	HighAge	20 years and over (%)	Cost and Operational	26.599	41.180	8.040	10.193
	CRU	Steel price index	Cost and Operational	136.969	234.883	74.917	44.764
Container	*Neworder*	New order ship (,000 TEU)	Market	1137.342	3314.011	60.479	872.586
	Trade	World Seaborne Container Trade (,000 TEU)	Market	111689.4	181128.2	46779.62	44759.81
	Fleet	Fleet capacity in (,000 TEU)	Market	9714.625	19743.70	2920.019	5447.255
	NBPI	New-building price index	Cost and Operational	89.369	123.565	69.017	16.667
	Freight	Containership Timecharter Rate Index	Market	76.238	152.000	35.000	31.607
	Earings	Clarksons Average Containership Earnings ($/Day)	Market	12000.143	25018.000	5070.000	5531.321
	Orderbook	Oderbook of container (,000 TEU)	Cost and Operational	3069.479	6626.198	626.928	1856.390
	Bunkerprices	380cst bunker prices, Rotterdam	Cost and Operational	289.051	639.635	67.621	191.154
	Demolition	Total container Demolition (,000 TEU)	Market	146.495	654.411	2.3460	181.601
	HighAge	20 years and over (%)	Cost and Operational	13.551	24.200	4.1400	6.866
	CRU	Steel price index	Cost and Operational	136.969	234.883	74.917	44.764

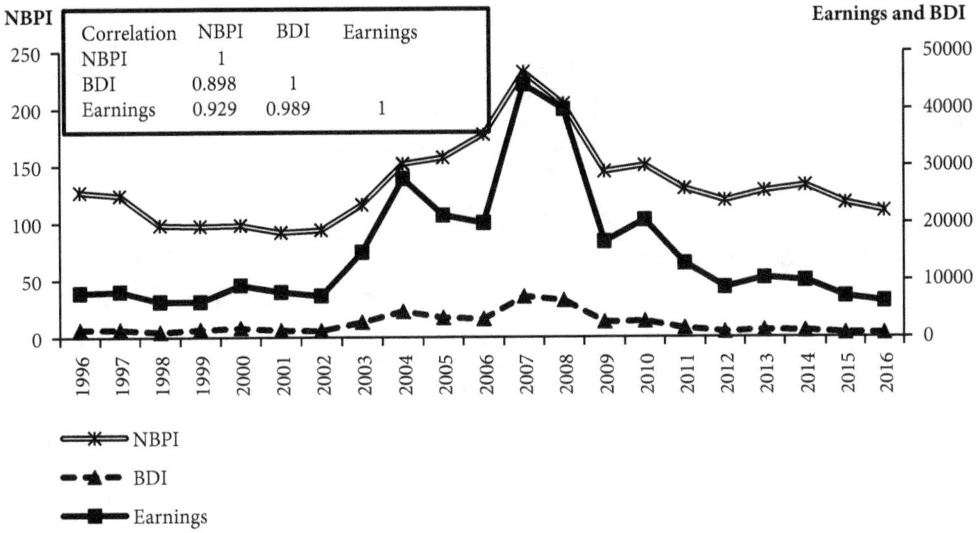

FIGURE 2: Correlations between new-building price, BDI, and earnings in the bulker market.

time series data for the shipping industry. As an aggregated model analysis, this study only considers the bulker and the container submarkets using yearly time series. The fleet size, demolition, order book, new order, and seaborne trade are from the SIN database. Considering the availability of the data, the sample is finally determined from 1996 to 2016. The Qianzhan database is a public data source which provides weekly, monthly, and yearly data for international, regional, national, and industrial aspects. In this study, the steel price index is obtained from the Qianzhan database. A detailed description of the variables can be found in Table 1.

According to the table, the ratio of new orders to the total fleet is 10.735% (45.378/422.703) for the bulker market and 11.708% (1137.342/9714.625) for the container market. Similarly, the order book ratio for the bulker market is 29.382% (ratio of order book to fleet) which is slightly lower than the container market of 31.596%. The ratio of demolitions to the fleet is 2.724% for the bulker market and 1.508% for the container market. This suggests the relatively active demolition behaviour in the bulker market. It can also be confirmed by the higher ratio of the older ships (*HighAge*) for the bulker ships (26.599%) than for the container ships (13.551%), as usually older ships are more inclined to be scraped than younger ones.

4.2. Simultaneous Equations Model. As discussed above, the three key variables influencing shipping fleet capacity are new orders, current fleet size, and demolitions. Considering the factors discussed in Section 3, we analyse the impact of new orders, current fleet and demolitions from two points of view: market and cost and operations (see detail in Table 1). Since the magnitudes of the values are different, the model is estimated separately for the bulker and container markets.

$$\begin{aligned} Neworder &= c_{11} + c_{12} \times Trade + c_{13} \times Fleet + c_{14} \\ &\quad \times NBPI + c_{15} \times Orderbook + \varepsilon_1 \end{aligned} \tag{1}$$

$$\begin{aligned} Fleet &= c_{21} + c_{22} \times Neworder\,(\text{-2 or} - 3) + c_{23} \\ &\quad \times NBPI + c_{24} \times bunker\ price + c_{25} \\ &\quad \times Trade + \varepsilon_2 \end{aligned} \tag{2}$$

$$\begin{aligned} Demolition &= c_{31} + c_{32} \times HighAge + c_{33} \times Laws + c_{34} \\ &\quad \times Fleet + c_{35} \times Neworder + c_{36} \\ &\quad \times CRU + \varepsilon_3 \end{aligned} \tag{3}$$

It is noteworthy that the correlations between the new-building price, freight, and average earnings are high (Figures 2 and 3). Therefore, only *NBPI* is included in the equations since it also represents the investment cost of a new ship. There is a lag for the *Neworder* variable in (2), which is determined by calculating the mean difference between the time placing a new-building order and the time it is delivered to the market. In this study, the mean shipbuilding lag is 2 and 3 for the bulker and container ship separately. In (3), the dummy variable, *Laws*, refers to industry-related scrapping guidelines. In recent years, international organizations such as the IMO, the ILO, and Greenpeace, as well as countries around the world, have given unprecedented attention to the shipbreaking industry and have developed guidelines to promote safety, health, and environmental protection in scrapping operations. For example, the ILO promulgated 'Safety and Health in Shipbreaking: Guidelines for Asian Countries and Turkey' in 2003 [46], and 'Guidelines on Ship Recycling' in the 23rd Assembly in November-December 2003 by Resolution A.962 [46, 47]. Therefore, the value of *Laws* is 1 in 2004 and after, while in previous years it is 0.

In this SEM system, *Neworder*, *Fleet*, and *Demolition* are endogenous variables which will be determined by all the other exogenous variables. In the system, the endogenous variable *Neworder* is an independent in the *Fleet* and *Demolition* equations, and the *Fleet* variable is an independent in the *Demolition* equation, so the system is interdependent.

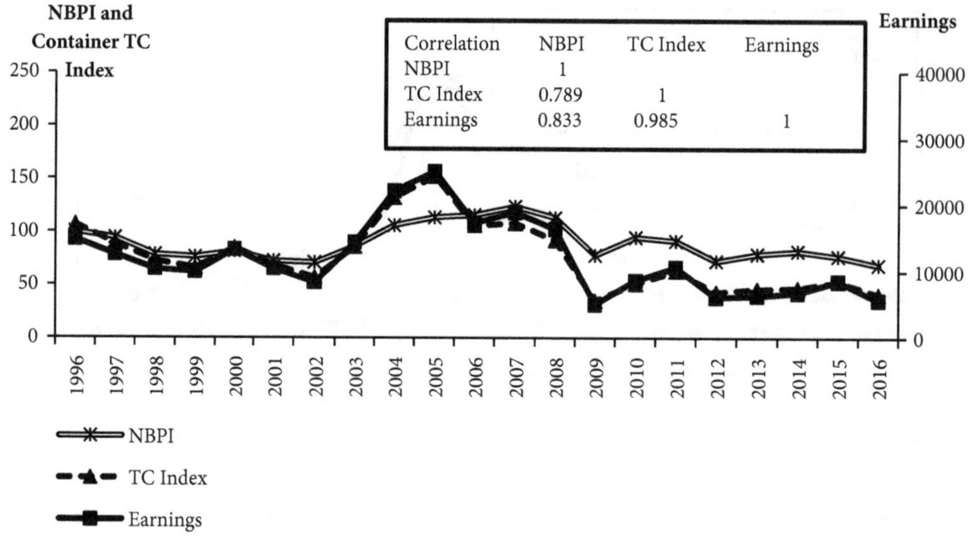

FIGURE 3: Correlations between new-building price, time charter index, and earnings in the container market.

This suggests that the residuals are correlated: cov(ε_1, ε_2)≠0, cov(ε_1, ε_3)≠0, cov(ε_2, ε_3)≠0. Put (1) to (2), and (1) and (2) to (3). It is obvious that cov(*Fleet*, ε_1)≠0, cov(*Demolition*, ε_1)≠0, and cov(*Demolition*, ε_2)≠0. Since the endogenous variables are correlated with the disturbances, the least squares estimators of the parameters with endogenous variables on the right-hand side are inconsistent.

A general method of obtaining consistent estimates is the method of instrumental variables (IV). All the exogenous and predetermined variables in the system are perfect instrumental variable for the estimation, because they are correlated with the endogenous variables as they appear in the equations, and they are independent to the error term as they are exogenous variables. The IV estimator will be consistent and have asymptotic covariance matrix. Based on the principle of instrumental variables, there are generally two approaches for estimation of SEM. One is limited information estimators which will estimate the structural parameters of each equation separately using all the information of the exogenous and predetermined variables in the whole system. Ordinary least squares estimator, instrumental variable estimator, two-stage least squares (2SLS) estimator, and limited information maximum likelihood (LIML) estimator all belong to this. But this type of estimator ignores information concerning that one endogenous variable may appear in other equations. It also ignores information that the error terms among the equations are correlated. Another type of estimation is full information estimator or system methods of estimation. It jointly estimates the equations in the model and get a more efficient estimation. System methods include three-stage least squares (3SLS) estimator and full information maximum likelihood (FIML) estimator. Detailed analyses can be obtained from Greene [48]. In this study, the 3SLS is adopted in the estimation where the exogenous variables and the predetermined endogenous variables in the system are used as instruments.

4.3. Empirical Results. Tables 2-3 are the estimated results from simultaneous equations for the bulker and container markets separately. The results show that most of the variables are significant, and the high adjusted R-squares suggest the goodness of the model fit.

4.3.1. Bulker Shipping Market. Table 2 is the estimated results for the bulker market. Equation (1) takes into account the effects of *seaborne trade, current fleet, new-building price,* and *order book*. However, the results show that only the *new-building price index* is significant with a positive coefficient. Generally speaking, new investment is usually negatively correlated with investment cost. Here the positive coefficient of the new-building price index suggests that new orders of ships increase with the increase of the new-building price. This can be explained by the high correlation (0.898) of the new-building price with the freight rate (BDI). Hence, the high new-building price implies high freight rate, which suggests a booming market. In this situation, shipping companies are eager to invest in new ships.

To further understand the results in (1), we illustrate the development of the *new order, order book, seaborne trade,* and *fleet* in the bulker market in Figure 4. The new order fluctuates regardless of the increasing trend in seaborne trade and fleet. It is interesting to see that ships on order are not a determinant factor for shipping companies making investment decisions. Although the order book stayed very high after 2008, there is no obvious decreasing trend in new orders and there were actually big jumps in 2010 and 2013. This indicates the irrationality of shipping company decisions to invest in the bulker market.

Equation (2) analyses the factors influencing fleet capacity in the bulker market. Since it usually takes two years to deliver a new-building to the bulker shipping market [3], we add the two-year lagged variable of *Neworder* to the regression. However, the coefficient is not significant.

TABLE 2: Estimated results of the SEM for bulker ship.

Simultaneous equations/Dependent	Variables	Coefficient	Std. Error	Prob.	Adjusted R-square	Durbin-Watson stat
Equation (1) *Neworder*	*Constant*	-75.511	42.749	0.085	0.612	1.972
	Trade	-0.020	0.079	0.807		
	Fleet	0.117	0.360	0.746		
	NBPI	**0.995**	**0.441**	**0.030**		
	Orderbook	0.014	0.079	0.861		
Equation (2) *Fleet*	***Constant***	**-159.629**	**48.180**	**0.002**	0.994	1.834
	Neworder(-2)[a]	0.022	0.079	0.783		
	NBPI	**-0.970**	**0.134**	**0.000**		
	Bunkerprice	**-0.090**	**0.036**	**0.018**		
	Trade	**0.232**	**0.011**	**0.000**		
	AR(1)	**0.632**	**0.176**	**0.001**		
Equation (3) *Demolition*	***Constant***	**-34.914**	**7.816**	**0.000**	0.828	2.089
	HighAge	**0.413**	**0.170**	**0.020**		
	Laws	**-11.522**	**3.009**	**0.001**		
	Fleet	**0.077**	**0.010**	**0.000**		
	Neworder	**-0.131**	**0.024**	**0.000**		
	CRU	**0.118**	**0.032**	**0.001**		
	AR(1)	**-0.471**	**0.234**	**0.051**		

Note: [a]Neworder (-2) is the two-year lagged variable of Neworder. The Durbin-Watson statistics suggest one-order serial correlation of the residuals in the equations, so *AR*(1) is added to deal this issue.

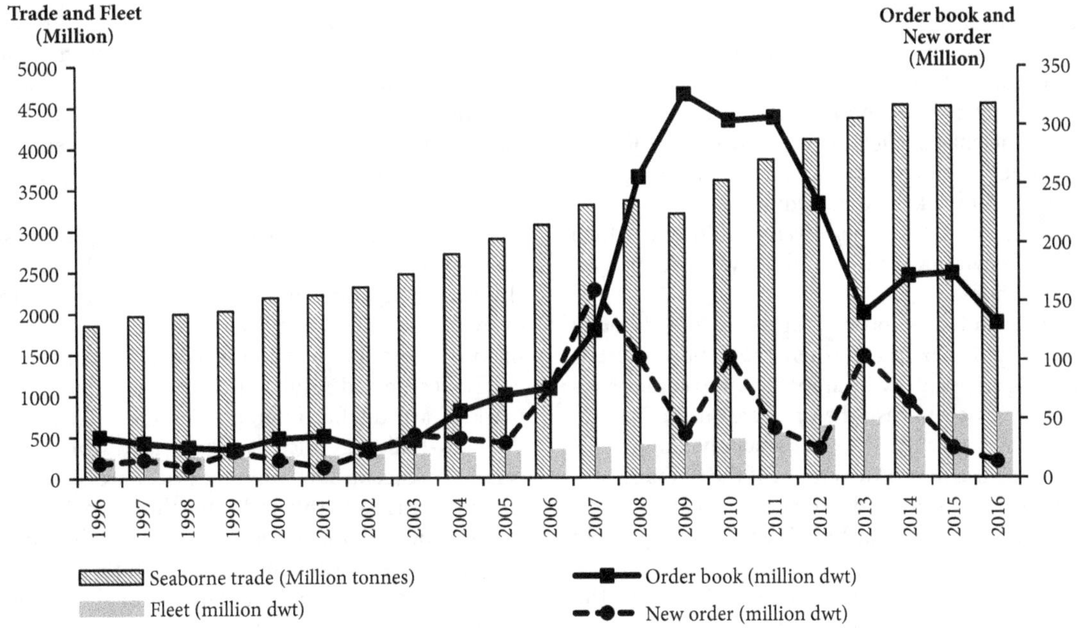

FIGURE 4: *New order, order book, seaborne trade,* and *fleet* development in the bulker market.

The coefficient for the new-building price, *NBPI*, is negatively significant. High new-building price not only suggests a boom in the shipping market but also represents the shipping company's investment cost. In a booming market, shipping companies rush to order new vessels, which pushes up prices. To keep pace with the demand for shipping transportation, some companies choose to invest in second-hand vessels as these can be put into the market immediately.

Bunker price represents the cost of operating the ship. Therefore, it is reasonable that the coefficient for bunker price is negatively significant.

Table 3: Estimated results of the SEM for container ship.

Simultaneous equations/Dependent	Variables	Coefficient	Std. Error	Prob.	Adjusted R-square	Durbin-Watson stat
Equation (1) Neworder	***Constant***	**-2862.771**	**436.579**	**0.000**	0.780	1.914
	Trade	**0.077**	**0.018**	**0.000**		
	Fleet	**-0.487**	**0.152**	**0.003**		
	NBPI	14.510	10.717	0.185		
	Orderbook	**-0.371**	**0.051**	**0.000**		
	AR(1)	**-0.766**	**0.162**	**0.000**		
Equation (2) Fleet	*Constant*	624.384	721.757	0.393	0.995	1.990
	Neworder(-3)[a]	0.067	0.078	0.395		
	NBPI	**-61.243**	**5.328**	**0.000**		
	BunkerPrice	-0.977	0.729	0.189		
	Trade	**0.129**	**0.004**	**0.000**		
	AR(1)	**0.487**	**0.218**	**0.032**		
Equation (3) Demolition	*Constant*	35.462	285.670	0.902	0.740	2.127
	HighAge	-0.223	10.615	0.983		
	Laws	15.786	89.884	0.862		
	Fleet	**0.034**	**0.012**	**0.006**		
	Neworder	**-0.072**	**0.030**	**0.021**		
	CRU	-1.086	0.741	0.152		

Note: [a]Neworder (-3) is the three-year lagged variable of Neworder.

The coefficient for the *trade* variable is positively significant. Obviously, with an increase in market demand, ship capacity supply will increase.

Equation (3) investigates the impacts on the scrapping market. The coefficient for *HighAge* is positively significant. Since the inspection of older ships by PSC authorities is more stringent, they are more likely to be demolished.

The coefficient for *Laws* is negatively significant, which is to say that the environmental laws do have an impact on shipbreaking activities.

Current fleet and new orders represent the capacity supply of the shipping market. When the supply exceeds demand, shipping companies will choose to scrap some obsolete or inefficient vessels to ease the pressure on the market. Conversely, when the demand exceeds supply, companies will choose to increase new orders to meet the demand and ship scrapping will be suspended. So the coefficient for *Fleet* is positively significant, but that for *Neworder* is negatively significant.

CRU represents the demand in the steel industry. The higher the steel price index, the larger the demand for steel. The positive coefficient for *CRU* suggests profit-motivated behaviour in the scrapping market. This result echoes to the main findings from Mikelis [30].

4.3.2. Container Shipping Market. The estimated results for the container market (Table 3) are different from those for the bulker market. In (1), the coefficient for *Trade* is positively significant and the coefficient for *Fleet* is negatively significant. This is reasonable because the higher the amount of maritime trade, the higher the demand for capacity. To meet the market demand, shipping companies will increase the number of new orders. On the contrary, when the capacity of the current fleet increases, shipping companies aiming to maximise profits will not choose to add new orders. Because an increase in new orders will lead to oversupply in the subsequent few years, the freight rate and profitability will decline.

The coefficient for *Orderbook* is negatively significant. This suggests that a high volume of order book will decrease shipping companies' investment in new ships as otherwise an overcapacity may result. This indicates the rational decision-making of the shipping companies in the container market.

The estimated results of (1) for the bulker and container markets are totally different. For the bulker market, only the new-building price is significant, while all the other variables (except new-building price) are significant for the container market. This indicates that capacity development in the container market is more rational than in the bulker market, as the investment decisions in the container market consider demand, supply, and operational conditions.

Similar to the results in Table 2, the estimates in (2) suggest that the capacity supply of the container market is mainly affected by trade and new-building price. It should be noted that it usually takes three years to build a container vessel [3], so the three-year lagged variable of *Neworder* is included in (2).

In (3), only the coefficients for *Fleet* and *Neworder* are significant. Similar to the results for the bulker market, the coefficient for *Fleet* is positive but is negative for *Neworder*. However, the coefficients for *HighAge* and *Laws* are insignificant. This is possibly due to the higher capital cost of a container ship, and it is more difficult to scrap a container ship just because of old age. In contrast, bulker ships are relatively small and cheap, and the demolition decision can easily be impacted by market and operational factors.

5. Summary and Conclusion

As one of the most important modes of transporting goods, shipping has made a significant contribution to globalisation. Through an analysis of the shipping market, we propose an econometric model focusing on the three key variables of its fleet supply: new orders, current fleet size, and demolitions. Our review of the literature indicates that the main factors influencing these variables are related to markets, costs, and operations. Considering the interdependence between these three endogenous variables, the 3SLS method is used in the regression analysis, leading to a better degree of fitting. Considering concerns about environmental protection in recent years, we add the regulations issued by international agencies to the factors influencing the demolition equation to examine the impact of these regulations on the shipbreaking industry.

The estimated results suggest the significant impact of market, cost, and operational factors on new orders, fleet development, and ship demolition. The results also indicate that new ship investment in the container market is conducted more rationally than in the bulker market, as market, cost, and operational factors all impact the investment decision in the container market. This is to be expected because there is a relatively larger requirement of capital investment for container ships. The result also suggests that shipping companies are more hesitant to engage in demolitions in the container market than in the bulker market.

In short, the model developed in this study is useful for providing information relevant to public policy and for decision-makers in private enterprises. Maritime organizations and environmental organizations may be able to use this information to regulate the shipbreaking industry so as to reduce environmental pollution. In addition, investors can use this information in ship financing decisions to minimise investment risks. Finally, the owner and the ship operator can use these results to design and implement business strategies to maximise their profits.

Conflicts of Interest

The authors declare that they have no conflicts of interest.

Acknowledgments

This research is supported by the National Natural Science Foundation of China (71673181, 71831008) and the National Social Science Fund of China (17BGL259).

References

[1] M. Stopford, "Maritime economics 3e," 2009, London: Routledge.

[2] UNCTAD, "Review of Maritime Transport, 2017," in *United Nations Publication*, United Nations, New York, NY, USA, 2017.

[3] Clarksons, "Shipping review database," 2016, Clarksons Research Ltd.

[4] P. Nielsen, L. Jiang, N. G. M. Rytter, and G. Chen, "An investigation of forecast horizon and observation fit's influence on an econometric rate forecast model in the liner shipping industry," *Maritime Policy & Management*, vol. 41, no. 7, pp. 667–682, 2014.

[5] M. G. Kavussanos, "Time varying risks among segments of the tanker freight markets," *Maritime Economics & Logistics*, vol. 5, no. 3, pp. 227–250, 2003.

[6] M. S. Haigh, N. K. Nomikos, and D. A. Bessler, "Integration and causality in international freight markets: Modeling with error correction and directed acyclic graphs," *Southern Economic Journal*, vol. 71, no. 1, pp. 145–162, 2004.

[7] A. W. Veenstra, "The term structure of ocean freight rates," *Maritime Policy & Management*, vol. 26, no. 3, pp. 279–293, 1999.

[8] A. M. Goulielmos, "What can we learn from 259 years of shipping cycles?" *International Journal of Shipping and Transport Logistics*, vol. 2, no. 2, pp. 125–150, 2010.

[9] A. Svendsen, *Sea Transport and Shipping Economics. Weltwirtschaftliches Archiv*, Republication for the Institute for Shipping and Logistics, Bremen, 1958.

[10] B. Volk, *The Dynamics Of Supply And Demand İn Tramp Shipping*, Fachhochschule University Of Applied Sciences: Launceston, Launceston, 2002.

[11] L. Fan and M. Luo, "Analyzing ship investment behaviour in liner shipping," *Maritime Policy & Management*, vol. 40, no. 6, pp. 511–533, 2013.

[12] V. D. Norman and T. Wergeland, *Nortank-A Simulation Model of the Freight Market for Large Tankers*, Centre for Applied Research, Norwegian School of Economics and Business Administration, Bergen, Norway, 1981.

[13] M. Beenstock and A. Vergottis, "An econometric model of the world market for dry cargo freight and shipping," *Applied Economics*, vol. 21, no. 3, pp. 339–356, 1989.

[14] M. Beenstock and A. Vergottis, *Econometric Modeling of World Shipping*, Chapman & Hall, London, UK, 1993.

[15] M. Luo, L. Fan, and L. Liu, "An econometric analysis for container shipping market," *Maritime Policy & Management*, vol. 36, no. 6, pp. 507–523, 2009.

[16] K. Cullinane, "A short-term adaptive forecasting model for biffex speculation: A box-jenkins approach," *Maritime Policy & Management*, vol. 19, no. 2, pp. 91–114, 1992.

[17] M. G. Kavussanos, "Comparisons of volatility in the dry-cargo ship sector: Spot versus time charters, and smaller versus larger vessels," *Journal of Transport Economics and Policy*, vol. 30, no. 1, pp. 67–82, 1996.

[18] A. W. Veenstra and P. H. Franses, "A co-integration approach to forecasting freight rates in the dry bulk shipping sector," *Transportation Research Part A: Policy and Practice*, vol. 31, no. 6, pp. 447–458, 1997.

[19] M. G. Kavussanos, I. D. Visvikis, and R. A. Batchelor, "Over-the-counter forward contracts and spot price volatility in shipping," *Transportation Research Part E: Logistics and Transportation Review*, vol. 40, no. 4, pp. 273–296, 2004.

[20] M. G. Kavussanos, I. D. Visvikis, and D. Menachof, "The unbiasedness hypothesis in the freight forward market: Evidence from cointegration tests," *Review of Derivatives Research*, vol. 7, no. 3, pp. 241–266, 2004.

[21] M. G. Kavussanos and I. D. Visvikis, "Shipping freight derivatives: A survey of recent evidence," *Maritime Policy & Management*, vol. 33, no. 3, pp. 233–255, 2006.

[22] R. Batchelor, A. Alizadeh, and I. Visvikis, "Forecasting spot and forward prices in the international freight market," *International Journal of Forecasting*, vol. 23, no. 1, pp. 101–114, 2007.

[23] M. Prokopczuk, "Pricing and hedging in the freight futures market," *Journal of Futures Markets*, vol. 31, no. 5, pp. 440–464, 2011.

[24] L. Dai, H. Hu, F. Chen, and J. Zheng, "The dynamics between newbuilding ship price volatility and freight volatility in dry bulk shipping market," *International Journal of Shipping and Transport Logistics*, vol. 7, no. 4, pp. 393–406, 2015.

[25] J. Yin, M. Luo, and L. Fan, "Dynamics and interactions between spot and forward freights in the dry bulk shipping market," *Maritime Policy & Management*, vol. 44, no. 2, pp. 271–288, 2017.

[26] Z. H. Munim and H.-J. Schramm, "Forecasting container shipping freight rates for the Far East-Northern Europe trade lane," *Maritime Economics & Logistics*, vol. 19, no. 1, pp. 106–125, 2017.

[27] V. Y. H. Lun and P. Marlow, "The impact of capacity on firm performance: A study of the liner shipping industry," *International Journal of Shipping and Transport Logistics*, vol. 3, no. 1, pp. 57–71, 2011.

[28] N. Matz-Lück, "Safe and sound scrapping of 'rusty buckets'? the 2009 Hong Kong ship recycling convention," *Review of European Community and International Environmental Law*, vol. 19, no. 1, pp. 95–103, 2010.

[29] I. L. Buxton, "The market for ship demolition," *Maritime Policy & Management*, vol. 18, no. 2, pp. 105–112, 1991.

[30] N. E. Mikelis, "A statistical overview of ship recycling," *WMU Journal of Maritime Affairs*, vol. 7, no. 1, pp. 227–239, 2008.

[31] S. Knapp, S. N. Kumar, and A. B. Remijn, "Econometric analysis of the ship demolition market," *Marine Policy*, vol. 32, no. 6, pp. 1023–1036, 2008.

[32] N. D. Kagkarakis, A. G. Merikas, and A. Merika, "Modelling and forecasting the demolition market in shipping," *Maritime Policy & Management*, vol. 43, no. 8, pp. 1021–1035, 2016.

[33] J. Yin and L. Fan, "Survival analysis of the world ship demolition market," *Transport Policy*, vol. 63, pp. 141–156, 2018.

[34] N. R. Fox, "An oligopoly model of ocean liner shipping," *Review of Industrial Organization*, vol. 9, no. 3, pp. 343–355, 1994.

[35] B. Franck and J.-C. Bunel, "Contestability, competition and regulation. The case of liner shipping," *International Journal of Industrial Organization*, vol. 9, no. 1, pp. 141–159, 1991.

[36] J. E. Davies, "Comepetition, contestability and the liner shipping industry," *Journal of Transport Economics and Policy*, vol. 20, no. 3, pp. 299–312, 1986.

[37] H. E. Haralambides, "Determinants of price and price stability in liner shipping," in *Workshop on The Industrial Organization of Shipping and Ports*, National University of Singapore, 2004.

[38] T. E. Notteboom, "Consolidation and contestability in the European container handling industry," *Maritime Policy & Management*, vol. 29, no. 3, pp. 257–269, 2002.

[39] E. Hirata, "Contestability of container liner shipping market in alliance era," *Asian Journal of Shipping and Logistics*, vol. 33, no. 1, pp. 27–32, 2017.

[40] A. Jugović, N. Komadina, and A. P. Hadžić, "Factors influencing the formation of freight rates on maritime shipping markets," *Scientific Journal of Maritime Research*, vol. 29, no. 1, pp. 23–29, 2015.

[41] L. D. Thien, *The Maritime Commons: Digital Repository of the World Maritime University*, World Maritime University, 2005.

[42] J. O. Anyanwu, "Analysis of the Interrelationships between the various shipping markets," *Industrial Engineering Letters*, vol. 3, no. 10, pp. 29–42, 2013.

[43] R. Adland and H. Jia, "Shipping market integration: The case of sticky newbuilding prices," *Maritime Economics & Logistics*, vol. 17, no. 4, pp. 389–398, 2015.

[44] J. J. Evans, "Replacement, obsolescence and modifications of ships," *Maritime Policy & Management*, vol. 16, no. 3, pp. 223–231, 1989.

[45] N. E. Mikelis, "Ship recycling markets and the impact of the Hong Kong convention," in *Proceedings of the International Conference on Ship Recycling*, World Maritime University, Malmo, Sweden, 2013.

[46] ILO, "Safety and health in shipbreaking: Guidelines for Asian countries and Turkey," 2004 Jul. 20, 2017; http://www.ilo.org/public/libdoc/ilo/2004/104B09_279_engl.pdf.

[47] IMO, "The development of the Hong Kong Convention," 2009 Jul. 20, 2017; http://www.imo.org/en/OurWork/Environment/ShipRecycling/Pages/Default.aspx.

[48] W. H. Greene, *Econometric Analysis*, Prentice Hall, Upper Saddle River, NJ, USA, 5th edition, 2008.

4

A Real-Time Obstacle Avoidance Method for Autonomous Vehicles using an Obstacle-Dependent Gaussian Potential Field

Jang-Ho Cho,[1] **Dong-Sung Pae**,[1] **Myo-Taeg Lim**,[1] **and Tae-Koo Kang**[2]

[1]*School of Electrical Engineering, Korea University, Seoul, Republic of Korea*
[2]*Department of Information and Telecommunication Engineering, Sangmyung University, Cheonan, Republic of Korea*

Correspondence should be addressed to Myo-Taeg Lim; mlim@korea.ac.kr and Tae-Koo Kang; tkkang@smu.ac.kr

Academic Editor: Jose E. Naranjo

A new obstacle avoidance method for autonomous vehicles called *obstacle-dependent Gaussian potential field* (ODG-PF) was designed and implemented. It detects obstacles and calculates the likelihood of collision with them. In this paper, we present a novel attractive field and repulsive field calculation method and direction decision approach. Simulations and the experiments were carried out and compared with other potential field-based obstacle avoidance methods. The results show that ODG-PF performed the best in most cases.

1. Introduction

Autonomous vehicle systems need path planning, indoor and outdoor localization, obstacle avoidance, object detection, and classification of cars, humans, pets, and traffic signs, signals, etc.

Path planning is one of the fundamental technologies in automotive vehicles. There are two categories in path planning, namely, global path planning and local path planning. Global path planning needs information from a geographic information system (GIS) and global localization, while local path planning needs relative position and obstacle avoidance. There are many algorithms for both global and local path planning. In global path planning, graphical methods such as the Voronoi algorithm [1–4], the cell decomposition algorithm [5–7], and fast marching algorithm [4, 8, 9] are used to develop a graph and then find the shortest path to the goal.

Obstacle avoidance is one of the essential technologies in local path planning and one of the critical technologies that guarantees human and vehicle safety. For decades, many researches have been performed on this theme and many methods have been developed, a few of which have been implemented in real systems. In order to avoid collision with obstacles, a robot needs not only to detect obstacles but also to recalculate the detouring path and to steer itself toward a safe and efficient path in real time.

Khatib [10] formulated the concept of artificial potential field method (PFM) for implementing an obstacle avoidance system. This method consisting of an attractive field and a repulsive field is simple and computationally light. The two different fields are calculated and summed up, after which the robot follows the direction of the calculated vector. Avoiding obstacles is achieved by the repulsive field and heading toward the goal by the attractive field. The concept of artificial potential fields can also be used as a global path planner [11].

The bug algorithm was introduced by Lumelsky and Stepanov [12], in which a vehicle runs along the wall of an obstacle whenever it meets one. If the direction toward the goal is available again, the vehicle leaves the obstacle and goes toward the goal.

Borenstein and Koren [13] presented a different method to avoid obstacles called the vector field histogram method (VFH), and Ulrich and Borenstein [14] improved the method by considering the robots size and choosing a safe and efficient angle (VFH+: later upgraded to VFH∗ [15]).

The dynamic window approach (DWA) using the vehicle's dynamics was proposed in [16]. This method consists of three velocity-angular velocity windows: the first window is the vehicles' minimum to maximum velocity-angular velocity

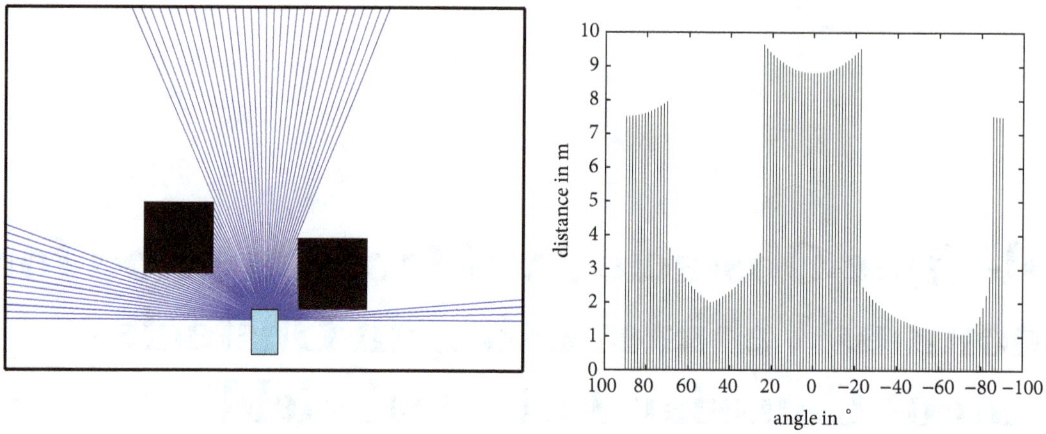

(a) (b)

FIGURE 1: (a) Illustration of the laser range finder and (b) corresponding sensor data (distance versus angle).

window, the second is an obstacle-free zone in the first window, and the last one is the available velocity-angular velocity window feasible from the current velocity-angular velocity pair. In the intersection of the three windows, DWA chooses a velocity-angular velocity pair maximizing the objective function.

Anderson et al. [17] used model predictive control (MPC) to develop an optimal control-based path planning method that uses a relatively new control method to predict vehicle movement. The main drawback of this approach is that it is basically a nonlinear optimal control problem, and even when the problem can be linearized, finding the optimal path in real time is not easy. This is highly dependent on CPU performance.

The nearness diagram (ND) method was proposed by Minguez and Montano [18], who classified real situations into five categories, each with a different decision-making procedure. Later on, this method was upgraded by several other researchers [19–21].

The follow-the-gap method (FGM) was developed by Sezer and Gokasan [22], in which the angle of maximum gap is found and then the weighted average of the angle and the angle toward the goal are calculated.

Park et al. [23] devised the advanced fuzzy potential field method (AFPFM) as an improved algorithm of conventional PFM by using fuzzy membership and TS fuzzy inference [24]. In this method, the attractive field of AFPFM is the same as that of the conventional PF, but the repulsive field calculation is carried out using TS fuzzy inference.

Many kinds of sensors are used to detect obstacles such as sonar, laser range finder, LIDAR, stereo vision, and 3D depth sensor. In this study, we used a laser range finder and introduced a new method to avoid obstacles that we call the obstacle-dependent Gaussian potential field (ODG-PF) method, which we compared with PFM, FGM, and AFPFM.

This article consists of 6 sections. Section 1 is the introduction. Algorithms for comparison with ODG-PF are mentioned in Section 2. Section 3 describes ODG-PF in detail. The results of simulations are in Section 4. Corresponding outdoor experiments carried out are in Section 5. The conclusion is in Section 6.

2. Related Work

All of the following methods use a range sensor. The blue lines in Figure 1(a) are infrared laser beams used to measure the distance. As depicted in Figure 1(b), the data from laser range finder are distances corresponding to predefined angles in the sensor.

These distance data are manipulated differently in different algorithms.

2.1. The Conventional Potential Field Method. Essentially, potential field-based methods use an artificial potential field consisting of an attractive field and a repulsive field. The attractive field attracts the vehicle toward the goal while the repulsive field repels the vehicle from obstacles. The following equations depict conventional potential field calculations:

$$f_{total} = f_{att} + f_{rep}, \quad (1)$$

$$f_{att} = k_{att} \frac{\boldsymbol{r}_{goal} - \boldsymbol{r}}{\left|\boldsymbol{r}_{goal} - \boldsymbol{r}\right|}, \quad (2)$$

$$f_{rep} = \begin{cases} -k_{rep} \sum_{i=1}^{n} \left(\frac{1}{d_i} - \frac{1}{d_{max}} \right) \boldsymbol{s}_i, & \text{if } d_i < d_{max} \\ 0, & \text{othewise} \end{cases} \quad (3)$$

where $\boldsymbol{s}_i = (\boldsymbol{r} - \boldsymbol{o}_i)/|\boldsymbol{r} - \boldsymbol{o}_i|$.

Equation (1) shows the total field calculation, while the attractive field that "attracts" the vehicle toward the goal is calculated by (2), where $\boldsymbol{r}_{goal}$ is the position vector of the goal point and $\boldsymbol{r}$ is the position vector of the vehicle. Equation (3) is the calculation for the repulsive field that pushes the vehicle away from obstacles in which $\boldsymbol{o}_i$ is the position vector of each

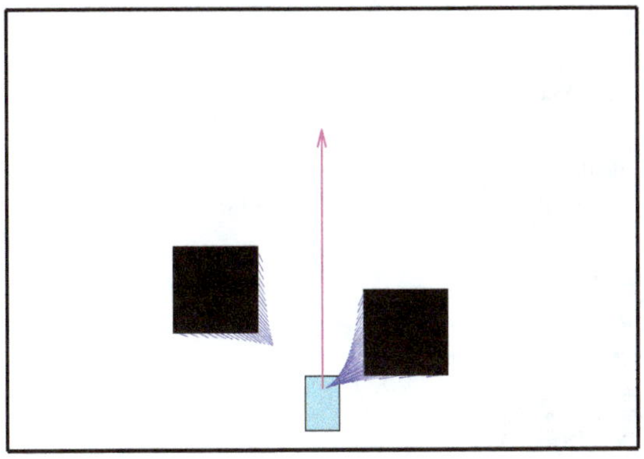

FIGURE 2: Attractive field and repulsive field in PFM.

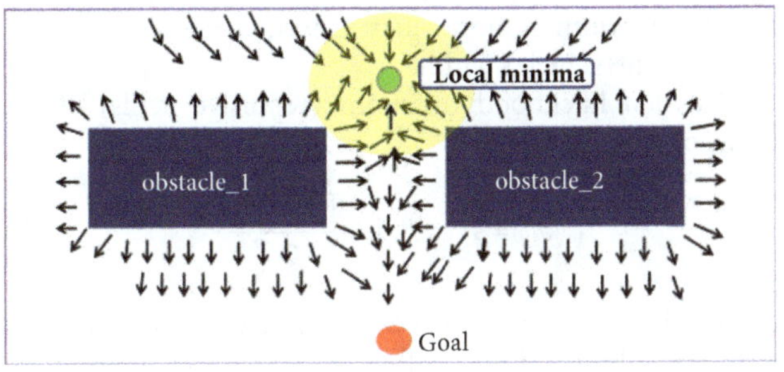

FIGURE 3: Local minima problem of conventional PFM.

obstacle. Figure 2 shows the attractive field and the repulsive field.

As depicted in Figure 3, the conventional potential field method has a drawback in that it becomes easily stuck at local minima [25].

2.2. The Follow-the-Gap Method. FGM is not based on an artificial potential field, but its guild angle decision is somewhat similar. Besides, its obstacle detecting and gap finding methods are somewhat like our ODG-PF.

The main idea of FGM is to find the maximum gap in front of the vehicle and calculate the middle angle of the gap (θ_{gap}) first. From the received data, we know where the obstacles are and we should enlarge the obstacles to a degree with regard to the width of the vehicle. By enlarging the obstacles, we can easily know whether a gap is good for the vehicle to go through or not. This method also attains the angle toward the goal for which the direction decision is carried out using the weighted average of the angle of maximum gap and the angle toward the goal.

In the example in Figure 1, the maximum gap is in -14° ~ 18° range and the medium angle (2°) becomes θ_{gap} as depicted in Figure 4. The guide angle is chosen by calculating the weighted average of θ_{gap} and θ_{goal} as follows:

$$\theta = \frac{(\alpha/d_{min})\,\theta_{gap} + \theta_{goal}}{\alpha/d_{min} + 1}, \quad (4)$$

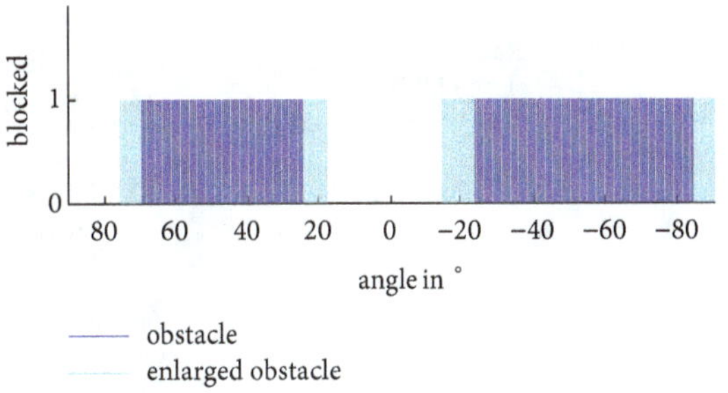

FIGURE 4: Finding the maximum gap from sensor data.

where d_{min} is the minimum distance from all of the distance data from the range sensors and α is a value that we should decide to avoid an obstacle and travel efficiently toward the goal. If it is small, the vehicle will travel more directly toward the goal but will be more likely to collide with obstacles whereas if it is large, the vehicle will travel more safely but the path toward the goal will not be so efficient. If the vehicle is near an obstacle, d_{min} will be small and $(\alpha/d_{min})\theta_{gap}$ will become greater than θ_{goal}, so the vehicle will travel more toward θ_{gap} rather than toward θ_{goal}. If the vehicle is far away from any obstacles, d_{min} will become large and $(\alpha/d_{min})\theta_{gap}$ will become relatively small, so the vehicle will travel more directly toward θ_{goal}.

2.3. The Advanced Fuzzy Potential Field Method. The repulsive force calculation in AFPFM is different from that in conventional PFM. The following equation shows the repulsive force calculation in AFPFM:

$$f_{rep} = -k_{rep}\left(\sum_{i=1}^{n}\frac{\varepsilon + d_i}{d_{max}}\right) \cdot \frac{\sum_{p=1}^{n} w_p\left(d_{max}/\left(\varepsilon + d_p\right)\right)\left(1/d_p - 1/d_{max}\right)\mathbf{s}_p}{\sum_{p=1}^{n} w_p}, \quad (5)$$

where d_{max} is the maximum range of the distance sensor, ε is the safety term considering the vehicle's outer hull, k_{rep} is the repulsive field coefficient with basically the same meaning in that of PFM, and w_p is a weight that, in TS fuzzy, is the minimum value of three membership functions u_{d_p}, u_{ϕ_p}, and u_{ψ_p}. The detailed calculation processes are explained in [23].

3. The Obstacle-Dependent Gaussian Potential Field

Here, we introduce ODG-PF. The main idea behind this method is that, after receiving distance data from the range sensor(s) (as shown in Figure 1(b)), we consider only the objects that are within the threshold range (2 m, for example), enlarge the obstacles with regard to the vehicle's width, and construct a Gaussian (repulsive) potential field from them. Next, we calculate the attractive field from the yaw angle information from an inertial measurement unit (IMU). The total field is made of these two fields, and, from it, we choose the angle with the minimum total field value. Figure 5 is the flowchart of the whole process.

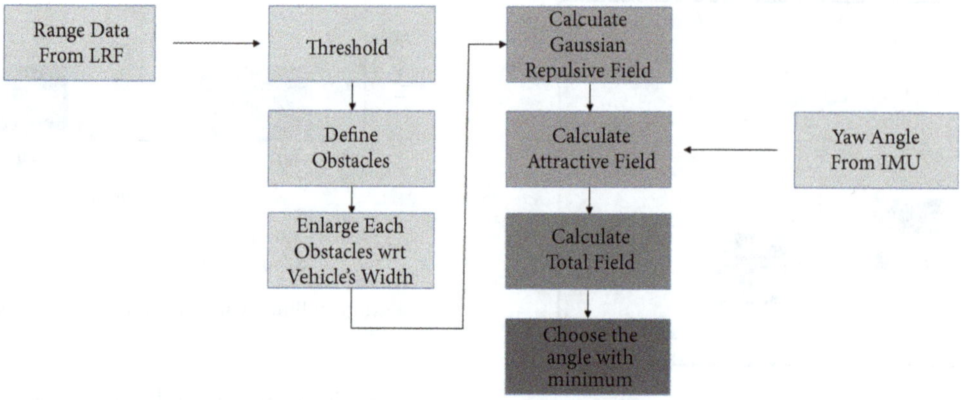

FIGURE 5: Flowchart of ODG-PF.

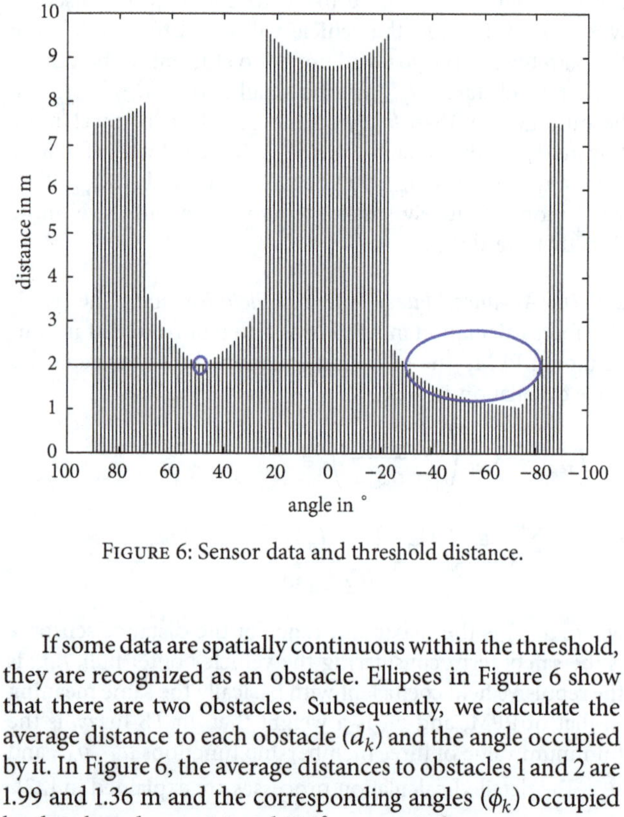

FIGURE 6: Sensor data and threshold distance.

If some data are spatially continuous within the threshold, they are recognized as an obstacle. Ellipses in Figure 6 show that there are two obstacles. Subsequently, we calculate the average distance to each obstacle (d_k) and the angle occupied by it. In Figure 6, the average distances to obstacles 1 and 2 are 1.99 and 1.36 m and the corresponding angles (ϕ_k) occupied by the obstacles are 1.5 and 51.5°, respectively.

In real systems, we should consider the vehicle's width, and, in many algorithms, the vehicle's width or size is added to the obstacle's size and the vehicle's size is shrunk to zero. Figure 7 shows the obstacle enlarging process.

If we consider the vehicle's width, we need to recalculate angle Φ_k as

$$\Phi_k = 2\sigma_k = 2 \cdot \operatorname{atan2}\left(d_k \tan\frac{\phi_k}{2} + \frac{w_{robot}}{2}, d_k\right), \quad (6)$$

where d_k is the average distance to the k_{th} obstacle.

As described in (6), Φ_k is obtained from each obstacle's average distance and the occupied angle of obstacle. The bright bars added to the left and right sides of the dark bar in Figure 7 are the enlargement of the obstacle after considering the vehicle's width. If more safety is needed, we can add more margin to the vehicle's width.

Gaussian likelihood functions (repulsive fields) of the obstacles are calculated as

$$f_k(\theta_i) = A_k \exp\left(-\frac{(\theta_k - \theta_i)^2}{2\sigma_k^2}\right) \quad (7)$$

where θ_k is the center angle of each obstacle. This is a function of θ_i. The subscript i means that the i_{th} data segment of the sensor data (i = 1...361) and it is the sequence number of each angle. Each Gaussian likelihood function becomes a component of the repulsive field. The coefficient A_k should be set in order that the Gaussian likelihood of each obstacle fully embraces the angle occupied by each obstacle. Subsequently,

$$\tilde{d}_k = A_k \exp\left(-\frac{(\theta_k - (\theta_k \pm \sigma_k))^2}{2\sigma_k^2}\right) = A_k \exp\left(-\frac{1}{2}\right). \quad (8)$$

Thus,

$$A_k = \tilde{d}_k \exp\left(\frac{1}{2}\right), \quad (9)$$

where $\tilde{d}_k = d_{max} - d_k$. d_{max} is the maximum detection range of the range sensor and σ_k is half of the angle occupied by the k_{th} obstacle.

In Figure 8(a), the angles occupied by and the Gaussian likelihoods of the obstacles are shown. This explains that the Gaussian likelihood of each obstacle fully embraces the occupied angle at $\tilde{d}_k$.

We obtain the repulsive field by summing up all of the Gaussian likelihood functions of obstacles into one function as

$$f_{rep}(\theta_i) = \sum_{k=1}^{n} A_k \exp\left(-\frac{(\theta_k - \theta_i)^2}{2\sigma_k^2}\right). \quad (10)$$

This is also a function of angle θ_i. The attractive field is calculated as

$$f_{att}(\theta_i) = \gamma\left|\theta_{goal} - \theta_i\right|. \quad (11)$$

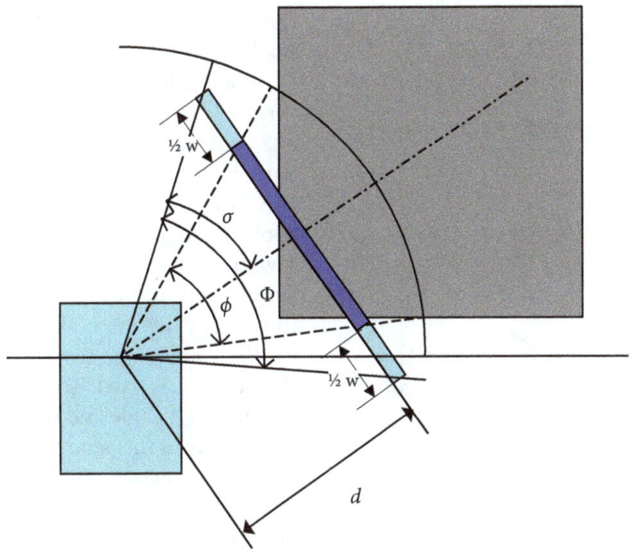

FIGURE 7: Enlarging the angle occupied by obstacles.

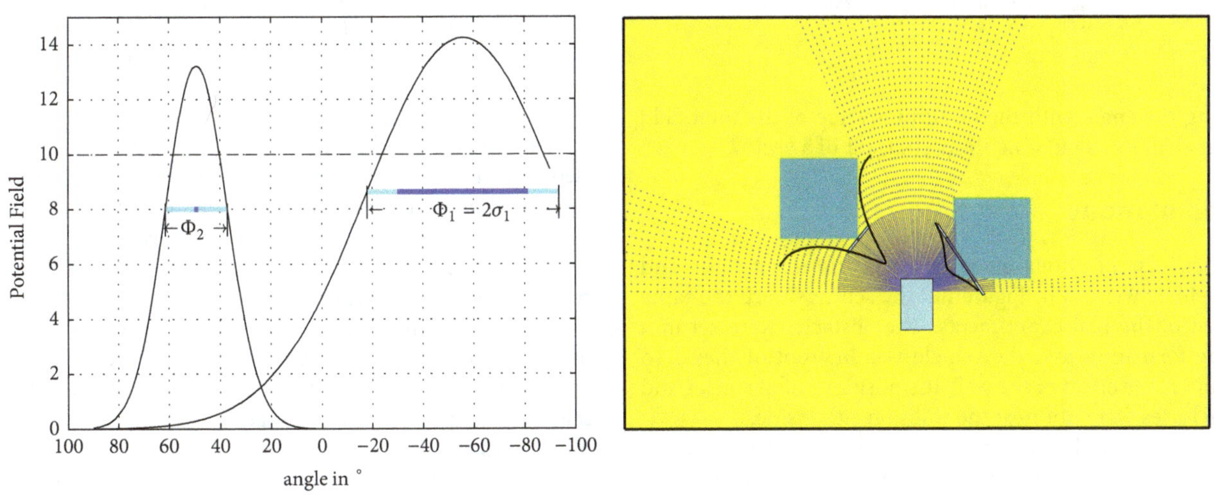

(a) (b)

FIGURE 8: (a) Gaussian likelihood of each obstacle and (b) illustration with vehicle and obstacles.

The total field is calculated by adding these two fields. Since these two fields are functions of θ_i, the total field is also a function of θ_i:

$$f_{total}(\theta_i) = f_{rep}(\theta_i) + f_{att}(\theta_i). \tag{12}$$

In (11), γ is the value that we should choose (it is like k_{att} in PFM and AFPFM). If γ is too small, the vehicle will avoid obstacles but the path will be inefficient whereas if it is too large, the vehicle will be more likely to collide with obstacles. The γ value was chosen as 5.0 by executing several simulations. We used this predetermined value in all the simulations and experiments. The calculation example of the fields is shown in Figure 9. Finally, the angle minimizing the total field is chosen as the heading angle. The small circle in Figure 9 points out this angle.

This method has some features that are not like in classical potential field-based methods, i.e., used by Ge and Cui [26]. First, instead of putting the range sensor data into equation, ODG-PF defines obstacles (occupied angle, average distance, etc.) from range sensor data and then calculates the repulsive fields of the obstacles. Second, the repulsive field and the attractive field of ODG-PF are functions of angle; that is to say, they are not vectors. In the classical potential field, the direction is decided by the direction of the total vector of the repulsive field and the attractive field. Thus, ODG-PF is very robust that even if there are small changes in the environment, the trajectory is not so much affected by them. Mathematically, the main difference between PFM and ODG-PF is that the attractive field and the repulsive field (as well as the total field) are angle functions while all of the fields in PFM are vector values. The direction decision is made by

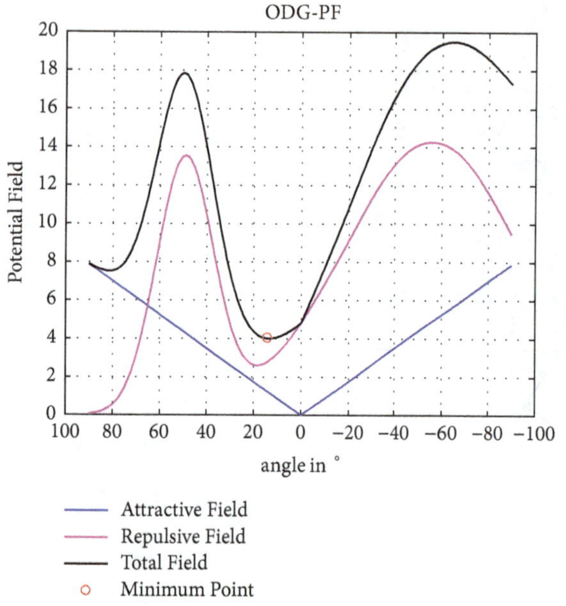

FIGURE 9: ODG-PF potential fields.

finding the angle with the minimum value of the total field rather than by calculating the arctangent of a vector.

4. Simulations

4.1. Settings. Positions of obstacles in the simulations and the vehicle were as in Figure 10. The settings were the same as that of the real experiments. The obstacles were set in a zigzag formation and the vehicle was in front of them. To minimize the effect of the wall, the positions of obstacles and the vehicles were not near the walls on either side.

We compared our novel ODG-PF method to conventional PFM, FGM, and AFPFM. PFM and AFPFM use artificial potential field and it was useful to compare them with the ODG potential field. FGM uses gaps between obstacles and the obstacle dependency of ODG-PF is comparable. All of the algorithms used in this study decide on the direction angle from the received range sensor data. From the sensor data in Figure 1(b), we were able to know where the maximum gap in FGM was and to calculate the repulsive fields of PFM, AFPFM, and ODG-PF.

Each method had some parameters that needed to be defined. For PFM and AFPFM, we fixed the repulsive field coefficient (k_{rep}) at 0.16 and carried out simulations with them while varying the attractive field coefficient (k_{att}) for simplicity.

For FGM, we needed to choose an appropriate α in (4), and so we set it to 0.5 m, which meant that angles toward the goal and the maximum gap had the same weights when the minimum distance from the obstacles and the vehicle was 0.5 m.

Actually, we should choose the maximum sensor distance for PFM and AFPFM. Our range sensor can detect object as far as 10m, but if we changed the d_{max} value in PFM and AFPFM, we would have had to change all of the fuzzy inference rules in AFPFM. For a fair comparison, we left it unchanged.

We needed to define two things for ODG-PF. One is the threshold distance which we set to 2.0 m in order to be easily adaptable to any low-speed, low-resolution sensors. Another parameter we needed to decide on was γ, which is like k_{att} in PFM and AFPFM but, as described in (7), is used to apply a penalty proportional to the deviation from the angle toward the goal. We set γ to 5.0, which was obtained by both simulations and experiments. ODG-PF is not very sensitive to γ, and, experimentally, it can be set to 4.0~6.0.

θ_{goal}, the angle toward the goal in all the algorithms in this study, was fixed. That is to say, there was no goal position but the fixed angle and the purpose of the obstacle avoidance algorithms in this study was to do their best to keep the fixed θ_{goal} without colliding with anything.

Thus, we needed to change the vector of the attractive field described in (2) to

$$f_{att} = k_{att} \begin{pmatrix} \cos\left(\theta_{goal} - \theta_{current}\right) \\ \sin\left(\theta_{goal} - \theta_{current}\right) \end{pmatrix}. \quad (13)$$

4.2. Simulation Results under a Static Environment. We carried out simulations with the algorithms using the same settings shown in Figure 10, the results of which are shown in Figure 11. Figure 11(a) exhibits the results of the PFM simulations. As predicted, k_{att} played a key role in the trade-off between safety and efficiency, and PFM did not perform efficiently in any of the cases. As already compared in [23], PFM was very sensitive to the repulsive field and the trajectories were easily affected even when the vehicle was relatively far away from the obstacles. A large k_{att} value made it less sensitive.

FGM performed better than PFM in that it tried to find the safest angle, as shown in Figure 11(b) and tended to follow the maximum gap if the vehicle came close to the obstacles, as described in (4). Anyway, FGM will be in danger of a collision if an obstacle is between the angle of the maximum gap (θ_{gap}) and the angle toward the goal (θ_{goal}). AFPFM was better than PFM in all cases, but when k_{att} was large, there were collisions, as shown in Figure 11(c). The value of k_{att} should be much larger (by 10 times, for simplicity) than PFM. The reason for this is that there is a coefficient in calculating repulsive field and there are also weights of TS Fuzzy in AFPFM which are a maximum of 15 times larger than those in PFM when an obstacle is placed close to the front of the vehicle.

The simulation of ODG-PF performed the best in all cases. We performed the same experiments several times and there were no collisions. It always avoided the obstacles successfully, and its heading angle finally returned to θ_{goal}, as shown in Figure 11(d).

PFM did not attain good results and the final deviation from the initial y position was the largest. When we set k_{att} to a large value, the deviation became smaller but, as shown in Figure 11(a), the risk of collision also became larger.

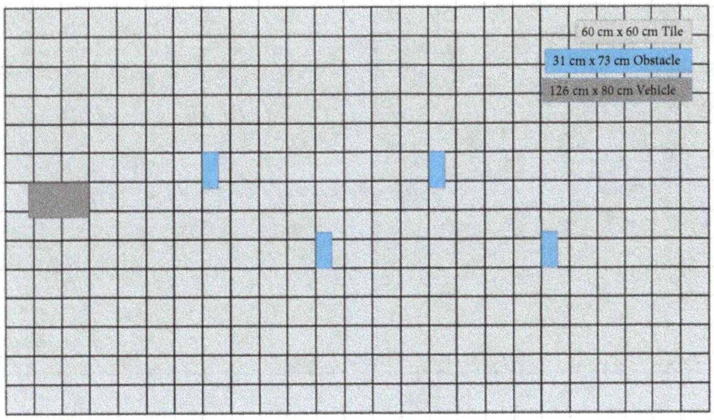

FIGURE 10: Mobile robot and obstacle settings.

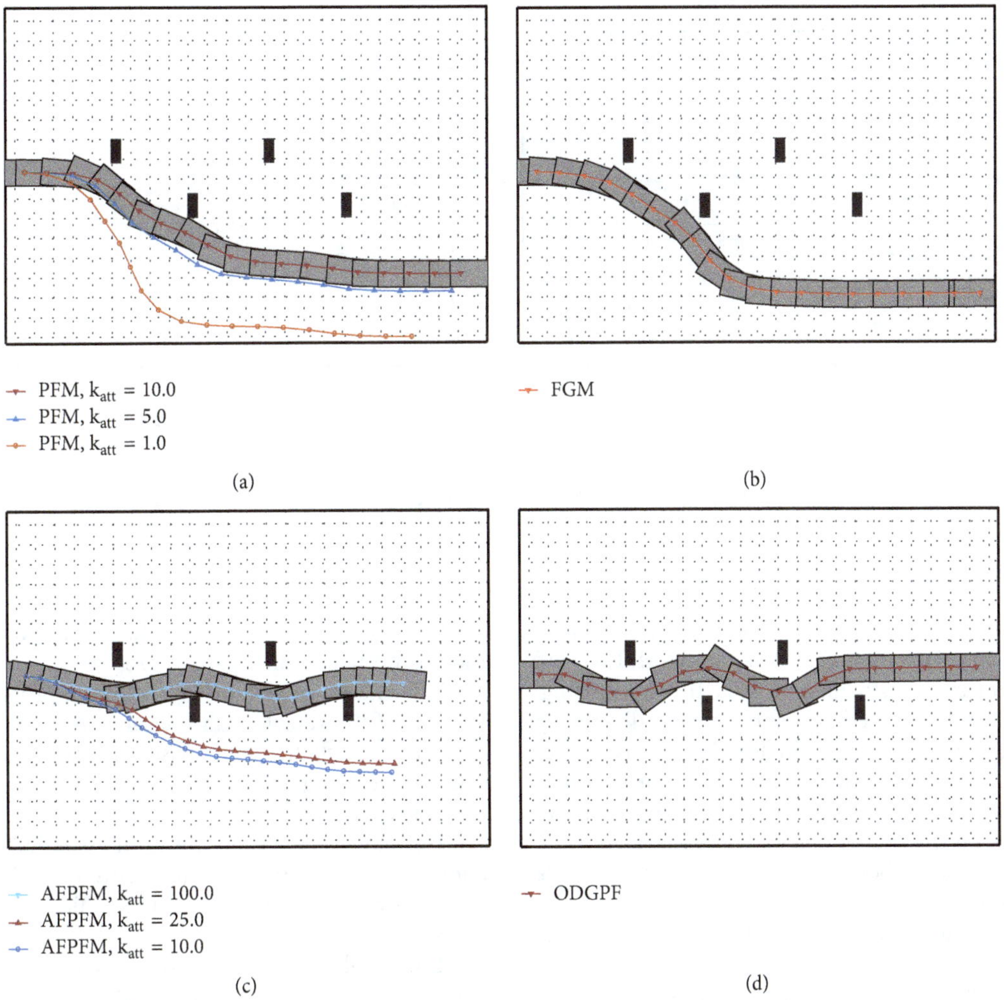

FIGURE 11: Simulation results under static environment: (a) PFM, (b) FGM, (c) AFPFM, and (d) ODG-PF.

The results for FGM were not too bad. When we set α to a smaller value, this caused collisions, and larger values caused bigger deviations.

The results of AFPFM show that it is dependent on k_{att} as PFM is, but they also show that this algorithm could be improved. In Figure 11(c), we can imagine that there would be an optimal k_{att} (or membership function) value that efficiently avoids obstacles without much deviation in simulation, but the optimal k_{att} value would vary on a case-by-case basis.

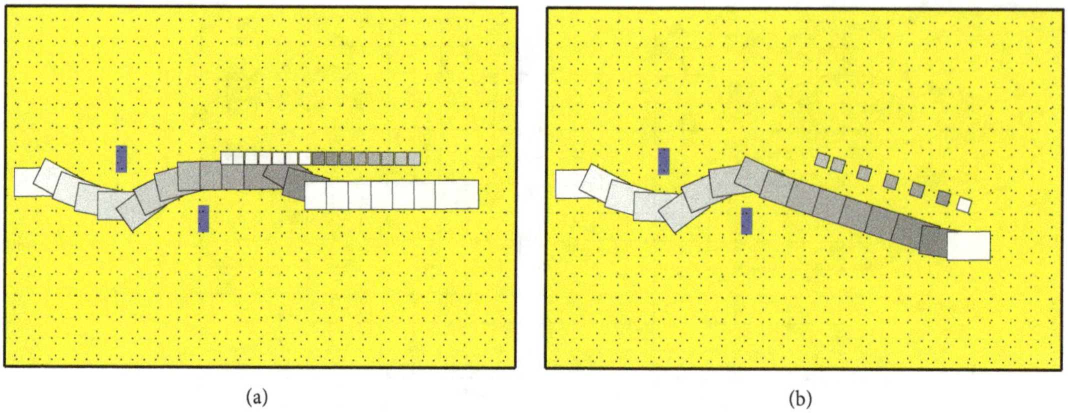

FIGURE 12: Simulation results under dynamic environments: (a) Scenario 1. (b) Scenario 2.

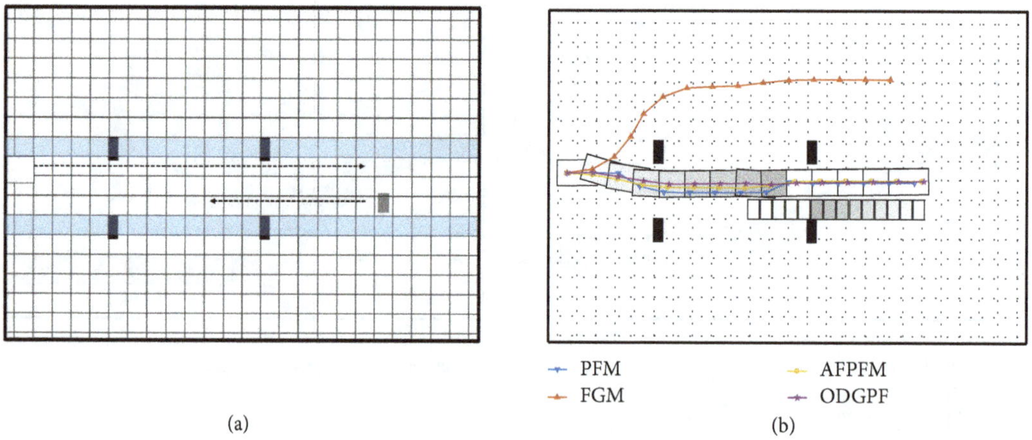

FIGURE 13: Dynamic environment Scenario 3: (a) scenario and (b) simulation results.

4.3. Simulation Results under a Dynamic Environment. Another merit of ODG-PF is that it can successfully avoid obstacles without any adjustment even when they are moving. Any obstacle avoidance method cannot avoid all moving obstacles because of the vehicle's physical limitations such as velocity, angular velocity, radius of rotation, sensor measurement range, etc. The success of avoiding moving obstacles depends heavily on the relative velocity between the moving obstacle and the vehicle, but it also depends on the obstacle avoidance algorithm.

We designed three moving obstacle scenarios. In Scenarios 1 and 2, we used the first two static obstacles used in the previous experiments and a moving obstacle, but the trajectory of the moving obstacle in Scenario 1 was different from that of Scenario 2.

Figure 12 shows the scenarios and simulations of moving obstacle avoidance in Scenarios 1 and 2. As shown in Figure 12(a), the vehicle using ODG-PF successfully avoided the moving obstacle. The colors of the vehicle and the moving obstacle became darker as they became closer and after they met (the point at which they were closest), their color returned to white. By using different colors, the reader can easily distinguish the movements of the vehicle and the moving obstacle with regard to time steps.

In Scenario 2, the moving obstacle approached from the left side of the vehicle with the same velocity as the vehicle (after the vehicle had gone past the second static obstacle). The vehicle using ODG-PF moved as if it was going along with the obstacle.

To compare the algorithms more strictly, we devised another experimental scenario. Figure 13(a) shows the static obstacles and the direction of the vehicle and the moving obstacle.

The simulation results of this moving obstacle scenario show that vehicles with PFM, AFPFM, and ODG-PF were able to avoid both the static obstacles and the moving obstacle. However, when we saw the details of the simulation results for PFM and AFPFM, their movements were somewhat unstable that the vehicle moved in a zigzag fashion.

5. Experiments

5.1. Hardware Specifications and Setup. Figure 14 shows the configuration of the system.

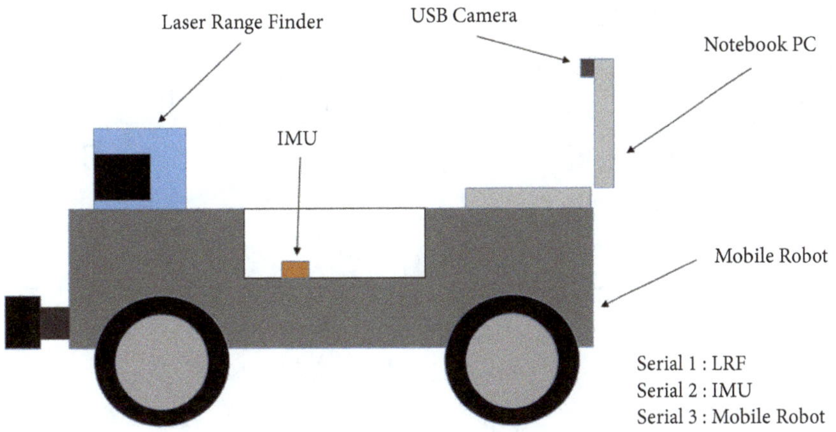

FIGURE 14: System configuration.

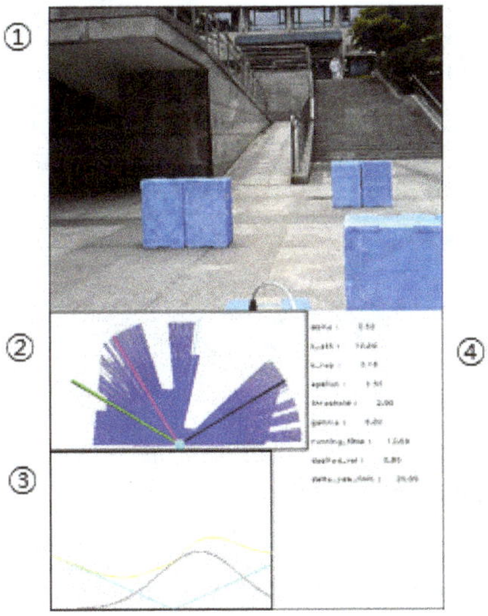

FIGURE 15: Program screen recorded by the PC on the robot.

The vehicle used was an Unmanned Solution™ ERP42v1 without a braking system. Its physical dimensions were 126 × 80 × 30 cm. The range sensor was a SICK™ LMS200 laser measurement system, which could sense the distances from obstacles within ±90° range at 0.5° resolution. We used Xsens™ MTI-100 IMU to calculate the yaw angle of the system using Madgwick's algorithm [27]. In all cases, the yaw angle of the starting point was 0°. A Hardkernel™ USB-CAM 720P USB camera with the resolution set to 640 x 480 was used to record the experiments. The CPU of the notebook PC was an Intel i5 (2 core, 4 thread, 2.67 GHz) and 4 GB of memory. The OS used was Windows 7. To implement the algorithms and show the graphs, Microsoft Visual C++ and OpenCV library were used. All of the data were recorded in real time. Figure 15 shows a snapshot of a recorded video. As we implemented all the algorithms in one PC program, the user should choose one of them before running the program.

The picture shows ① the front view from the USB camera, ② graphical sensor data, ③ calculated data (ODG-PF in this example), and ④ the parameter values used. Thus, recorded videos can be used to analyze the cause of failure later.

FIGURE 16: Outdoor experimental environment.

5.2. Outdoor Experimental Environment. The experiments were performed at the back of the Innovation Hall at Korea University. One obstacle consisted of two 56 × 36.5 × 30 cm expanded polystyrene boxes. Basically, the positions of the obstacles and the vehicle were the same as those in the simulation experiments. The outdoor experimental environment is shown in Figure 16.

5.3. Experiment Results under a Static Environment. The trajectory of each experiment was calculated later from the saved data using

$$x(n+1) = x(n) + v \cdot \cos\theta \cdot \Delta t, \tag{14}$$

$$y(n+1) = y(n) + v \cdot \sin\theta \cdot \Delta t. \tag{15}$$

The value of k_{att} for AFPFM was initially set at 100 because we used this value in the indoor experiments. For conventional PFM, we initially used k_{att}= 10 for the indoor experiments.

Figure 17 is the trajectories of the experiments. The time interval between markers in all the figures of trajectories is 0.8s. The velocity was set to constant but, in the beginning of

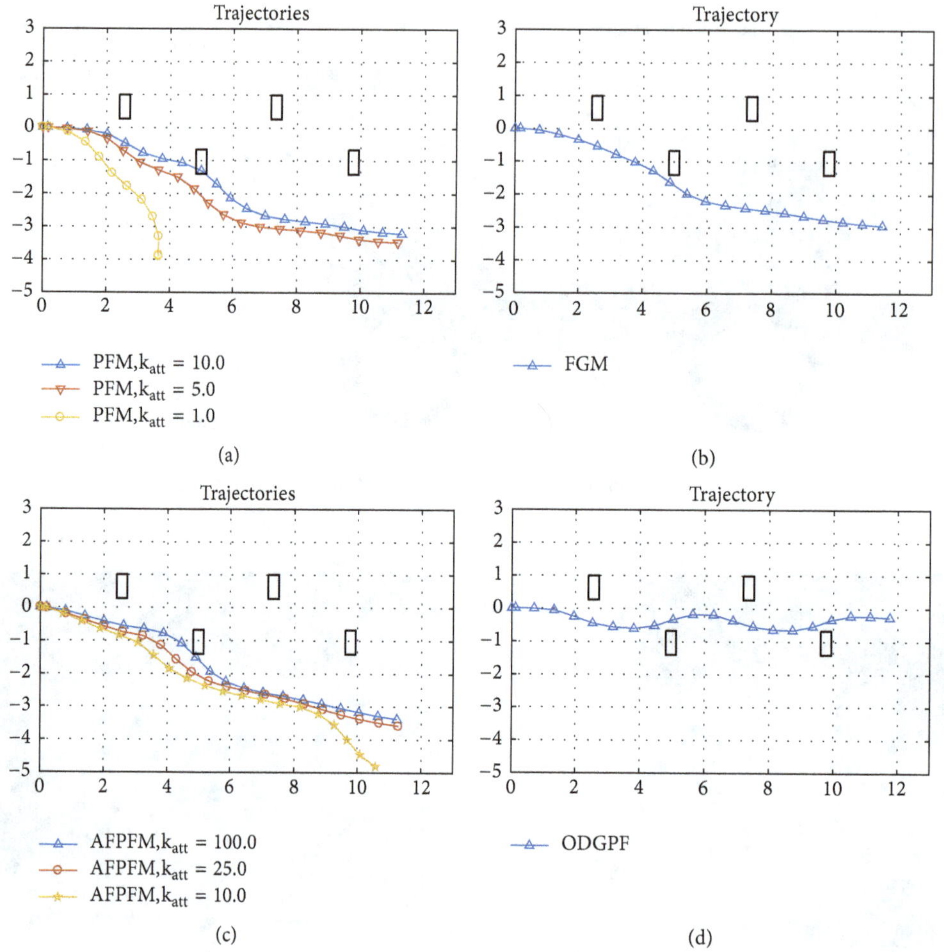

FIGURE 17: Trajectories of the vehicle with (a) PFM, (b) FGM, (c) AFPFM, and (d) ODG-PF.

each navigation, the robot was accelerated from zero speed. The sampling time of IMU is 0.01s and it was used as the reference clock.

For the PFM with k_{att}= 10 (Figure 17(a)), and AFPFM with k_{att}= 100 (Figure 17(c)) cases, the vehicle collided with the second obstacle.

As occurred in the simulations, Figure 17(a) shows that PFM did not perform well and collided with obstacles when k_{att} was large and strayed from the desired angle when k_{att} was small. When k_{att} was too small, it strayed from the goal angle, and it collided with the second obstacle when $k_{att} = 10$.

The vehicle with FGM performed very well when it is needed to avoid obstacles in complex environments. There were no collisions in several experiments, although it did not attain an efficient trajectory.

Figure 17(c) shows the performance of the vehicle with AFPFM. When k_{att} was large ($k_{att} = 100$), the vehicle collided with the second obstacle, which was not the same results as in the simulation, indicating that there might have been a problem in the vehicle's dynamic model. Moreover, there was a pedestrian at the end of the experiment with k_{att}= 10.

ODG-PF performed the best in all cases. We carried out the same experiments several times and there were no collisions. It always managed to avoid the obstacles, as shown in Figure 17(d).

The experimental results consistently show that ODG-PF efficiently and successfully avoided obstacles. Other algorithms were able to avoid obstacles but the deviations from the original y position were large.

5.4. Experiment Results under the Dynamic Environment. As simulated in Figure 13, we performed three scenarios for moving obstacle avoidance.

In Scenarios 1 and 2, the positions of two static obstacles were the same as for static obstacle avoidance and we performed experiments with all of the aforementioned algorithms. The experimental results of vehicles with PFM, AFPFM, and FGM show that their trajectories were not much different from those of the static obstacle cases because they were far away from the moving obstacle. As shown in Figures 17–19, the speed of mobile robot was constant except in the beginning of each experiment, where the mobile robot was accelerated from zero speed to the specified speed. We will upgrade ODG-PF to consider the acceleration of the mobile robot in the future work.

In Scenario 3, we analyzed all of the algorithms using different static obstacle positions and scenarios.

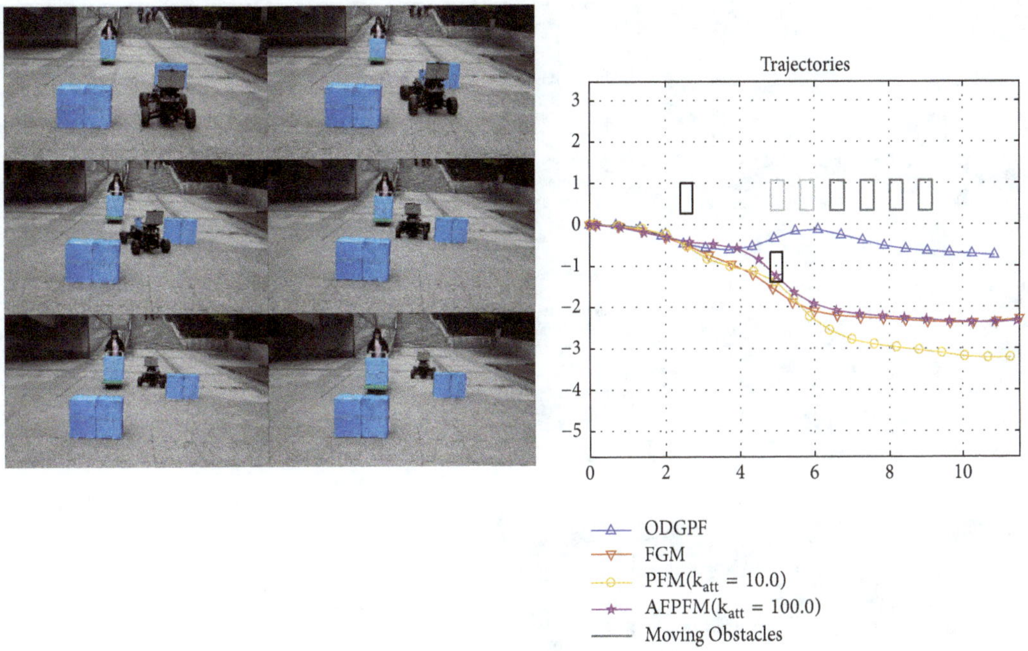

(a) (b)

FIGURE 18: Experimental results of the vehicle with ODG-PF for Scenario 1.

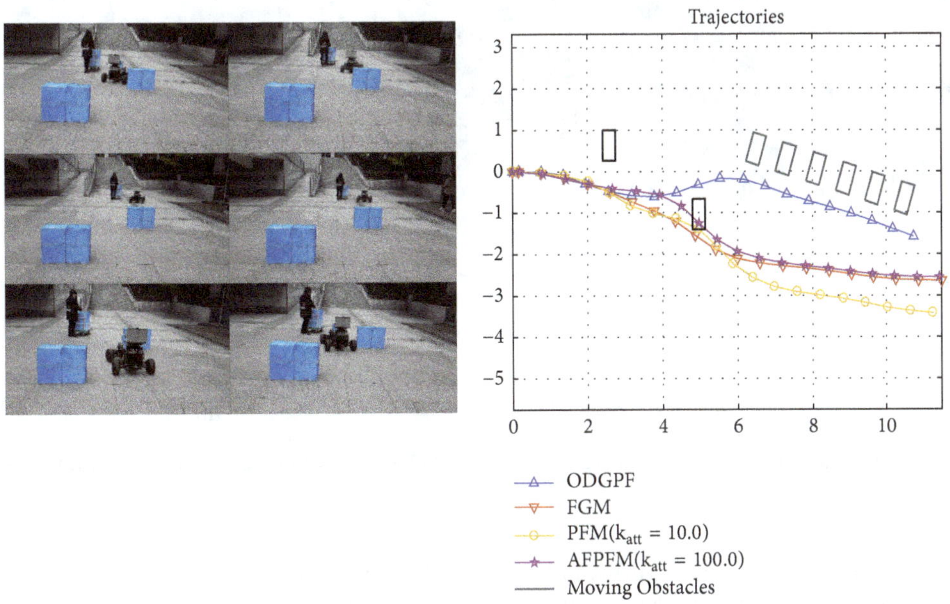

(a) (b)

FIGURE 19: Experimental results of the vehicle with ODG-PF for Scenario 2.

Figure 18 show some images from the Scenario 1 experiment and the trajectory. The vehicle's velocity was 0.8 m/s in both the simulations and the experiments.

Figure 19 show some images of the experiments and the trajectory for Scenario 2. We can see that they also agreed with the simulation results shown in Figure 13(b).

As in Scenarios 1 and 2, we performed experiments with the vehicle and all of the aforementioned algorithms for moving obstacle Scenario 3. As shown in Figure 13(b), all of the algorithm except for FGM successfully avoided both the static obstacles and the moving obstacles in the simulations. This scenario is harsh because at the end of the experiments, the gap between a static obstacle and the moving obstacle was only slightly wider than the vehicle's width. In the simulations, we noted that all the algorithms successfully passed through the gap when it was relatively

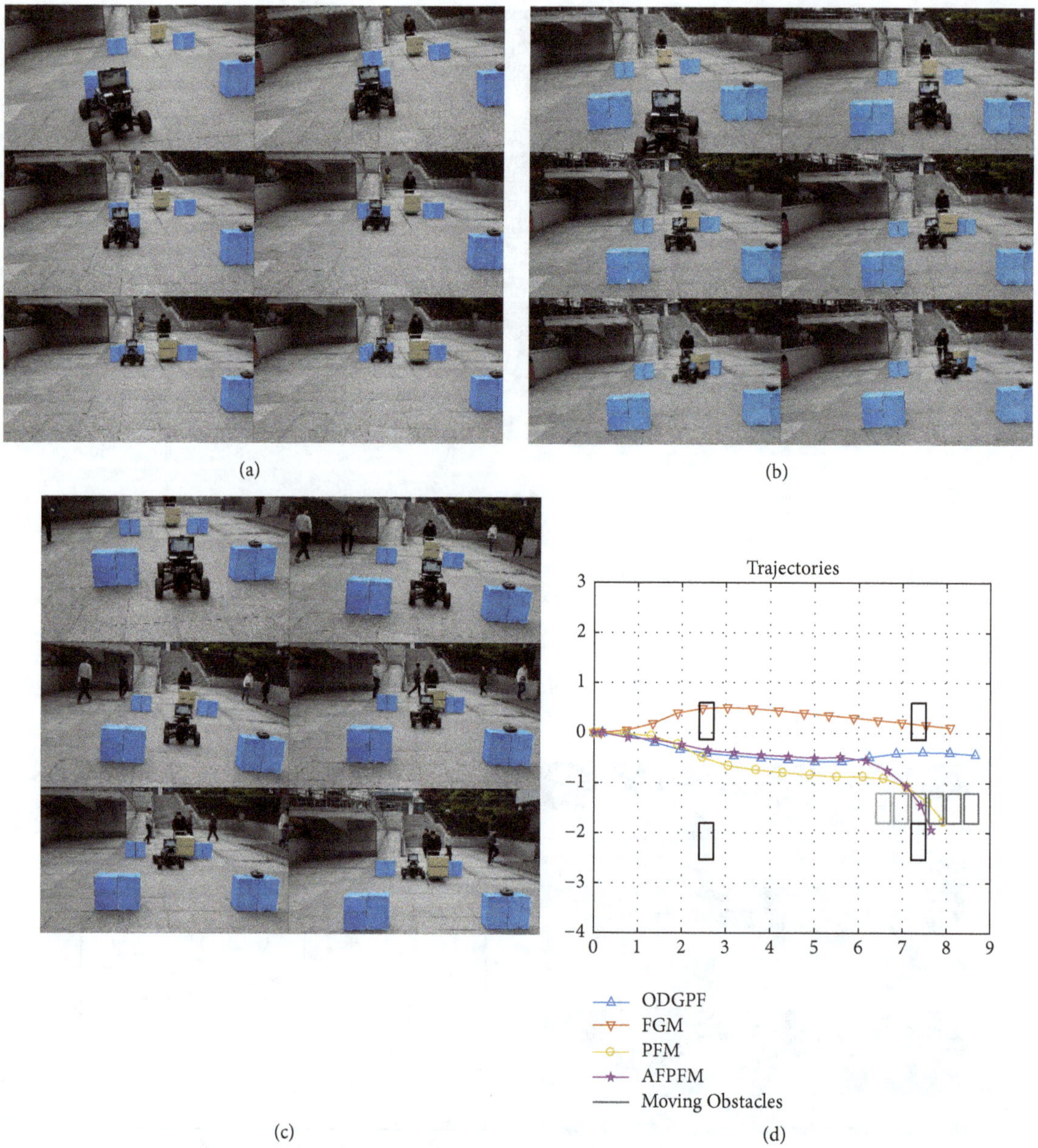

FIGURE 20: Experimental results for Scenario 3: (a) FGM, (b) AFPFM (k_{att}=100), (c) ODG-PF, and (d) trajectories.

wide. We shrunk the gap in the simulations to a degree where one of them started to move differently. In Figure 13(b), the vehicle with FGM went toward the maximum gap, which was to the left of the vehicle when it started to move. We also performed experiments with a wider gap with which all the algorithms worked well.

Figure 20 shows some images from Scenario 3 experiments and trajectories. In results different from the simulation, only ODG-PF successfully passed through the gap between the static obstacles and the moving obstacle. The reasons for success and failure are discussed later.

For PFM and AFPFM, the direction is decided by the arctangent value of the total field vector. As described in (1), the total field is the sum of the attractive field and the repulsive field. The repulsive field becomes larger when the distance to the obstacles becomes smaller. At a certain point, the repulsive field starts to overwhelm the attractive field and the arctangent value directs the vehicle backwards. We analyzed the reason for failure, which is explained in the next section. The mobile robot was not able to change direction abruptly, which resulted in a collision.

Nevertheless, the vehicle avoided both the static obstacles and the moving obstacle closely but successfully with ODG-PF. The Gaussian repulsive field of the ODG-PF described obstacles efficiently and did not fail to detect them.

Figure 20(d) shows the trajectories of the vehicle using different obstacle avoidance algorithms. Besides the ODG-PF case, the experimental results were not the same as the simulation results, and so we think that the vehicle model used in the simulation might have had some errors in it.

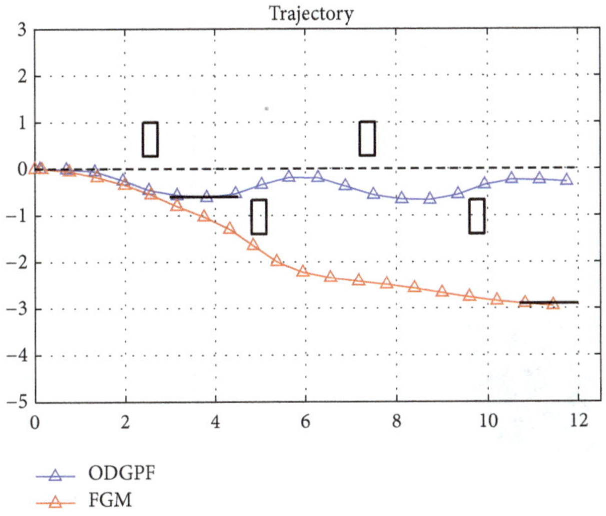

FIGURE 21: Maximum deviation from the initial direction line.

The experimental trajectory of FGM was different from that of the simulation, which we think is due to the vehicle model. As in the simulation result shown in Figure 13(b), FGM tried to follow the maximum gap.

5.5. Evaluation and Discussion. To evaluate the results of the different algorithms, we should consider whether the vehicle with each algorithm collided with any obstacles or not. If the vehicle with a particular algorithm did not collide with any obstacles, we should evaluate the algorithm with some criteria. When considering the efficiency (depicted in Figure 21), the maximum deviation from the initial direction line can be a good criterion.

Table 1 contains the collision results and the maximum deviation from the initial line for each algorithm. Avoiding obstacles is more important than efficiency because this is a matter of safety. Efficiency matters only if there are no collisions with obstacles.

As is evident from the results in Table 1, ODG-PF avoided obstacles with outstanding efficiency and AFPFM was more efficient than conventional PFM.

The experiments with moving obstacles show that only ODG-PF successfully avoided the moving obstacles in all the 3 scenarios. The experimental results of Scenario 3 show that, in a tough situation, only ODG-PF successfully avoided both static and moving obstacles.

Failure analyses show why other methods could not avoid the obstacles and there were different reasons for each of these. In PFM and AFPFM, the total field is a vector sum of the repulsive field and the attractive field. The vector direction (recommended angle of the vehicle by PFM or AFPFM) is calculated using the arctangent of the vector. As shown in Figure 22, if the attractive field is not large enough to push the repulsive field, the direction of the vector sum will be at the back of the vehicle. The problem is that using a large attractive field is not the solution because the repulsive field is inversely proportional to the distances to obstacles.

TABLE 1: Experimental results of static obstacle avoidance.

No.	Algorithm	Parameter	Collision	Max. Deviation
1	PFM	k_{att}= 10	○	3.2 m
2	PFM	k_{att}= 5	-	3.7 m
3	PFM	k_{att}= 1	-	4.6 m
4	FGM	α= 0.5	-	3.4 m
5	AFPFM	k_{att}= 100	○	3.2 m
6	AFPFM	k_{att}= 25	-	3.0 m
7	AFPFM	k_{att}= 10	-	4.3 m
8	ODG-PF	γ= 5.0	-	**0.6 m**

Furthermore, vehicles have some physical constraints such as turning radii, maximum wheel angles, and body sizes. We cannot abruptly adjust the vehicle's direction to go backwards. Because turning rate is limited, turns trying to avoid obstacle can often lead to positions that are more dangerous.

As depicted in Figure 23, the guide angle in FGM is calculated as in (4). The drawback of this calculation is that the direction of guide angle can be toward an obstacle when it is between the maximum gap angle and the goal angle because the guide angle is the weighted average of the two angles.

6. Conclusions

In this paper, we introduce a novel method for obstacle avoidance, ODG-PF, which avoids obstacles very efficiently and safely. Although AFPFM is an improvement that avoids the local minima problem, its attractive field coefficient should be adjusted when the environment is changed. As with conventional PFM, when the vehicle is in a narrow environment, the repulsive field becomes strong; thus k_{att} value needs to be large. Otherwise, the robot will be stuck at a local minimum or go toward the goal very inefficiently.

Simulations and experiments showed that vehicle movements with ODG-PF were very stable. It seems that this stable movement is because of two reasons. One reason is that it detects and defines obstacles first rather than directly calculating the repulsive field from the distance data, and the other is that it finds the angle with minimum values from the total field function. Thus, ODG-PF does not have the local minima problem that causes a vehicle to become stuck at certain points.

One more merit of ODG-PF is that it avoids both static and moving obstacles without any adjustments. In the performed moving obstacle scenarios, the vehicle using it did not collide with any of the static or moving obstacles and instead avoided them using the same method. This is also quite a good point in that it does not need to carry out any time-consuming activities such as image processing or computer vision processing. ODG-PF can be easily implanted in a real-time system because the overhead when calculating it is relatively light.

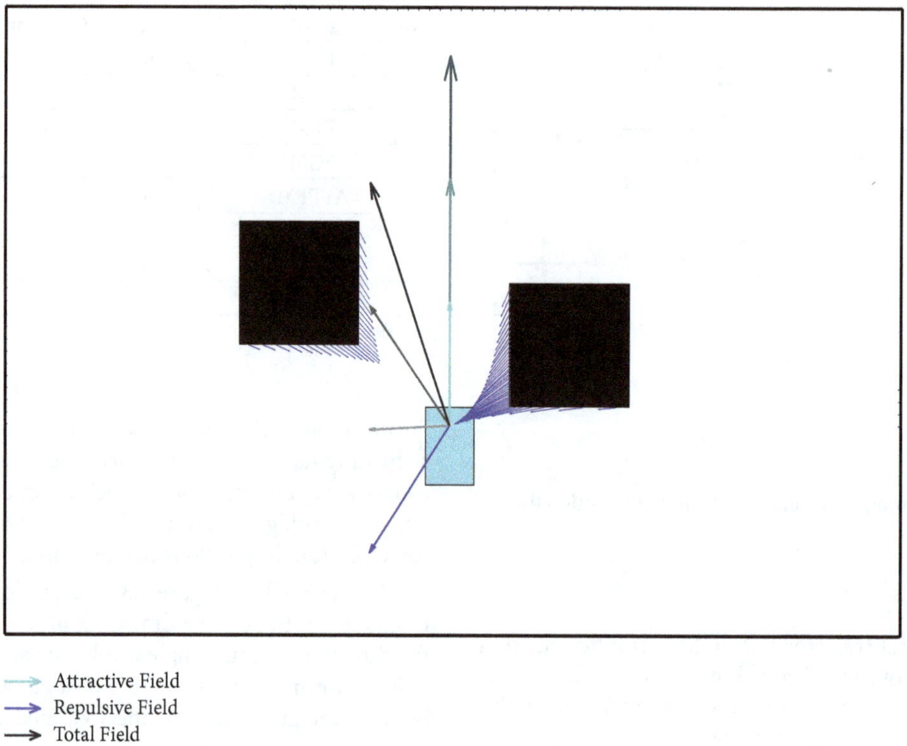

FIGURE 22: Total field of PFM and AFPFM with varying k_{att}.

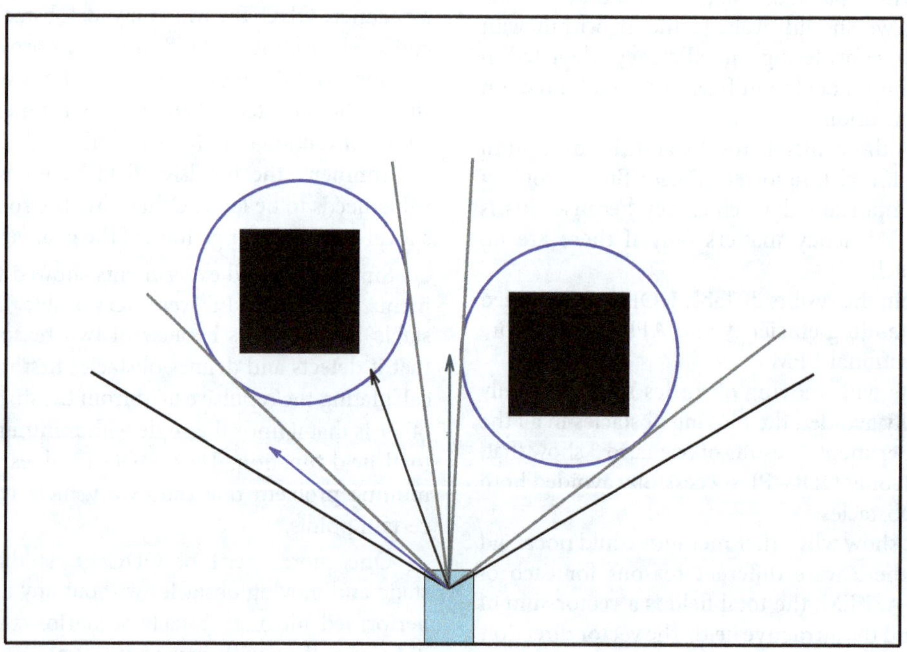

FIGURE 23: Guide angle calculation in FGM.

Conflicts of Interest

The authors declare that there are no conflicts of interest regarding the publication of this paper.

Acknowledgments

This work was supported by the Basic Science Research Program through the National Research Foundation of Korea (NRF) funded by the Ministry of Education (NRF-2016R1D1A1B01016071 and NRF-2016R1D1A1B 03936281) and also in part by the Brain Korea 21 Plus Project in 2018.

References

[1] H. Choset, *Sensor based motion planning: The hierarchical generalized Voronoi graph*, California Institute of Technology, Pasadena, USA, 1996.

[2] E. Masehian and M. R. Amin-Naseri, "A voronoi diagram-visibility graph-potencial field compound algorith for robot path planning," *Journal of Robotic Systems*, vol. 21, no. 6, pp. 275–300, 2004.

[3] P. Bhattacharya and M. L. Gavrilova, "Voronoi diagram in optimal path planning," in *Proceedings of the 4th International Symposium on Voronoi Diagrams in Science and Engineering (ISVD '07)*, pp. 38–47, July 2007.

[4] S. Garrido, L. Moreno, M. Abderrahim, and F. Martin, "Path Planning for Mobile Robot Navigation using Voronoi Diagram and Fast Marching," in *Proceedings of the 2006 IEEE/RSJ International Conference on Intelligent Robots and Systems*, pp. 2376–2381, Beijing, China, October 2006.

[5] T. Arney, "An efficient solution to autonomous path planning by approximate cell decomposition," in *Proceedings of the 2007 3rd International Conference on Information and Automation for Sustainability, ICIAFS*, pp. 88–93, Australia, December 2007.

[6] R. V. Cowlagi and P. Tsiotras, "Beyond quadtrees: Cell decompositions for path planning using wavelet transforms," in *Proceedings of the 46th IEEE Conference on Decision and Control 2007, CDC*, pp. 1392–1397, USA, December 2007.

[7] G. I. Delgado, R. O. Casallas, and G. E. Jacinto, "Path planning in static scenarios using image processing and cell decomposition," in *Proceedings of the 2014 IEEE International Autumn Meeting on Power, Electronics and Computing, ROPEC 2014*, Mexico, November 2014.

[8] P. Melchior, B. Orsoni, O. Lavialle, A. Poty, and A. Oustaloup, "Consideration of obstacle danger level in path planning using A* and fast-marching optimisation: comparative study," *Signal Processing*, vol. 83, no. 11, pp. 2387–2396, 2003.

[9] S. Garrido, L. Moreno, and D. Blanco, "Exploration of a cluttered environment using Voronoi Transform and Fast Marching," *Robotics and Autonomous Systems*, vol. 56, no. 12, pp. 1069–1081, 2008.

[10] O. Khatib, "Real-time obstacle avoidance for manipulators and mobile robots," in *Proceedings of the IEEE International Conference on Robotics and Automation (ICRA '85)*, vol. 2, pp. 500–505, 1985.

[11] A. Oustaloup, B. Orsoni, P. Melchior, and H. Linarès, "Path planning by fractional differentiation," *Robotica*, vol. 21, no. 1, pp. 59–69, 2003.

[12] V. J. Lumelsky and A. A. Stepanov, "Path-planning strategies for a point mobile automaton moving amidst unknown obstacles of arbitrary shape," *Algorithmica*, vol. 2, no. 1-4, pp. 403–430, 1987.

[13] J. Borenstein and Y. Koren, "The vector field histogram—fast obstacle avoidance for mobile robots," *IEEE Transactions on Robotics and Automation*, vol. 7, no. 3, pp. 278–288, 1991.

[14] I. Ulrich and J. Borenstein, "VFH+: reliable obstacle avoidance for fast mobile robots," in *Proceedings of the IEEE International Conference on Robotics and Automation (ICRA '98)*, pp. 1572–1577, May 1998.

[15] I. Ulrich and J. Borenstein, "VFH*: local obstacle avoidance with look-ahead verification," in *Proceedings of the IEEE International Conference on Robotics and Automation (ICRA '00)*, pp. 2505–2511, April 2000.

[16] D. Fox, W. Burgard, and S. Thrun, "The dynamic window approach to collision avoidance," *IEEE Robotics and Automation Magazine*, vol. 4, no. 1, pp. 23–33, 1997.

[17] S. J. Anderson, S. C. Peters, T. E. Pilutti, and K. Iagnemma, "An optimal-control-based framework for trajectory planning, threat assessment, and semi-autonomous control of passenger vehicles in hazard avoidance Scenarios," *International Journal of Vehicle Autonomous Systems*, vol. 8, no. 2–4, pp. 190–216, 2010.

[18] J. Minguez and L. Montano, "Nearness diagram navigation (ND): A new real time collision avoidance approach," in *Proceedings of the 2000 IEEE/RSJ International Conference on Intelligent Robots and Systems*, pp. 2094–2100, Japan, November 2000.

[19] J. Minguez, J. Osuna, and L. Montano, "A "divide and conquer" strategy Based on situations to achieve reactive collision avoidance in troublesome scenarios," in *Proceedings of the Proceedings- 2004 IEEE International Conference on Robotics and Automation*, pp. 3855–3862, USA, May 2004.

[20] J. W. Durham and F. Bullo, "Smooth nearness-diagram navigation," in *Proceedings of the 2008 IEEE/RSJ International Conference on Intelligent Robots and Systems, IROS*, pp. 690–695, France, September 2008.

[21] M. Mujahed and H. Jaddu, "Smooth and safe Nearness-Diagram (SSND) Navigation for autonomous mobile robots," *Advanced Materials Research*, vol. 403-408, pp. 4718–4726, 2012.

[22] V. Sezer and M. Gokasan, "A novel obstacle avoidance algorithm: "Follow the gap method"," *Robotics and Autonomous Systems*, vol. 60, no. 9, pp. 1123–1134, 2012.

[23] J. W. Park, H. J. Kwak, Y. C. Kang, and D. W. Kim, "Advanced Fuzzy Potential Field Method for Mobile Robot Obstacle Avoidance," *Computational Intelligence and Neuroscience*, vol. 2016, Article ID 6047906, 13 pages, 2016.

[24] T. Takagi and M. Sugeno, "Fuzzy identification of systems and its applications to modeling and control," *IEEE Transactions on Systems, Man, and Cybernetics*, vol. 15, no. 1, pp. 116–132, 1985.

[25] Y. Koren and J. Borenstein, "Potential field methods and their inherent limitations for mobile robot navigation," in *Proceedings of the IEEE International Conference on Robotics and Automation*, pp. 1398–1404, April 1991.

[26] S. S. Ge and Y. J. Cui, "Dynamic motion planning for mobile robots using potential field method," in *Proceedings of the 8th IEEE Mediterranean Conference on Control and Automation*, 2000.

[27] S. O. Madgwick, AHRS algorithms and calibration solutions to facilitate new applications using low-cost MEMS, Electronic Thesis, University of Bristol: Bristol, UK 2014.

5

Development and Testing of Intelligent Alcohol Transportation Security System

Velaphi Msomi and Shandene E. A. Solomons

Mechanical Engineering Department, Engineering Faculty, Cape Peninsula University of Technology, P.O. Box 1906, Bellville 7535, South Africa

Correspondence should be addressed to Velaphi Msomi; msomiv@gmail.com

Academic Editor: Aboelmaged Noureldin

The development and testing of intelligent liquid transportation security system are being reported in this paper. The targeted fluid to be secured was ethanol alcohol and this was due to the theft cases occurring during the transportation of this product from the supplier to the customer. The system was developed such that only the radar level sensor (VEGAPULS 62) might be in contact with the fluid and the rest of the system remained outside the liquid carrying container to be secured. The system was developed such that it reports any abnormal liquid level drop through short message service (SMS). The functioning of the developed system was tested through the use of 1040 L Intermediate Bulk Container (IBC) filled with water which was hauled for about 1.5 km. The liquid theft was simulated and the system sent two SMS. The first SMS reported the beginning of water level drop and the second one reported the ending of water level drop. The second SMS reported the amount of liquid that was taken out of the container.

1. Introduction

Transportation of bulk goods forms a large part of the economic sector. Logistic companies have large fleets carrying goods to various destinations. Whilst these operations can be controlled on-site, once in transit it becomes difficult to monitor the goods being carried leaving them vulnerable to theft. Many tracking system solutions have been developed but most of them were designed to monitor the quality of the solid food being transported [1]. These food quality monitoring systems detect the presence of various gases which might be threat to the quality of food inside the packages. They do not provide the quantity being monitored during the transportation. Oil and gas industries are the ones which have done work on leak detection and monitoring during transportation [2]. It should be noted that the mostly used oil and gas transportation mode is the underground pipe system. Alkhaledi et al. [3] presented gas leak detection system called fault tree analysis and this technique identifies the exact point of leakage. Elaoud et al. [4] developed an oil leakage detection technique which is based on transient negative pressure waves. Scott and Barrufet [5] developed a complex method which uses real-time transient model to detect the small pipes leakages. Real-time pipeline leak detection method based on volume balancing was developed by Doorhy [6]. The method has capabilities to detect small pipeline leakages but it takes considerable time to detect them.

The monitoring systems employed in oil and gas industry deal with fluid loss which affects the economy. The objective of developing and employing such systems is to maintain the quality and the quantity of goods being transported. It is based on this objective that the oil and gas based literature is included in this work because it relates to the objective that this work is trying to achieve. This paper reports on the development and testing of the intelligent liquid monitoring system for road transportation mode. The targeted liquid to be secured is ethanol alcohol. The objective of developing such system is to eliminate or reduce theft of this product occurring during its transportation. The system is designed such that it monitors liquid level during transportation. It has a capability to report the amount of liquid taken out of the transporting container. The system reports to the supplier through short message service (SMS). All the SMS sent to the supplier get duplicated and stored to the server that can be accessed by the supplier. This duplication of SMS serves as a backup in case of cellphone theft or damage.

2. Solution Specifications

In general, the development of a system has to meet certain specifications; hence this section is included.

The development and the implementation of any safety device require that the developed device must comply with certain standard to avoid problems. There are number of explosion proof design standards available commercially which evaluate the explosiveness of the devices or systems being developed [7]. All the components used in constructing our system were evaluated for their explosiveness, that is, looking at the explosive temperature for each component against the explosive range specified by Jespen [7]. Tank level was the critical variable to be measured as it was the focal key performance indicator (KPI) of this work. Various container configurations exist which means the system had to be flexible to accommodate different configurations and sizes. The level sensor to be used for this application had to be biocompatible and not reactive since the targeted liquid to be secured was for human consumption. The level sensor had to provide real-time continuous readings even if the liquid quality and temperature vary. System location point on the carrying container and the system casing were also the critical factors as they had an impact on satellite connectivity. Power failure backup was also a requirement for our system. The battery system was used as power backup plan in case of the main power supply failure.

3. Methodology

This section describes all the steps followed in the construction of the system such that the objectives described in the solution specification are satisfied. The system was constructed through the following major components.

3.1. Radar Level Sensor (VEGAPULS 62). The VEGAPULS 62 level sensor is a noncontact sensor which uses microwaves to detect the level of the liquid. It is designed to measure different types of liquid and does not react with the measured liquid. The maximum measuring range of this is 38 m. The sensor can operate under a very harsh environmental condition, that is, temperature range of –196°C to 450°C. More specifications could be found on the sensor specifications sheet.

3.2. Siemens LOGO! CMR2040. LOGO! CMR2040 is the electronic device which is used for wireless communication. CMR is an abbreviation of communications module radio. The satellite and the sensor information captured by LOGO! CMR2040 got monitored through S7-1200 programmable logic controller (PLC). S7-1200 PLC is a device used in controlling decentralized system. More information about these devices is freely available from websites. The general schematic overview diagram for the connection and wireless communication is shown in Figure 1. The switch symbol in

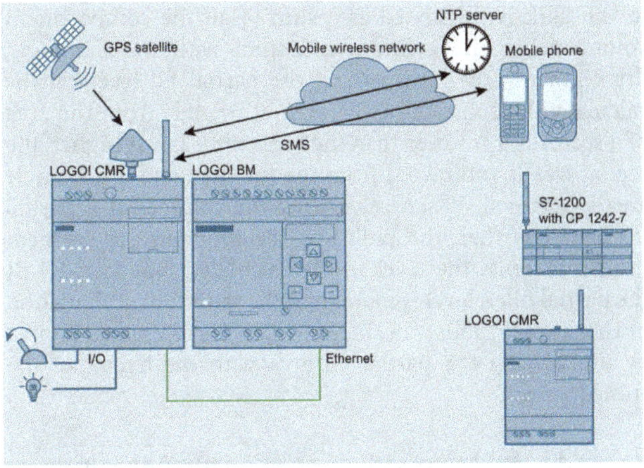

FIGURE 1: Schematic diagram for LOGO! CMR2040 wireless communication (extracted from the product manual sheet).

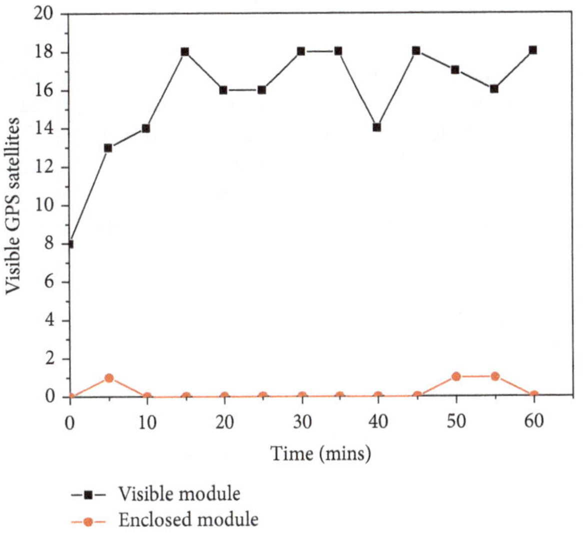

FIGURE 2: Satellite visibility over an hour.

the Figure 1 is reserved for the input signal (in our case the level sensor) whilst the bulb symbol is for activation of the notifications (like alarms or red flashing lights). The wireless communication between the system installed in the liquid carrying container and the mobile phone was through the Siemens LOGO! CMR2040 device. The receiver reads the latitude and longitude coordinates into its onboard memory in degrees, minutes, and seconds format. It also stored the altitude, signal state, and amount of visible satellites in use up to 18 (see Figure 2). This data is transferred from the LOGO! to the S7-1200 programmable logic controller (PLC) for monitoring via Profinet protocol for route referencing and liquid level monitoring.

It should be noted that the liquid carrying container dimensions need to be recorded during the input control data for accurate readings. In our case, the cylindrical configurations were used in design calculations because this is how most alcohol containers are configured. The total volume in the tank is calculated as a sum of all the compartment volumes. For the cylindrical section with the flat ends, the segment area representing the partial fill level can be calculated by subtracting the area of triangle from the area of a sector if it is lower than the radius [8]. If it is higher, the sector area is subtracted from the full area and the triangle area is retained. When calculating the volume for a partial fill level less than the radius of the cylinder, the segment itself represents the level and is calculated based on (1). If the partial filled level is more than the radius, the calculation of the segment follows (2). The volume is calculated simply by multiplying the partial fill area with the length of the compartment.

$$A_{\text{sm}} = R^2 \arccos\left(1 - \frac{h}{R}\right) - (R - h)\sqrt{2Rh - h^2}$$

$$A_{h>R} = \pi R^2 - R^2 \arccos\left(1 - \frac{h}{R}\right) + (R - h_{\text{norm}})\sqrt{2Rh - h^2}. \quad (1)$$

For the end compartments with the 2 : 1 elliptical heads, the cylindrical section, which includes the straight flange distance in its length, is also calculated with (1) and (2). A single head's wetted area (S) is calculated using (2). It should be noted that the wetted area depends on variables R and frictional liquid level (F), where F is calculated using (3) and ε is equal to 0.866 for 2 : 1 elliptical heads.

$$S = \frac{\pi R^2}{2}\left\{(F - 0.5)\sqrt{1 + 12(F - 0.5)^2} + 1 + \frac{1}{4\varepsilon}\ln\left[\frac{4\varepsilon(F - 0.5) + \sqrt{1 + 12(F - 0.5)^2}}{2 - \sqrt{3}}\right]\right\}. \quad (2)$$

Fractional liquid level is equal to partial fill height divided by internal cylinder diameter. The elliptical head calculation is based on (3). The cylindrical volume including the straight flange length plus the elliptical head volume gives the total volume of the compartment.

$$V = 2R^3\frac{\pi}{12}\left[3\left(\frac{h}{2R}\right)^2 - 2\left(\frac{h}{2R}\right)^3\right] \quad (3)$$

It was deemed necessary to check the performance of the GPS module when its antenna was enclosed inside the system encasing and when it was outside. Figure 2 shows the results for the comparison test performed between the enclosed GPS module antenna and the line of sight GPS module antenna. The purpose of enclosing the GPS module was a way of protecting the system from possible theft or damage. It is clear from the graph (Figure 2) that the idea of enclosing the system will compromise the accuracy of the feedback from the system. This suggests that there will be a lot of signal loss from the system.

The system can be operated using the user interface screen (S7-1200 PLC screen). The screen is designed to be

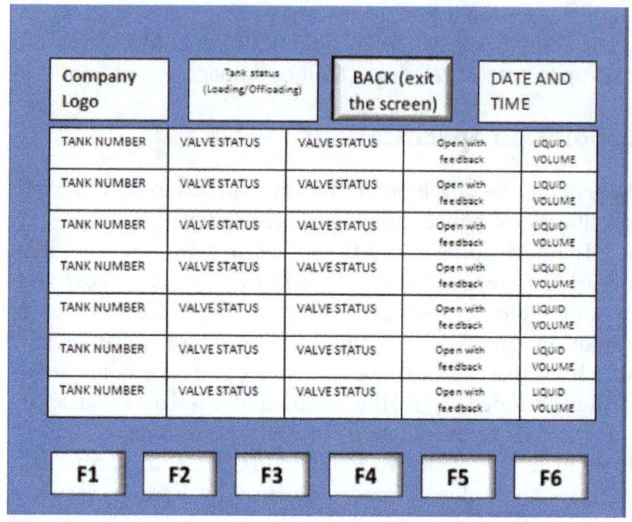

FIGURE 3: Operator interface process control.

a touch screen and the access to the screen is protected by a password. A database with all the relative client location coordinates is stored in the PLC. These variables may be configured on the system but are secure with password protection. The location of the liquid carrying container is continuously cross-referenced against the database to determine whether it is in range of a stored location. Once the tanker is in a range, the system will set the location as active. Since the earth is an oblate spheroid, the distance between two points on earth must be measured on the surface of the sphere. Equation (4) is used to determine the large-circle distance which was useful in determining the data that could be used to locate the position of the liquid carrying container.

$$d = 2r \cdot \arcsin\sqrt{\text{harvesin}(\Delta\phi) + \cos\phi_1\cos\phi_2\,\text{harvesin}(\Delta\lambda)}. \quad (4)$$

When one wants to access data for the system through screen shown in Figure 3, one-time pin (OTP) needs to be used. The user receives an SMS with OTP and its expiry information (see Figure 4). Upon successful entry of the OTP, the user may then access the process control functions where the valves can be controlled for filling and offloading (see Figure 3).

The control system contains two 12 VDC 12 Ah batteries as backup supply to continue powering the unit even if the main supply is cut. This allows monitoring of the main power supply using a current transducer to measure the current drawn and a diode ensuring that the current from the batteries does not flow backward to the main power supply. The logic controller and its communications module need to be powered so as to send notification of a loss of power from the truck whilst on route and ensure all the compartment levels are being monitored. If in battery mode whilst out of range of a saved location, the system sends an SMS notification. The control panel door is monitored by a proximity sensor to check if the enclosure's door has been opened which will allow for disconnection of power to the controllers. If so, an SMS notification is also sent.

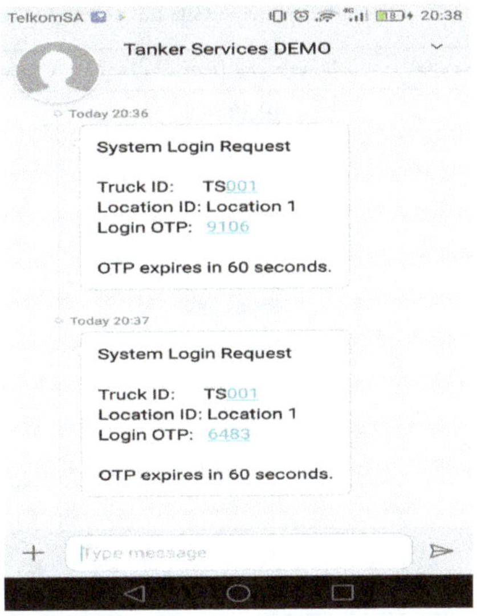

FIGURE 4: One-time pin SMS notification.

FIGURE 5: System installation test.

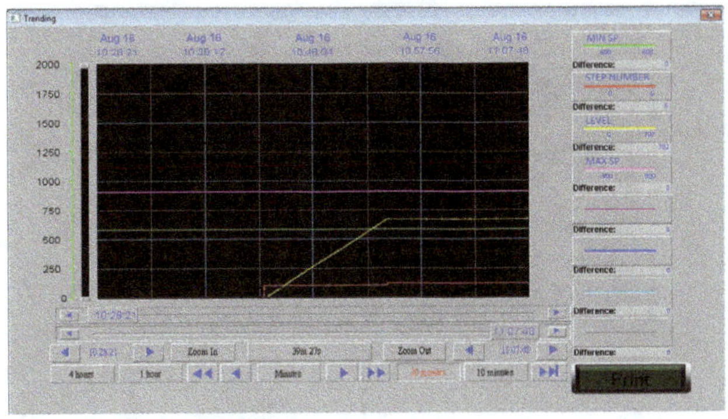

FIGURE 6: IBC filling up profile.

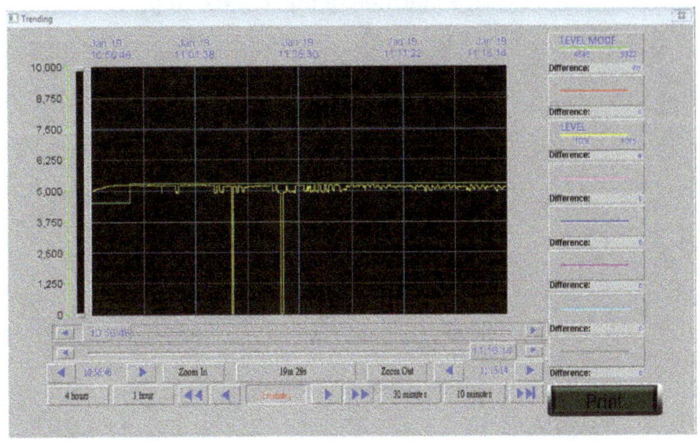

FIGURE 7: Instantaneous liquid level readings.

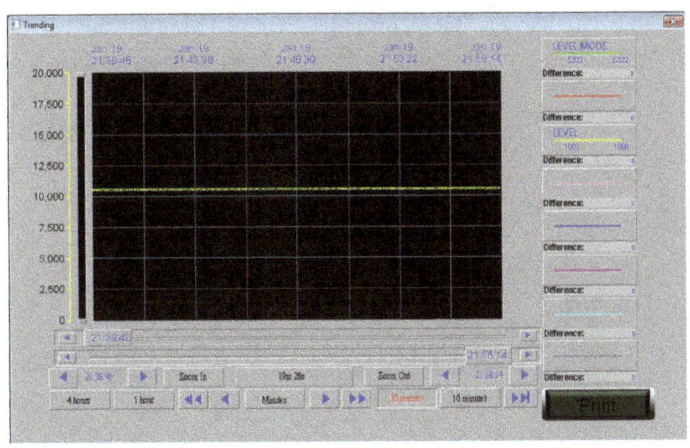

FIGURE 8: Dampened liquid level fluctuation readings.

4. Results and Discussion

This section is looking at the performance of the system when it is installed. For the purpose of testing the system, a 1040-liter IBC filled with water was placed on a trailer and hauled for about 2 km to test the reading stability (see Figure 5).

The filling up of the container profile gets recorded and stored in the server which can be accessed by the supplier. Figure 6 shows IBC being filled with water for testing purposes. There are a number of crucial inputs that the system require before the filled container leaves the supplier. Those inputs include the amount of liquid in the container (indicated by yellow line), the maximum setpoint that the fluid can reach during fluctuation (purple straight line), and the minimum setpoint that the fluid can reach during fluctuation (lime straight line). The liquid fluctuation is assumed to be emanating from the unevenness of the road surface when the vehicle carrying liquid is in motion. Figure 7 shows the liquid level fluctuation with two sharp peaks resulting from excessive vehicle breaking. The liquid level fluctuation became a problem in accurately monitoring the level drop and hence the introduction of averaging function block which was programmed through PLC. The dampening of the erratic instantaneous readings is shown in Figure 8.

Figure 9 demonstrated the simulation of the abnormal liquid level drop. The yellow line shows the emptying of the liquid from the carrying container (IBC). It should be noted that there was a stable liquid level before dropping. This indicates that the liquid carrying container was not in motion. Figure 10 is the demonstration of the SMS notification received by the supplier. The SMS carries the information about the vehicle carrying the liquid container, the GPS coordinates, and the amount of liquid inside the container. The demonstration SMS giving notification of the ending of the abnormal activity is shown in Figure 11. The second

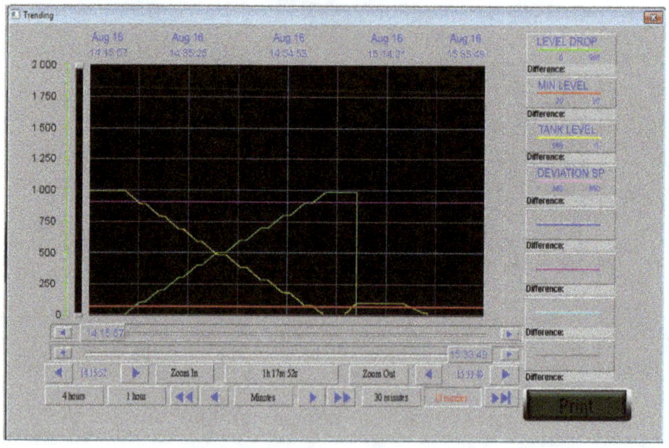

FIGURE 9: Simulation of liquid level drop.

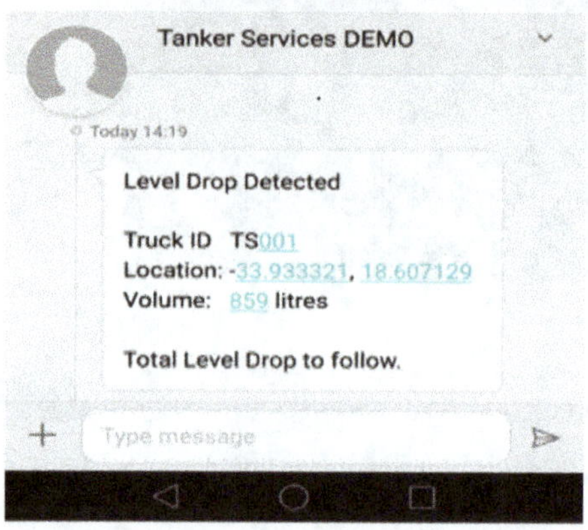

FIGURE 10: Liquid level drop detection SMS notification.

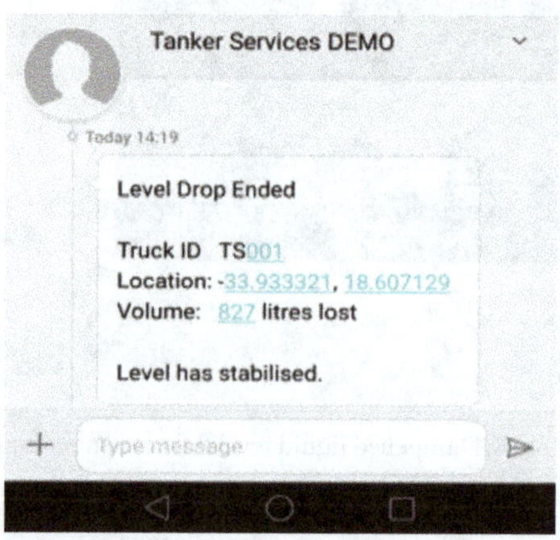

FIGURE 11: Liquid level drop detection SMS notification.

SMS records location, vehicle information, and the amount of liquid taken out of the container.

A test was done to compare the circular distance measurement (see (4)) function programmed into the PLC to a measuring wheel's distance. The results are shown in Table 1.

The haversine formula's accuracy is more than suitable for the application coming in at an average of 1.18% deviation from 0 to 100 m, which amounts to 0.354 m at 30 m. The maximum current consumption from the system is listed in Table 2.

TABLE 1: Measuring wheel versus haversine formula.

Measuring wheel	Haversine formula	Visible satellites
10 m	10.38 m	8
20 m	20.09 m	8
30 m	29.96 m	9
40 m	40.33 m	8
50 m	50.08 m	10
60 m	60.80 m	11
70 m	70.23 m	11
80 m	81.66 m	9
90 m	91.29 m	8
100 m	101.68 m	9

TABLE 2: System maximum current consumption.

Item	Max current
CPU1211C DC/DC/DC PLC	2.9 A
SM-1231 PLC ANALOG CARD	0.5 A
KTP700	0.23 A
LOGO! OBA8	3 A
LOGO! CMR2040	0.3 A
PHOENIX ETHERNET HUB	0.5 A
PHOENIX SAFETY BARRIERS × 2	0.2 A
VEGAPULS62 × 7	0.154 A
MSFG 24-EX SOLENOID VALVE × 7	0.17 A
Total	*7.954 A*

The total current consumption adds up to 7.954 A. This means that with a 12 Ah batter capacity the system can stay powered for at least another 1 hour and 30 min.

5. Conclusion

The level monitoring system was successfully constructed and tested. Each segment has been fulfilled and level monitoring with alarm notifications has been proved to be possible with the suggested setup. The incorporation of the GPS module for both positioning and route control is made possible through the system and adds tremendous value to user access control and level alarm notification stamping. The system also has a backup battery that allows the system to continue monitoring even when the power from the truck has been interrupted. This assists with system tampering along with control panel door status.

Conflicts of Interest

The authors declare that there are no conflicts of interest that may arise from this work.

Acknowledgments

The authors would like to thank Wynand Kotze (Automation Works) for the project approval and assistance with equipment for proof of concept, Reinhold de Villiers (Automation Works) for assistance with Tank Volume measurement, and Theuns Kotze (Automation Works) for assisting with continuing with tanker trial.

References

[1] F. Bibi, C. Guillaume, N. Gontard, and B. Sorli, "A review: RFID technology having sensing aptitudes for food industry and their contribution to tracking and monitoring of food products," *Trends in Food Science & Technology*, vol. 62, pp. 91–103, 2017.

[2] P.-S. Murvay and I. Silea, "A survey on gas leak detection and localization techniques," *Journal of Loss Prevention in the Process Industries*, vol. 25, no. 6, pp. 966–973, 2012.

[3] K. Alkhaledi, S. Alrushaid, J. Almansouri, and A. Alrashed, "Using fault tree analysis in the Al-Ahmadi town gas leak incidents," *Safety Science*, vol. 79, pp. 184–192, 2015.

[4] S. Elaoud, L. Hadj-Taïeb, and E. Hadj-Taïeb, "Leak detection of hydrogen-natural gas mixtures in pipes using the characteristics method of specified time intervals," *Journal of Loss Prevention in the Process Industries*, vol. 23, no. 5, pp. 637–645, 2010.

[5] S. Scott and M. Barrufet, "Worldwide assessment of industry Leak detection capabilities for single and multiphase pipelines," Tech. Rep., 2003.

[6] J. Doorhy, "Real-time pipeline leak detection and location using volume balancing," *Pipeline Gas Journal*, vol. 238, no. 2, pp. 65-66, 2011.

[7] T. Jespen, "ATEX – Explosive Atmospheres: Risk Assessment, Control and Compliance," *Springer International Publishing*, pp. 46–52, 2016.

[8] A. Barderas and B. Roadea, "How to calculate the volume of partially full tanks," in *Proceedings of the International Journal of Research in Engineering and Technology*, pp. 2–7, 2016.

6

Predicting Work Zone Collision Probabilities via Clustering: Application in Optimal Deployment of Highway Response Teams

Przemysław Sekuła[1,2] **Zachary Vander Laan**[1]
Kaveh Farokhi Sadabadi,[1] **and Mirosław J. Skibniewski**[1,3,4]

[1] *Department of Civil and Environmental Engineering, University of Maryland, College Park, MD, USA*
[2] *Faculty of Informatics and Communication, University of Economics in Katowice, Katowice, Poland*
[3] *Institute for Theoretical and Applied Informatics, Polish Academy of Sciences, Gliwice, Poland*
[4] *Chaoyang University of Technology, Taichung, Taiwan*

Correspondence should be addressed to Przemysław Sekuła; psekula@umd.edu

Academic Editor: Richard S. Tay

This paper proposes a clustering approach to predict the probability of a collision occurring in the proximity of planned road maintenance operations (i.e., work zones). The proposed method is applied to over 54,000 short-term work zones in the state of Maryland and demonstrates an ability to predict work zone collision probabilities. One of the key applications of this work is using the predicted probabilities at the operational level to help allocate highway response teams. To this end, a two-stage stochastic program is used to locate response vehicles on the Maryland highway network in order to minimize expected response times.

1. Introduction

Work zone collisions accounted for approximately 1.2 to 1.7 percent of all 2006-2013 crashes in the United States, amounting to 67,523 incidents in 2013 [1]. These collisions resulted in numerous injuries (e.g., over 47,000 in 2013), as well as fatalities that occurred in about 1 percent of the cases. Mohan and Gautam [2] performed a thorough analysis of work zone accidents in the US between 1995 and 1997 and estimated the direct costs resulting from worker and passenger injuries to be approximately $6.2 billion, which composes only part of the overall cost. Although this value may be somewhat lower now given that annual work zone collisions have mirrored the gradual downward trend of overall collisions [1], there are still enormous safety and financial benefits that can be achieved by avoiding or lessening the severity of such accidents. Here we consider two ways in which transportation agencies may mitigate the impact of work zone collisions:

(i) *Optimally allocating response teams.* Transportation agencies could allocate response vehicles in accordance with anticipated risks at work zones and thereby reduce the time it takes to reach the scene of an accident. This would improve their ability to assist injured motorists, clear blocked traffic lanes, and help prevent additional secondary accidents.

(ii) *Adjusting work zone characteristics.* Transportation agencies could identify risky work zone projects in the planning phase and adjust work zone characteristics in order to improve their safety and operational performance.

Both of these applications depend on the ability to *quantify how work zone characteristics affect collision occurrence risk in work zones,* which is precisely the focus of this paper. We briefly describe how this fits into the broader context of work zone research and focus in particular on two types of approaches which may be employed to model this relationship: statistical modeling and machine learning.

Existing work zone research tends to be oriented towards mobility or safety applications. The mobility-oriented research generally focuses on the effect of work zones on traffic capacity, queuing delays, and other performance metrics [3–8], as well as optimizing controllable work zone parameters [9–14]. Research focused on safety commonly describes how work zones affect collision frequency [15–18], severity [19–23], or both [24–28]. In particular, this paper examines how work zone characteristics (e.g., duration, time of day, and number of lanes) impact the likelihood of work zone collisions occurring. Here we review the two prevailing methodologies used to quantify this relationship, considering each approach separately.

(i) *Statistical Modeling.* The majority of literature in this area uses regression techniques to model the relationship between explanatory variables describing work zones and the frequency with which incidents occur, employing traditional maximum likelihood and Bayesian methods to estimate the parameters. However, least squares regression is not ideal for representing the relationship between work zone characteristics and crash frequency, because it assumes normally distributed errors and the dependent variable (i.e., number of crashes) is not continuous [29]. While isolated examples of linear regression models exist in the literature (e.g., [30]), researchers have found that discrete models with Poisson-distributed responses are better suited for modeling crash frequency. Accordingly, the Poisson and related models (e.g., Negative Binomial (NB), Poisson-lognormal, and zero-inflated Poisson and NB) are more prevalent in the literature [29]. The Poisson model's primary deficiency is that it has poor estimation performance when the variance is greater than the mean (i.e., overdispersion), which the NB model addresses by specifying a gamma distribution for the error term [29]. There are numerous applications of the NB model [31–36], focusing on different aspects of crash frequency modeling (e.g., location within a work zone, time of day, duration, fatal/injury/property damage crash classifications). For example, Venugopal and Tarko [32] estimate work zone crash frequency based on traffic levels, work zone lengths, work duration, on/off ramps, and classification of work zones, observing that there is not a statistically significant difference in crash rate for day and nighttime work zones. From a slightly different perspective, Srinivasan et al. [33] use the NB models to predict crash modification factors, while Yang et al. [37] incorporate concepts of measurement error in the NB modeling framework. However, NB models are susceptible to underdispersion for small datasets, which Poisson-lognormal models seek to counter [24, 25, 38]. Finally, zero-inflated models attempt to address the fact that work zone datasets do not contain many incidents (i.e., the response variable primarily takes the value of zero), which can cause overdispersion. The zero-inflated approach generally uses two states: one where accidents can occur and the other where they cannot. Binary choice models (e.g., binary probit or logit) are often used to determine which system the state is in, and Poisson and NB approaches are used to model crash frequency for the state in which accidents are allowed [39–43].

Another class of advanced statistical techniques includes random parameter [44–47], generalized additive [48, 49], Markov switching [50–52], and hierarchical models [53, 54]. We do not spend extensive time discussing these methods here, primarily because existing literature in this area focuses on non-work zone crash frequency modeling applications, which is not the focus of this paper.

The models described thus far are primarily useful for predicting the number of crashes that will occur in a work zone over a period of time. As Yang, Ozbay, Xie, and Bartin [55] point out that sometimes it is useful to frame the problem in terms of crash probability (also referred to as crash risk) instead, where the goal is to predict the probability that a work zone will encounter a collision based on its characteristics. In this case, the response variable is binary (i.e., accident or no accident) rather than a positive integer (i.e., number of collisions). Logistic regression is a popular approach for handling binary responses, examples of which include Al-Ghamdi [56], Harb et al. [57], and Bham et al. [58]. However, when modeling crash risk for short work zone durations, the estimates are often biased because there are rarely crashes during the specified time period [55]. Being able to estimate crash probabilities for short durations is one of the core objectives of this paper and has not been studied extensively in the literature. To the best of our knowledge, the only paper which addresses this problem for short durations is Yang, Ozbay, Xie, and Bartin [55], which implements a rare event logistic regression model to account for the biases. They illustrate this approach by considering 466 work zones along a 25-mile section of the New Jersey Turnpike.

(ii) *Machine Learning.* An alternative approach to model relationships between work zone characteristics and incident risks is through machine learning techniques, including neural and Bayesian neural networks, support vector machines, and clustering. However, existing work zone research employing these techniques tends to focus on characterizing work zone capacity, delays, incident duration, and other metrics rather than quantifying crash risks (e.g., [59–64]). For example, Jiang and Adeli [61] use neural network models to assess work zone capacity, concluding that they can estimate the WZ capacity with less than 10% error with this approach. In terms of work zone safety, the literature deals primarily with crash counts (not crash probability) for generic roadway incidents that do not consider the specific characteristics of work zones. Examples of this are Xie et al. [65] and Abdulhafedh [66], both of which compare neural network approaches to traditional NB models, noting that the neural networks show superior predictive capability in their case studies.

Li et al. [67] perform a similar comparison using support vector machines (SVM) and find that the SVM approach better estimates crash frequency in comparison to an NB model. They also point out that although neural networks and SVMs provide similar prediction capabilities, the SVM approach does not overfit the data. From a clustering approach, Ma and Kockelman [20] group road segments with similar characteristics and use a linear regression model to estimate crash frequency within each cluster. Other clustering applications include roadway crashes [68, 69] and safety projects [70], but to our knowledge current literature does not include papers which cluster work zones based their characteristics in order to infer crash risk.

Despite many different modeling approaches and perspectives, the majority of literature focuses on either work zones or crash frequency independently, with comparatively little research dedicated to modeling collision frequency/risk based on work zone characteristics. Within this niche body of research, most papers conclude that increased vehicle demand, work zone length, and duration all increase the number of collisions that occur at a work zone over a period of time [71]. However, we are more interested in determining the probability of a collision rather than the number of collisions that occur. This poses problems when considering short time durations with classical statistical approaches due to the fact that accidents occur so infrequently. The one paper that addresses this issue (i.e., the rare event logistic regression approach by Yang, Ozbay, Xie, and Bartin [55]) showed good performance in a case study, although the model was tested on a single corridor with less than 500 work zones. While machine learning models appear promising based on closely related studies, none of them have been applied specifically to estimating collision probabilities based on work zone characteristics. Accordingly, we focus on this underresearched area and present a clustering approach to estimate the probabilities of work zone collisions, which is useful for determining where to best locate response teams and identifying work zone characteristics that can be modified to reduce collision probabilities. We make the following contributions:

(i) We propose a scalable, unsupervised learning approach to predict the probability of a collision occurring at a short-term work zone. In contrast to most classical regression approaches which do not explicitly calculate crash risk/probability and are not appropriate for short-term durations, this machine learning approach clusters all work zones based on salient characteristics, calculates a collision probability for each cluster, and assigns new work zones to existing ones with similar features. Furthermore, it yields collision predictions without having access to current traffic volumes or work zone lengths and would easily scale to larger work zone datasets (e.g., see [72] for a clustering application involving 1 billion data points).

(ii) We present an integrated approach which combines both work zone collision risk predictions and actionable response recommendations, which is intuitive and applicable for practitioners. This approach takes historic information about work zones and returns the optimal allocation of highway response teams, which could be readily implemented by an agency such as the Coordinated Highways Action Response Team [73], an example of which is illustrated in the case study using the previously described approach and a stochastic optimization model.

The remainder of this paper is organized in the following manner. We begin by discussing the historical work zone and collision dataset that is used for model development in subsequent sections. Next, we describe clustering methods and explain how we determine the number of clusters and quantify model performance. We then apply this clustering framework to the work zone data set, noting how different approaches affect the model performance. Afterwards, we discuss example applications, focusing on optimally locating highway response vehicles by using the work zone incident probabilities as an input to a stochastic optimization model. Finally, we draw conclusions and suggest future steps to extend the research.

2. Materials and Methods

2.1. Data. Data describing work zones (WZs) and collisions that take place in their proximity (i.e., within 1 mile distance) were collected from the Regional Integrated Transportation Information System [77] and Coordinated Highways Action Response Team [73], whose joint objective is to improve operations of Maryland's highway system. We focus on those WZs that were set up and cleared in less than one day, which account for about 99% of all the WZs observed during 2010-2015. The resulting dataset considered in our analysis includes 54,463 WZs (Figure 1) and 380 WZ collisions. In most cases, a WZ had one or two lanes closed, and those closures took place along either the main or shoulder lanes (Figure 2). Furthermore, maintenance work lasted for 6 hours on average (Figure 3). In order to minimize negative effects on mobility and safety, only about 7% of the maintenance work was conducted during the peak-periods (7-9 AM and 4-6 PM). It is worth noting that about 72% of the work was done in daylight (Figure 3) and that WZs were typically set up on weekdays (Figure 4)

The observed 380 WZ collisions occurred at a very small subset of all the recorded WZs. There were hardly any WZs with more than one collision; however, a few observed as many as three collisions (Figure 5). The highest number of WZ collisions took place on Thursdays, while far fewer occurred on Saturdays and Sundays (Figure 6), which is unsurprising due to lower maintenance activity over the weekends (Figure 4). The seasonal distribution of WZ collisions (Figure 6) implies that most took place in summer and fall, which is expected due to higher maintenance activities during these seasons (Figure 4). Additionally, Figure 7 shows a visual representation of Average Annual

TABLE 1: Input variables considered in the analysis of WZ collisions.

(1) Day of the week when the WZ was opened (nominal)	(7) Duration of the WZ at night (continuous)
(2) Season of the year when the WZ was operational (nominal)	(8) Total number of lanes (ordinal)
(3) Road class (nominal: MD, I, US)	(9) Number of main lanes closed (ordinal)
(4) Duration of the WZ during the peak hours (continuous)	(10) Number of shoulder lanes closed (ordinal)
(5) Duration of the WZ during the off-peak hours (continuous)	(11) Number of median lanes closed (ordinal)
(6) Duration of the WZ in daylight (continuous)	(12) Average annual daily traffic (AADT) (ordinal)

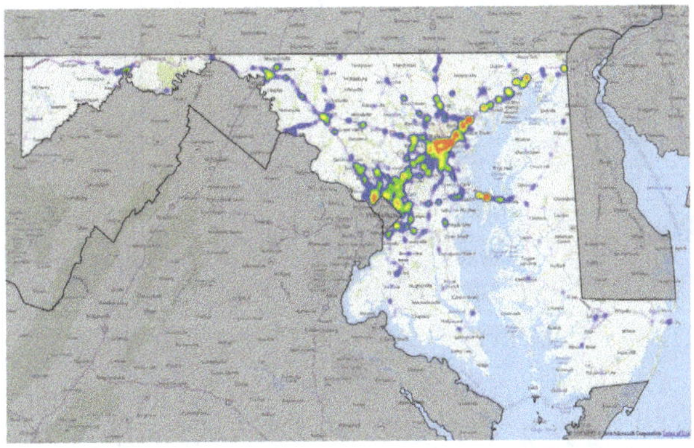

FIGURE 1: Heat map showing locations of over 54 thousand WZs that took place in Maryland during 2010-2015. The pie chart indicates that majority of WZs were setup along MD and Interstate roads.

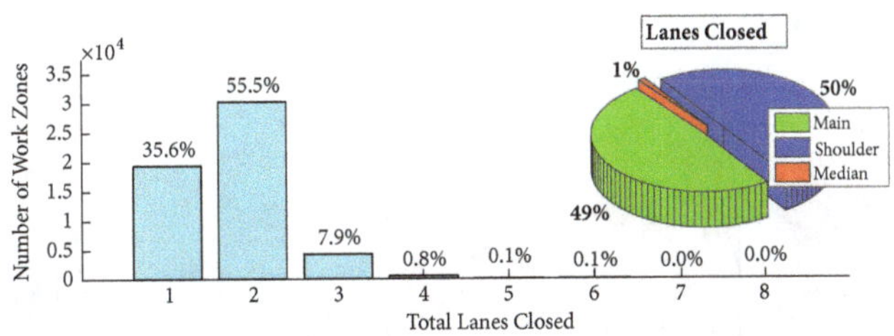

FIGURE 2: Most WZs would have one or two lanes closed, while closures of more than four lanes were very rare and typically occurred close to toll plazas. For the most part, closures affected main and shoulder lanes.

Daily Traffic (AADT) on the Maryland road network in 2013, indicating that a wide range of AADT values are observed, and highlighting the spatial distribution of high-volume roads. Based on this descriptive analysis of available data, we selected twelve input variables (Table 1) that might affect safety levels in the proximity of a WZ and thereby influence occurrence of a collision. The correlation between a subset of these variables and the number of WZ collisions is provided in Figure 8 and indicates relatively weak correlation between the WZ collisions and other variables.

2.1.1. Data Processing. Data preprocessing can significantly improve performance of the clustering algorithm presented in the following section. Categorical variables were modeled using their binary counterparts, and the peak and day duration of a WZ were divided by its overall duration because we are interested in predicting the one-hour probability of a collision. The off-peak and night durations were removed from the model as redundant information (e.g., relative peak and relative off-peak add up to one), and all the variables were normalized in order to enhance performance of the clustering approach presented in the next section. The WZ features used for clustering are presented in Table 2.

Note that WZ lengths were not available for the 54,000 WZs considered in this study, but since WZ lengths have been shown to influence crash frequency [71], we expect that including this variable in the future would improve the model accuracy. Additionally, a variety of other temporal variables would likely enhance model performance (e.g., speed variance and weather), but these characteristics would be unavailable ahead of time, and the proposed approach focuses on predicting the crash probability at a future work zone. Along these lines, it may be worth considering using hourly volume profiles [78] instead of AADT, as it would allow the model to account for the average hourly volumes during the specific work hours (e.g., 7 AM to 9 AM). Of course, like the AADT, the volume profiles would only help give a rough idea about the future volumes, because these volumes would be affected by the presence of the work zone.

2.2. Methodology. The objective is to use historical data in order to predict the probability of a collision occurring within

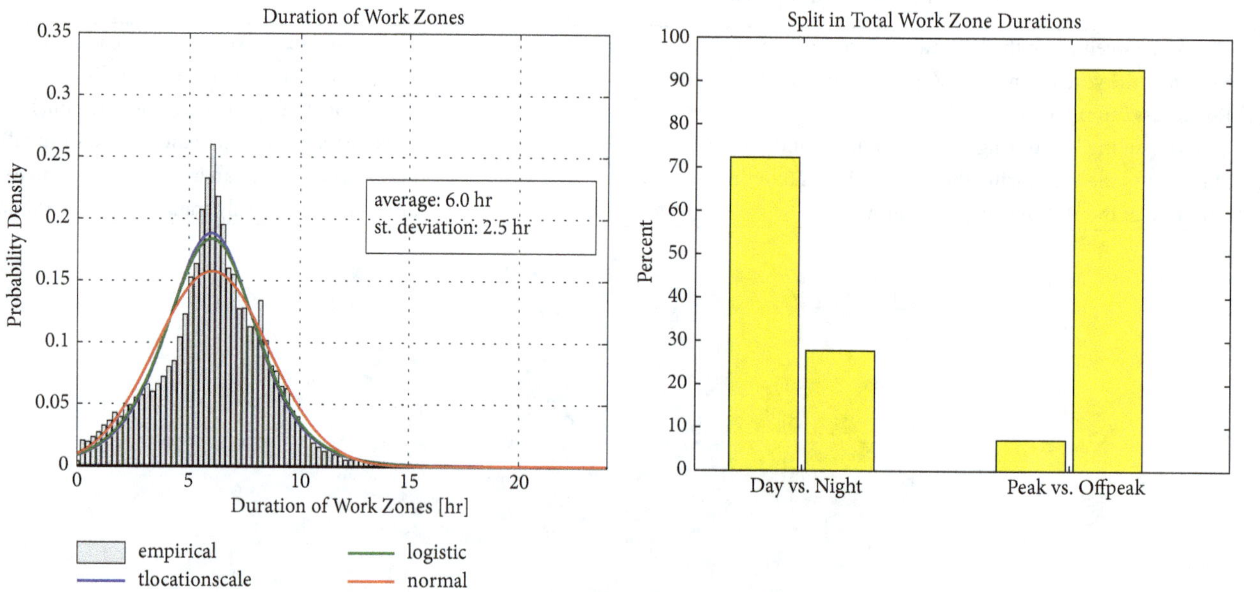

FIGURE 3: Duration of WZs and split in total work zone hours in day/night and peak/off-peak. Seventeen continuous distributions were fitted to WZ duration [74], and the probability density functions of those distributions that best fit in terms of Bayesian information criterion are shown in the figure on the left. Sunrise and sunset times for the dates of maintenance work and geographic locations of WZs are computed using MATLAB toolbox from SCRIPPS Institute of Oceanography [75].

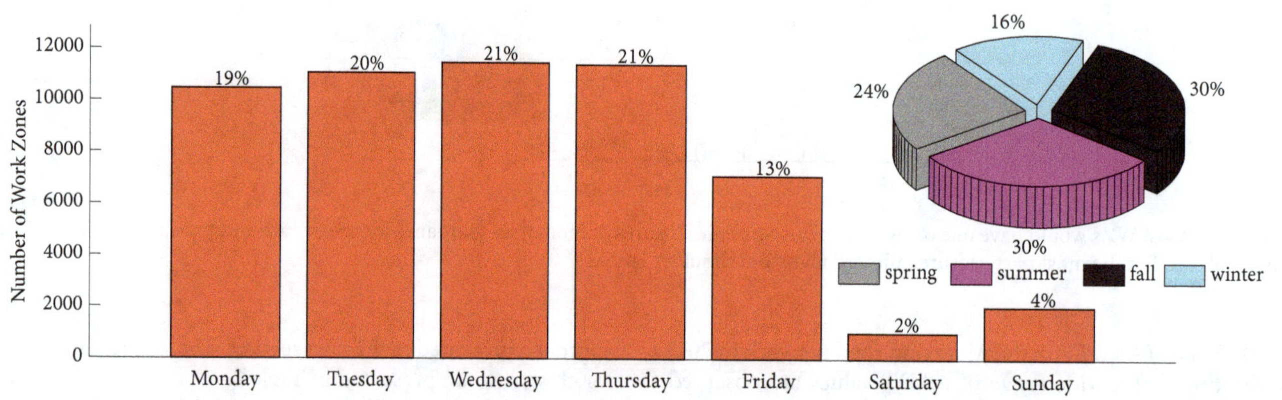

FIGURE 4: Most maintenance work was carried out on weekdays and during summer and fall.

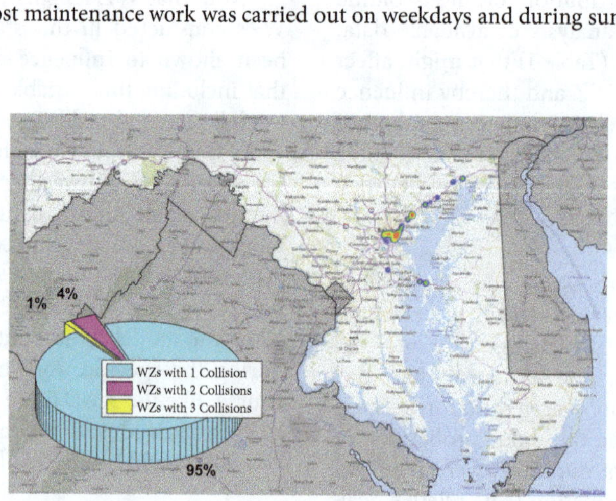

FIGURE 5: Heat map showing locations of 359 WZs within which 380 collisions occurred. The pie chart indicates that very few WZs had more than one collision.

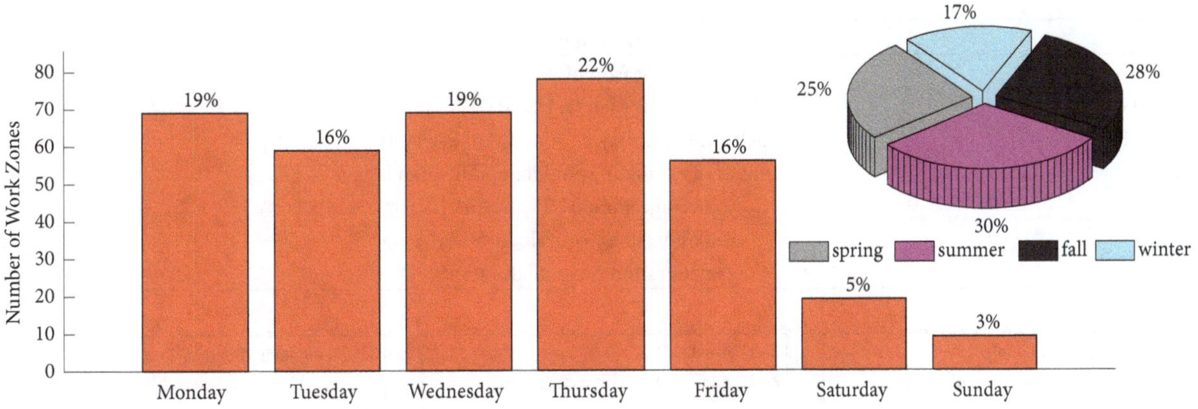

FIGURE 6: Most collisions took place on weekdays and during summer and fall, when high maintenance activity was observed.

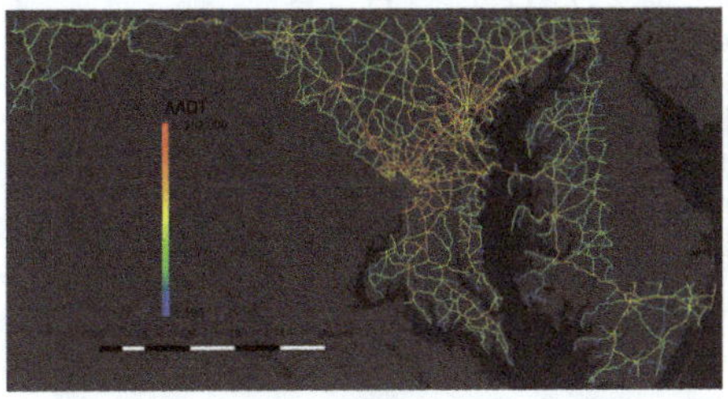

FIGURE 7: Visual representation of AADT in 2013 for the road links in Maryland.

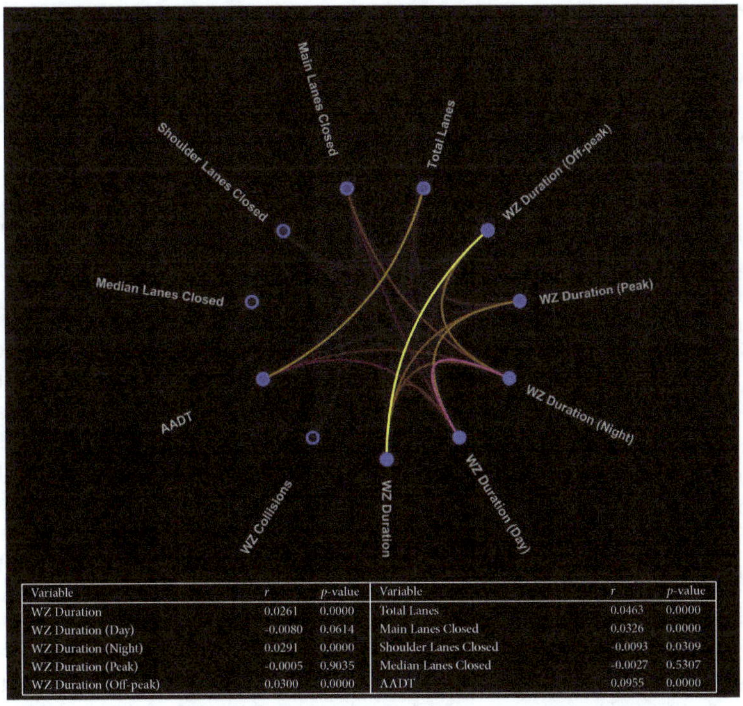

Variable	r	p-value	Variable	r	p-value
WZ Duration	0.0261	0.0000	Total Lanes	0.0463	0.0000
WZ Duration (Day)	-0.0080	0.0614	Main Lanes Closed	0.0326	0.0000
WZ Duration (Night)	0.0291	0.0000	Shoulder Lanes Closed	-0.0093	0.0309
WZ Duration (Peak)	-0.0005	0.9035	Median Lanes Closed	-0.0027	0.5307
WZ Duration (Off-peak)	0.0300	0.0000	AADT	0.0955	0.0000

FIGURE 8: Numeric values indicate correlation between (continuous and ordinal) input variables and number of WZ collisions. Yellow and magenta denote positive and negative correlation, respectively. Brighter links indicate greater correlation, while less visible links imply low correlation. Unfilled tokens indicate that a variable has low total absolute correlation with other covariates [76].

TABLE 2: Input variables used for WZ clustering.

Variable	Description	Columns
(1) Season	Season of year	3
(2) Road type	Type of road (I, US, MD)	3
(3) Weekend	Equals 1 for Saturday or Sunday, otherwise 0	1
(4) Lanes	The number of all, main, shoulder and median lanes	1
(5) AADT	Average annual daily traffic	1
(6) Peak	Peak duration / overall duration	1
(7) Day	Day duration / overall duration	1

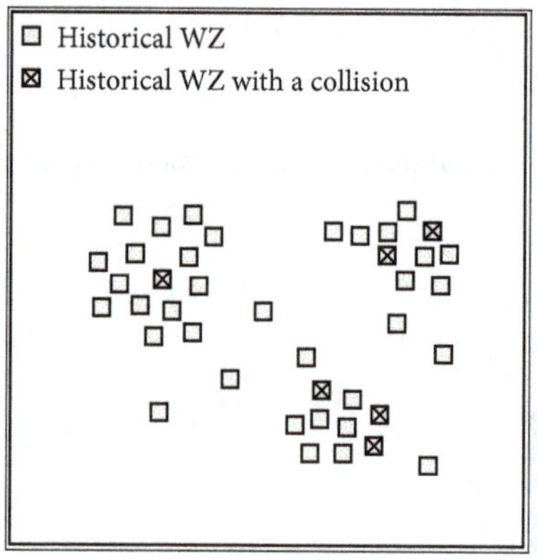

(a) Historical (unclustered) work zones

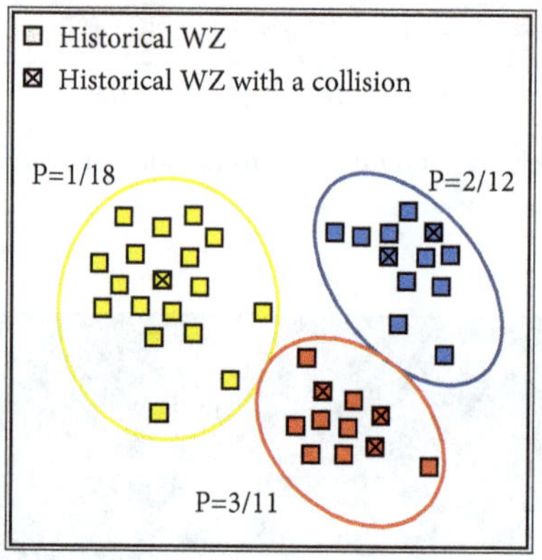

(b) Clustered WZs and corresponding collision probabilities

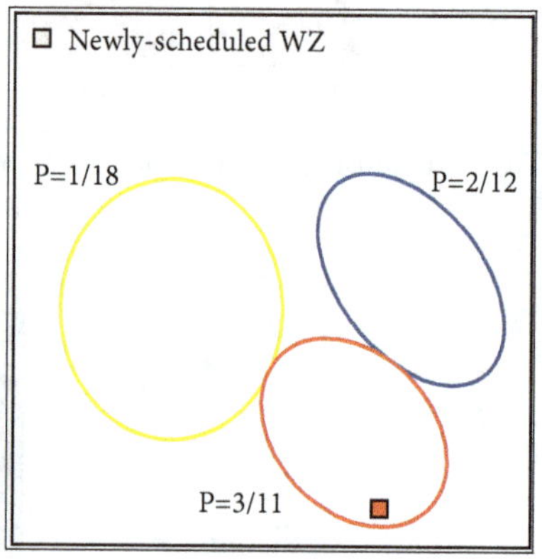

(c) Newly scheduled WZ is attributed to the closest cluster

FIGURE 9: Graphical description of the methodology via a trivial example including 2 dimensions, 3 clusters, and 41 data points (i.e., WZs). After clustering historical WZs, for each cluster, we compute the number of WZs with at least one collision over the total number of WZs within the cluster. A newly scheduled WZ is attributed to the cluster with the most similar features. The predicted probability of a collision occurring at this WZ corresponds to the ratio computed for the cluster it was attributed to.

a future WZ (e.g., road maintenance work scheduled for the next week), based on the underlying assumption that WZs with similar features should have comparable safety levels. Accordingly, we take the set of 54,000 WZs observed in the past (Section 2.1) and partition it into clusters of WZs with similar features (Table 2). After grouping WZs with similar characteristics, we calculate the collision probability for each cluster by counting the number of WZs where at least one collision occurred and dividing it by the total number of WZs within that cluster. Thus, the probability of a collision occurring at a future WZ can be predicted by assigning it to the closest cluster (in terms of its features) and looking up that cluster' observed collision probability, a process that is illustrated in Figure 9. The relation between predicted overall collision probability denoted by P and one-hour predicted collision probability delivered by the model (i.e., P_h) is computed as

$$P = 1 - (1 - P_h)^D, \quad (1)$$

where D is duration of the observed WZ.

The historical data can be partitioned into clusters of WZs with similar features via classical k-means clustering [79, 80]. Given a set of observations $(\mathbf{x}_1, \ldots, \mathbf{x}_n)$, where observation $\mathbf{x}_i$ is a d-dimensional vector of variables describing ith WZ, the k-means clustering seeks to partition these observations into k sets $\mathbf{S} = \{S_1, \ldots, S_k\}$ so as to minimize the within-cluster sum of squares. More formally, WZs are clustered by solving an optimization problem

$$\underset{\mathbf{S}}{\arg\min} \quad \sum_{l=1}^{k} \sum_{\mathbf{x} \in S_l} \|\mathbf{x} - \boldsymbol{\mu}_l\|^2, \quad (2)$$

where $\boldsymbol{\mu}_l$ is the mean of points in S_l. Since finding an exact solution to the above problem is NP-hard, many efficient heuristic techniques have been devised over the past 50 years. We apply k-means++ algorithm [81], which employs a randomized seeding technique to select initial cluster centers in a way that reduces the chance of the algorithm getting stuck in a suboptimal solution (Algorithm 1). However, even the k-means++ algorithm is not guaranteed to find an optimal solution in instances that include many dimensions and data points. To alleviate this problem, we perform clustering 100 times using different initial cluster centers selected by the k-means++ technique and choose the one with the smallest cost function for further analysis.

To evaluate the model, the dataset is first separated into training and testing datasets. Upon selecting the number of clusters (discussed below), the k-means++ algorithm is used to cluster the training data into k discrete groups. Each point in the test dataset is then associated with its most similar cluster from the training dataset in order to compute the

```
Procedure k-means++
  Select centers for k clusters by using randomized seeding technique
  repeat
    Assign each point to the closest cluster center
    Compute new cluster centers using assigned points
  until Convergence (cluster centers did not change during the last step)
```

ALGORITHM 1: Outline of the k-means++.

forecasted probability of a collision. These points are then ordered from lowest to highest probability (i.e., decreasing in safety) and separated into a specified number of groups, referred to as quantiles. For quantile i, the mean of the forecasted probability can be calculated (denoted F_i), which can be compared to the actual probability of collision (found by computing the ratio of work zones from quantile i which had at least one collision to the number of work zones in i, denoted A_i). More specifically, the symmetric mean absolute percentage error ($SMAPE$) is used to quantify the error between the actual and forecasted probabilities across all quantiles, which is computed as

$$SMAPE = \frac{1}{\tilde{n}}\sum_{i=1}^{\tilde{n}} \frac{|F_i - A_i|}{|F_i| + |A_i|}, \quad (3)$$

where $\tilde{n}$ is the number of quantiles.

In order to determine the number of clusters to use, we employ three different methods: Elbow, Silhouette, and Cross-Validation. In the Elbow method, the number of clusters is increased until the improvement in the objective function becomes marginal. The Silhouette method [82] requires computing each point's similarity to both its own cluster and the others. This measure is computed as

$$s_i = \frac{b_i - a_i}{\max(a_i, b_i)}, \quad (4)$$

where a_i represents the average dissimilarity of data point i with all other data within the same cluster, and b_i is the lowest average dissimilarity of the data point i to any cluster of which i is not a member. A value of s_i close to 1 implies that the point was appropriately clustered; values close to 0 mean that the points are on the border of two clusters, whereas negative values imply misclustering. The third approach is based on cross-validation, where the training dataset is divided into training and cross-validation sets. The model is then trained using the new, reduced training dataset, and evaluated by using the cross-validation dataset. After selecting the number of clusters, the model is retrained using the entire (undivided) training set.

Additionally, these three methods are helpful in determining the relevant model scenarios. In the base scenario, we assume that all the variables are of equal importance and thus are normalized from 0 to 1. To explore the influence of individual variables, we inflate their normalized values and study how this affects the clustering results. Thus, the Elbow, Silhouette, and Cross-Validation methods help provide insight into the number of clusters and appropriate variables to use.

3. Results and Discussion

The proposed clustering approach may overfit the data (i.e., memorize data points rather than detect patterns). In order to check for possible overfitting, it is helpful to train models with various sizes of training data and then check the behavior of the test error. When overfitting occurs, the error should significantly decrease as the training dataset size increases, but if the model is able to generalize well (i.e., does not overfit), further changes in training dataset size should not result in a meaningful reduction of test error. The results for the base scenario are presented in Figure 10 and imply that the model does not overfit the data for training sets greater than 50% of the entire dataset. Thus, the employed clustering approach may be used to predict WZ collision probabilities.

3.1. Number of Clusters. Selecting the number of clusters is essential for obtaining satisfactory results. After some preliminary experiments, the lower and upper bound on the number of clusters were set to 8 and 21, respectively. It should be noted that if a model includes few clusters, the WZs in each cluster may be quite diverse. This results in relatively small differences among mean collision probabilities in each cluster. Consequently, even if the accuracy of clustering is very high, the model cannot be employed for predictions. On the other hand, having many clusters should (in theory) improve the prediction accuracy. However, if some clusters do not include enough data points with collisions (due to relatively few collisions in the entire dataset), then the estimated collision probabilities may be inaccurate for these clusters. Therefore the upper bound on the number of clusters should correspond to the size of the training set and the number of collisions in the entire dataset.

As argued before, the authors considered three methods to select the number of clusters: Elbow, Silhouette, and Cross-Validation. The accuracy of each was estimated using a relative difference between the error of the model indicated by the method and error of the best model selected using a test set. The results for each method were computed for different scenarios and for different training dataset sizes (due to the overfitting analysis only the results for training dataset sizes greater or equal of 50% of entire dataset size were taken into consideration), which are summarized in Table 3.

TABLE 3: Aggregated accuracy metrics of various methods for selecting the best model (smaller numbers indicate better accuracy).

	Silhouette	Cross-Validation	Elbow
Mean	0.36365	0.40575	0.44222
Std. dev.	0.28405	0.42113	0.43249

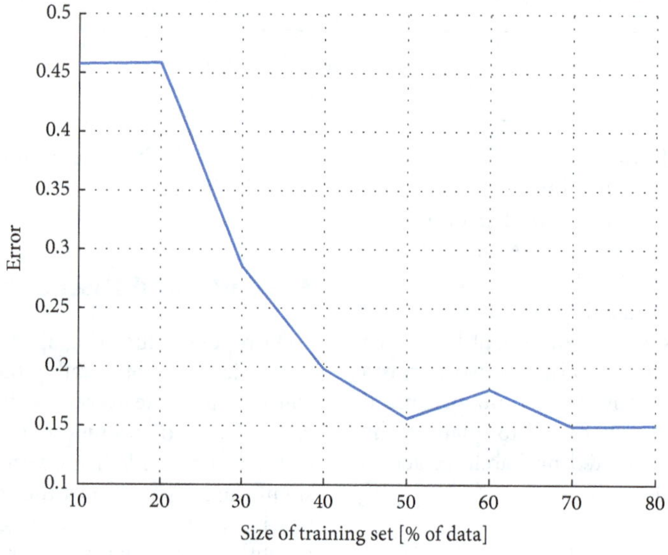

FIGURE 10: Predicted and actual mean collision probability for the base scenario.

The results indicate that the Silhouette method has the best accuracy and smallest standard deviation amongst the three measures, meaning that it is less prone to changes of scenario or training dataset size and, consequently, more reliable. Accordingly, the Silhouette method was chosen to determine the number of clusters.

3.2. Choosing a Scenario. The base scenario assumes that all the features are equal, but in actuality some may have a larger impact on the predicted values and others. In order to verify this, different scenarios were created and tested. In each scenario some features are more (or less) important than others, and the proximity in dimensions corresponding to these features has a greater impact on the attribution of WZs to certain clusters. A list of tested scenarios is presented in Table 4, while errors associated with the performance of the best model in each scenario are shown in Figure 11.

Scenarios 1-7 imply that AADT can deteriorate model accuracy. Specifically, increasing the importance of AADT inflates errors regardless of the data size. Conversely, scenarios 4, 6, and 7, which correspond to lanes, day/night, and peak/off-peak, are characterized by relatively smaller errors. Scenarios 8 and 9 and to some extent 11 show that AADT has no significant impact on model accuracy. In fact, reducing the significance of AADT or completely removing it does not affect the accuracy. This is unsurprising because AADT represents an average annual measure, which may be a poor indicator of traffic volumes during particular hours of road maintenance work. Moreover, scenarios 10-12 indicate that WZ hours of operations are highly relevant, as models that emphasize day/night and peak/off-peak values provide more accurate predictions. Finally, scenarios 13 and 14 show that the number of lanes, day/night, and peak/off-peak data have a crucial impact on prediction accuracy. However, performance of all these models is comparable, so we use the Silhouette method to select both the number of clusters and the best scenario.

3.3. The Best Model. The error for the model whose specifications are determined using the Silhouette method for 3 quantiles is 2.95%, indicating a high level of model accuracy. The accuracy for each quantile is shown in Figure 12(a), while the *SMAPE* errors based on the number of quantiles for the selected model are shown in Figure 12(b). While the number of quantiles (the resolution of the model) is increasing, the *SMAPE* errors also increase, meaning accuracy of the model is reduced. It is worth noticing that for 4, 5, 6 and 7 quantiles the *SMAPE* errors are almost constant (from 9.84% to 10.88% with 10.5% mean), thus implying that if 11% error is acceptable, it is possible to increase the resolution of the model up to 7 quantiles. When the resolution of the model increases above 7 quantiles, the error grows to an unacceptable level as a result of an insufficient amount of training data. Since the predicted probability of collision in each quantile is computed by dividing the number of work zones with a collision occurrence by the total number of WZs in the quantile, when there are not enough WZs in a quantile, even a single collision can significantly change predictions.

TABLE 4: List of tested scenarios.

Scenario	Description
0	Base scenario, all features equally important
1	Importance of seasons increased
2	Importance of road type increased
3	Importance of weekday/weekend increased
4	Importance of number of lanes increased
5	Importance of AADT increased
6	Importance of peak/off-peak increased
7	Importance of day/night increased
8	Importance of AADT decreased
9	AADT data removed
10	Importance of peak/off-peak and day/night increased
11	Importance of peak/off-peak and day/night increased, AADT data removed
12	Uses only peak/off-peak and day/night data
13	Importance of number of lanes, peak/off-peak and day/night increased
14	Uses only number of lanes, peak/off-peak and day/night data

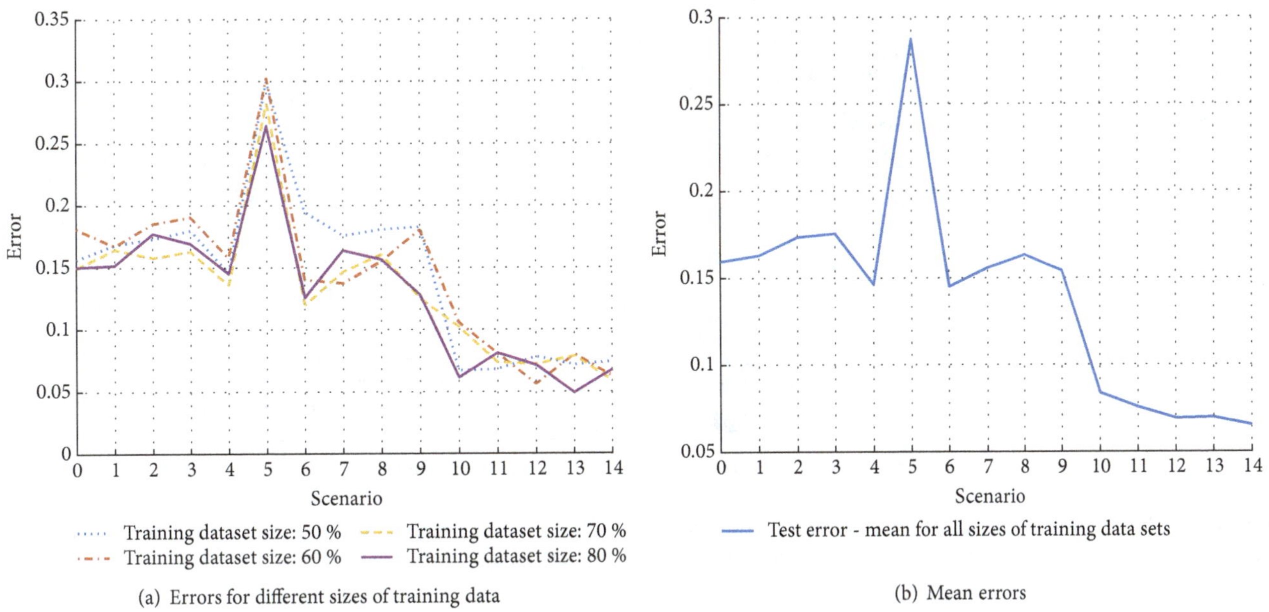

FIGURE 11: *SMAPE*-based errors associated with the best model in each scenario.

Accordingly, it is possible to increase the accuracy of the model for higher resolutions by increasing the size of the training dataset.

3.4. Example Application. Using the previously proposed clustering approach to predict the probability of a collision occurring within a WZ, we now provide an illustrative application of the proposed model. This hypothetical case study pertains to the jurisdiction of the Coordinated Highways Action Response Team, whose objective is to improve operations of Maryland's highway system. Suppose that this agency has a list of planned maintenance work for the following day and is interested in deploying a fixed number of response units to tackle collisions that may happen within these WZs. Clearly, we can assign these WZs to the clusters derived in the previous section and consequently estimate collision probability for each of the WZs scheduled for the following day (Figure 9). Once these probabilities are computed, the allocation of response units becomes a stochastic facility location problem, which has been tackled extensively in the operations research literature [83, 84]. In order to solve this response team allocation problem based on collision probabilities obtained from clustering, we formulate a two-stage stochastic program [85], the details of which are described in the Appendix.

In this illustrative example we randomly sample the 40 WZs that were used to test the clustering methodology and pretend they represent the maintenance work scheduled for

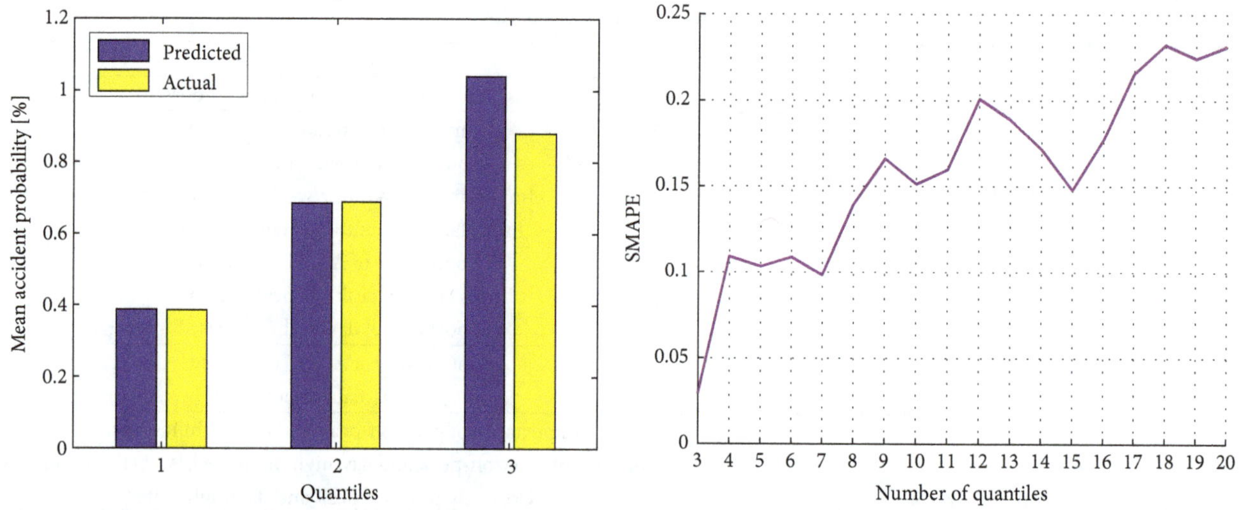

(a) Accuracy for 3 quantiles

(b) *SMAPE* errors regarding the number of quantiles

FIGURE 12: Accuracy and *SMAPE* errors for the selected model.

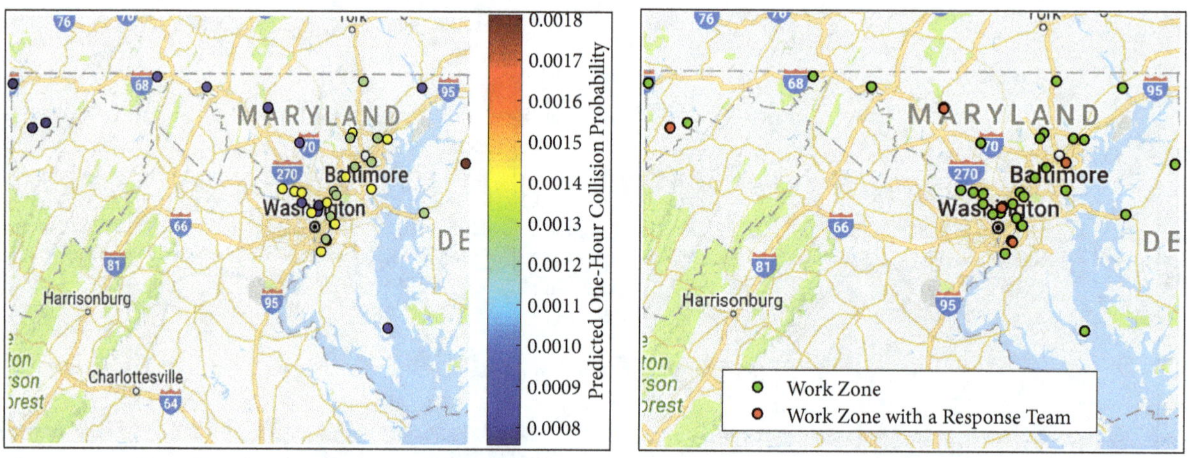

(a) Collision probabilities predicted via clustering

(b) Optimal response team allocation

FIGURE 13: Collision probabilities at 40 future WZs are estimated via the proposed clustering procedure. These probabilities are used as an input for the two-stage stochastic model to optimally allocate 5 highway response teams.

the following day. Consequently, we assign the 40 WZs to clusters built based on historical WZs (2010-2015), in order to compute the collision probabilities associated with each of the 40 newly-scheduled WZs (Figure 13(a)). Finally, using the optimization model provided in the Appendix, we determine the optimal location of the response units on the Maryland highway network for the observed time period. Figure 13(b) visualizes the optimal allocation when 5 response units are available, and additional allocations are provided in the Appendix (Figure 14) as the number of available response vehicles is perturbed from this value. The optimal allocation can be updated as the situation in the field evolves (e.g., one WZ is cleared and another on is set up), by simply reapplying the model given a new set of inputs (i.e., new configuration of WZs and corresponding collision probabilities). Alternatively, the proposed optimization model can be extended into a multiperiod stochastic program [85], which would better address dynamic relocation of response vehicles. This extended model would also require the collision probabilities estimated in this paper as inputs.

In addition to optimal allocation of highway response units, the proposed clustering method can be used to determine or adjust WZ parameters. Specifically, the presented model could help modify WZ parameters (e.g., lanes closed, day/night, and peak/off-peak duration) in order to meet certain safety levels (e.g., keep the collision probability below a specified threshold). For example, the easternmost WZ in Figure 13(b) has higher collision probability than others, so

- Work Zone
- Work Zone with a Response Team

(a) 3 response units

- Work Zone
- Work Zone with a Response Team

(b) 4 response units

- Work Zone
- Work Zone with a Response Team

(c) 6 response units

- Work Zone
- Work Zone with a Response Team

(d) 7 response units

- Work Zone
- Work Zone with a Response Team

(e) 8 response units

- Work Zone
- Work Zone with a Response Team

(f) 9 response units

FIGURE 14: Sensitivity analysis showing the optimal allocation of response teams when the number of available units is perturbed from $m = 5$.

its features may be modified to improve its safety level. A government agency might also introduce regulations to keep accident probabilities below certain level or provide financial incentives for those contractors who perform roadwork while maintaining high safety levels.

4. Conclusions

This paper proposed a clustering approach to predict the probability of a collision occurring in the proximity of planned maintenance work, which is important for allocation of highway response units. We presented the first application of clustering in the analysis of WZ collisions, which involved a large dataset of over 54,000 WZs in Maryland. The model showed good prediction accuracy, and its potential application was illustrated by optimally allocating response units in the Maryland highway network. Namely, collision probabilities determined via clustering were used as inputs to a two-stage stochastic program to optimally deploy highway response teams. Additionally, the proposed clustering approach can be used to adjust features of WZs to meet specified safety levels.

The proposed clustering method has certain limitations corresponding to the number of quantiles, so in some cases it may be used for classification of WZs rather than to predict the exact collision probabilities. Including more data would allow for additional quantiles to be used (i.e., preparation of higher-resolution models) and including additional WZ features may be useful as well. It would also be interesting to test clustering algorithms other than the k-means++, such as k-medoids or c-means and compare their performance. For the allocation of highway response teams, one could consider a multiperiod extension of the stochastic model used in this paper.

Appendix

The problem of allocating highway response units is modeled as a two-stage stochastic integer program, with the goal to locate m units in a way that would minimize the expected response time to WZs where collisions occur. In the first stage, we need to decide the location of response units, which may be placed at any of the WZ locations (or in their proximity). In the second stage, we assign these response units to WZs depending on the collision occurrences in each scenario and observe response times. More formally, let I denote a set of WZs indexed by i and j. Let ξ_i be a random parameter indicating the probability of a collision occurring at WZ $i \in I$, which was estimated using the proposed clustering approach. A particular realization of the aforementioned random parameters is denoted by $\omega \in \Omega$. Define x_i as an integer variable indicating the number of response units located in the proximity of WZ $i \in I$. We define $J(\omega) \subseteq I$ as a set of WZ locations where collisions occurred in a particular realization of random parameters $\omega \in \Omega$. Let $y_{ij}(\omega)$ be a binary variable which equals 1 if a response unit located at WZ $i \in I$ is dispatched to WZ $j \in J(\omega)$, given a particular realization of random parameters $\omega \in \Omega$. Let $d_{ij}(\omega)$ be the distance (or travel time) from WZ $i \in I$ to WZ $j \in J(\omega)$. We also define $\mathbf{x} = \{x_i \mid i \in I\}$ and $\boldsymbol{\xi} = \{\xi_i \mid i \in I\}$ as vectors of $|I|$ elements.

We formulate the first-stage problem as

$$\min_{x_i \in \mathbb{Z}^+} \quad \mathbb{E}\left[Q\left(\boldsymbol{x}, \boldsymbol{\xi}\right)\right] \tag{A.1}$$

$$\text{s.t.} \quad \sum_{i \in I} x_i \leq m, \tag{A.2}$$

where $Q(\boldsymbol{x}, \boldsymbol{\xi}(\omega))$ is the solution to the following second-stage problem:

$$Q\left(\boldsymbol{x}, \boldsymbol{\xi}\left(\omega\right)\right) = \min_{y_{ij}(\omega) \in \{0,1\}} \quad \sum_{i \in I} \sum_{j \in J(\omega)} y_{ij}\left(\omega\right) d_{ij}\left(\omega\right) \tag{A.3}$$

$$\text{s.t.} \quad \sum_{j \in J(\omega)} y_{ij}\left(\omega\right) \leq x_i \quad \forall i \in I \tag{A.4}$$

$$\sum_{i \in I} y_{ij}\left(\omega\right) = 1 \quad \forall j \in J\left(\omega\right) \tag{A.5}$$

As argued above, the first-stage problem seeks to allocate teams in order to minimize the expected response time subject to vehicle availability constraint. The response time for a particular realization of random parameters is computed in the second stage. Constraints (A.4) ensure that vehicles are dispatched only from locations where they are available, while (A.5) guarantee that a response team is directed to each WZ collision. To ensure feasibility of the second state, we let $m \geq |J(\omega)|$, $\forall \omega \in \Omega$, which implies that the number of response teams is greater or equal than the total number of locations at which collisions occur simultaneously. This condition, however, can be relaxed by extending the model to consider priorities in serving collision locations. Finally, the proposed two-stage stochastic model is implemented in GAMS and solved in the extensive form using CPLEX solver. We observe computation times of several seconds for the case study involving 40 WZs, 5 response teams and 1,000 scenarios (i.e., 1,000 realizations of $\boldsymbol{\xi}$). The corresponding optimal allocation of response teams is provided in Figure 13(b), with additional allocations shown in Figure 14 for different numbers of available response teams.

Disclosure

This research did not receive specific funding, but was performed as part of employment at the Center for Advanced Transportation Technology, University of Maryland.

Conflicts of Interest

The authors declare that they have no conflicts of interest.

References

[1] FHWA, "Facts and statistics - work zone safety," 2016, http://www.ops.fhwa.dot.gov/wz/resources/facts_stats/safety.htm.

[2] S. B. Mohan and P. Gautam, "Cost of highway work zone injuries," *Practice Periodical on Structural Design and Construction*, vol. 7, no. 2, pp. 68–73, 2002.

[3] A. G. Beacher, M. D. Fontaine, and N. J. Garber, "Evaluation of the late merge work zone traffic control strategy," Technical Report FHWA/VTRC 05-R6, 2004.

[4] M. Papageorgiou, I. Papamichail, A. D. Spiliopoulou, and A. F. Lentzakis, "Real-time merging traffic control with applications to toll plaza and work zone management," *Transportation Research Part C: Emerging Technologies*, vol. 16, no. 5, pp. 535–553, 2008.

[5] N. J. Fudala and M. D. Fontaine, "Work zone variable speed limit systems: Effectiveness and system design issues," Tech. Rep., 27 Nicholas J Fudala and Michael D Fontaine. Work zone variable speed limit systems, Effectiveness and system design issues. Technical Report FHWA/VTRC 10-R20, 2010.

[6] J. Weng and Q. Meng, "Modeling speed-flow relationship and merging behavior in work zone merging areas," *Transportation Research Part C: Emerging Technologies*, vol. 19, no. 6, pp. 985–996, 2011.

[7] J. Weng and Q. Meng, "Estimating capacity and traffic delay in work zones: an overview," *Transportation Research Part C: Emerging Technologies*, vol. 35, pp. 34–45, 2013.

[8] B. Du, I. Steven, and J. Y. Chien, "Feasibility of shoulder use for highway work zone optimization," *Journal of Traffic and Transportation Engineering*, vol. 1, no. 4, pp. 235–246, 2014.

[9] A. Tarko, S. Kanipakapatnam, and J. Wasson, "Modeling and Optimization of the Indiana Lane Merge Control System on Approaches to Freeway Work Zones, Part I," Purdue University FHWA/IN/JTRP-97/12-1, 1998.

[10] S. Chien and P. Schonfeld, "Optimal work zone lengths for four-lane highways," *Journal of Transportation Engineering*, vol. 127, no. 2, pp. 124–131, 2001.

[11] S. Chien, Y. Tang, and P. Schonfeld, "Optimizing work zones for two-lane highway maintenance projects," *Journal of Transportation Engineering*, vol. 128, no. 2, pp. 145–155, 2002.

[12] Q. Meng and J. Weng, "Optimal subwork zone length and project start time for short-term daytime work zones from the contractor's perspective," *Transportation Research Part C: Emerging Technologies*, vol. 29, pp. 72–83, 2013.

[13] A. Tympakianaki, A. Spiliopoulou, A. Kouvelas, I. Papamichail, M. Papageorgiou, and Y. Wang, "Real-time merging traffic control for throughput maximization at motorway work zones," *Transportation Research Part C: Emerging Technologies*, vol. 44, pp. 242–252, 2014.

[14] G. Casal, D. Santamarina, and M. E. Vázquez-Méndez, "Optimization of horizontal alignment geometry in road design and reconstruction," *Transportation Research Part C: Emerging Technologies*, vol. 74, pp. 261–274, 2017.

[15] J. Wang, W. E. Hughes, F. M. Council, and J. F. Paniati, "Investigation of highway work zone crashes: What we know and what we don't know," *Transportation Research Record*, no. 1529, pp. 54–62, 1996.

[16] J. Daniel, K. Dixon, and D. Jared, "Analysis of fatal crashes in Georgia work zone," *Transportation Research Record*, no. 1715, pp. 18–23, 2000.

[17] N. J. Garber and M. Zhao, "Distribution and characteristics of crashes at different work zone locations in Virginia," *Transportation Research Record*, no. 1794, pp. 19–28, 2002.

[18] S. D. Schrock, G. L. Ullman, A. S. Cothron, E. Kraus, and A. P. Voigt, "An analysis of fatal work zone crashes in Texas," Technical Report FHWA/TX-05/0-4028-1, 2004.

[19] H. T. Abdelwahab and M. A. Abdel-Aty, "Predicting injury severity levels in traffic crashes: a modeling comparison," *Journal of Transportation Engineering*, vol. 130, no. 2, pp. 204–210, 2004.

[20] J. Ma and K. Kockelman, "Crash frequency and severity modeling using clustered data from Washington state," in *Proceedings of the ITSC 2006: 2006 IEEE Intelligent Transportation Systems Conference*, pp. 1621–1626, Canada, September 2006.

[21] Y. Li and Y. Bai, "Highway work zone risk factors and their impact on crash severity," *Journal of Transportation Engineering*, vol. 135, no. 10, pp. 694–701, 2009.

[22] G. Khan, A. R. Bill, and D. A. Noyce, "Exploring the feasibility of classification trees versus ordinal discrete choice models for analyzing crash severity," *Transportation Research Part C: Emerging Technologies*, vol. 50, pp. 86–96, 2015.

[23] A. J. Khattak and F. Targa, "Injury severity and total harm in truck-involved work zone crashes," *Transportation Research Record*, no. 1877, pp. 106–116, 2004.

[24] E. S. Park and D. Lord, "Multivariate poisson-lognormal models for jointly modeling crash frequency by severity," *Transportation Research Record*, no. 2019, pp. 1–6, 2007.

[25] J. Ma, K. M. Kockelman, and P. Damien, "A multivariate Poisson-lognormal regression model for prediction of crash counts by severity, using Bayesian methods," *Accident Analysis & Prevention*, vol. 40, no. 3, pp. 964–975, 2008.

[26] Q. Meng, J. Weng, and X. Qu, "A probabilistic quantitative risk assessment model for the long-term work zone crashes," *Accident Analysis & Prevention*, vol. 42, no. 6, pp. 1866–1877, 2010.

[27] C. Xu, A. P. Tarko, W. Wang, and P. Liu, "Predicting crash likelihood and severity on freeways with real-time loop detector data," *Accident Analysis & Prevention*, vol. 57, pp. 30–39, 2013.

[28] Y.-J. Kweon, I.-K. Lim, and M. D. Fontaine, "Work zone safety performance measures for Virginia," Technical Report FHWA/VTRC 16-R10, 2016.

[29] D. Lord and F. Mannering, "The statistical analysis of crash-frequency data: a review and assessment of methodological alternatives," *Transportation Research Part A: Policy and Practice*, vol. 44, no. 5, pp. 291–305, 2010.

[30] R. Pal and K. C. Sinha, "Analysis of crash rates at Interstate work zones in Indiana," *Transportation Research Record*, no. 1529, pp. 43–53, 1996.

[31] A. J. Khattak, A. J. Khattak, and F. M. Council, "Effects of work zone presence on injury and non-injury crashes," *Accident Analysis & Prevention*, vol. 34, no. 1, pp. 19–29, 2002.

[32] S. Venugopal and A. Tarko, "Safety models for rural freeway work zones," *Transportation Research Record*, no. 1715, pp. 1–9, 2000.

[33] R. Srinivasan, G. Ullman, M. Finley, and F. Council, "Use of empirical bayesian methods to estimate crash modification factors for daytime versus nighttime work zones," *Transportation Research Record*, no. 2241, pp. 29–38, 2011.

[34] E. Chen and A. Tarko, "Analysis of crash frequency in work zones with focus on police enforcement," *Transportation Research Record*, no. 2280, pp. 127–134, 2012.

[35] O. Ozturk, K. Ozbay, H. Yang, and B. Bartin, "Crash frequency modeling for highway construction zones," in *Transportation Research Board 92nd Annual Meeting*, pp. 13–4555, 2013.

[36] D. Eustace, A. Aylo, and W. Y. Mergia, "Crash frequency analysis of left-side merging and diverging areas on urban freeway segments - A case study of I-75 through downtown Dayton,

Ohio," *Transportation Research Part C: Emerging Technologies*, vol. 50, pp. 78–85, 2015.

[37] H. Yang, K. Ozbay, O. Ozturk, and M. Yildirimoglu, "Modeling work zone crash frequency by quantifying measurement errors in work zone length," *Accident Analysis & Prevention*, vol. 55, pp. 192–201, 2013.

[38] K. El-Basyouny and T. Sayed, "Collision prediction models using multivariate Poisson-lognormal regression," *Accident Analysis & Prevention*, vol. 41, no. 4, pp. 820–828, 2009.

[39] D. Lord, S. P. Washington, and J. N. Ivan, "Further notes on the application of zero-inflated models in highway safety," *Accident Analysis & Prevention*, vol. 39, no. 1, pp. 53–57, 2007.

[40] A. K. Sharma and V. S. Landge, "Zero inflated negative binomial for modeling heavy vehicle crash rate on Indian rural highway," *Internet Journal of Advances in Engineering & Technology*, vol. 5, no. 2, pp. 292–301, 2013.

[41] C. Dong, D. B. Clarke, X. Yan, A. Khattak, and B. Huang, "Multivariate random-parameters zero-inflated negative binomial regression model: an application to estimate crash frequencies at intersections," *Accident Analysis & Prevention*, vol. 70, pp. 320–329, 2014.

[42] C. Dong, S. H. Richards, D. B. Clarke, X. Zhou, and Z. Ma, "Examining signalized intersection crash frequency using multivariate zero-inflated Poisson regression," *Safety Science*, vol. 70, pp. 63–69, 2014.

[43] F. Chen, S. Chen, and X. Ma, "Crash frequency modeling using real-time environmental and traffic data and unbalanced panel data models," *International Journal of Environmental Research and Public Health*, vol. 13, no. 6, article no. 609, 2016.

[44] J. C. Milton, V. N. Shankar, and F. L. Mannering, "Highway accident severities and the mixed logit model: an exploratory empirical analysis," *Accident Analysis & Prevention*, vol. 40, no. 1, pp. 260–266, 2008.

[45] P. C. Anastasopoulos and F. L. Mannering, "A note on modeling vehicle accident frequencies with random-parameters count models," *Accident Analysis & Prevention*, vol. 41, no. 1, pp. 153–159, 2009.

[46] E. Chen and A. P. Tarko, "Modeling safety of highway work zones with random parameters and random effects models," *Analytic Methods in Accident Research*, vol. 1, pp. 86–95, 2014.

[47] Z. Zeng, W. Zhu, R. Ke et al., "A generalized nonlinear model-based mixed multinomial logit approach for crash data analysis," *Accident Analysis & Prevention*, vol. 99, pp. 51–65, 2017.

[48] Y. Xie and Y. Zhang, "Crash frequency analysis with generalized additive models," *Transportation Research Record*, no. 2061, pp. 39–45, 2008.

[49] Y. Zhang, Y. Xie, and L. Li, "Crash frequency analysis of different types of urban roadway segments using generalized additive model," *Journal of Safety Research*, vol. 43, no. 2, pp. 107–114, 2012.

[50] N. V. Malyshkina, F. L. Mannering, and A. P. Tarko, "Markov switching negative binomial models: an application to vehicle accident frequencies," *Accident Analysis & Prevention*, vol. 41, no. 2, pp. 217–226, 2009.

[51] N. V. Malyshkina and F. L. Mannering, "Zero-state Markov switching count-data models: An empirical assessment," *Accident Analysis & Prevention*, vol. 42, no. 1, pp. 122–130, 2010.

[52] Y. Xiong, J. L. Tobias, and F. L. Mannering, "The analysis of vehicle crash injury-severity data: A Markov switching approach with road-segment heterogeneity," *Transportation Research Part B: Methodological*, vol. 67, pp. 109–128, 2014.

[53] X. Qin, J. N. Ivan, N. Ravishanker, and J. Liu, "Hierarchical Bayesian estimation of safety performance functions for two-lane highways using Markov chain Monte Carlo modeling," *Journal of Transportation Engineering*, vol. 131, no. 5, pp. 345–351, 2005.

[54] D.-G. Kim, Y. Lee, S. Washington, and K. Choi, "Modeling crash outcome probabilities at rural intersections: Application of hierarchical binomial logistic models," *Accident Analysis & Prevention*, vol. 39, no. 1, pp. 125–134, 2007.

[55] H. Yang, K. Ozbay, K. Xie, and B. Bartin, "Modeling crash risk of highway work zones with relatively short durations," in *In Transportation Research Board 94th Annual Meeting*, Washington, DC, USA, 2015.

[56] A. S. Al-Ghamdi, "Using logistic regression to estimate the influence of accident factors on accident severity," *Accident Analysis & Prevention*, vol. 34, no. 6, pp. 729–741, 2002.

[57] R. Harb, E. Radwan, X. Yan, A. Pande, and M. Abdel-Aty, "Freeway work-zone crash analysis and risk identification using multiple and conditional logistic regression," *Journal of Transportation Engineering*, vol. 134, no. 5, pp. 203–214, 2008.

[58] G. H. Bham, B. S. Javvadi, and U. R. R. Manepalli, "Multinomial Logistic Regression Model for Single-Vehicle and Multivehicle Collisions on Urban U.S. Highways in Arkansas," *Journal of Transportation Engineering*, vol. 138, no. 6, pp. 786–797, 2012.

[59] H. Adeli and X. Jiang, "Neuro-fuzzy logic model for freeway work zone capacity estimation," *Journal of Transportation Engineering*, vol. 129, no. 5, pp. 484–493, 2003.

[60] A. Karim and H. Adeli, "Radial basis function neural network for work zone capacity and queue estimation," *Journal of Transportation Engineering*, vol. 129, no. 5, pp. 494–503, 2003.

[61] X. Jiang and H. Adeli, "Clustering-neural network models for freeway work zone capacity estimation.," *International Journal of Neural Systems*, vol. 14, no. 3, pp. 147–163, 2004.

[62] S. Ghosh-Dastidar and H. Adeli, "Neural network-wavelet microsimulation model for delay and queue length estimation at freeway work zones," *Journal of Transportation Engineering*, vol. 132, no. 4, pp. 331–341, 2006.

[63] S. Demiroluk and K. Ozbay, "Adaptive learning in bayesian networks for incident duration prediction," *Transportation Research Record*, vol. 2460, no. 1, pp. 77–85, 2014.

[64] B. Du, S. Chien, J. Lee, L. Spasovic, and K. Mouskos, "Artificial neural network model for estimating temporal and spatial freeway work zone delay using probe-vehicle data," *Transportation Research Record*, vol. 2573, pp. 164–171, 2016.

[65] Y. Xie, D. Lord, and Y. Zhang, "Predicting motor vehicle collisions using Bayesian neural network models: an empirical analysis," *Accident Analysis & Prevention*, vol. 39, no. 5, pp. 922–933, 2007.

[66] A. Abdulhafedh, "Crash Frequency Analysis," *Journal of Transportation Technologies*, vol. 06, no. 04, pp. 169–180, 2016.

[67] X. Li, D. Lord, Y. Zhang, and Y. Xie, "Predicting motor vehicle crashes using Support Vector Machine models," *Accident Analysis & Prevention*, vol. 40, no. 4, pp. 1611–1618, 2008.

[68] S. Y. Sohn, "Quality function deployment applied to local traffic accident reduction," *Accident Analysis & Prevention*, vol. 31, no. 6, pp. 751–761, 1999.

[69] T. F. Golob and W. W. Recker, "A method for relating type of crash to traffic flow characteristics on urban freeways," *Transportation Research Part A: Policy and Practice*, vol. 38, no. 1, pp. 53–80, 2004.

[70] S. C. Wong, B. S. Y. Leung, B. P. Y. Loo, W. T. Hung, and H. K. Lo, "A qualitative assessment methodology for road safety policy strategies," *Accident Analysis & Prevention*, vol. 36, no. 2, pp. 281–293, 2004.

[71] H. Yang, K. Ozbay, O. Ozturk, and K. Xie, "Work Zone Safety Analysis and Modeling: A State-of-the-Art Review," *Traffic Injury Prevention*, vol. 16, no. 4, pp. 387–396, 2015.

[72] R. Wu, B. Zhang, and M. Hsu, "Clustering billions of data points using GPUs," in *Proceedings of the Combined Workshops on UnConventional High Performance Computing Workshop Plus Memory Access Workshop, UCHPC-MAW '09, Co-located with the 2009 ACM International Conference on Computing Frontiers, CF'09*, pp. 1–5, Italy, May 2009.

[73] CHART, "Coordinated highways action response team," 2016, http://www.chart.state.md.us/.

[74] M. Sheppard, "Allfitdist," MATLAB Central File Exchange, 2012.

[75] R. Pawlowicz, "Function adapted by the SCRIPPS Institute for Oceanography," 2009.

[76] O. Komarov, "MATLAB Central File Exchange," *Retrieved*, pp. 01-02, 2013.

[77] L. P. Michael, R. B. Jeffrey, and A. Steffes, "Overview and status of regional integrated transportation information system in the national capital region," in *Transportation Research Board 87th Annual Meeting*, pp. 08–1299, 2008.

[78] D. Schrank, B. Eisele, T. Lomax, and J. Bak, "2015 urban mobility scorecard," 2015, http://d2dtl5nnlpfr0r.cloudfront.net/tti.tamu.edu/documents/mobility-scorecard-2015.pdf.

[79] H. Steinhaus, "Sur la division des corps matériels en parties," *Bulletin de l'Academie Polonaise des Sciences*, vol. 4, no. 12, pp. 801–804, 1957 (French).

[80] E. W. Forgy, "Cluster analysis of multivariate data: Efficiency versus interpretability of classifications," *Biometrics*, vol. 21, Article ID 768769, pp. 768-769, 1965.

[81] D. Arthur and S. Vassilvitskii, "k-means++: the advantages of careful seeding," in *Proceedings of the ACM-SIAM Symposium on Discrete Algorithms*, pp. 1027–1035, 2007.

[82] P. J. Rousseeuw, "Silhouettes: a graphical aid to the interpretation and validation of cluster analysis," *Journal of Computational and Applied Mathematics*, vol. 20, pp. 53–65, 1987.

[83] L. V. Snyder, "Facility location under uncertainty: a review," *IIE Transactions*, vol. 38, no. 7, pp. 547–564, 2006.

[84] M. T. Melo, S. Nickel, and F. Saldanha-da-Gama, "Facility location and supply chain management—a review," *European Journal of Operational Research*, vol. 196, no. 2, pp. 401–412, 2009.

[85] J. R. Birge and F. V. Louveaux, *Introduction to Stochastic Programming*, Springer, Berlin, Germany, 2011.

7

Passenger Mobility in a Discontinuous Space: Modelling Access/Egress to Maritime Barrier in a Case Study

G. Birgillito, C. Rindone, and A. Vitetta

Dipartimento di ingegneria dell'Informazione, delle Infrastrutture e dell'Energia Sostenibile, Università Mediterranea di Reggio Calabria, Reggio Calabria, Italy

Correspondence should be addressed to A. Vitetta; vitetta@unirc.it

Academic Editor: Gonçalo Homem de Almeida Correia

The present study analyses a transport system in a discontinuous space. Classical specifications of transport models cannot be applied directly when evaluating the influence of territorial discontinuity and related barriers on user behaviour. Adjustments are required for this specific case because studies relative to discontinuous spaces are limited. The influence concerns the different travel components (access, barrier, on board, and egress) and could impact travel choices (departure time, destination, mode, and path). In this paper, the models refer to access and egress components with a focus on the mode of travel choice level. The paper focusses on the influence of discontinuity, introducing some adjustments to the classical demand models used to simulate discontinuity crossing. The main variables influencing the user's choice, and their relative weight in discontinuous space, are investigated. These elements are fundamental for any planning and design procedures to improve the quality of mobility. In the paper, the case study of the Strait of Messina in southern Italy is analysed. In this case, the barrier is constituted by the sea that physically separates the two shores. In this work, only strait crossings via hydrofoil, from Reggio Calabria to Messina, are considered in random sample interviews.

1. Introduction

In this paper, problems related to temporal and spatial discontinuities in a transport system are studied. The issue is common to many European spatial planning and transport policies that aim to remove bottlenecks and barriers limiting the mobility of people and goods. For instance, the Trans-European Networks for Transport (TEN-T) [1] help "close gaps, remove bottlenecks, and eliminate technical barriers that exist between the transport networks of EU Member States" [2]. Moreover, planned TEN-T interventions pursue the objectives of removing bottlenecks and filling missing links. However, many academic research have not traditionally focussed on transport bottlenecks [3]. Considering the existing financial and environmental constraints, construction of new infrastructures is not the only solution. As an example, the importance of, and the problems related to, bottlenecks in public transport operations is highlighted by van Oort et al. [4]. As a consequence, a new approach to study and solve bottlenecks and barrier problems is needed.

In this context, it is crucial to develop travel demand models for transportation planning, as they can be used to estimate present and future demand [5–8], also considering recent research developments in the type of choice models [9].

In transport systems, behavioural demand models are typically studied (specified, calibrated, and validated) in continuous territories when simulating travellers' choices. In a territory with a discontinuity, two continuous areas are divided by a physical element (e.g., sea, lake, or river) and cannot be travelled using a single mode in a continuous time (i.e., pedestrian, bicycle, or private car on road). Users who travel only within the continuous area have many transport alternatives at their disposal (individual, collective, semicollective, and multimodal). However, for trips between two continuous areas, they are forced to use only multimode collective or semicollective alternatives with discontinuous services (i.e., pedestrian, bicycle, or private car on discontinuity ship service). The discontinuity may impact one or more travel choice levels (departure time, destination, mode, and path). There is a great deal of literature on behaviour demand models for simulations of travel users' choices in a continuous space [10, 11]. However, similar studies considering the presence of a discontinuity in the spaces are limited.

This research is in the field of demand models related to a discontinuous space. There are three main components in the travel choice for a discontinuous space: *access* (i.e., reaching the terminal from the point of origin), *barrier* (i.e., waiting at the terminal for the service), *on board* crossing the discontinuity (i.e., moving from the terminal located in the area of origin to the terminal located in the destination area), and *egress* (i.e., reaching the final destination from the terminal). In this context, adjustments to the classical specification of the models for the simulation of user choices are required for each travel component (i.e., users could have different choice set alternatives in the same trip in the three components).

This paper considers models with reference to the mode adopted for the access and egress component. The main goals of the paper are to consider the influence of discontinuity on the travel choice; to introduce adjustments to the classical demand model to take account of the mode choice for access and egress given a service crossing the discontinuity; and to apply the model (specification, calibration, and validation) in a case study based on a survey. This paper is innovative in that it proposes models representing passenger mobility in the presence of a space discontinuity.

The problem is (a) relevant and (b) general. (a) It is relevant because many urbanised areas in the real world are separated by a discontinuity, and traditional and consolidated models cannot be directly applied to design and planning. The innovation also relates to the specification, calibration, and validation of a specific model for the discontinuity. It cannot be simulated by a traditional approach adopted for a waiting trip component in transit system [12]. Instead, it has to be considered as a model evaluated in the context of an entire trip. (b) It is general because it considers the problem of access and egress models with experimentation in a real case where a territory serves as a natural laboratory. The case study examined is the Strait of Messina Area (Italy), including the metropolitan cities of Messina (Sicily) and Reggio Calabria (Calabria). This area represents a whole because of its unique geographic characteristics (minimum strait width of just 3.2 km) and consequently the close relationships between the two shores of the Strait. The presence of activities in both cities generates significant flows of people, despite poor transportation system integration and the considerable amount of time spent in this short-medium distance mobility. The Strait Area also involves long distance trips to and from Sicily. Both freight and passenger flows in the area are significant. This work focusses on the second.

In the study, one segment of strait-crossing passenger mobility is focussed. The goal of the present work is to model the mode choice only for users having previously decided to cross the Strait of Messina by hydrofoil on the Reggio Calabria–Messina route. For achieving this goal, a survey was conducted, and the obtained data are used to calibrate and validate a behavioural model.

The paper has the following structure. Section 2 presents a brief literature review. Section 3 presents the proposed method and illustrates the problem of mobility with barrier (Section 3.1). The method to analyse the problem is proposed for the current (Section 3.2.1) and scenario configurations (Section 3.2.2) in order to support decision makers and planners (Section 3.3). In Section 4, the proposed method is applied to the case study to analyse passenger mobility with a barrier constituted by the Strait of Messina, located in southern Italy (Sections 4.1 and 4.2). The applications concern current (Section 4.2.1) and scenario configurations (Section 4.2.2), producing results useful for decision makers and planners (Section 4.3). At last, Section 5 offers conclusions with the main results and considerations.

2. Literature Review

A transport system includes a supply component, including infrastructures and services offered in continuous or discontinuous spaces, and a demand component, referring to user travel needs. These two components interact, generating flows in continuous or discontinuous transport networks.

Referring to the continuous space context, and adopting a network approach, some elements of the supply component (e.g., car, pedestrian, etc.) are generally represented as continuous systems [10]. Other elements (e.g., transit systems) have to be represented as discontinuous systems [13]. In this context, discontinuity in infrastructures and services becomes a constraint [14, 15].

Referring to the discontinuous space context, some authors have considered territorial discontinuity as physical constraints that act as an absolute or relative barrier to movements and exchanges [16–18]. These barriers are related to the presence of physical, and/or topographical features (e.g., a sea, lake, or river, etc.) [19]. The impacts of a discontinuous space in a transport system are significant for island accessibility. In fact, on islands, discontinuity of space results in poor accessibility, since they can only be accessed by sea or air transport services [20, 21].

This condition implies the presence of specific infrastructures (e.g., a port) and services (e.g., maritime services) that allow users to overcome the barriers. Rodrigue et al. [22] distinguish between absolute barriers, where modal change is constrained, and relative barriers, where path choice is constrained. Both cases imply a friction on movement.

Barriers influence accessibility and mobility in relation to the interaction between territorial and transport systems [23]. A specific issue is related to island accessibility, where the discontinuity of space makes private transport an unavailable alternative with respect to continental areas [24]. In these specific cases, access and egress distances influence modal choice [25]. Regarding these problems, two interesting works concern the Bosphorus Strait crossing in Istanbul. The first describes the effects of the introduction of the BRT system that crosses the Bosphorus Strait in Istanbul, including the consequent modal shift for trips between the European and Asian shores of the strait [26]. The second provides an overview of economic, social, and environmental impacts of Europe-Asia crossing transportation policies on Istanbul over

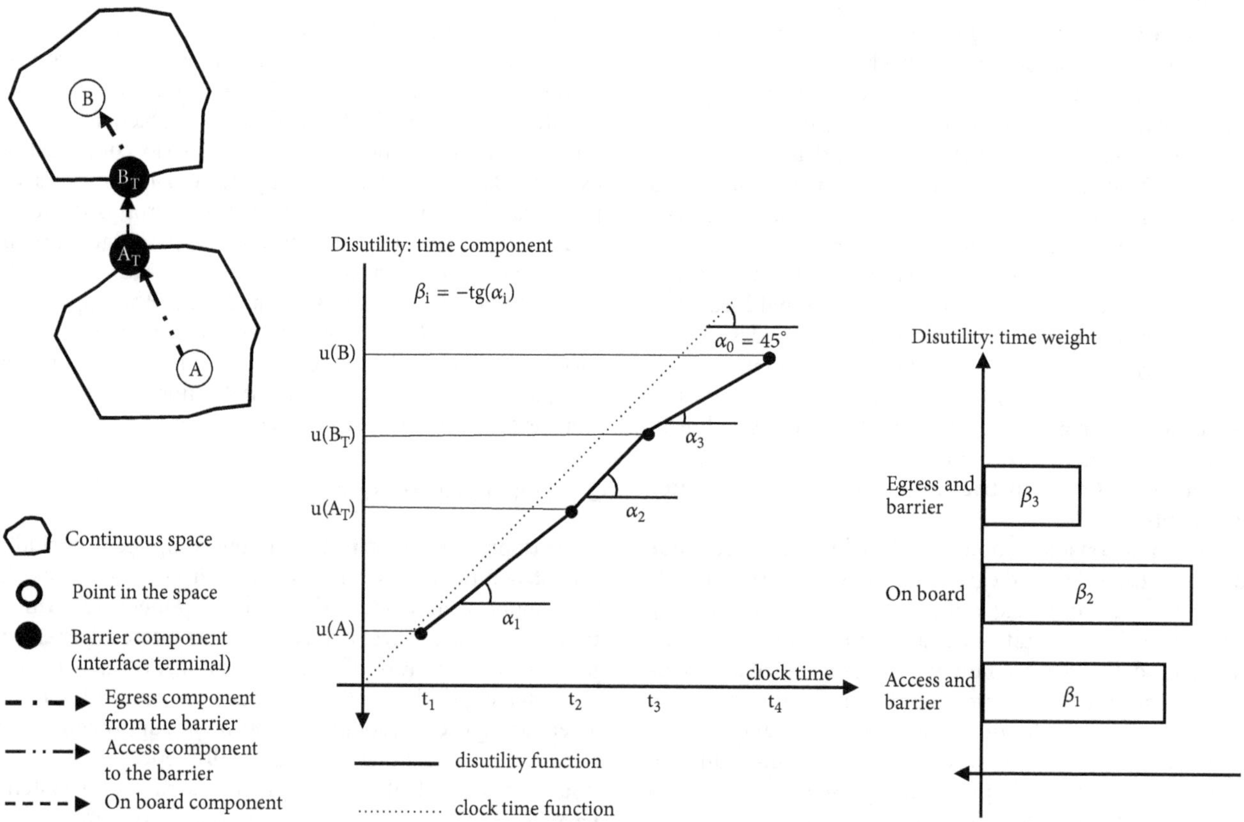

(a) Discontinuous space (b) Perceived disutility function: time component (c) Perceived disutility function: time weight

FIGURE 1: Travel in discontinuous space: Disutility function.

time [27]. Thus, there appears to be a lack of travel demand modelling for discontinuous territories. The schematisation of a discontinuous transportation network generally includes terminal nodes that are connected to their respective network and allow intermodal transport in order to overcome physical discontinuity.

The problem of discontinuity and barriers and their implications on user behaviours on travel choices are not extensively explored in the literature. In applications, simulation modelling of the discontinuity of space is generally represented as discontinuous transport systems in a continuous space. Consolidated approaches propose models to represent discontinuous transport services (e.g. transit services) adopting frequency-based [28] or schedule-based approaches [29, 30]. However, in most cases, the principal hypothesis is that discontinuous transport services are offered in a continuous space. The presence of discontinuity in the space limits available travel alternatives, forcing users to use at least two transport modes. Few studies explicitly consider the effects on user behaviour and travel choices in the presence of a discontinuity in the spaces.

For these reasons, models estimating current and future transport demand in a discontinuous space have to be constructed (specified, calibrated, and validated) [10]. The modelling approach adopted in this paper is derived from a classical random utility theory. The objective here is to clarify the implications of space discontinuity in transport modelling. For this reason, classical tools like Revealed Preferences (RP) and Stated Preferences (SP) surveys on a sample of users are adopted to support construction of the simulation model [31].

3. Modelling Discontinuity

3.1. Mobility with Barrier. The problem is how to represent a transportation system (supply, demand, and interactions) in a discontinuous space. The main difference with respect to a trip in a continuous space is the lack of alternatives to complete the trip using a single mode of transport between two points in the discontinuous space. Also, in a continuous space, there are transport supply alternatives with discontinuity (e.g., transit services). In any case, there is always at least one alternative without interruptions in a continuous space.

The diagram in Figure 1 shows an example of discontinuity effects in a transport system. In the example, the trips between A and A_T and between B_T and B can be completed in the continuous space. In these types of trips, there are no interruptions (or barriers) and there is at least one mode of transport (for instance, pedestrian mode) to complete them. In the same diagram, the trip between A and B presents a discontinuity component. To complete this trip, users are forced to change the transport mode and are constrained by the service schedule. They must use discontinuous services linking interface terminals between continuous and discontinuous spaces (such as two ports). More precisely, to travel between two continuous spaces, users are constrained by schedules and travel times of transport services connecting A_T and B_T. The individual transport modes for traveling in a continuous space (i.e., pedestrian, bicycle, or private car on road) are not available to overcome the discontinuity in the space (between A_T and B_T).

In this case, the trip has the following components:

(i) an *access component* representing the part of the trip from the point of origin (A in Figure 1(a)) to the departure terminal in the same space (A_T in Figure 1(a)). Available transport modes and services for this trip can be continuous or discontinuous, characterising a user disutility specification (e.g., in Figure 1(b), disutility is relative to the time variable only);

(ii) a *barrier component* representing the part of the trip within the barrier (A_T in Figure 1(a)) where some operations must be performed in the departure phase in order to use the transit services covering the discontinuity (e.g., check-in, waiting, ticket acquisition, etc.) and in the arrival phase (e.g., discharge, etc.). These terminals are constrained points in the continuous space belonging to the origin (A in Figure 1(a)) and the continuous space belonging to the destination (B in Figure 1(a));

(iii) an *on-board component* representing the part of the trip between the departure and arrival terminals (A_T and B_T in Figure 1(a)). The two terminals are linked by transport services that are discontinuous in time, characterised by a speed and a capacity (Figure 1(a));

(iv) an *egress component* representing the part of the trip from the arrival terminal point (B_T in Figure 1(a)) to the destination point (B in Figure 1(a)) in the same continuous space. Available transport modes and services for this trip can be continuous or discontinuous as in the access component (Figure 1(a)).

The barrier is a bottleneck in the transport system. Adopting a behavioural approach, a generic user perceives each trip component in a different way. It is possible to estimate the disutilities for each trip component. For instance, Figure 1(b) shows the disutility function: the time component for a user moving from A to B is represented as an example. The function varies with time. The user perceives each time component with a different weight (β_i), that is, the slope of the correspondent part of the disutility function. To obtain the β_i values, a survey is needed to specify and calibrate a behavioural model simulating travel choices in a discontinuous space. In the example, the *on-board*, *egress* and *barrier* components have the maximum and the minimum weight values, respectively (Figure 1(c)).

3.2. Proposed Modelling Approach. The proposed method aims to analyse users' perceptions of the discontinuity in terms of quantitative variables describing travel disutility. The characteristics of the current configuration of transport systems (in continuous and discontinuous spaces) are inputs of the method. These are composed of infrastructures (linear and punctual), services (timetables, running times, fares, etc.), and territory activities (residential, production, commercial, etc.). The problem can be studied in two different approaches.

(i) The first (*big data analytics*) consists in collecting data on the RP of users from different sources in real time or in off-time. The collected data allow planners to elaborate statistical indicators for the current discontinuity configuration. From the collected data, some statistical results, relative to historical configuration, can be obtained for each temporal period, users category, and travel choice level (i.e., O/D matrices, trajectories, and level of service attributes). From this historical data traditional travel demand statistical models could be constructed. The forecasted effects related to interventions (or scenario configurations) to solve the discontinuity problems refer only to historical and current configuration. For instance, effects of introduction of a new travel alternative, not present in the current situation, are estimated with a statistical significance referred to the current configuration.

(ii) The second (*transport models*) is an evolution of the first. The data and statistics obtained in the first way are inputs to model the current and scenario configurations. Indicators to evaluate users' RP behaviour on discontinuity can also be obtained for scenarios completely different from the present configuration. Moreover, data and their processing allow planners to design scenario configurations, to obtain feedback from users through SP survey data, and to simulate potential effects produced by interventions to solve discontinuity. With respect to the first approach, the statistical significance increases.

Indicators of user perceptions of the discontinuity in scenario configurations can be simulated by following the second way (this is not possible with the first). The proposed method supports transport planning in the selection of transport control variables (travel times, fares, costs, etc.) that influence travel choices in the presence of a discontinuity in a transport system. Figure 2 summarises the contents of the two ways. The structure of Sections 3 and 4 is also shown in relation to the entire proposed and applied method.

3.2.1. Current Configuration Modelling. The first phase aims to simulate current users' perceptions of a discontinuity by identifying quantitative variables that influence travel choices and related disutilities. To construct a simulation model (specification, calibration, and validation) an RP survey must be conducted. Descriptive and inferential statistical analysis is performed to estimate travel choice probabilities in relation to the variables and parameters representing travel disutility (e.g., times and costs). This allows analysts to identify variables representing users' perceptions in the current scenario.

Behavioural models are based on random utility theory and attempts to reproduce users' choice behaviour [32]. A

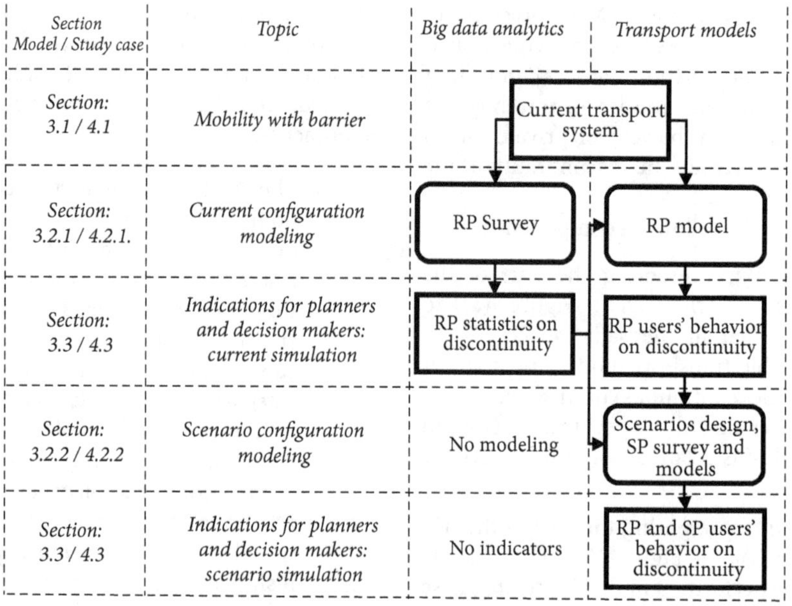

FIGURE 2: The proposed approach.

proper understanding of users' travel choices is essential in order to evaluate the ex ante effectiveness of possible transportation policies or services. The main assumption is that every individual is a rational decision-maker who maximises utility or minimises disutility. The user chooses an alternative from a finite choice set of mutually exclusive available alternatives. Each user assigns a perceived utility u^{RP} to each alternative m and selects the one that maximises utility.

For users, the perceived utility u is assumed to be a random variable with a known probability function. Its expected value is known as systematic utility v, which is generally assumed to be a linear combination of attributes, weighted by parameters (or coefficients) β. Different random utility model forms can be derived by assuming different joint probability distribution functions for the perceived utility U.

Note that an estimation of the parameters β could be obtained adopting big data analytics and traditional travel demand statistical models. However, parameters have a statistical significance referred only to current situation.

It is assumed that the travel disutility of the trip with mode m from a generic point A to a generic point B ($u^{RP}{}_{m}$) has the following functional dependency:

$$u^{RP}{}_{m} = \varphi\left(\mathbf{v}^{RPc}{}_{m}, \mathbf{v}^{RPd}{}_{m}, \boldsymbol{\beta}^{RPc}{}_{m}, \boldsymbol{\beta}^{RPd}{}_{m}\right) \quad (1)$$

where

$\mathbf{v}^{RPc}{}_{m}$, is the vector of variables representing users' perceptions in the continuous space;

$\boldsymbol{\beta}^{RPc}{}_{m}$, is the vector of parameters weighting variables in the continuous space;

$\mathbf{v}^{RPd}{}_{m}$, is the vector of variables representing users' perceptions in the discontinuous space;

$\boldsymbol{\beta}^{RPd}{}_{m}$, is the vector of parameters weighting variables in the discontinuous space.

The utility is not known with certainty by the analyst; thus $\mathbf{u}^{RP}$ might be represented in general by a vector of a random variable containing the utility relative to all modes. The expected value of this random variable (known as systematic utility) is assumed to be equal to $v^{RP}{}_{m}$:

$$v^{RP}{}_{m} = E\left(u^{RP}{}_{m}\right) \quad (2)$$

It can be assumed that it is the sum of the expected utility of variables representing users' perceptions in the continuous space ($\boldsymbol{\beta}^{RPc}{}_{m}{}^{(T)}\,\mathbf{v}^{RPc}{}_{m}$) and discontinuous space ($\boldsymbol{\beta}^{RPd}{}_{m}{}^{(T)}\,\mathbf{v}^{RPd}{}_{m}$):

$$v^{RPd}{}_{m} = \boldsymbol{\beta}^{RPc}{}_{m}{}^{(T)}\,\mathbf{v}^{RPc}{}_{m} + \boldsymbol{\beta}^{RPd}{}_{m}{}^{(T)}\,\mathbf{v}^{RPd}{}_{m} \quad (3)$$

where (T) stands for the transposed vector. Note that the systematic utility of the alternative is assumed to be the sum of two linear combinations of attributes weighted by parameters (or coefficients) β. The parameters are distinct for continuous and discontinuous spaces.

Starting from these disutilities it is possible to estimate travel choice probabilities for the current scenario. Under this hypothesis, although the selected alternative cannot be predicted, the probability that the decision-maker will select the alternative from his choice set can be expressed. The Multinomial Logit Model (MNL) is the simplest random utility model. Its specific assumptions are as follows: the perceived utilities are distributed as random Gumbel variables with scale parameter θ; they are identically and independently distributed across alternatives (covariance of random residuals of any two alternatives is null); and they are identically and independently distributed across observations/individuals [33]. Referring to the same trip (A to B) and to the current configuration, the MNL probability relative to the modal alternative m is [32]

$$p^{RP}{}_{m} = \psi\left(\mathbf{u}^{RP}, \theta\right) = \frac{\exp\left(v^{RP}{}_{m}/\theta\right)}{\Sigma_{m'} \exp\left(v^{RP}{}_{m'}/\theta\right)} \qquad (4)$$

where

ψ is the function that evaluates the probability for each alternative m for a given probability density function;

θ is the parameter of the probability density function.

Note that, considering a Gumbel probability density function,

(i) the function ψ can be specified in a closed form;

(ii) adding a linear specification of the expected value of $\mathbf{u}^{RP}$, the parameter θ can be included in the β; parameters β, including the parameter θ, can be calibrated using the RP survey with the maximum likelihood method.

In the case studied in this paper, the focus is on two aspects relating to the mode choice for access and egress. The first concerns the construction of the choice set that could be different for access and egress in the same trip. The second concerns the specification of the models that could be different in relation to the travel components. These adjustments are necessary to consider the discontinuity effects.

3.2.2. Scenario Configuration Modelling. By considering the current transport scenario and users' perception of the variables, the second phase aims to simulate users' perception of a discontinuity in relation to a set of planned transport scenarios. They are represented by infrastructure and service characteristics in order to reduce user disutilities related to discontinuity. Then an SP survey and a simulation model have to be performed. Each scenario is represented by a set of variations of the variables identified in the first phase. Planned scenarios are presented to a sample of travel users as potential future transport alternatives. The SP survey and model allow analysts to quantify the users' perception of the planned scenarios proposed to overcome discontinuity. For this reason, descriptive and inferential statistics are recalculated in order to estimate travel choice probabilities for each scenario when varying quantitative variables representing planned scenarios.

Considering a similar specification adopted in Section 3.2.1 (in all the notations of the variables, SP is adopted instead of RP), referring to the same trip (A to B) and to scenario configuration, the probability for the modal alternative m is

$$p^{SP}{}_{m} = \psi\left(\mathbf{u}^{SP},\ \theta\right) \qquad (5)$$

Parameters β, including parameter θ, can be calibrated using an SP survey with the maximum likelihood method.

3.3. Indications for Planners and Decision Makers. The two described approaches (big data analytics and transport model) can support planners and decision makers in solving problems related to the presence of a transport system in a discontinuous space. However, as already pointed out in Section 3.2, the possible indications that can be obtained using the two ways are very different.

Using *big data analytics*, planners and decision makers have to limit statistics on the current configuration to RP. Estimations of quantitative effects deriving from interventions on infrastructures and transport services to solve discontinuity with scenarios different from the current configuration are not available.

The *transport model* supports the transportation planning process in order to identify and quantify to what extent factors influence travel choices in continuous and discontinuous spaces. Factor quantification is essential to define and obtain an ex ante evaluation of effects produced by planned scenarios.

Modelling of discontinuity effects in the current and planned configurations allows planners to consider user needs and then to increase the sustainability of the proposed transport infrastructures and services. Quantitative effects estimated by planners support decision makers in planning, programming, and realising the proposed solutions. In particular, modelling and quantification of discontinuity effects support decision making process for the definition of the priority solutions order.

4. The Case Study

The aim of the present study is to develop an access/egress model for the Strait of Messina crossing for users who traversed the Strait by hydrofoil on the Reggio Calabria–Messina route. The structure of Section 4 is the same as Section 3. The Strait is a territorial discontinuity in the TEN-T Scandinavian Mediterranean Corridor.

4.1. Mobility with Barrier. The case study area is the Strait of Messina Area that includes the metropolitan cities of Messina (Sicily) and Reggio Calabria (Calabria). From a mobility point of view, the cities have strong relationships with each other and with their suburbs (short-medium distance mobility). The Strait Area also involves long distance trips to and from Sicily.

According to a recent estimate, total mobility in the Strait Area involves a flow of about 216,000 users/day on an average work day, most of whom (62%) travel by private car, and 38% by transit, including transport services in the Strait; work and school/university travel predominates (90%). Private automobile is the favourite transport mode (86%) for work-related commutes, while transit is preferred (74%) for study travel [34].

The mutual attraction between the urban areas of Messina and Reggio Calabria is expressed by the estimated number of strait crossings of about 21,000 users/day in both directions. Of these, about 61% are within the Strait Area, 18% originate or end inside the Strait Area, and 21% both originate and end outside the Strait Area [34]. In the port areas of Messina and Reggio Calabria, loaded and unloaded freight total 6.6 tons/year and 4.7 tons/year, respectively [35]. Approximately 3,000 passengers/day have been counted on an average working day in the Reggio Calabria harbour node, for the Messina–Reggio Calabria and vice versa for the hydrofoil service alone. This relationship is slightly imbalanced towards the Sicilian side, which is more attractive than Calabria: 59% of trips are from Calabria to Sicily, while 41% are in the opposite direction [34]. Furthermore, 37.3% of trips are for work purposes and 26.1% for school or university purposes. Private motor transport is the favourite mode (56%) to access the harbour [34]. Poor integration in the transportation supply system and the physical discontinuity represented by the sea result in long travel times, with 82% of strait crossings by hydrofoil on the Messina–Reggio Calabria route taking between one and three hours [36].

Ferries are one of the available strait crossing modes for trains, motorised vehicles, and passengers, covering the Messina-Villa San Giovanni (RC) route. The other available strait-crossing mode is the hydrofoil service, a high-speed boat for passengers only. It covers the Messina-Reggio Calabria and Messina–Villa San Giovanni (RC) routes in both directions.

Of the set of available alternatives (ferries and hydrofoils) for crossing the discontinuity, the hydrofoil alternative connecting Messina (belonging to Sicilia region) and Reggio Calabria (belonging to Calabria region) is selected. This selection derives from the following considerations: Messina and Reggio Calabria are the main cities in the area, the majority of the population lives in these two centres, the alternative is specialised for passenger service, and the service is not available at night.

Due to the physical discontinuity causing different transportation conditions on the two shores, mode choice is analysed separately for the two parts of each trip, before and after the crossing: the first part (*access*), from the origin of the trip to Reggio Calabria harbour, and the second part (*egress*), from Messina harbour to the destination. Trips in access and egress have different characteristics and choice sets. So, two kinds of mode choice models have been developed for access and for egress, using the theory and models for travel demand presented in Section 3.2.

Three planned scenarios are considered: Scenario 1: integrated ticketing, time coordination and integrated ticketing for intermodal transit in the Strait Area involving urban public transport in Messina and Reggio Calabria cities and hydrofoil service; Scenario 2: direct bus, a bus service directly connecting Reggio Calabria to Messina, without transfers, including strait crossing by ferry via Villa San Giovanni; Scenario 3: confirmed, to confirm the chosen travel alternative in access and egress.

4.2. Proposed Modelling Approach. The current configuration is studied starting from an RP survey, while the scenario configurations are studied starting from an SP survey. Starting from the survey, RP and SP models are specified, calibrated, and validated.

It should be highlighted that the survey aims to represent only one segment of strait-crossing mobility. In addition, it should be underlined that, because of the period during which the survey was conducted (June 2017) and because of the cyclical variations of travel demand, high-school student trips are not considered, although their share in the total number of this kind of trip is significant during the school year. Almost all users (90.6%) are commuters (workers and university students).

Strata were identified with each scheduled hydrofoil trip in a day and the number of interviews conducted for each *stratum* is proportional to its size. The number of definitively validated interviews is 245, while an average flow of 1,236 passengers/day was counted during the same period. Thus, an actual sampling rate of 19.8% was estimated. The main information asked of interviewees was as follows: origin and destination of the trip, trip purpose, hydrofoil departure time, departure time from place of origin and estimated arrival time at destination, transportation modes used in the travel chain to reach Reggio Calabria harbour (*access*) and to reach the destination from Messina harbour (*egress*), sex, age, employment, home-based trip, and frequency of the same trip. Lastly, an SP question was asked.

4.2.1. Current Configuration Modelling. An RP survey was conducted over three days in June 2017 at the hydrofoil service terminal in Reggio Calabria harbour, meaning that only users of this service travelling from Reggio Calabria to Messina were interviewed. The survey received the support of ATAM, the public transport company in Reggio Calabria. Once created, the database [37] was used as input in the model development process.

The method used is Simple Stratified Random Sampling, in which the population is divided into mutually exclusive groups of users (*strata*), whose number is known, and from which simple random samples of elements are drawn [10].

From the RP survey statistical data analysis, almost all of the investigated trips originated in the city of Reggio Calabria (86.1%). Similarly, most of the trips had Messina as their destination (80.4%). For simplification purposes, the chosen transportation modes were grouped in three main categories (walk, motorised, and transit), based on their characteristics. Reggio Calabria harbour is reached primarily via private vehicle, while egress in Messina harbour is distributed almost equally among modes.

The mode choice models developed in the present work are the behavioural model (Random Utility Models), and MNL, with different specifications for access and egress. The mode choice alternatives considered in the choice set are walking, motorised, and transit, with the same meaning as expressed above. Systematic utility functions of the three alternatives are a linear combination of the level of service attributes (travel times, monetary costs) and alternative specific individual attributes (home-based trips, walking distance, and trip purpose). Table 1 shows the model specification and all estimated parameters for the developed models and their statistical tests [36].

First, informal tests were conducted. All estimated parameters fulfilled sign conditions and presented expected values. The parameter for passengers whose home was the point of origin of the trip ($\beta_{\text{o\_home}}$), referring only to the motorised mode, presented a positive value for the access model and a negative value for the egress model, reflecting mostly the availability of the motorised mode in access and

TABLE 1: Calibrated RP models.

	Name	U.o.m. attribute	Access model: Walk	Access model: Motorized	Access model: Transit	Estimated* Value (t-Student)	Egress model: Walk	Egress model: Motorized	Egress model: Transit	Estimated* Value (t-Student)
β_{time}	Travel time	Minutes	X	X	X	-0.0629 (-5.50)	X	X	X	-0.0422 (-4.02)
β_{cost}	Monetary cost	Euro		X	X	-0.3303 (-5.62)		X	X	-0.2728 (-5.36)
β_{o_home}	1 if home as origin	Dummy		X		1.338 (4.63)		X		-1.814 (-5.11)
$\beta_{dist2km}$	1 if walking distance ≤ 2 km	Dummy	X			1.139 (5.27)	X			1.454 (3.45)
β_{uni}	1 if home-university purpose	Dummy							X	1.021 (3.45)
β_{work}	1 if home-work purpose	Dummy						X		1.767 (6.13)
Log-likelihood						-149.64				-169.52
ρ^2						0.437				0.365
McFadden ρ^2						0.316				0.365
χ^2						138.41				194.86
p.value						2.22E-16				2.22E-16
VOT						11.4 €/h				9.3 €/h

its unavailability in egress, on the other side of the Strait. All obtained VOTs (Value of Times, see, e.g., [38]) were in line with scientific literature.

Later, formal statistical tests were performed to verify certain hypotheses on parameter estimation. In the developed models, Student t-values were significant for a confidence level of 0.95. Likelihood Ratio values for estimated parameters exceeded the 95th percentile of the corresponding χ^2 variable. Therefore, the null hypothesis that the true model has $*$ parameters (null or ASA) can be rejected with a very low probability of error. For the access model, ρ^2 and McFadden pseudo ρ^2 presented the values of 0.437 and 0.316, respectively, while both their values were 0.365 for the egress model.

Therefore, it can be stated that the presented models have a good fit and a fair ability to reproduce the mode choices in access and egress made by a sample of users crossing the Strait of Messina from Reggio Calabria to Messina by hydrofoil.

4.2.2. Scenario Configuration Modelling. SP surveys investigate individuals' preferences for services or policies that do not currently exist [31]. The choice contexts proposed to users refer to hypothetical design scenarios. By conducting SP surveys, new and currently unavailable choice alternatives and new model attributes can be introduced, relevant attribute variability can be checked, and generally more information can be obtained. However, with SP surveys, some deviations in data analysis and in calibrated models are introduced, derived from the user's possible different behaviour between the stated and the actual choice in the new scenario when realised, because of a lack of relevant attributes, justification bias, and so on [10]. Among the questions, users were asked to indicate which option they would choose for the same trip in the scenario configurations.

From the SP survey statistical data analysis, the integrated ticketing alternative with time coordination is the preferred one (60.8%). Indeed, about 60% of interviewees stated they would be willing to switch to public transit if time coordination and integrated ticketing policies were implemented. About 25% would continue to use their motorised vehicles, while 12.8% of motorised users in access and 18.3% in egress would choose the direct bus service.

The present section models the possible modal shift in the hypothetical scenarios presented to the users in the SP question, for those choosing motorised mode in access in the current context (152 users, 62.8%). The mode choice model is MNL. The alternative scenarios considered in the choice set are as follows: 1 – integrated ticketing; 2 – direct bus RC – Villa S.G. – ME; 3 – confirmed. They refer to both the first portion of the trip (*access*) and the Strait crossing phase by hydrofoil in the first and third alternatives, and by ferry in the second one. The SP question relates to the three scenario travel alternatives (1, 2, and 3 in Table 2) represented in terms of time and cost. Among the alternatives, Alternative 3 confirms the current configuration, while the other two alternatives are easily identifiable by users because they introduce variations to the current configuration. In addition, most users cross the strait regularly and are familiar with the transport system in the area. These aspects should reduce the bias in the SP survey.

Table 2 shows the model specification and all estimated parameters for the developed model and their statistical tests. Two models are proposed: the difference between them

TABLE 2: Calibrated SP models.

	Name	*U.o.m. attribute*	*(1) Integrated ticketing*	*(2) Direct bus (RC – Villa S.G. – ME)*	*(3) Confirmed*	*Estimated value (t-Student)*	
β_{time}	Travel time	Minutes	X	X	X	-0.0268 (-2.26)	
β_{timeB}	Travel time for barrier	Minutes	X	X	X		-0.0091 (-0.37)
β_{timeA}	Travel time for access	Minutes	X	X	X		-0.0400 (-2.44)
β_{cost}	Monetary cost	Euro	X	X	X	-0.1229 (-0.87)	-0.1288 (-0.91)
β_{o_home}	1 if home as origin	Dummy	X			0.6374 (2.21)	0.5591 (1.83)
β_{centre}	1 if origin in city centre	Dummy	X			0.5521 (1.91)	0.4122 (1.14)
β_{uni}	1 if home-university purpose	Dummy	X			0.900 (2.22)	0.9190 (2.26)
Log-likelihood						-130.71	-129.2
ρ^2						0.217	0.226
McFadden ρ^2						0.063	0.073
χ^2						17.43	20.45
p.value						5.8E-4	1.4E-4
VOT						13.1 €/h	4.26 €/h 18.64 €/h

concerns the travel time that is considered aggregated or disaggregated for *access* and *barrier* trip components.

All estimated parameters fulfil sign conditions and present expected values. The weight of access and barrier time is significantly different. The other parameters refer to the integrated ticketing alternative and present positive values, expressing the positive relation between the users with the characteristics represented by the parameters and their availability to shift to public transport mode.

Student t-values for almost all the estimated parameters are significant for a confidence level of 0.95. Only the parameters related to monetary cost (β_{cost}) and travel time for the barrier (β_{timeB}) are not significant. Indeed, motorised users do not seem overly concerned by the monetary costs in this context, but care mostly about the time spent travelling. Furthermore, the time component to access the terminal is much more significant than the time spent overcoming the barrier. The Likelihood Ratio value for estimated parameters exceeds the 95th percentile of the corresponding χ^2 variable. Therefore, the null hypothesis that the true model has $*$ parameters (null or ASA) can be rejected with a very low probability of error. Also, the VOTs agree with the scientific literature with significant different values for access and barrier. Goodness of fit index ρ^2- is not very high (0.217 or 0.226), probably because the hypothetical scenarios, which include a new travel mode alternative and new policies for intermodal public transport, are not very clear to the interviewed users. Thus, the answers may be scattered and illogical. However, as highlighted above, these deviations are part and parcel of an SP survey.

4.3. Indications for Planners and Decision Makers. The results of the statistical analysis obtained following the *big data analytics* approach provide limited indications for planners and decision makers. Statistics on RP travel choices are available only for the current configuration. There are no provisional estimations for scenario configurations.

The results obtained following the *transport models* approach provide simulations of user behaviours in the current and scenario configurations (Sections 4.2.1 and 4.2.2). Several indications for planners and decision makers are reported below.

A first indication concerns users' perception of travel time with respect to the access and the barrier trip components. Access time is significant in terms of the Student t-value, while the barrier time is not significant. This confirms that users take access time into high consideration, due to service discontinuity and the absence of any other continuous alternative. Another indication concerns the monetary cost that is not significant in terms of Student t-value in both cases.

Figure 3 summarises the disutility of travel time for the access and barrier components. Note that the access time component has a weighting about four times greater than the barrier time component (Figure 3(c)).

These results can be used to define and evaluate alternative transport scenarios and specific interventions aimed to limit space discontinuity effects. In particular,

(i) the relevant role of travel time, compared to other disutility variables, indicates that interventions to reduce it are needed;

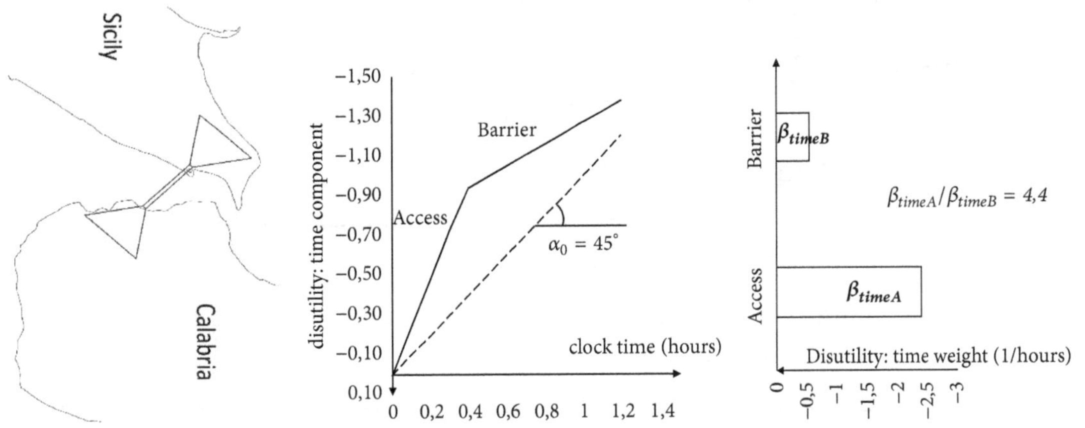

(a) Discontinuous space (b) Perceived disutility function: time component (c) Perceived disutility function: time weight

FIGURE 3: Travel in discontinuous space: The case study.

(ii) the relevant role of access time, compared to barrier time, indicates the need of interventions to improve infrastructures and services connecting the terminal to its catchment area.

Figure 4 reports choice probability as a function of an attribute variation. The figure has three rows and two columns. Each row relates to a single possible scenario; the first column is for the travel time variation and the second column for the monetary cost variation only in the scenario reported in the row. For a fixed value of the attribute's percentage variation, the observed probability slope

(i) (comparing columns) is more elastic comparing percentage variation in travel time with respect to monetary cost. This comparison provides information on the importance users assign to the time variation with respect to the cost variation.

(ii) (comparing rows) assumes the maximum value in the integrated ticketing scenario and is similar to the direct bus scenario. This result provides information about the importance of realising interventions encouraging integrated ticketing to travel in the study area.

(iii) (observing third row) is slightly elastic to time and cost in the current configuration (*confirmed*). This result indicates that users potentially prefer new scenarios with respect to the current configuration.

5. Concluding Remarks and Further Developments

This paper presents an analysis of transport systems in a discontinuous territory. A specific focus concerns users' perception of discontinuity in travel disutility. The proposed method studies the problem in two approaches:

(i) The first is based on statistical analyses in the context of a *big data analytics* approach. The results and information for planners and decision makers are limited to analysing the current configuration.

(ii) The second is based on the *transport model* approach constructed using collected data obtained in the first way. The results and information for planners and decision makers are available for both the current and scenario configurations.

The proposed approach has been experimented to study mobility in a territory in southern Italy characterised by territorial discontinuity created by the Strait of Messina. For this study area, a survey was conducted, from which some statistical results have been obtained. These data support transport model specification and calibration. In particular, two mode choice models have been developed. The application focusses only on strait crossings by hydrofoil on the specific route from Reggio Calabria to Messina. The physical discontinuity generates different transportation conditions on either shore. For this reason, two different transport mode choice models are developed for the two parts of a generic trip between the two cities through the urban harbours. The first represents trips from the point of origin to Reggio Calabria harbour (*access*), and the second represents trips from Messina harbour to the destination (*egress*). The choice set includes walking, motorised, and transit modes.

The models proved fully satisfactory, since all parameters are significant and their validation tests provide good results. Moreover, VOT, ρ^2, and McFadden's pseudo ρ^2 also presented values in line with the scientific literature. Conducting a second survey on the same population of users, with a different sample, and checking frequencies of choice could be a further step to verify the models' ability to reproduce user choices.

The results can be considered as the first step of the research. The results reported in the paper are derived from numerous tests in the specification-calibration-validation process. Many trial and error specifications were tested, also considering the same specification for the utility of the three travel components as a whole. These tests do not provide

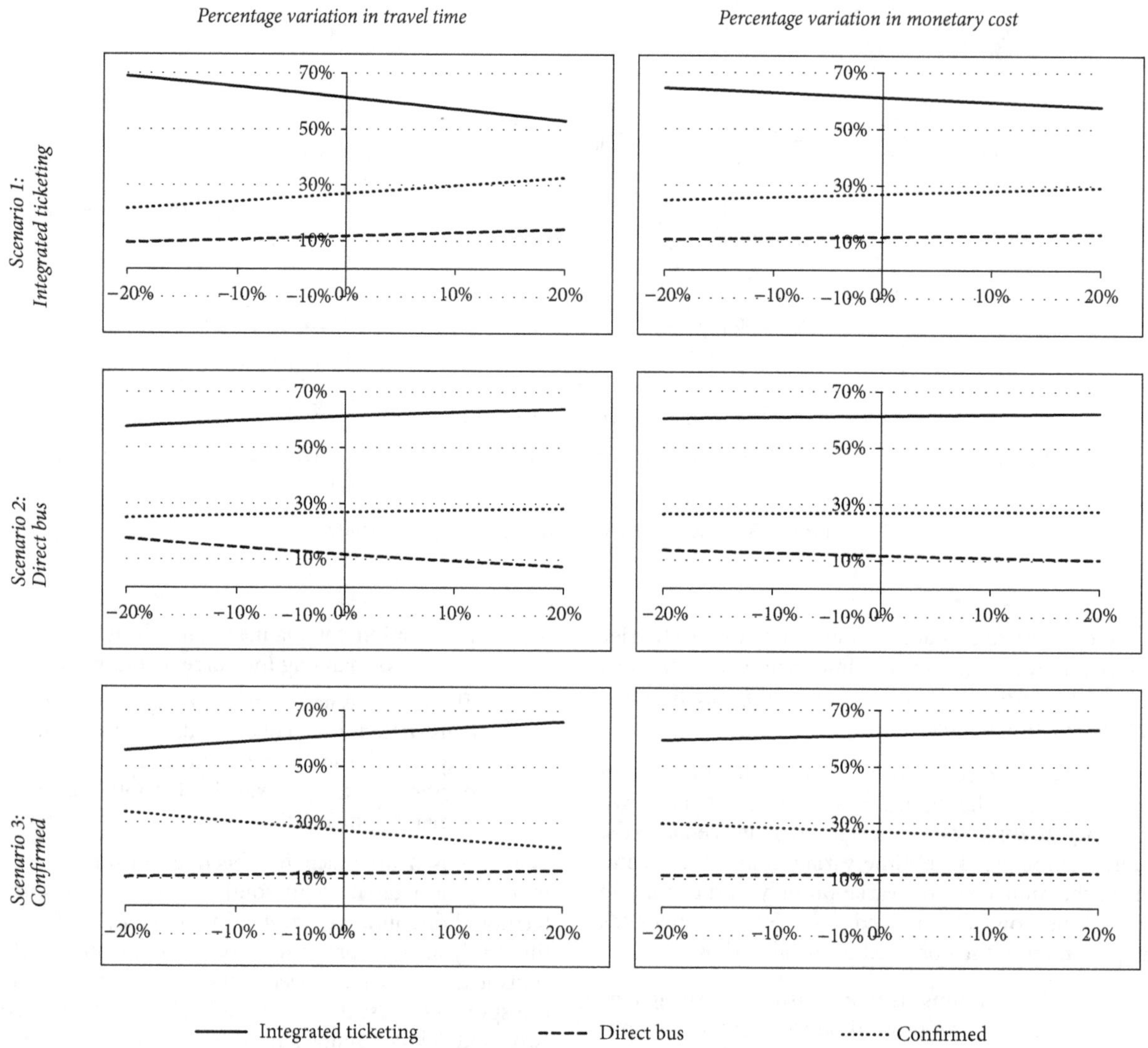

FIGURE 4: SP choice probability with respect to the variation of travel time or cost in one alternative.

significant statistical results, only when two different access and egress models are adopted do the statistical results give good results.

All of the results obtained confirm that user behaviour is different when traveling in a discontinuous space with respect to a trip in a continuous space. The differences have to be considered by planners and decision makers when defining interventions to limit discontinuity effects. The results obtained for the case study show the relative level of importance users assign to different trip components. These results indicate that it is essential to define interventions to improve intermodality and mobility integration for discontinuous territories. For instance, SP data analysis highlighted that hydrofoil users seem very interested in integrated ticketing and time coordination policies for public transport in the Strait Area. A mode choice model, including strait crossings by hydrofoil and by ferry in both directions, could support transport planning in the study area by estimating a modal shift for each alternative scenario.

Two main lessons were learned. (a) In a real context, discontinuity in space has impacts on connectivity; it creates time (e.g., schedule, availability during the day) and space constraints (e.g., few terminals and unavailability of individual transport mode). (b) In the modelling context, adjustments simulating discontinuity effects are required in the classical demand model.

The results show some differences between user behaviour in the presence of a discontinuity in space. Referring to the specific case study, the role of travel time —and in particular access time, spent on reaching the terminal— emerges among disutility variables. This is one of the main peculiarities of the research topic.

Further developments regard extensions of the proposed method and relative applications. In relation to the model extension, a larger sample could be considered to simulate the entire trip chain and other demand components. Effects of discontinuity on others travel choice levels different from mode choice (departure time, destination, mode, and path) have to be studied. Additionally, the effects of attributes (i.e., costs, frequency) need to be investigated in order to estimate the measurement of discontinuity on travel behaviour.

These future research directions depend on the availability of other data that could be obtained by designing and implementing a mobility monitoring system for the study area. The monitoring centre could also be useful for operators and system managers to help them better design and plan the transport system. The problem can be studied in the context of emerging Mobility as a Service (MaaS).

The problem related to model transferability needs to be studied in detail, considering that each area with discontinuity has specific characteristics that are very often different from other areas. In relation to applications, an extension of the survey campaign to the entire area, involving short, medium, and long-distance trips, could be conducted in the study area in order to develop more comprehensive mode choice models. These can be very helpful in assessing possible future transportation policies, since they allow modal choice simulation in design scenarios developed on the basis of imposed goals and restrictions.

Conflicts of Interest

The authors declare that they have no conflicts of interest.

Acknowledgments

Some numerical results are developed by Giovanni Birgillito during the course of Master on Transport System Management and Logistics at the Mediterranea University of Reggio Calabria. This research is partially supported by the project GRE.ENE.LOG., funded by Calabria Region via POR CALABRIA FESR - FSE 2014 - 2020, Prog. 2894, Prot. SIAR 52115, date 17/02/2017.

References

[1] European Commission, "Regulation (EU) No 1315/2013 of the European Parliament and of the council of 11 December 2013 on Union guidelines for the development of the trans-European transport network and repealing Decision No 661/2010/EU," 2013.

[2] European Commission, "Regulation (EU) no 1316/2013 of the European parliament and of the council of 11 December 2013 establishing the Connecting Europe Facility, amending Regulation (EU) No 913/2010 and repealing Regulations (EC) No 680/2007 and (EC) No 67/2010," 2013.

[3] P. A. Witte, B. W. Wiegmans, F. G. van Oort, and T. J. M. Spit, "Chokepoints in corridors: Perspectives on bottlenecks in the European transport network," *Research in Transportation Business and Management*, vol. 5, pp. 57–66, 2012.

[4] N. van Oort, D. Sparing, T. Brands, and R. M. P. Goverde, "Data driven improvements in public transport: the Dutch example," *Public Transport*, vol. 7, no. 3, pp. 369–389, 2015.

[5] M. Salah Mahmoud, K. M. Nurul Habib, and A. Shalaby, "Survey of cross-regional intermodal passenger travel (Script): A joint RP/SP survey integrated with multimodal trip planning," *Transportation Research Record Journal of the Transportation Research Board*, pp. 108–118, 2015.

[6] M. S. Mahmoud, K. N. Habib, and A. Shalaby, "Application of RP/SP Data to the joint estimation of mode choice models: lessons learned from an empirical investigation into cross-regional commuting trips in the Greater Toronto and Hamilton Area," in *Proceedings of the Transportation Research Board 95th Annual Meeting*, 2016.

[7] F. Russo, C. Rindone, and P. Panuccio, "European plans for the smart city: from theories and rules to logistics test case," *European Planning Studies*, vol. 24, no. 9, pp. 1709–1726, 2016.

[8] G. Musolino, C. Rindone, A. Polimeni, and A. Vitetta, "Planning urban distribution center location with variable restocking demand scenarios: General methodology and testing in a medium-size town," *Transport Policy*, 2018.

[9] A. Vitetta, "A quantum utility model for route choice in transport systems," *Travel Behaviour and Society*, vol. 3, pp. 29–37, 2016.

[10] E. Cascetta, *Transportation Systems Analysis: Models and Applications*, Springer, NY, USA, 2009.

[11] J. D. Ortúzar and L. G. Willumsen, *Modelling Transport*, Wiley, USA, 4th edition, 2011.

[12] N. H. M. Wilson and A. Nuzzolo, *Schedule-Based Modeling of Transportation Networks: Theory and applications*, Springer Science, 2009.

[13] A. Nuzzolo, F. Russo, and U. Crisalli, *Transit Network Modelling. The Schedule-Based Dynamic Approach*, Franco Angeli, Milan, Italy, 2002.

[14] F. Russo, *Sistemi di trasporto merci. Approcci quantitativi per il supporto alle decisioni di pianificazione strategica tattica ed operativa a scala nazionale*, Franco Angeli, Milan, Italy, 2005.

[15] F. Russo, C. Rindone, L. Amodeo, D. Lamari, V. Placido, and M. Praticò, "Freight transportation in European logistics: A comparison of Northern and Southern range gateways," in *Proceedings of the 13th Annual International Industrial Simulation Conference 2015, ISC 2015*, pp. 55–62, Spain, 2015.

[16] D. Chapman, D. Pratt, P. Larkham, and I. Dickins, "Concepts and definitions of corridors: Evidence from England's Midlands," *Journal of Transport Geography*, vol. 11, no. 3, pp. 179–191, 2003.

[17] D. Musolino, "Characteristics and effects of twin cities integration: the case of Reggio Calabria and Messina, 'walled cities' in Southern Italy," *Regional Science. Policy & Practice*, 2018.

[18] D. Musolino, "The mental maps of Italian entrepreneurs: a quali-quantitative approach," *Journal of Cultural Geography*, vol. 35, no. 2, pp. 251–273, 2018.

[19] H. S. Thorsen and I. Thorsen, "Effects of transportation barriers on geographic asymmetries in labour markets," *Research in Transportation Economics*, vol. 63, pp. 27–37, 2017.

[20] I. Spilanis, T. Kizos, and P. Petsioti, "Accessibility of peripheral regions: Evidence from Aegean Islands (Greece)," *Island Studies Journal*, vol. 7, no. 2, pp. 199–214, 2012.

[21] R. Fernandes and P. Pinho, "The distinctive nature of spatial development on small islands," *Progress in Planning*, vol. 112, pp. 1–18, 2017.

[22] J. P. Rodrigue, C. Comtois, and B. Slack, *The Geography of Transport Systems*, Routledge, London, UK, NY, USA, 2006.

[23] K. T. Geurs and B. van Wee, "Accessibility evaluation of land-use and transport strategies: Review and research directions," *Journal of Transport Geography*, vol. 12, no. 2, pp. 127–140, 2004.

[24] S. Karampela, T. Kizos, and I. Spilanis, "Accessibility of islands: Towards a new geography based on transportation modes and choices," *Island Studies Journal*, vol. 9, no. 2, pp. 293–306, 2014.

[25] A. C. van den Enden and C. W. Lobuizen, "The influence of access and egress distance on modal choice of train versus automobile in the context of urban and regional planning and locational decisions," in *Proceedings of the Transportation Planning Research Colloquium*, vol. II, pp. 14–16, Zandvoort, The Netherlands, 1983.

[26] P. Alpkokin and M. Ergun, "Istanbul Metrobüs: first intercontinental bus rapid transit," *Journal of Transport Geography*, vol. 24, pp. 58–66, 2012.

[27] P. Alpkokin, J. A. Black, S. Iyinam, and A. S. Kesten, "Historical analysis of economic, social and environmental impacts of the Europe-Asia crossings in Istanbul," *International Journal of Sustainable Transportation*, vol. 10, no. 2, pp. 65–75, 2016.

[28] H. Spiess and M. Florian, "Optimal strategies: a new assignment model for transit networks," *Transportation Research Part B: Methodological*, vol. 23, no. 2, pp. 83–102, 1989.

[29] C. O. Tong and S. C. Wong, "A stochastic transit assignment model using a dynamic schedule-based network," *Transportation Research Part B: Methodological*, vol. 33, no. 2, pp. 107–121, 1999.

[30] A. Vitetta, A. Cartisano, and A. Comi, "Application for Comparing Frequency and Schedule-Based Approaches in the Simulation of a Low Frequency Transit System," in *Schedule-Based Dynamic Transit Modeling: theory and applications*, N. H. M. Wilson and A. Nuzzolo, Eds., vol. 28 of *Operations Research/Computer Science Interfaces Series*, Springer, Boston, MA, 2004.

[31] D. A. Hensher, "Stated preference analysis of travel choices: the state of practice," *Transportation*, vol. 21, no. 2, pp. 107–133, 1994.

[32] M. Ben Akiva and S. R. Lerman, *Discrete Choice Analysis Theory and Application to Travel Demand*, MIT Press, Cambridge, 1985.

[33] F. Koppelman and C. Bhat, *A Self Instructing Course in Mode Choice Modelling: Multinomial and Nested Logit Models*, U.S. Department of Transportation, Federal Transit Administration, 2006.

[34] G. Delfino, C. Rindone, A. Vitetta, and D. Iannò, *Stretto di Messina: uno studio della mobilità intermodale per i passeggeri*, Alfagi Edizioni, Villa S.G. (RC), Italy, 2011.

[35] MIT – Ministero delle Infrastrutture e dei Trasporti, "Conto Nazionale delle Infrastrutture e dei Trasporti. Anni 2016-2017," Istituto Poligrafico e Zecca dello Stato S.p.A., Rome, Italy, 2018.

[36] G. Birgillito, *Transport Models for Discontinuous Territory: An Application to the Strait of Messina Area, Master Thesis*, DIIES, Università degli Studi Mediterranea di Reggio Calabria, 2017.

[37] ATAM SpA, "Database of survey on passenger mobility in the Strait Area," 2017.

[38] Y.-Y. Tseng and E. T. Verhoef, "Value of time by time of day: A stated-preference study," *Transportation Research Part B: Methodological*, vol. 42, no. 7-8, pp. 607–618, 2008.

8

Random Forests-Based Operational Status Perception Model in Extra-Long Highway Tunnels with Longitudinal Ventilation

Chao Qian,[1] Jianxun Chen,[2] Yanbin Luo,[2] and Shuguang Li[1]

[1]*School of Electronic & Control Engineering, Chang'an University, Xi'an 710064, China*
[2]*School of Highway, Chang'an University, Xi'an 710064, China*

Correspondence should be addressed to Jianxun Chen; chenjx1969@chd.edu.cn

Academic Editor: Juan C. Cano

An increasing number of extra-long highway tunnels have been built and put into operation around the world, but the quantified segmentation criteria for evaluating the in-tunnel operational status have not yet been enacted up till the present moment. Meanwhile, ventilation facilities could not satisfy the dynamic requirements of fresh air demand under fast spatial-temporal variation of traffic conditions and operating environment. In this study, the operational data collected from an extra-long highway tunnel were deeply analyzed using big data technology. By combining traffic flow and environmental monitoring data, a data-driven perception model based on the Random Forests was structured. The prediction results show that the proposed model provides better performances as compared to contrast models, indicating it had better ability to adapt to the dynamic changes of in-tunnel operational status while realizing accurate prediction. The designed intelligent control strategies of ventilation facilities and traffic operation applying for different operational status would provide a theoretical basis and data support for lifting the level of intelligent control as well as promoting energy saving and consumption reducing in extra-long highway tunnels.

1. Introduction

By the end of 2016, 815 extra-long highway tunnels with a total length of 3622.7 km were built in China [1]. Owing to the influence of traffic volume and fleet composition, vehicle emissions accumulate sequentially. These emissions are difficult to disperse, especially in the case of extra-long highway tunnels with high traffic loads and frequent traffic congestions. Tunnel ventilation has become the primary problem during operation periods.

For road tunnels, there exist several, very different approaches to ventilation concepts [2]. They have common objectives, opposite in nature: (a) the pollution levels within admissible margins and (b) the energy consumption for ventilation facilities to fulfill objective (a) should be minimal. Under some circumstances, it is difficult to meet both objectives concurrently by using simple ventilation control algorithms [3]. Thus many advanced control methodologies have been proposed in recent decades.

Appropriate and accurate ventilation control systems can not only decrease energy consumption and save operation cost but also provide drivers with a comfortable and safe driving environment. Standard linear feed-backward control was applied in early ventilation automatic control schemes such as PI or PID. However, these conventional control schemes reach their limits of applicability as soon as nonlinear effects become increasingly dominant. Funabashi et al. (1991) and Koyama et al. (1993), respectively, proposed ventilation control systems for longitudinal ventilation road tunnels with nonlinear programming and fuzzy control applications [4, 5]. Chen et al. (1998) designed a fuzzy logic control model for prediction of pollutant concentrations and adjustment of jet fans [6]. Chu et al. (2008) demonstrated a genetic algorithm in combination with fuzzy control to maintain an adequate level of the pollutants and minimize power consumption [7]. Bogdan et al. (2008) developed a model predictive and fuzzy control algorithm for a longitudinal ventilation system [8]. The predictive controller estimates fresh air requirements (depending on traffic and weather conditions) and calculates the number of necessary jet fans, while the fuzzy controller compares measured and admissible levels of pollutants and adjusts a predicted number of jet fans to keep the pollutant levels within predefined boundaries. Euler-Rolle et al. (2017) applied a model based nonlinear dynamic feedforward control in the longitudinal tunnel ventilation to enhance standard feedback control and improve the closed-loop behavior [9]. However, all these contributions focused rather on the specific pollutants control than on the overall control and dynamic characteristics of in-tunnel operational status. Unchanging ventilation mode and unreasonable control strategy lead to enormous energy consumption and economic loss [10].

The in-tunnel operational status can be considered as a result generated by the combined action of four transportation elements, including the driver, vehicle, road, and environment. Li et al. (2015) focused on the diffusion properties of CO, NO, and $PM_{2.5}$ influenced by in-tunnel traffic force [11]. Yamada et al. (2016) and Martin et al. (2016) concentrated on the impact of in-tunnel tunnel environment (e.g., NO_2 level and particle number concentrations) on the driver and the passenger [12, 13]. Up till the present moment, the quantified segmentation criteria for evaluating the operational status in extra-long highway tunnels have not been enacted. Meanwhile, the analysis and mining of the in-tunnel operational status by deeply combining the real-time traffic flow and environmental information have also seldom been studied.

The decision tree is a classical classification algorithm, which is essentially a data recursive partitioning process based on a series of rules. Since the single decision tree has some drawbacks, such as low precision and overfitting, the ensemble learning method, which summates simple machine learning algorithms to produce better predictive performance than could be achieved by the most sophisticated solutions, has become popular in research in the field of machine learning. Practitioners created various solutions to improve a decision tree by replicating it many times and averaging results. For classification task, the ensemble can be used as a voting system, choosing the most frequent response class as an output for all its replications.

Aiming at finding the best way to replicate the trees in an ensemble, Breiman (1996) tested the effects of bootstrap sampling (sampling with replacement), which not only leaves out some noise but also creates more variation in the ensembles, improving the results. This technique is called "**b**ootstrap **agg**regat**ing**" and use the acronym **bagging** [14]. Noticing that results of an ensemble of trees improved when the trees differ significantly from each other, Breiman (2001) proposed a new ensemble model, Random Forests (RF), which add a layer of randomness to bagging [15, 16]. Random Forests change how the classification or regression trees are constructed by constructing each tree using a different bootstrap sample of the data, which turns out to perform very well compared with many other classifiers, including discriminant analysis, support vector machines, and neural networks,and is robust against overfitting [17].

The main goal of this work was to fuse the in-tunnel traffic flow data (such as fleet segmentation and traffic volume) and ambient air data (such as the concentrations of toxic gas and particular matter and air velocity) based on big data technology and to build a Random Forests-based perception model realizing accurate prediction of the in-tunnel operational status.

2. Material and Methods

2.1. Operational Monitoring Data. The Xi'an–Hanzhong Expressway (Xihan Expressway) is one of the most critical sections of the G5 Beijing–Kunming Expressway (a part of the China National Expressway Network, commonly known as the Jingkun Expressway), which connects north and southwest China, in Shaanxi province. A critical controlling project in the Xihan Expressway, the Qin Mountains tunnel group (Figure 1), comprises three extra-long highway tunnels, No. 1 tunnel, No. 2 tunnel, and No. 3 tunnel, passing through the Qin Mountains. The mountains are the most important geographical entities that divide northern and southern China.

No. 1 tunnel is a twin-bore tunnel with unidirectional traffic in each bore. The tunnel comprises southbound (SB) and northbound (NB) tunnels, with each direction having two lanes for motor vehicles. Figure 2 depicts the overall structure of the SB tunnel. In total, 11 lay-bys (emergency parking bays), numbered from ESA-1 to ESA-11, have been built along the length of the tunnel. The ventilation mode is longitudinal and is powered by 30 jet fans; an inclined shaft is reserved, and the air supply and exhaust system with additional axial fans had not yet been equipped. Since it was constructed and opened to traffic in 2007, the traffic has consistently increased. Among all vehicles, heavy-good vehicles (HGV) have shown the most notable increase.

In this study, four lay-bys (ESA-1, ESA-4, ESA-8, and ESA-11) were selected as the monitoring or data-collection sites in the driving direction. A real-time monitoring experiment for the operational environment was performed from Nov 27, 2016, to Dec 3, 2016. Through the experiment, raw monitoring data of the operational status were obtained. The details of data and monitoring instruments are listed in Table 1.

Raw monitoring data were preprocessed at statistical or resampling intervals of 15 min. That is, traffic flow data were converted to cumulative values every 15 min. The other monitoring data were calculated as average values for each 15 min interval. Finally, the statistical dataset of the operational environment was obtained.

The proportions of passenger cars (PC), light-duty vehicles (LDV), and HGV were 29.46%, 3.21%, and 67.32%, respectively. LDV had the lowest proportion, which smoothly changed; hence, its impact on the in-tunnel operational status could be ignored. The Pearson correlation coefficients indicated that PC had weak positive correlations with CO and NO_2. Consequently, the impact of PC on the in-tunnel operational status could also be ignored. Finally, only HGV was retained in the traffic flow data. Profiles of pollutant concentration in the driving direction exhibited a triangular distribution characteristic, increasing consistently from the tunnel entrance to the tunnel exit; this characteristic is consistent with the conventional wisdom of longitudinal ventilation systems. In conclusion, five types of data collected

Table 1: Monitoring instruments and data.

Data category	Data name	Instrument and model	Unit	Statistical interval
Ambient air	CO	ThermoFisher Model 48i CO Analyzer	ppm	15-min average value
	NO_2	ThermoFisher Model 42i NO_x Analyzer	ppm	
	Air velocity	AZ Instrument 9871	m/s	
	$PM_{2.5}$	ThermoFisher Model 5030i Particulate Monitor	mg/m^3	
Traffic flow	PC	Laser vehicle detector	vehicle/15 min	15-min accumulative value
	LDV		vehicle/15 min	
	HGV		vehicle/15 min	

Figure 1: Qin Mountains tunnel group.

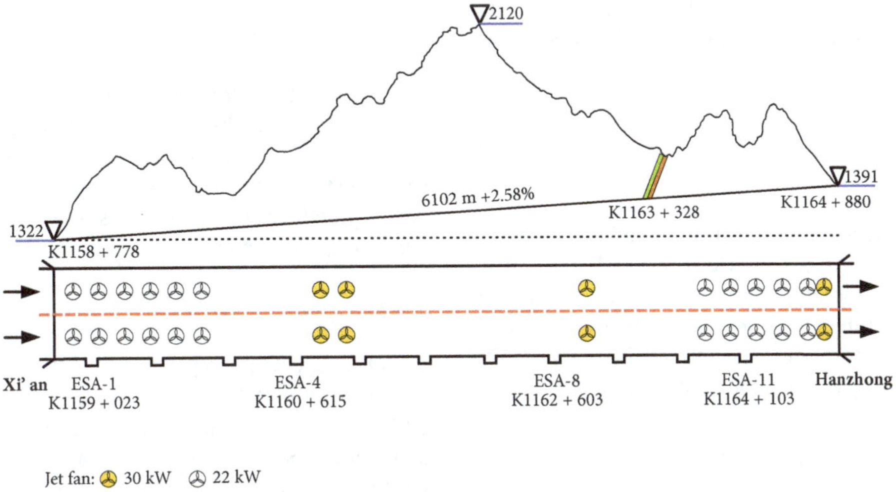

Figure 2: Overall structure of Qin Mountains No. 1 tunnel in the southbound direction.

from ESA-11, the monitoring site with the highest degree of pollution, were selected as the sample dataset; these data were the CO, NO_2, air velocity, $PM_{2.5}$, and HGV data.

2.2. Clustering Method. A five-dimensional space was obtained from the sample dataset. Clustering analysis for the operational status is the task of grouping the sample dataset in such a way that status data in the same group (called a cluster) are more similar (in some sense or another) to each other than to those in other clusters. In centroid-based clustering, the task can be summarized as finding the cluster centers and assigning the sample data to the nearest cluster center such that the squared distances from the cluster are minimized and thus obtaining a classification method for multiclass operational statuses.

Fuzzy C-Means (FCM) clustering is a fuzzy clustering algorithm based on an objective function; this algorithm was developed by Dunn [18] and improved by Bezdek [19]. Given its advantages in big data applications, FCM clustering was chosen in this study. Consider that the ith sample data $x_i = (x_{i1}, x_{i2}, x_{i3}, x_{i4}, x_{i5})$ denote a five-dimensional monitoring result, namely, the values of CO, NO_2, air velocity, $PM_{2.5}$, and HGV. The sample dataset containing N measured values is denoted by X. Then X can be expressed by a $N \times 5$ matrix, as shown in the following:

$$X = \begin{bmatrix} x_{11} & x_{12} & x_{13} & x_{14} & x_{15} \\ x_{21} & x_{22} & x_{23} & x_{24} & x_{25} \\ \vdots & \vdots & \vdots & \vdots & \vdots \\ x_{N1} & x_{N2} & x_{N3} & x_{N4} & x_{N5} \end{bmatrix} \quad (1)$$

The FCM aims to minimize the following objective function:

$$\sum_{v=1}^{k}\sum_{i=1}^{N} u_{iv}^2 \|x_i - m_v\|^2 = \sum_{v=1}^{k}\sum_{i=1}^{N} u_{iv}^2 \sum_{f=1}^{5} \left(x_{if} - m_{vf}\right)^2 \quad (2)$$

where k is a preset number of operational status, i.e., cluster numbers; v is the sequence number of a cluster; m_v is the center of the cluster v; u_{iv}^2 stands for the unknown membership of sample x_i in cluster v with a membership exponent 2 to determine the level of cluster fuzziness; $\|x_i - m_v\|^2$ denotes the squared Euclidean distance between x_i and m_v; f is the sequence number of five-dimensional space; and m_{v1}, m_{v2}, m_{v3}, m_{v4}, and m_{v5} represent the values of cluster center m_v corresponding to CO, NO_2, air velocity, $PM_{2.5}$, and HGV, respectively. Cluster center m_{vf} can be calculated by the following equation:

$$m_{vf} = \frac{\sum_{i=1}^{N} u_{iv}^2 x_{if}}{\sum_{i=1}^{N} u_{iv}^2} \quad (3)$$

Kaufman and Rousseeuw (2008) proposed a new fuzzy clustering algorithm FANNY based on FCM [20]. The FANNY algorithm has some definite advantages over FCM: lower sensitivity to outliers or otherwise erroneous data and better recognition of nonspherical clusters. In the FANNY algorithm, the following equation is derived from (2):

$$\min \sum_{v=1}^{k} \frac{\sum_{i=1}^{N}\sum_{j=1}^{N} u_{iv}^2 u_{jv}^2 d\left(x_i, x_j\right)}{2\sum_{j=1}^{N} u_{jv}^2} \quad (4)$$

where $d(x_i, x_j)$ represents the given distances (or dissimilarities) between samples x_i and x_j; Euclidean distance is in common use. Each pair is encountered twice because $d(x_j, x_i)$ also occurs, and the factor 2 in the denominator compensates for this duplicity. The membership function is subject to the following constraints:

$$u_{iv} \geq 0, \quad i = 1, \cdots, N;$$
$$\sum_{v=1}^{k} u_{iv} = 1, \quad i = 1, \cdots, N. \quad (5)$$

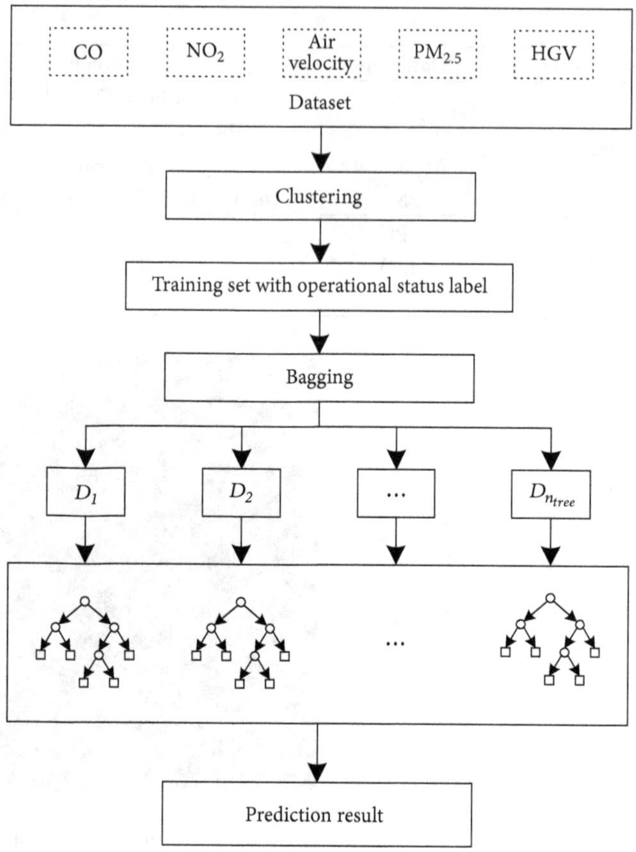

FIGURE 3: Diagram of Random Forests combining with clustering analysis.

The optimization problem is solved as shown in (4) to calculate and obtain the membership coefficients of all samples in every cluster $u_{iv} (1 \leq i \leq N, 1 \leq v \leq k)$ and each cluster center m_v. Thus, each sample is assigned to the cluster in which it has the largest membership, and the fuzzy clustering is completed.

2.3. Perception Model

Definition 1 (the perception of in-tunnel operational status). Given a training set $T = \{(x_1, y_1), \cdots, (x_N, y_N)\} \in (X^5 \times Y)^N$, $x_i \in X^5$ is the ith sample in the training set and it includes the values of CO, NO_2, air velocity, $PM_{2.5}$, and HGV; $y_i \in Y = \{c_1, c_2, c_3, c_4\}$ corresponds to one of the four operational statuses of the ith sample—lightly polluted, moderately polluted, heavily polluted, and severely polluted; and $i = 1, \cdots, N$ is the serial number of the training set. According to algorithmic modeling [21], the target is to find a function $f(x) : X^5 \longrightarrow Y$—an algorithm that operates on X^5 to predict the responses of in-tunnel operational status Y.

The ensemble of the Random Forests combining with clustering analysis is shown in Figure 3, in which clustering results of operational status are taken as inputs of Random Forests-based perception model. For perception model, first of all, bootstrap samples of size n_{tree} with replacement from the training set are taken, and a new series of training subsets are formed by the bagging technique. Then, randomly

TABLE 2: Confusion matrix for in-tunnel operational status.

Actual status	Perception status				Total
	Lightly polluted (c_1)	*Moderately polluted* (c_2)	*Heavily polluted* (c_3)	*Severely polluted* (c_4)	
Lightly polluted (c_1)	$n_{1,1}$	$n_{1,2}$	$n_{1,3}$	$n_{1,4}$	$N_{1,.}$
Moderately polluted (c_2)	$n_{2,1}$	$n_{2,2}$	$n_{2,3}$	$n_{2,4}$	$N_{2,.}$
Heavily polluted (c_3)	$n_{3,1}$	$n_{3,2}$	$n_{3,3}$	$n_{3,4}$	$N_{3,.}$
Severely polluted (c_4)	$n_{4,1}$	$n_{4,2}$	$n_{4,3}$	$n_{4,4}$	$N_{4,.}$
Total	$N_{.,1}$	$N_{.,2}$	$N_{.,3}$	$N_{.,4}$	N

select partial features in training subset for finding the best split variable whenever splitting the sample in a tree and create a complete tree using the bootstrapped examples. Next, compute the performance of each tree using examples that were not chosen in the bootstrap phase (out-of-bag data). Finally, calculate a vote on new cases when completing all the trees in the ensemble. Declare for each of them the winning class as a prediction.

2.4. Modeling Approach. There are the following two crucial parameters in Random Forests modeling, namely, n_{tree} and m_{try}:

(1) n_{tree}—the number of trees to grow;

(2) m_{try}—the number of variables randomly sampled as candidates at each split.

Herein, n_{tree} determines the overall scale of the whole Random Forests, and m_{try} defines the structure of a single decision tree. In other words, n_{tree} and m_{try} determine the construction of the Random Forests at macroscopic and microcosmic levels, respectively.

In R, the *randomForest* package provides an interface to the Breiman and Cutler's Fortran programs of Random Forests, and *randomForest()* function implements the algorithm for classification and regression [22]. The function prototype is as follows:

randomForest (*formula, data, mtry, ntree, na.action*)

in which *formula* describes the model to be fitted; *data* is a data frame containing the variables in the model; *mtry* is the number of variables randomly sampled; *ntree* is the number of decision trees; *na.action* specifies the action to be taken if NAs are found.

Since the bootstrap performs sampling with replacement from the training set, its probability to be chosen as the out-of-bag (OOB) sample is $(1-1/N)^N$. For large N, the number of OOB samples is expected to be a fraction $e^{-1} \approx 0.368$ of the training set. It means each decision tree is grown by using approximately $1-e^{-1} \approx 63.2\%$ of the training samples, leaving $e^{-1} \approx 36.8\%$ as the OOB samples. Since the OOB part of the data has not been used in tree construction, it can be used to estimate the ensemble prediction performance in the following way.

Let D_b^{OOB} be the OOB part of the data for the bth tree. Then use the bth tree to predict D_b^{OOB}. Since each training sample x_i is in an OOB sample set, on the average approximately $e^{-1} \approx 36.8\%$ of the time the ensemble prediction $\widehat{Y}^{OOB}(x_i)$ can be calculated by aggregating only its OOB predictions. Calculate an estimate of the error rate (ER) for classification by

$$ER \approx ER^{OOB} = n^{-1}\sum_{i=1}^{n} I\left(\widehat{Y}^{OOB}(x_i) \neq Y_i\right) \quad (6)$$

where $I(\cdot)$ is the indicator function.

2.5. Evaluation Metric. A status set $Y = \{c_1, c_2, c_3, c_4\}$ is used to denote the four classes of the in-tunnel operational statuses—lightly polluted, moderately polluted, heavily polluted, and severely polluted. Then the confusion matrix (as shown in Table 2) is chosen to describe the classification performance.

In Table 2, $n_{i,j}$ denotes the number of actual statuses identified as c_j by the classification model. The confusion matrix reflects the distribution of status set Y, among which the jth column reflects the precision of c_j and ith row reflects the recall (also known as sensitivity) of c_i. Thus, for the particular operational status, e.g., c_j, the precision (P_{c_j}) and recall (R_{c_j}) are calculated by the following.

$$P_{c_j} = \frac{n_{j,j}}{N_{.,j}} \times 100\% \quad (7)$$

$$R_{c_j} = \frac{n_{j,j}}{N_{j,.}} \times 100\% \quad (8)$$

Besides, the other evaluation metric is the harmonic average of the precision and recall and is called F-measure (F). It is calculated as follows:

$$F_{c_j} = \frac{2P_{c_j}R_{c_j}}{P_{c_j} + R_{c_j}} \times 100\% \quad (9)$$

3. Results

3.1. Classification of Operational Status. Determining the optimal number of clusters is a fundamental issue in clustering analysis. In this study, this value was estimated by the optimum average silhouette width [23]. Suppose a data set is partitioned into k clusters, the silhouette width of sample x_i is then defined as

$$S_k(x_i) = \frac{B_k(x_i) - A_k(x_i)}{\max(A_k(x_i), B_k(x_i))} \quad (10)$$

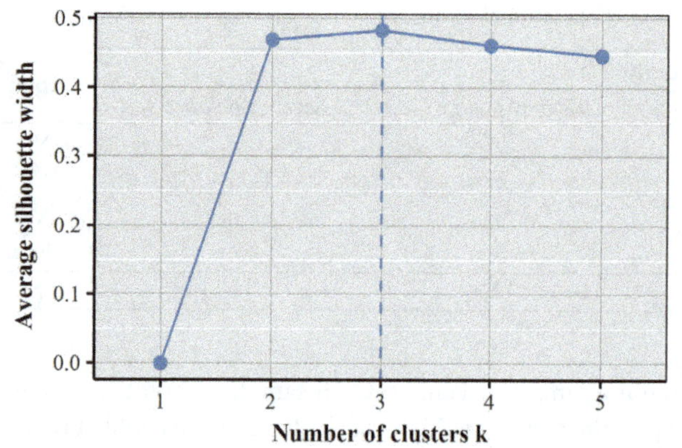

FIGURE 4: Optimal number of clusters; higher average silhouette widths are preferred.

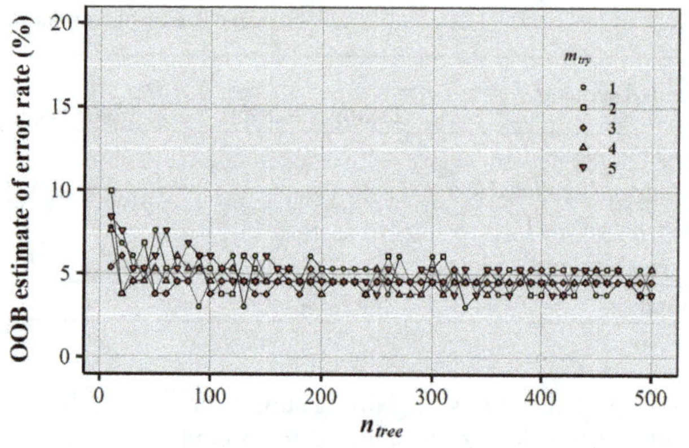

FIGURE 5: Influence of the combined parameters (n_{tree} and m_{try}) on OOB error.

where $A(x_i)$ is the average dissimilarity between x_i and all other samples in the cluster to which x_i belongs. Similarly, $B(x_i)$ is the minimum average dissimilarity between x_i and all other clusters to which x_i does not belong. The average silhouette method computes the average silhouette width (S_k) of all N samples for different values of k:

$$S_k = \frac{1}{N}\sum_{i=1}^{N} S_k(x_i), \quad k = 1, 2, 3 \cdots \tag{11}$$

The optimal number of clusters k is the one that maximizes the average silhouette width over a range of possible values for k.

The average silhouette widths with $k = 1, 2, 3, 4, 5$ for this study are shown in Figure 4. The silhouette plot shows that the k value of 3 corresponded to the maximum width, so the optimal number of in-tunnel operational status is 3 for the actual monitoring dataset. One of the four in-tunnel operational statuses did not appear in the experiment, so the next step is to verify which status was missing.

By applying FANNY algorithm to the preprocess data, three cluster centers are obtained using the following equation.

$$m_{vf} = \begin{bmatrix} 0.82 & 0.081 & 3.2 & 0.28 & 0 \\ 5 & 1.1 & 5.9 & 0.46 & 56 \\ 14 & 1.8 & 6.7 & 0.72 & 115 \end{bmatrix} \tag{12}$$

In (12), the five elements in each row represent the values of cluster centers in the following order: CO (ppm), NO_2 (ppm), air velocity (m/s), $PM_{2.5}$ (mg/m^3), and HGV (veh/15 min). The number of HGV is 0 in the first cluster, and the NO_2 concentrations in the second and third cluster exceed the PIARC standard (1 ppm) [24]. Thus, the three rows represent the cluster centers of lightly polluted, heavily polluted, and severely polluted statuses. The moderately polluted status did not appear when the traffic volume increased slightly.

3.2. *Modeling of Status Perception*

3.2.1. Optimal Combined Parameter. Before tuning the parameters in the Random Forests, the in-tunnel operational dataset is divided into a training set and a test set in the ratio 7:3. The former was used for parameter tuning and variable importance calculation. The latter was used for model evaluation. The minimum OOB ER principle is considered to be the reference to optimize the combination of parameters n_{tree} and m_{try}. R implementation code performed on a desktop PC running Windows 10, with a 3.6 GHz Intel i7 quad-core CPU and 16 GB RAM is shown as in Algorithm 1.

Assuming $n_{tree} = 10, 20, \cdots, 500$ and $m_{try} = 1, 2, 3, 4, 5$, 250 combinations of n_{tree} and m_{try} are run iteratively, and the relation between the combined parameters and OOB estimate of ER is obtained as shown in Figure 5.

Figure 5 shows that OOB ER was largely influenced by parameter n_{tree}; the error decreased with increasing n_{tree}, making the perception results more accurate. However, the time consumed for each iteration remained on the order of milliseconds, and, hence, the calculation time could be ignored. When $n_{tree} > 200$, OOB ERs tended to converge. The parameter m_{try} had less impact on the OOB ER. The results further validated that the Random Forests would be less likely to overfit, and the classification error would converge with an increasing number of decision trees. Consider the classification accuracy; an optimal combined parameter was identified as $n_{tree} = 500$ and $m_{try} = 1$, corresponding to 4.55% as an unbiased estimation of the OOB ER. The R code of status perception modeling is shown below.

```
Status.rf <- randomForest(Status ~ ., data = trainingset, mtry = 1, ntree = 500, na.action = na.omit)

print(Status.rf)
```

3.2.2. Importance Measurement of Variables. Another important feature of the Random Forests is the measurement of variable importance, which allows ranking variables regarding the importance and optimizing the variable subset, thus avoiding the problems created by dimensionality and reducing the computational complexity. There are two indexes to measure variable importance: mean decrease accuracy (MDA) and mean decrease Gini-index (MDG). The former is defined as the average decrease between the percentage of votes for the correct class in the untouched OOB data and

```
# division of training set and test set
set.seed(100)
ind <- sample(2, nrow(Mydataset), replace = TRUE, prob = c(0.7,0.3))
trainingset <- Mydataset[ind==1, ] # training set accounts for 70%
testset      <- Mydataset[ind==2, ] # test set accounts for 30%
# combined parameters in Random Forests
library(randomForest)
library(caret)
M <- ncol(trainingset)
ntree <- 10*c(1:50)
result <- data.frame()
# model training
set.seed(100)
for(m in 1:(M-1)){
  for(n in ntree){
    fit.rf <- randomForest(Status ~., data = trainingset, mtry = m, ntree = n, na.action = na.omit)
    OOB.ER <- 1-confusionMatrix(as.table(fit.rf$confusion[,c(-4)]))$overall["Accuracy"]
    result <- rbind(result, data.frame(m, n, OOB.ER))
  }
}
print(result)
```

Algorithm 1

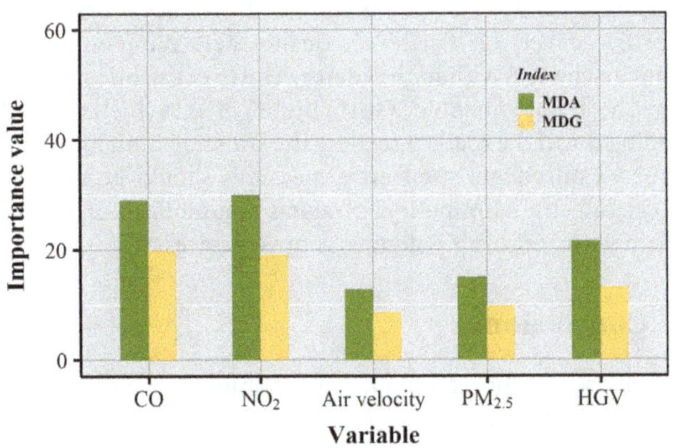

Figure 6: Comparison of importance of different variables.

the percentage of votes for the correct class in the variable permuted OOB data averaged over all trees. The latter is defined as the total decrease in the Gini-index from splitting on the variable averaged over all trees [22, 25]. The bigger the MDA and MDG, the more important is the variable. The optimal combined parameter in the training set was applied, and the importance indexes of each variable were calculated, as shown in Figure 6.

As seen from Figure 6, during the dynamic evolution process of in-tunnel operational status, the importance order of variables from largest to smallest is as follows: NO_2, CO, HGV, $PM_{2.5}$, and air velocity. NO_2 and CO, the two main types of gaseous pollutants, are the primary factors that affect the changes in the in-tunnel operational status.

3.3. Perception Results of Operational Status. In this study, the Naïve Bayes, Support Vector Machine (SVM), and Random Forests-based perception model were applied to predict operational status in the test set. Evaluation metrics for these three models are listed in Table 3.

Naïve Bayes classifier assumes that the value of a particular feature is independent of the value of any other feature, which is always invalid in practice. The naive design and apparently oversimplified assumption affect classification performance of Naïve Bayes. SVM can efficiently perform a nonlinear classification using what is called the kernel trick, implicitly mapping their inputs into high-dimensional feature spaces. However, a significant practical question, the selection of the kernel function parameters, is still not entirely solved. Table 3 indicates that the precision, recall, and F-measure for the Random Forests-based model were better than those for the Naïve Bayes or SVM model. For further calculation, the average precision, recall, and F-measure in the Naïve Bayes model were 94.85%, 86.79%, and 90.19%, respectively. In the SVM-based model, the average precision, recall, and F-measure were 96.20%, 90.83%, and 93.24%, respectively. In contrast, the average precision, recall, and F-measure in the Random Forests-based model were 98.83%, 95.52%, and 97.07%, respectively. The results validated that the Random Forests-based perception model offers the best performance among the three models, indicating its better adaptability to the dynamic changes of the operational status in extra-long highway tunnels.

4. Discussion

4.1. Optimal Number of Clusters. Determining the number of clusters in a dataset, a quantity often labeled k, is a frequent problem in data clustering and is a distinct issue from the process of actually solving the clustering problem. The correct choice of k is often ambiguous, with interpretations depending on the shape and the scale of distribution of points in a dataset and the desired clustering resolution of the user.

TABLE 3: Comparison of evaluation metric for different perception models.

Operational status	Naïve Bayes			SVM			Random Forests		
	P (%)	*R* (%)	*F* (%)	*P* (%)	*R* (%)	*F* (%)	*P* (%)	*R* (%)	*F* (%)
Lightly polluted	100	80	88.89	100	85	91.89	100	90.00	94.74
Heavily polluted	88.71	98.21	93.22	92.44	98.21	95.24	96.49	100	98.21
Severely polluted	95.83	82.14	88.64	96.15	89.29	92.59	100	96.55	98.25

In this study, the silhouette method was chosen for assessing the natural number of operational statuses. Frankly, the determination of $k = 3$ was of a little subjectivity; $k = 2$ or $k = 3$, after all, was only slightly smaller than it. In consequence, long-term accumulation of in-tunnel operational monitoring data is crucial for the rational classification of operational status.

4.2. Management and Control Strategy for Ventilation and Traffic Flow. The perception model can be used to determine the real-time in-tunnel operational status by using a combination of pollutant concentration and traffic volume monitoring results. In the SB direction of Qin Mountains No. 1 tunnel, the percentages of the operational environment with heavily polluted and severely polluted statuses were 59% and 31%, respectively. The lightly polluted status contributed less than 10% of the operational environment. The box-plots presenting the distributions of CO, NO_2, air velocity, $PM_{2.5}$, and HGV under different statuses are shown in Figures 7(a)–7(e).

Although the moderately polluted status did not appear, the fluctuation range of CO (Figure 7(a)), NO_2 (Figure 7(b)), and $PM_{2.5}$ (Figure 7(d)) exhibited a tendency to intensify with the deterioration of the in-tunnel operational status, which was basically the same as the tendency for HGV. Figure 7(c) shows that there was a minimal number of HGV in regions with the lightly polluted status. Natural ventilation mode was used during that period, and, hence, the air velocities underwent a rather significant fluctuation influenced by the movement of vehicles (piston effect). For heavily and severely polluted statuses, all jet fans were turned on, and the air velocities were relatively stable (Figure 7(e)). Even so, the concentration of NO_2 still exceeded the PIARC standard.

The classification of in-tunnel operational statuses provides a scientific way to develop strategies for intelligent ventilation and traffic management and control. For lightly polluted status, consider switching off the fans and depending only on natural ventilation. For moderately polluted status, consider switching on the fans with a variable-frequency drive (VFD) to save energy consumption. For heavily polluted status, consider operating the jet fans at the fully open position and activating the axial fans in the inclined shaft in a timely manner. For severely polluted status, the in-tunnel air quality is terrible and the tunnel is filled with smog and smoke, threatening driving safety; therefore, all jet fans and axial fans should be fully operated. If the tunnel is operated under the severely polluted status for extended periods of time, temporary traffic control measures should be executed to ensure driving safety [26], for instance, limiting HGV powered by diesel engines passing through the tunnel or diverting them upstream of the tunnel.

4.3. Impact on Ecology and Environmental Management. The ecology and environmental impact of transportation are significant because transportation is a major consumer of energy and burns most of the world's petroleum. According to the annual report of Chinese Ministry of Environmental Protection, more than 246 million vehicles emitted 45.47 million tons of pollutants in China in 2014 [27]. Vehicle emissions have become one of the principal sources of air pollution and a significant cause of dust–haze and photochemical smog. Reducing transportation emissions will produce considerable positive effects on Earth's air quality, acid rain, smog, and climate change. Although stricter vehicle emission standards have been implemented, a vast number of old vehicles are still rolling down the road, exceeding the emission limit by several times. Consequently, effective measures should be made to accelerate the elimination of aging automobiles or retrofit them with approved pollutant control devices.

5. Conclusions

In this study, the operational monitoring data in an extra-long highway tunnel were analyzed in detail using big data technology. By combining monitoring results of CO, NO_2, air velocity, $PM_{2.5}$, and HGV, a data-driven model for in-tunnel operational status perception was structured. The major conclusions are as follows.

By applying the FANNY algorithm, the optimal number of clusters for obtaining the in-tunnel operational status was determined following the principle of maximum average silhouette width. Owing to the restriction of the total experimental duration, the clustering results did not contain all four operational statuses. Unfortunately, the moderately polluted status was not observed. The next step is to perform long-term monitoring of the in-tunnel operational environment and obtain massive data, thus realizing more scientific and reasonable classification of in-tunnel operational statuses.

A Random Forests-based perception model was built for determining in-tunnel operational status. Taking the perception accuracy into consideration primarily, an optimal combined parameter of the Random Forests was identified. Prediction results indicated that the proposed model was better than the contrast models and had the better adaptability to dynamic changes of operational status in extra-long highway tunnels, thus realizing accurate predictions.

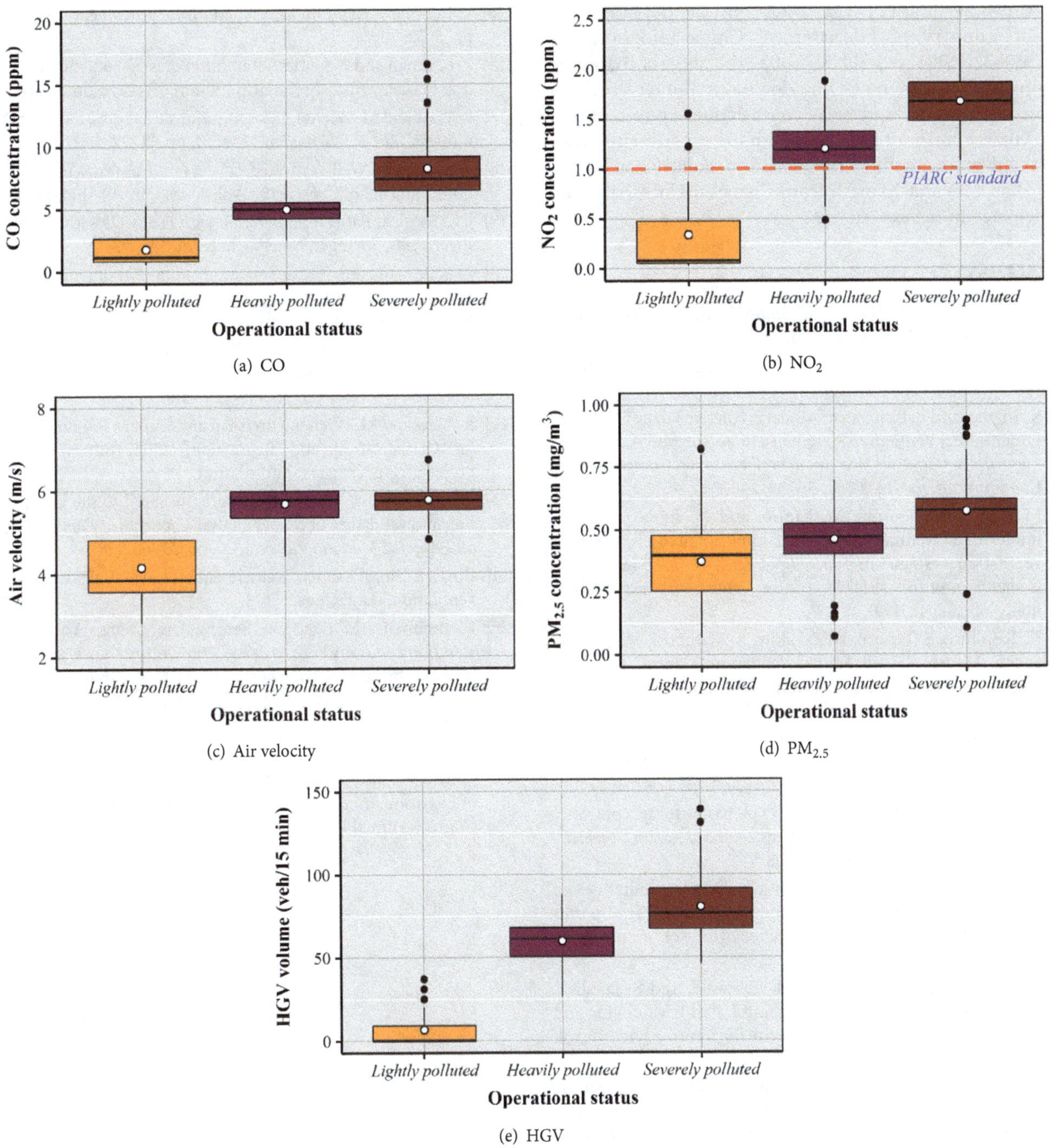

FIGURE 7: Distributions of particular variables under different statuses.

The distribution of individual variable under different operational statuses were analyzed. The management and control strategies for ventilation and traffic flow under lightly polluted, heavily polluted, and severely polluted statuses were discussed. These strategies could help improve the operation and management level of extra-long highway tunnels and provide a scientific method to realize energy saving and emission reduction.

Conflicts of Interest

The authors declare that they have no conflicts of interest.

Acknowledgments

This work was supported by the National Natural Science Foundation of China under Grant no. 51678063 and

Fundamental Research Funds for the Central Universities of Ministry of Education of China under Grants nos. 310832161006 and 310821173102. The authors thank the administration center of Qin Mountains tunnel group of Shaanxi provincial expressway construction group corporation Xihan branch for assistance with the experiments. They also thank graduate students Penglei Sun, Yuhui Zhai, Wei Li, Tongzhan Liu, and Xiang Ji from Chang'an University who participated in the experiments.

References

[1] Ministry of Transport of China, *Statistical Bulletin: Transportation Industry in 2016*, Beijing, China, 2017.

[2] PIARC, "Road tunnels: operational strategies for emergency ventilation," La Défense, 2011.

[3] S. Bogdan and B. Birgmajer, "Model Predictive Fuzzy Control of Longitudinal Ventilation System in a Road Tunnel," *Automatika - Journal for Control, Measurement, Electronics, Computing and Communications*, vol. 47, pp. 39–48, 2006.

[4] M. Funabashi, I. Aoki, M. Yahiro, and H. Inoue, "A fuzzy model based control scheme and its application to a road tunnel ventilation system," in *Proceedings IECON '91 International Conference on Industrial Electrlnics, Control and Instrumentation*, Kobe, Japan, 1991.

[5] K. Toshihiro, Y. Tatsuro, W. Takahiro, S. Masanori, M. Miyako, and E. Hisashi, "Road Tunnel Ventilation Control Based on Nonlinear Programming and Fuzzy Control," *IEEJ Transactions on Industry Applications*, vol. 113, no. 2, pp. 160–168, 1993.

[6] P.-H. Chen, J.-H. Lai, and C.-T. Lin, "Application of fuzzy control to a road tunnel ventilation system," *Fuzzy Sets and Systems*, vol. 100, no. 1-3, pp. 9–28, 1998.

[7] B. Chu, D. Kim, D. Hong et al., "GA-based fuzzy controller design for tunnel ventilation systems," *Automation in Construction*, vol. 17, no. 2, pp. 130–136, 2008.

[8] S. Bogdan, B. Birgmajer, and Z. Kovačić, "Model predictive and fuzzy control of a road tunnel ventilation system," *Transportation Research Part C: Emerging Technologies*, vol. 16, no. 5, pp. 574–592, 2008.

[9] N. Euler-Rolle, M. Fuhrmann, M. Reinwald, and S. Jakubek, "Longitudinal tunnel ventilation control. Part 1: Modelling and dynamic feedforward control," *Control Engineering Practice*, vol. 63, pp. 91–103, 2017.

[10] C. Guo, M. Wang, L. Yang, Z. Sun, Y. Zhang, and J. Xu, "A review of energy consumption and saving in extra-long tunnel operation ventilation in China," *Renewable & Sustainable Energy Reviews*, vol. 53, pp. 1558–1569, 2016.

[11] Q. Li, C. Chen, Y. Deng et al., "Influence of traffic force on pollutant dispersion of CO, NO and particle matter (PM2.5) measured in an urban tunnel in Changsha, China," *Tunnelling and Underground Space Technology*, vol. 49, pp. 400–407, 2015.

[12] H. Yamada, R. Hayashi, and K. Tonokura, "Simultaneous measurements of on-road/in-vehicle nanoparticles and NOx while driving: Actual situations, passenger exposure and secondary formations," *Science of the Total Environment*, vol. 563-564, pp. 944–955, 2016.

[13] A. N. Martin, P. G. Boulter, D. Roddis et al., "In-vehicle nitrogen dioxide concentrations in road tunnels," *Atmospheric Environment*, vol. 144, pp. 234–248, 2016.

[14] L. Breiman, "Bagging predictors," *Machine Learning*, vol. 24, no. 2, pp. 123–140, 1996.

[15] L. Breiman, "Random forests," *Machine Learning*, vol. 45, no. 1, pp. 5–32, 2001.

[16] L. Breiman and A. Cutler, "Random Forests," 2001, https://www.stat.berkeley.edu/~breiman/RandomForests/cc_home.htm.

[17] A. Liaw and M. Wiener, "Classification and regression by randomforest," *The R Journal*, vol. 2, no. 3, pp. 18–22, 2002.

[18] J. C. Dunn, "Well-separated clusters and optimal fuzzy partitions," *Journal of Cybernetics*, vol. 4, no. 1, pp. 95–104, 1974.

[19] J. C. Bezdek, *Pattern Recognition with Fuzzy Objective Function Algorithms*, Springer, New York, NY, USA, 1981.

[20] L. Kaufman and P. Rousseeuw, *Finding Groups in Data: An Introduction to Cluster Analysis*, John Wiley & Sons, Hoboken, NJ, USA, 2008.

[21] L. Breiman, "Statistical modeling: the two cultures," *Statistical Science. A Review Journal of the Institute of Mathematical Statistics*, vol. 16, no. 3, pp. 199–231, 2001.

[22] A. Liaw and M. Wiener, "Breiman and Cutler's Random Forests for Classification and Regression," 2015, https://CRAN.R-project.org/package=randomForest.

[23] P. J. Rousseeuw, "Silhouettes: a graphical aid to the interpretation and validation of cluster analysis," *Journal of Computational and Applied Mathematics*, vol. 20, pp. 53–65, 1987.

[24] PIARC, "Road tunnels: vehicle emissions and air demand for ventilation," La Défense, 2012.

[25] L. Breiman, "Manual On Setting Up, Using, And Understanding Random Forests V3.1," 2002, https://www.stat.berkeley.edu/~breiman/Using_random_forests_V3.1.pdf.

[26] PIARC, "Integrated approach to road tunnel safety," La Défense, 2007.

[27] Ministry of Environmental Protection of China, *China Vehicle Emission Control Annual Report*, Beijing, China, 2015, Ministry of Environmental Protection of China, China Vehicle Emission Control Annual Report.

An Optimization Problem for Quadcopter Reference Flight Trajectory Generation

Karam Eliker,[1] Guoqing Zhang,[1] Said Grouni,[2] and Weidong Zhang[1,3]

[1]*Department of Automation, Shanghai Jiao Tong University, Shanghai 200240, China*
[2]*Faculty of Sciences Engineering, University of Boumerdes, 35000 Boumerdes, Algeria*
[3]*School of Computer Engineering and Science, Shanghai University, Shanghai 200444, China*

Correspondence should be addressed to Weidong Zhang; wdzhang@sjtu.edu.cn

Academic Editor: Howard Li

This paper deals with the reference flight trajectory generation and planning problems for quadcopter Unmanned Aerial Vehicle (UAV). The reference flight trajectory is defined as the composition of path and motion functions. Both of them are generated by using quintic B-spline functions. Based on differential flatness approach, the quadcopter dynamical constraints are satisfied instantaneously by computing the induced aerodynamical moments and lift force. The optimal reference flight trajectory, with respect to the mission requirements and imposed constraints, is reached by manipulating the control points' vectors of B-spline functions via a nonlinear constrained optimization method. The mission requirements are defined as a set of waypoints with their respective scheduled flight timetable. A minimum-energy cost function is developed to minimize the consumed energy and induced efforts by reference flight trajectory. For the need of the optimal overfly-times schedule, the overfly times with respect to the defined constraints and performance criteria are calculated. Numerical simulation results show the feasibility and effectiveness of the proposed optimization method.

1. Introduction

With a vast proliferation of Manned Aerial Vehicles (MAVs) and Unmanned Aerial Vehicles (UAVs), world air transportation traffic has witnessed a continuous increase causing a saturation in the airspace. The formal approach adopted by researchers to design the UAVs is to separate the avionics systems into three distinct and independent systems, which are Guidance, Navigation, and Control (GNC). By using this structure, UAVs could perform a variety of planning missions in either military or civilian fields. However, there are many incidents like a collision with civil aviation aircraft, invasion of privacy, and threats to security reported by [1] Federal Aviation Administration (FAA). These incidents are because of the lack of transportation rules and regulations supporting UAVs and the weakness current structure UAVs. Therefore, the need to integrate the UAVs in National Airspace System (NAS) is highly recommended.

To achieve this challenge, an adequate flight trajectory planner that could be embedded on quadcopter UAV should be developed. The flight trajectory planner could reduce uncertainty and increase predictability for Air Traffic Management (ATM). Planning missions in terms of waypoints with overfly times at which the UAV is scheduled to overfly could be a challenging problem for trajectory generation. Most of the previous works have focused on 3D trajectory planning problem and consider only waypoints without any overfly-times constraints. By contrast, this paper proposes a new optimization method to generate the reference flight trajectory that could fulfill some related requirements of NAS. The proposed flight trajectory planning should generate a trajectory that complies with the considered quadcopter dynamical constraints and meets the assigned missions. The assigned missions contain a set of waypoints with their respective scheduled flight timetable. For the necessity of optimal overfly-times schedule, the overfly times can be calculated in optimal way while considering the assigned constraints and performance criteria. By generating such reference flight trajectory, different tasks such as delivering and reconnaissance missions at acceptable overfly times could be performed. Moreover, the traffic managements for UAV could be enhanced by managing the scheduled flight timetable of waypoints.

Several papers of 3D trajectory planning problems while considering waypoints as physical points in the workspace have been done. On one hand, the authors in [2–7] proposed an optimization approach to solve Time Optimal Motion Planning (TOMP) between two configuration points by using a Nonlinear Programming (NLP) method. On the other hand, the authors in [8–10] have proposed an algorithm based on Pontryagin's minimum principle to solve TOMP between two states. In [11], the authors presented an approach that could link separate trajectories to form smooth analytically defined paths while satisfying a set of waypoint constraints. Moreover, in [12–14], a minimum snap cost function was designed to generate smooth trajectory while satisfying a sequence of consecutive waypoints. In [15], a quasi-optimal trajectory generation was developed to satisfy constraints concerning waypoints and their overfly times while minimizing the trajectory length.

Moreover, some papers have dealt with the minimum-energy problem. In [16], a quasi-optimal trajectory planner was proposed to generate trajectory between two points. The objective function was defined as a fuel consumption problem by considering the cost function as an average velocity. The authors in [17] supposed that the consumed energy was proportional to the rotors velocities. In [18], the authors designed a trajectory planning function for a small-scale helicopter in autorotation which could minimize the battery power consumption and avoid bang-bang type solutions. Nonlinear Programming problem was used to compute the optimal polynomial coefficients of the trajectory while minimizing the cost function written as the rate of all forces and moments applied on helicopter. In [19], a minimum-energy trajectory generation approach was proposed by defining the cost function as a function of the voltage and current across the four rotors. The similarity among these papers was the reference trajectory planning problems, defined as longitudinal, lateral, and altitude displacements, as NLP problem between two configuration points, while the difference was related to the parameterization of the reference trajectory.

The objective of this paper is the reference trajectory generation and planning problems for quadrotor UAV. The reference flight trajectory is defined as a composition of path and monotonically increasing motion functions. Both of them are generated by using quintic B-spline functions. Based on differential flatness approach, the quadcopter dynamical constraints are checked instantaneously by computing the induced aerodynamical roll and pitch, heading moments, and lift force, respectively. The optimal reference trajectory, which consists of longitudinal, lateral, and altitude displacements, in addition to yaw angle, is reached by manipulating the control points' vectors of B-spline functions via a nonlinear constrained optimization method while taking into account the mission requirements and defined constraints. The mission requirements are defined as a sequence of waypoints with their respective scheduled flight timetable, and each waypoint could represent a particular posture. A minimum-energy cost function is designed to enhance the performance criteria regarding trajectory length, induced efforts, consumed energy, and trajectory smoothness. For the optimal overfly-times schedule requirement, the overfly times are calculated by the proposed optimization method with regard to the interested constraints and performance criteria. Several configurations are studied and numerical simulation results show the feasibility and effectiveness of the proposed optimization method.

2. Quadcopter Flight Dynamics

The quadcopter UAV is a vertical take-off and landing (VTOL) aircraft. Its main characteristics are the ability of hovering, vertical take-off, and landing in any terrain. The displacement and rotation motions are achieved by adjusting the angular speed of each rotor. To design the reference flight trajectory, a simplified quadcopter flight dynamics are used to reduce the computation burden and to facilitate the optimization problem implementation. By considering that the quadcopter UAV performs angular motions of low amplitude and omitting the drag force, aerodynamical frictions, and gyroscopic effects moments, the simplified dynamic model of quadcopter UAV, in contrast to full quadcopter dynamics presented in [21, 22], is given by

$$\ddot{\phi} = \frac{1}{I_x} U_\phi \tag{1a}$$

$$\ddot{\theta} = \frac{1}{I_y} U_\theta \tag{1b}$$

$$\ddot{\psi} = \frac{1}{I_z} U_\psi \tag{1c}$$

$$\ddot{x} = \frac{U_z}{m} (\cos\phi \sin\theta \cos\psi + \sin\phi \sin\psi) \tag{1d}$$

$$\ddot{y} = \frac{U_z}{m} (\cos\phi \sin\theta \sin\psi - \sin\phi \cos\psi) \tag{1e}$$

$$\ddot{z} = -g + \frac{U_z}{m} (\cos\phi \cos\theta), \tag{1f}$$

where x, y, z are longitudinal, lateral, and altitude displacements and ϕ, θ, ψ are roll, pitch, and yaw angles in the Earth-fixed frame, respectively. g is the gravity acceleration and m is the mass of quadcopter UAV. I_x, I_y, I_z are roll, pitch, and yaw inertia moments, respectively. The lift force U_z and the aerodynamical roll U_ϕ, pitch U_θ, and heading U_ψ moments developed by the quadcopter UAV are expressed with respect to the rotors velocities:

$$\begin{pmatrix} U_z \\ U_\phi \\ U_\theta \\ U_\psi \end{pmatrix} = \begin{pmatrix} C_L & C_L & C_L & C_L \\ 0 & dC_L & 0 & -dC_L \\ -dC_L & 0 & dC_L & 0 \\ -C_D & C_D & -C_D & C_D \end{pmatrix} \begin{pmatrix} \omega_1^2 \\ \omega_2^2 \\ \omega_3^2 \\ \omega_4^2 \end{pmatrix}, \tag{2}$$

where C_L and C_D are lift and drag force coefficients, respectively. d is the distance between the center mass of quadcopter and the axis of the propeller. ω_i is the rotor i velocity, $i = \{1, 2, 3, 4\}$. From (1d), (1e), and (1f), roll (ϕ) and pitch (θ) angles are expressed as follows:

$$\phi = \arcsin\left(\frac{\ddot{x}\sin\psi - \ddot{y}\cos\psi}{\sqrt{\ddot{x}^2 + \ddot{y}^2 + (\ddot{z} + g)^2}}\right) \tag{3a}$$

$$\theta = \arctan\left(\frac{\ddot{x}\cos\psi + \ddot{y}\sin\psi}{\ddot{z} + g}\right) \tag{3b}$$

Using (1a)–(1f), the aerodynamical roll, pitch, and heading moments and lift force, which are induced by the reference flight trajectory, are expressed as follows:

$$U_\phi = I_x\ddot{\phi} \tag{4a}$$

$$U_\theta = I_y\ddot{\theta} \tag{4b}$$

$$U_\psi = I_z\ddot{\psi} \tag{4c}$$

$$U_z = m\sqrt{\ddot{x}^2 + \ddot{y}^2 + (\ddot{z} + g)^2} \tag{4d}$$

Based on (3a) and (3b), it can be concluded that the analytical expressions of $\dot{\phi}$, $\dot{\theta}$, $\ddot{\phi}$, and $\ddot{\theta}$ are functions of the first, second, third, and fourth derivatives of the vector components (x, y, z, ψ).

3. Optimization Problem Formulation

The optimization problem is concerned with the reference flight trajectory that satisfies waypoints constraints at specific schedule flight timetable. Since the quadcopter UAV can perform six Degrees of Freedom (DOF), the reference flight trajectory could contain six components:

$$\Im(t) = (x_d(t), y_d(t), z_d(t), \phi_d(t), \theta_d(t), \psi_d(t)) \tag{5}$$

However, the quadcopter UAV is an underactuated system. Only longitudinal, lateral, and altitude displacements (x, y, z) and yaw angle (ψ) can be directly controlled. Roll (ϕ) and pitch (θ) angles are systemically computed with respect to quadcopter dynamics. Differential flatness approach has shown interesting performance regarding the trajectory planning problem as demonstrated in [4, 5, 8, 12–14, 16–18]. Using the differential flatness method, the state space variables and control inputs can be defined in terms of flat output vector. Furthermore, the computation load on optimization could be reduced considerably. Equations (1a)–(1f) are differentially flat if there exists a flat output vector $\boldsymbol{\eta} \in \mathscr{R}^m$ in the following form [23]:

$$\boldsymbol{\eta} = \zeta\left(\mathbf{x}, \mathbf{u}, \dot{\mathbf{u}}, \ddot{\mathbf{u}}, \ldots, \mathbf{u}^{(r)}\right) \tag{6}$$

such that

$$\mathbf{x} = \alpha\left(\boldsymbol{\eta}, \dot{\boldsymbol{\eta}}, \ldots, \dot{\boldsymbol{\eta}}^{(q)}\right) \tag{7a}$$

$$\mathbf{u} = \beta\left(\boldsymbol{\eta}, \dot{\boldsymbol{\eta}}, \ldots, \dot{\boldsymbol{\eta}}^{(q)}\right), \tag{7b}$$

where $\mathbf{x}$ and $\mathbf{u}$ are state space and control input vectors, respectively. From (1a)–(1f), (3a)-(3b), and (4a)–(4d), it can be concluded that the vector (x, y, z, ψ) represents the flat output $\eta(t)$ vector. Therefore, the optimization problem can be treated from the flat output $\eta(t)$ vector and expressed as NLP problem in the following form:

$$\min \quad J(\eta(t)) \quad \text{for } t \in [0, T] \tag{8a}$$

$$\text{such that:} \quad h(\eta(t)) = 0 \tag{8b}$$

$$g(\eta(t)) \leqslant 0, \tag{8c}$$

where (8b) and (8c) denote a set of equality and inequality constraints, respectively. $J(\eta(t))$ is the objective function.

3.1. Mission Requirements and Constraints. The considered constraints are classified as follows.

(a) Mission Constraints. The mission constraints are defined as a set of conditions as follows:

(i) The quadcopter UAV posture:

$$\Im(0) = \Im_{\text{init}} \tag{9a}$$

$$\Im(t_i) = \Im_i, \quad i = 1, \ldots, n_{\text{waypoint}} \tag{9b}$$

$$0 < t_1 < t_2 < \cdots < t_{n_{\text{waypoint}}} < T < T_{\text{aut}} \tag{9c}$$

$$\Im(T) = \Im_{\text{fin}}, \tag{9d}$$

where T_{aut} and T are autonomy flight and final times, respectively. n_{waypoint} is the number of waypoints.

(ii) The quadcopter UAV scheduled flight timetable:

$$t_i^* - \delta t \leqslant t_i \leqslant t_i^* + \delta t, \tag{10}$$

where t_i^* is the predefined overfly time and δt is an allowed range of overfly time.

(iii) The quadcopter UAV velocity:

$$\dot{\Im}(0) = \overrightarrow{0}, \tag{11a}$$

$$\dot{\Im}(T) = \overrightarrow{0} \tag{11b}$$

(b) Workspace and Flight Envelope Constraints

(i) The workspace limits:

$$\phi_{\min} \leqslant \phi_d(t) \leqslant \phi_{\max} \tag{12a}$$

$$\theta_{\min} \leqslant \theta_d(t) \leqslant \theta_{\max} \tag{12b}$$

$$\psi_{\min} \leqslant \psi_d(t) \leqslant \psi_{\max} \tag{12c}$$

$$x_{\min} \leqslant x_d(t) \leqslant x_{\max} \tag{12d}$$

$$y_{\min} \leqslant y_d(t) \leqslant y_{\max} \tag{12e}$$

$$z_{\min} \leqslant z_d(t) \leqslant z_{\max} \tag{12f}$$

(ii) The flight envelope limits: the flight envelope is a set of flight conditions defined by the quadcopter height and airspeed.

$$z_{\min} \leqslant z_d(t) \leqslant z_{\max} \tag{13a}$$

$$\dot{\psi}_{\min} \leqslant \dot{\psi}_d(t) \leqslant \dot{\psi}_{\max} \tag{13b}$$

$$0 \leqslant V_a \leqslant V_{\mathrm{MO}}, \tag{13c}$$

where V_{MO} is the maximum operating speed.

(c) Induced Efforts Constraints. From (2), the induced efforts can be expressed as follows:

$$0 \leqslant U_z \leqslant 4C_L\omega_{\max}^2 \tag{14a}$$

$$-dC_L\omega_{\max}^2 \leqslant U_\phi \leqslant dC_L\omega_{\max}^2 \tag{14b}$$

$$-dC_L\omega_{\max}^2 \leqslant U_\theta \leqslant dC_L\omega_{\max}^2 \tag{14c}$$

$$-2C_D\omega_{\max}^2 \leqslant U_\psi \leqslant 2C_D\omega_{\max}^2 \tag{14d}$$

Since the quadcopter rotors are identical, $\omega_{\max}$ denotes the maximum rotor velocity.

3.2. Reference Flight Trajectory Parameterization. The reference trajectory can be defined as a composition [3, 6, 15, 24] of path and motion functions. Based on this definition, the velocity, acceleration, jerk, and snap profiles of reference flight trajectory could be adjusted to make them compliant with the assigned constraints. The reference flight trajectory $\mathfrak{I}(t) = (x_d(t), y_d(t), z_d(t), \psi_d(t))$ is written as follows:

$$\mathfrak{I}(t) = P(\lambda) \circ \lambda(t) = P(\lambda(t)) \tag{15a}$$

$$\dot{\mathfrak{I}}(t) = \dot{\lambda}(t)\frac{dP}{d\lambda}(\lambda(t)) \tag{15b}$$

$$\ddot{\mathfrak{I}}(t) = \ddot{\lambda}(t)\frac{dP}{d\lambda}(\lambda(t)) + \dot{\lambda}^2(t)\frac{d^2P}{d\lambda^2}(\lambda(t)) \tag{15c}$$

$$\dddot{\mathfrak{I}}(t) = \dddot{\lambda}(t)\frac{dP}{d\lambda}(\lambda(t)) + 3\dot{\lambda}(t)\ddot{\lambda}(t)\frac{d^2P}{d\lambda^2}(\lambda(t)) + \dot{\lambda}^3(t)\frac{d^3P}{d\lambda^3}(\lambda(t)) \tag{15d}$$

$$\begin{aligned}\mathfrak{I}^{(4)}(t) = {} & \lambda^{(4)}(t)\frac{dP}{d\lambda}(\lambda(t)) \\ & + 4\dot{\lambda}(t)\dddot{\lambda}(t)\frac{d^2P}{d\lambda^2}(\lambda(t)) \\ & + 3\ddot{\lambda}^2(t)\frac{d^2P}{d\lambda^2}(\lambda(t)) \\ & + 6\dot{\lambda}^2(t)\ddot{\lambda}(t)\frac{d^3P}{d\lambda^3}(\lambda(t)) \\ & + \dot{\lambda}^4(t)\frac{d^4P}{d\lambda^4}(\lambda(t)),\end{aligned} \tag{15e}$$

where the path $P(\lambda)$ is a set of time-independent configurations followed by the quadcopter UAV to fly from an initial configuration P_{init} to a final one P_{fin}. The motion $\lambda(t)$ is a positive scalar function that defines the way in which the path is achieved with respect to the time, $\lambda \in [0,1]$. In order to construct the path $P(\lambda)$ and motion $\lambda(t)$ functions, a parametric function has to be selected. For this purpose, B-spline function is used for its advantage to support the manipulation of the reference trajectory as functions of control points and knot vectors. Basically, B-spline function is demonstrated as follows [24]:

$$S(u) = \sum_{i=1}^{N} c_i B_i^p(u) \tag{16}$$

with

$$u_0 \leq \cdots \leq u_{j-1} \leq u_j \leq u_{j+1} \leq \cdots \leq u_{n_{\mathrm{knot}}}, \tag{17}$$

where c_i is a control point and N is a control points number. u is a knot vector and n_{knot} is the number of knots. B_i^p is a basis function of degree p. For this purpose, the knot vectors are selected as uniform sequences. To construct the path and motion functions, the following conditions are defined:

(a) Path Function $P(\lambda)$

(i) Posture condition: it satisfies the condition indicated in (9a)–(9d):

$$P(\lambda = 0) = P_{\mathrm{init}} \tag{18a}$$

$$P(\lambda_i) = P_i, \quad i = 1, \ldots, n_{\mathrm{waypoint}} \tag{18b}$$

$$P(\lambda = 1) = P_{\mathrm{fin}} \tag{18c}$$

(ii) Velocity condition: it ensures the condition of (11a) and (11b):

$$\dot{P}(\lambda = 0) = \vec{0}, \tag{19a}$$

$$\dot{P}(\lambda = 1) = \vec{0} \tag{19b}$$

(c) Motion Function $\lambda(t)$

(i) The first derivative of $\lambda(t)$:

$$\frac{d\lambda}{dt} \geqslant 0, \quad \forall t \in [0,T] \tag{20}$$

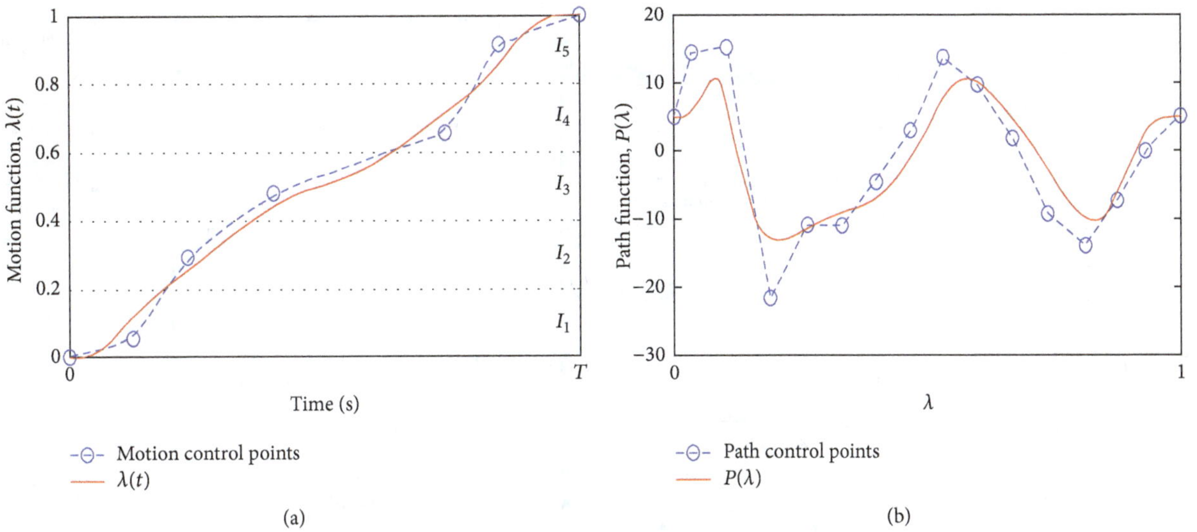

FIGURE 1: An illustration example for (a) motion $\lambda(t)$ and (b) path $P(\lambda)$ functions, respectively.

(ii) The motion must satisfy the following conditions:

$$\lambda(0) = 0 \tag{21a}$$

$$\dot{\lambda}(0) = 0 \tag{21b}$$

$$\ddot{\lambda}(0) = 0 \tag{21c}$$

$$\dddot{\lambda}(0) = 0 \tag{21d}$$

$$\lambda(T) = 1 \tag{21e}$$

$$\dot{\lambda}(T) = 0 \tag{21f}$$

$$\ddot{\lambda}(T) = 0 \tag{21g}$$

$$\dddot{\lambda}(T) = 0 \tag{21h}$$

(iii) The overfly-time conditions:

$$\lambda(t_i) = \lambda_i, \quad i = 1, \ldots, n_{\text{waypoint}} \tag{22}$$

From (3a)-(3b) and (15a)–(15e), the continuity of the second-order derivatives of the desired roll $\phi_d(t)$ and pitch $\theta_d(t)$ angles depends on the continuity of the fourth derivative of the path and motion functions. Therefore, the parametric function of $P(\lambda)$ is modeled as a quintic B-spline function using $(N_p + 2)$ control points number, where $N_p \geqslant 5$. The (N_p) control points number represents a set of path control points that will be manipulated by the proposed optimization method. The (2) control points number represents the fixed control points that satisfy the condition expressed in (18a) and (18c). The parametric function of $\lambda(t)$ is modeled as a quintic B-spline function using $(N_m + 8)$ control points number. The (N_m) control points number is a set of motion control points and the (8) control points number is a set of fixed control points that meet conditions defined by (11a)-(11b) and (21a)–(21h). In addition, the motion control points' vector is uniformly distributed along the time scale, guarantying that each motion control point belongs to an increasing bounded interval to satisfy the condition of (20). Figure 1 illustrates an example for path and monotonically increasing motion functions, respectively.

4. Performance Criterion

For rotor-craft systems, the electrical power is converted to mechanical power in the form of thrust. The developed mechanical power could be described as the product of the torque τ_i and angular velocity ω_i. Therefore, the power generated from each rotor i could be computed as follows:

$$P_i = \tau_i \omega_i = \left(C_D \omega_i^2\right) \omega_i \tag{23}$$

By using the first term $C_D\omega_i^2$ in (23), the proposed cost function is expressed as a function of rotor velocities and yaw dynamics:

$$J = \int_0^T \left[\left(\sum_{i=1}^{4} C_D \omega_i^2 \right)^2 + \ddot{\psi}^2 \right] dt \tag{24a}$$

$$= \int_0^T \left[C_D^2 \left(\omega_1^2 + \omega_2^2 + \omega_3^2 + \omega_4^2\right)^2 + \ddot{\psi}^2 \right] dt \tag{24b}$$

$$= \int_0^T \left[\beta \left(\ddot{x}^2 + \ddot{y}^2 + (\ddot{z} + g)^2\right) + \ddot{\psi}^2 \right] dt \tag{24c}$$

with

$$\beta = \left(C_D \frac{m}{C_L} \right)^2, \tag{25}$$

where the term $\ddot{\psi}^2$ is used to minimize the induced aerodynamical heading moment and the second term $\beta(\ddot{x}^2 + \ddot{y}^2 + (\ddot{z} +$

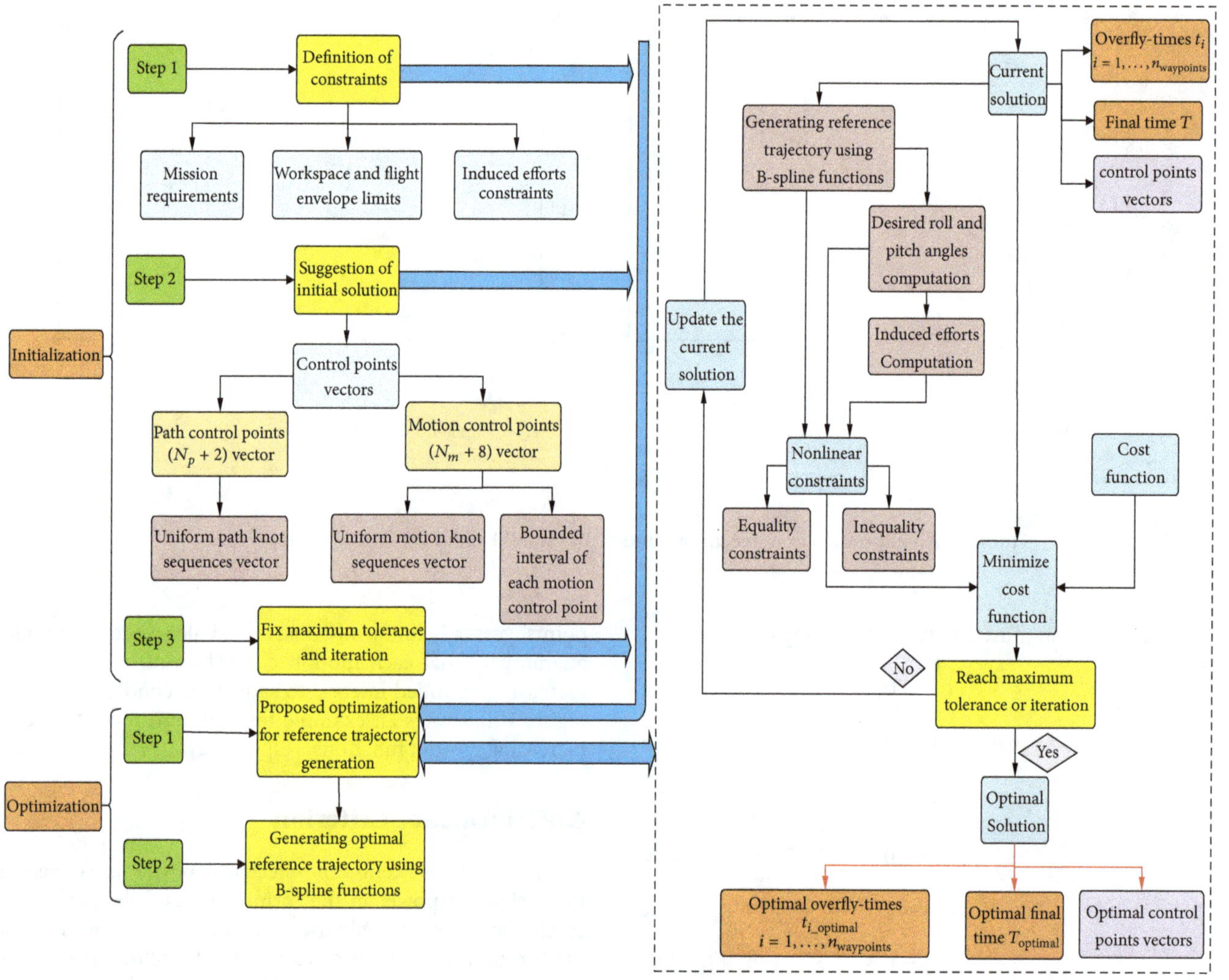

FIGURE 2: Proposed optimization method procedure.

$g)^2$) is introduced in the proposed cost function to minimize the induced efforts and consumed energy. This yields

$$J = \int_0^1 \left[\beta \left(\left(\frac{d^2 z_1}{d\lambda^2} \right)^2 + \left(\frac{d^2 z_2}{d\lambda^2} \right)^2 + \left(\frac{d^2 z_3}{d\lambda^2} + g \right)^2 \right) + \left(\frac{d^2 z_4}{d\lambda^2} \right)^2 \right] d\lambda \quad (26)$$

The synoptic procedure of optimization method is shown in Figure 2. For fixed scheduled flight timetable problem, only the positions of the control points' vectors are manipulated. The final solution will be the optimal control points' vectors. For this problem, the overfly-time constraints defined should be selected properly. For the requirements of the optimal overfly-times schedule, the overfly times are calculated by the proposed optimization problem with respect to the constraints of interest and performance criteria.

5. Numerical Simulations

5.1. Flight Trajectory 1. Let us consider the following configuration in Table 1. All simulations are done using N_m = 12 and N_p = 14. The constraints are listed in Table 3. The real parameters identified in [20] are used to test the proposed optimization method (see Table 2). A numerical solver in Matlab [25] via *"fmincon"* based upon Sequential Quadratic Programming (SQP) is used to solve the optimization method. To demonstrate the effectiveness of proposed cost function in (26), different cost functions defined in [15, 17] are used. Figure 3 displays the 2D and 3D reference flight trajectory with respect to the requirements of Table 1.

TABLE 1: Waypoints configuration with their scheduled flight timetable.

Waypoint	t (s)	x (m)	y (m)	z (m)	ψ (rad)
Initial	0	5	−5	0	0
1	10	10	0	20	−0.1745
2	25	0	15.708	20	0.3491
3	43	−10	0	20	0.5236
4	68	0	−15.708	20	0.1745
5	80	10	0	20	0
Final	95	5	5	0	0

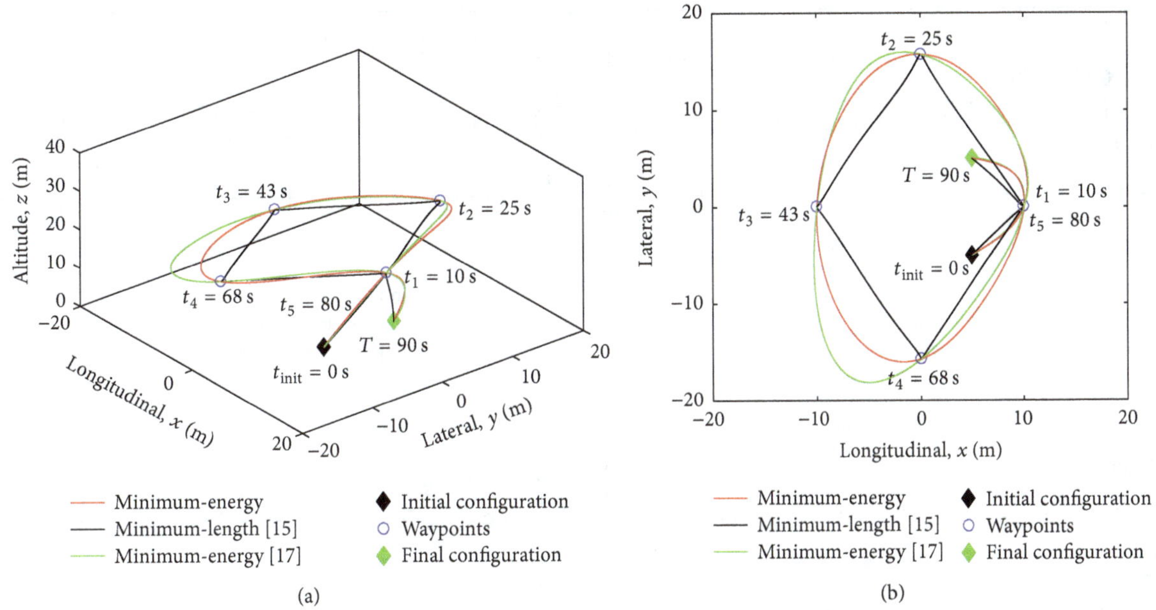

FIGURE 3: Optimal 3D and 2D reference flight trajectory.

TABLE 2: Quadcopter UAV parameters [20].

Parameter	Value
I_x	$0.0059\ \text{kg·m}^2$
I_y	$0.0059\ \text{kg·m}^2$
I_z	$0.0107\ \text{kg·m}^2$
m	0.694 kg
C_D	1.18×10^{-7} N·m/rad/s
C_L	3.8162×10^{-6} N/rad/s
d	0.18 m

TABLE 3: Considered constraints.

Constraint types	Constraints	Constraint ranges
Workspace	Longitudinal displacement (m)	$-30 \leqslant x \leqslant 30$
	Lateral displacement (m)	$-40 \leqslant y \leqslant 40$
	Altitude (m)	$0 \leqslant z \leqslant 120$
Flight envelope	Roll angle (rad)	$-0.5236 \leqslant \phi \leqslant 0.5236$
	Pitch angle (rad)	$-0.6109 \leqslant \theta \leqslant 0.6109$
	Yaw angle (rad)	$-0.7854 \leqslant \psi \leqslant 0.7854$
	Height (m)	$0 \leqslant z \leqslant 120$
	Yaw rate (rad/s)	$-1.7454 \leqslant \dot{\psi} \leqslant 1.7454$
	Airspeed (m/s)	$0 \leqslant V_a \leqslant 8$
Actuators	Maximum rotor velocity (rad/s)	$0 \leqslant \omega_{max} \leqslant 1047$

It could be noted that the proposed algorithm can deal with different cost functions. Moreover, the smoothness of the generated reference flight trajectory depends on the considered cost functions. Figure 4 shows the generated

TABLE 4: Comparison results of trajectory length and fitness value.

	Cost function type	Trajectory length (m)	Fitness value
1	Minimum-energy	129.16	1.1174×10^3
2	Minimum-length [15]	117.05	1.3609×10^4
3	Minimum-energy [17]	136.25	1.2241×10^4

longitudinal, lateral, and altitude displacements in addition to yaw angle. Figures 6 and 7 illustrate the induced efforts developed by each cost function. It could be seen from the proposed cost function in (26) that the mean values of the induced efforts are considerably less than those related to the cost functions in [15, 17]. Figure 5 displays the results given by each cost function in regards of trajectory length and consumed energy. The fitness value developed by the proposed cost function, which reflects consumed energy, is less important than other cost functions. The proposed cost function shows an interesting performance in terms of trajectory length, induced efforts, consumed energy, and smoothness, as demonstrated in Tables 4 and 5.

5.2. Flight Trajectory 2. Let us consider the configuration in Table 6. Simulations are done by considering three scenarios. In the first one, the waypoints overfly times are fixed. In the second one, the waypoints overfly times are freely computed by the proposed optimization method. However, in the last case, an allowed range of overfly time ($\delta t = 5$ s) is used.

Figure 8 displays the 3D and 2D generated reference flight trajectory for the three scenarios. Even if the imposed waypoints overfly times defined in Table 6 are fixed without any prior knowledge on optimality aspect, the proposed optimization method will show its ability to generate the reference trajectory of desired configuration despite its weakness in

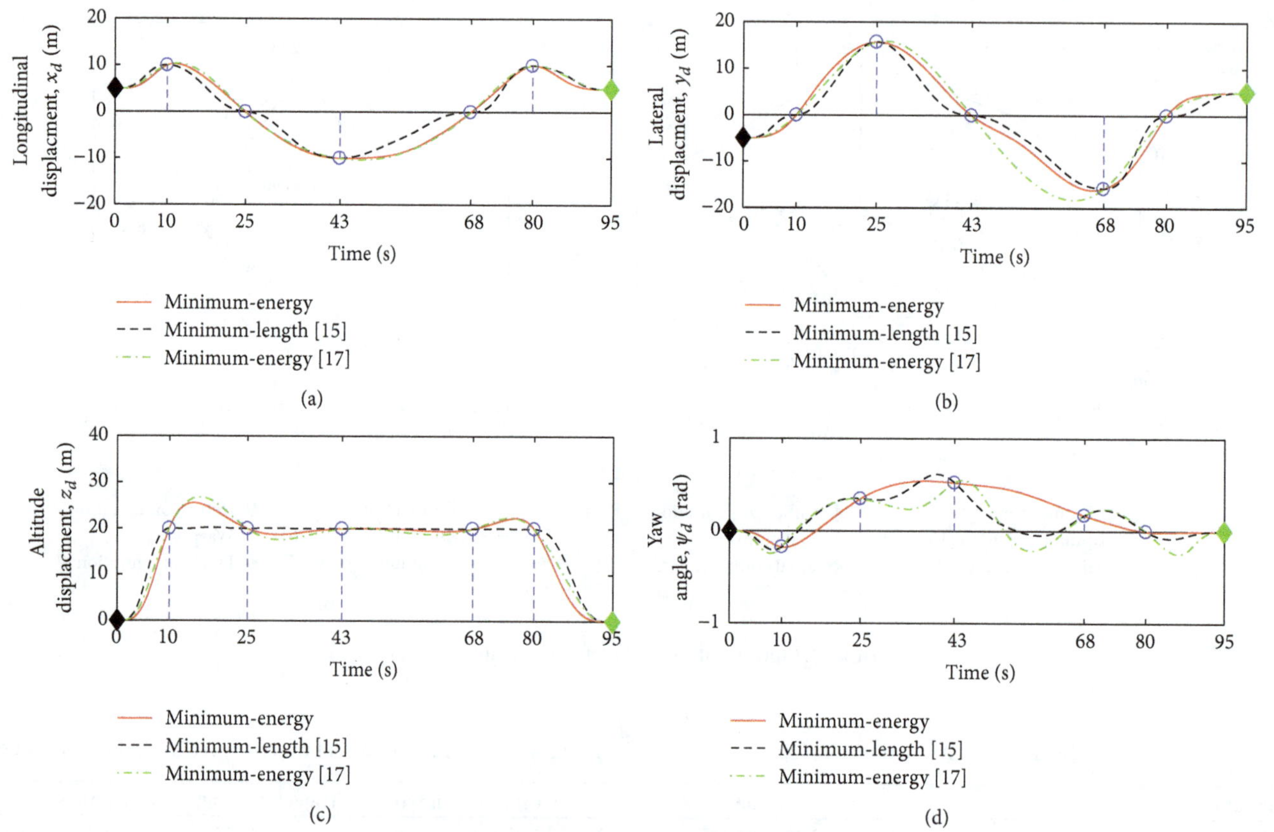

FIGURE 4: Desired (a) longitudinal, (b) lateral, and (c) altitude displacements and (d) heading angle, respectively.

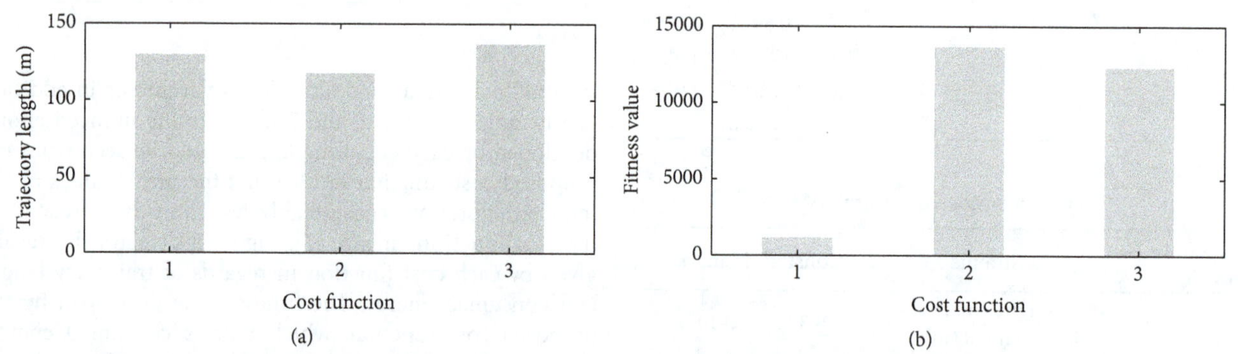

FIGURE 5: Performance criterion: (a) trajectory length and (b) fitness value, respectively.

terms of induced efforts, length trajectory, and consumed energy, as demonstrated in Figures 9 and 11. Table 8 shows the optimal overfly-times schedule computed in the second and third scenarios.

Figures 10 and 11 illustrate the induced lift force, aerodynamical moments, and their mean values computations with respect to each scenario. Based on the proposed optimization problem, the overfly times are computed optimally and consequently; the induced efforts, trajectory length, and consumed energy could be considerably reduced. Tables 7 and 9 show this aspect.

5.3. Flight Trajectory 3. Let us consider the same configuration in Table 6. In this study, different δt is used and T_{aut} = 20 min is fixed. Figure 12 displays the 3D and 2D generated reference flight trajectory and Table 10 shows the optimal overfly-times schedule computed for each selected δt. Figures 13, 14, and 15 display the trajectory length, consumed energy, and induced efforts developed by selecting δt. Tables 11 and 12 display the performance criteria assessment. It can be seen that while δt increased, the performance criteria showed enhancing results.

6. Conclusion

An optimization method for reference flight trajectory generation has been proposed, which allows the generation of feasible and realistic flight plan for quadcopter UAV and manages

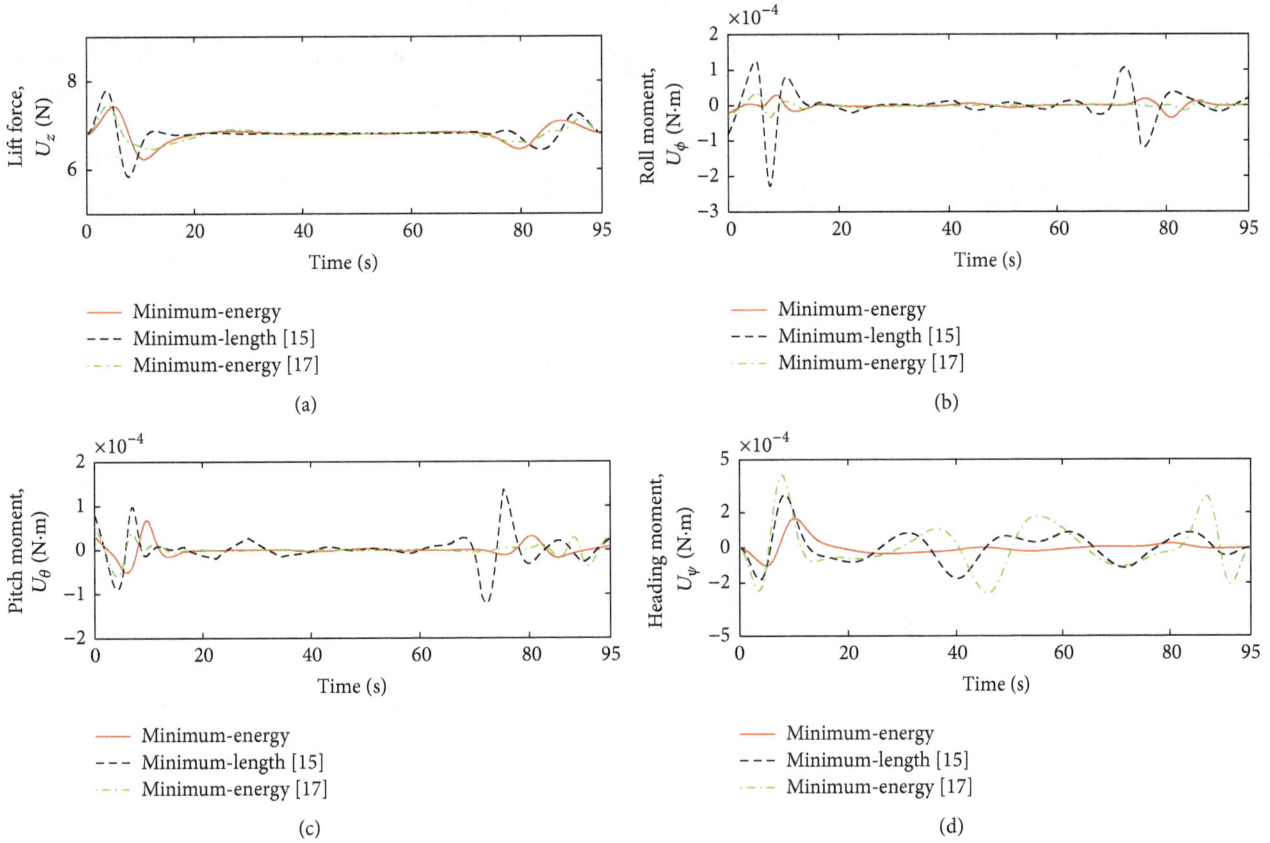

FIGURE 6: Induced efforts: (a) lift force and aerodynamical (b) roll and (c) pitch and (d) heading moments, respectively.

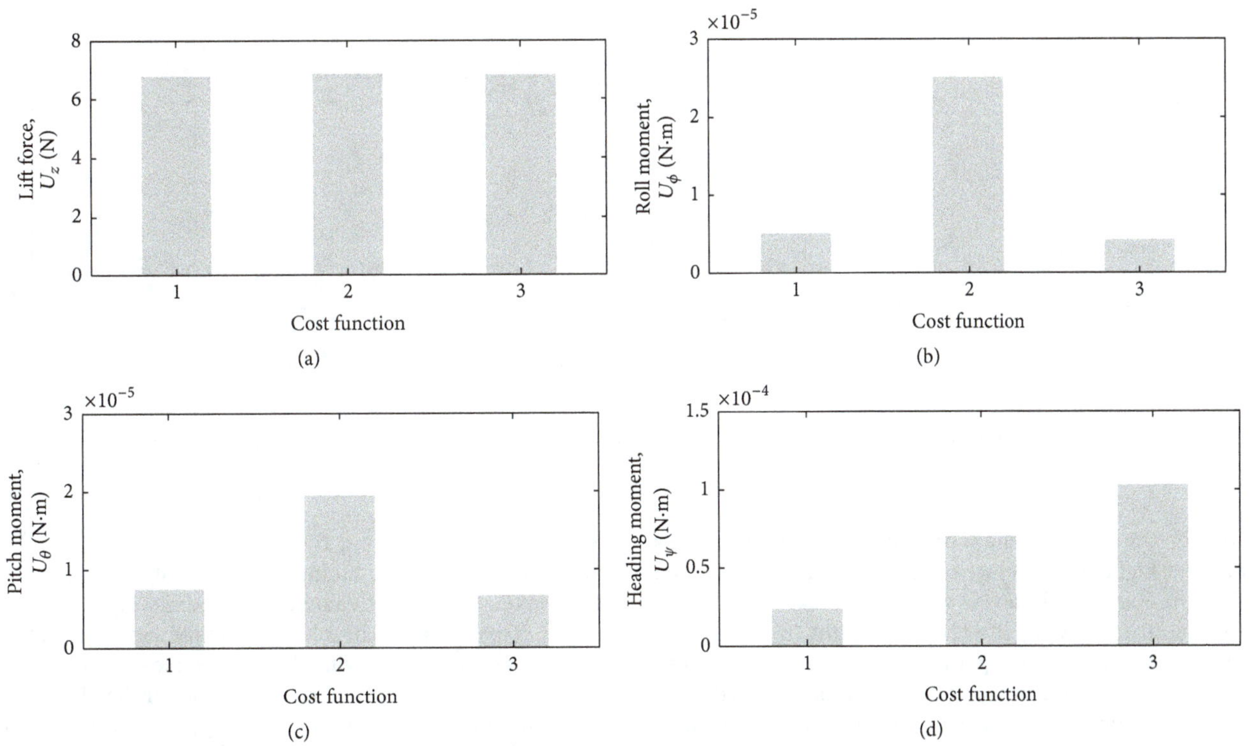

FIGURE 7: Mean values of (a) lift force and aerodynamical (b) roll and (c) pitch and (d) heading moments, respectively.

Table 5: Comparison results of induced efforts.

	Cost function type	Lift force (N)	Roll moment (N·m)	Pitch moment (N·m)	Heading moment (N·m)
1	Minimum-energy	6.8092	5.0493×10^{-6}	7.4343×10^{-6}	2.3455×10^{-5}
2	Minimum-length [15]	6.811	2.4896×10^{-5}	1.9401×10^{-5}	6.9616×10^{-5}
3	Minimum-energy [17]	6.8089	4.1299×10^{-6}	6.5995×10^{-6}	1.0246×10^{-4}

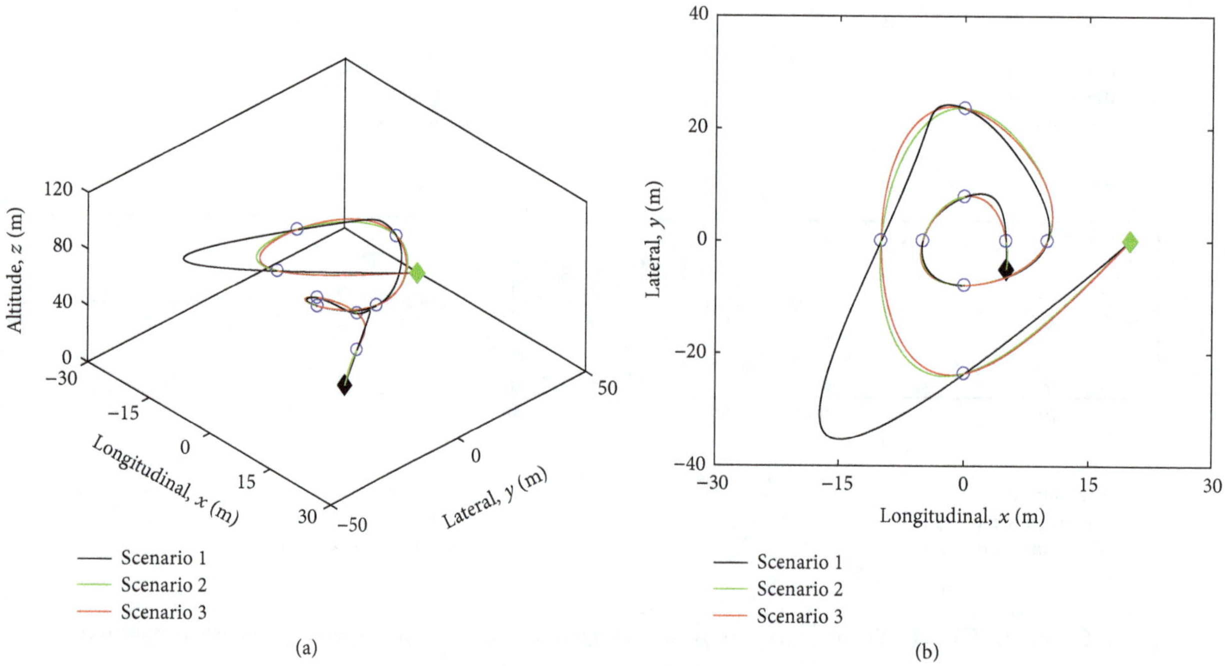

Figure 8: Optimal 3D and 2D reference flight trajectory.

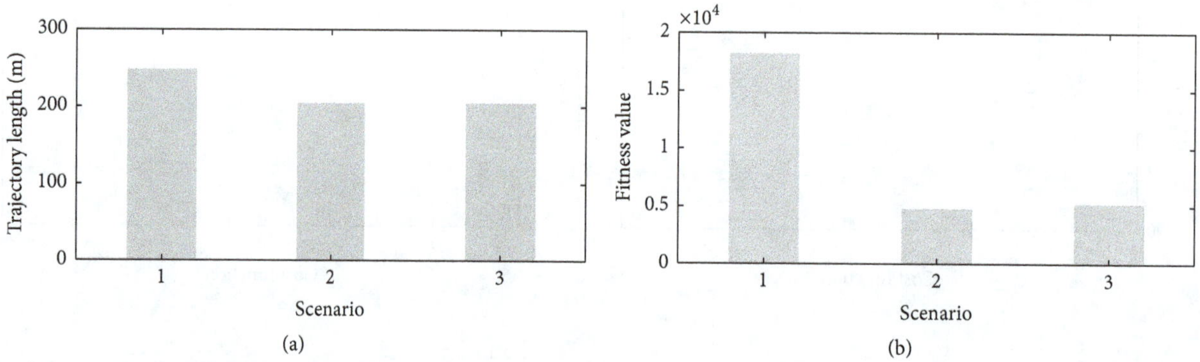

Figure 9: Performance criterion: (a) trajectory length and (b) fitness value, respectively.

adequately the scheduled flight timetable of waypoints. In this paper, the flight trajectory generation problem was treated as NLP problem and solved via a nonlinear constrained optimization technique. The reference flight trajectory was defined as a composition of path and motion functions. Both of them were generated by using B-spline functions. Based on differential flatness approach, the quadcopter UAV dynamical constraints were checked instantaneously by computing the induced aerodynamical roll, pitch, and heading moments and lift force, respectively. The optimal reference trajectory was reached by manipulating the control points' vectors of B-spline functions while taking into account the quadcopter mission requirements, dynamical constraints, flight envelope, and workspace limits. The mission requirements were defined as a series of waypoints with their respective scheduled flight timetable. A minimum-energy cost function has been proposed to minimize the consumed energy and induced efforts along the path. For requirement of optimal overfly-times schedule, the overfly times were computed in an optimal way regarding the defined constraints and performance criteria.

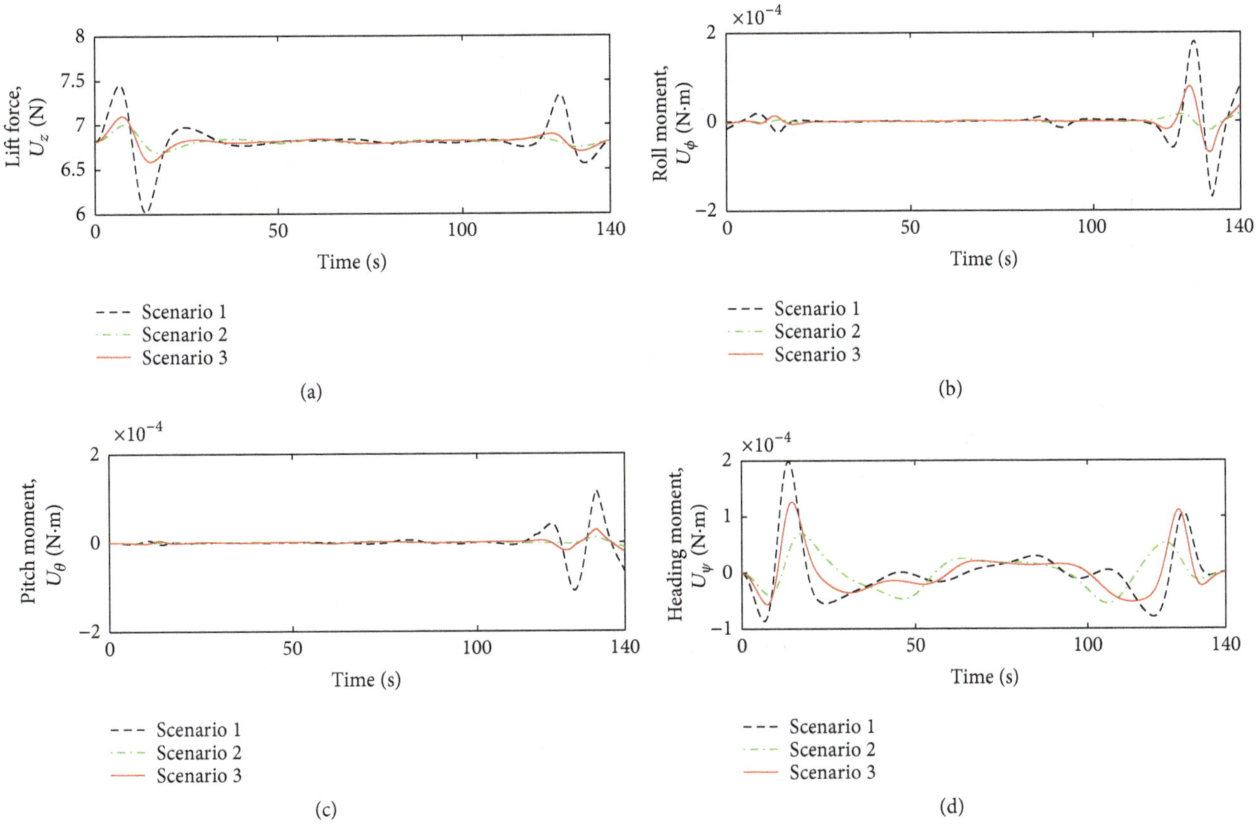

FIGURE 10: Induced efforts: (a) lift force and aerodynamical (b) roll and (c) pitch and (d) heading moments, respectively.

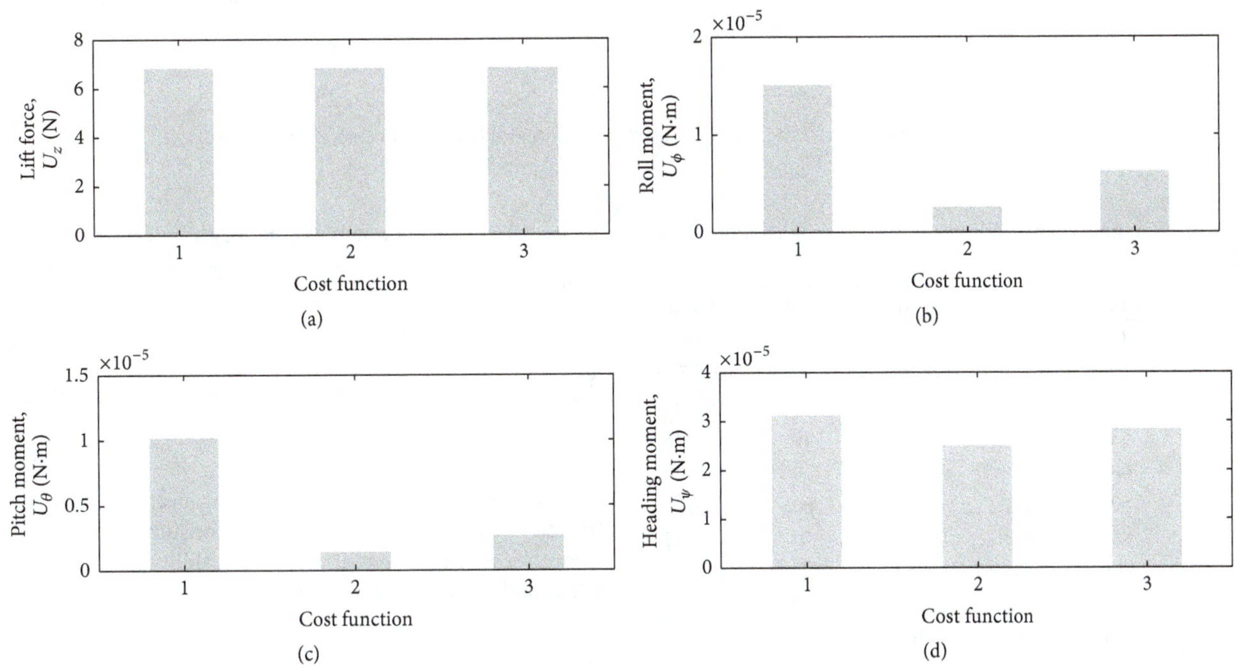

FIGURE 11: Mean values of (a) lift force and aerodynamical (b) roll and (c) pitch and (d) heading moments, respectively.

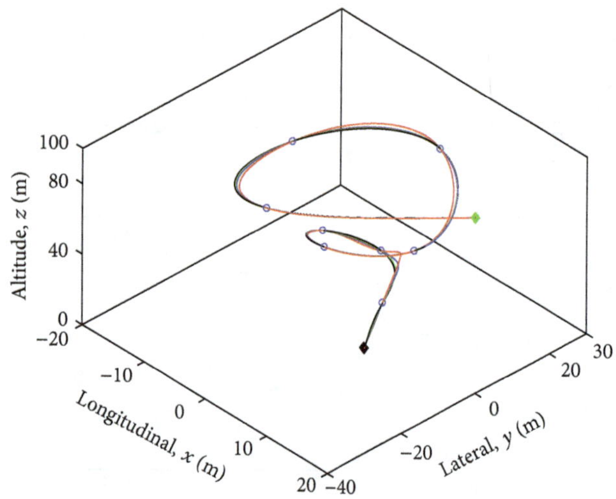

(a)

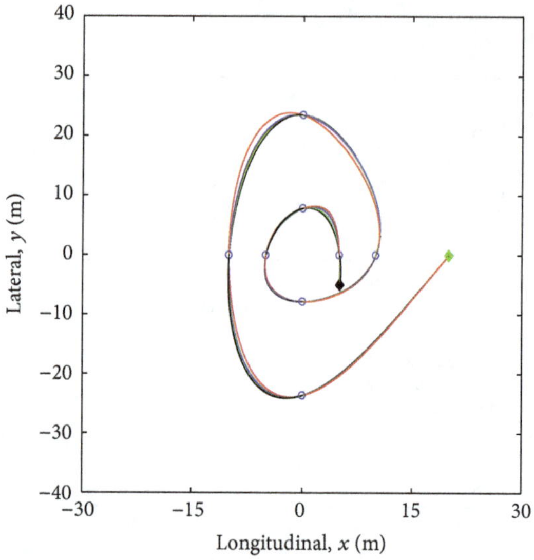

(b)

FIGURE 12: Optimal 3D and 2D reference flight trajectory.

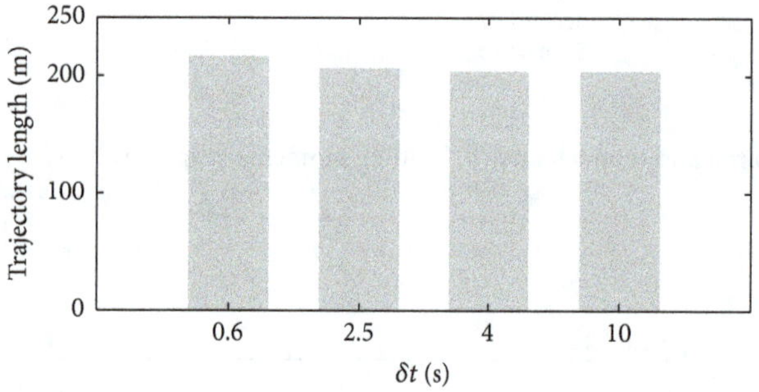

(a)

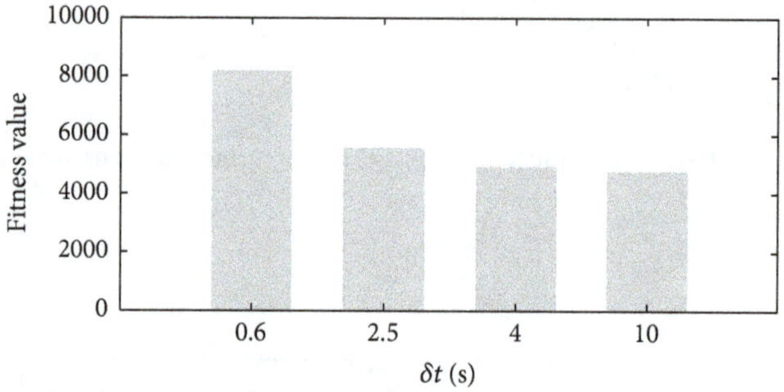

(b)

FIGURE 13: Performance criterion: (a) trajectory length and (b) fitness value, respectively.

TABLE 6: Waypoints configuration with their scheduled flight timetable.

Waypoint	t (s)	x (m)	y (m)	z (m)	ψ (rad)
Initial	0	5	−5	0	0
1	10	5	0	20	−0.1745
2	25	0	7.854	30	0.3491
3	43	−5	0	40	0.5236
4	68	0	−7.854	50	0.1745
5	80	10	0	60	0
6	95	0	23.5619	70	0.1745
7	118	−10	0	80	0.4363
8	130	0	−23.5619	90	0
Final	140	20	0	100	0

Numerical simulations in this study have shown the feasibility and effectiveness of the proposed optimization method to generate an optimal reference flight trajectory in terms of trajectory length, induced efforts, consumed energy, and smoothness. Moreover, it shows its ability to manage the scheduled flight timetable of waypoints.

TABLE 7: Comparison results of trajectory length and fitness value.

Scenario	Trajectory length (m)	Fitness value
1	247.85	1.8187×10^4
2	203.77	4.7394×10^3
3	203.95	5.1734×10^3

Conflicts of Interest

The authors declare that there are no conflicts of interest regarding the publication of this paper.

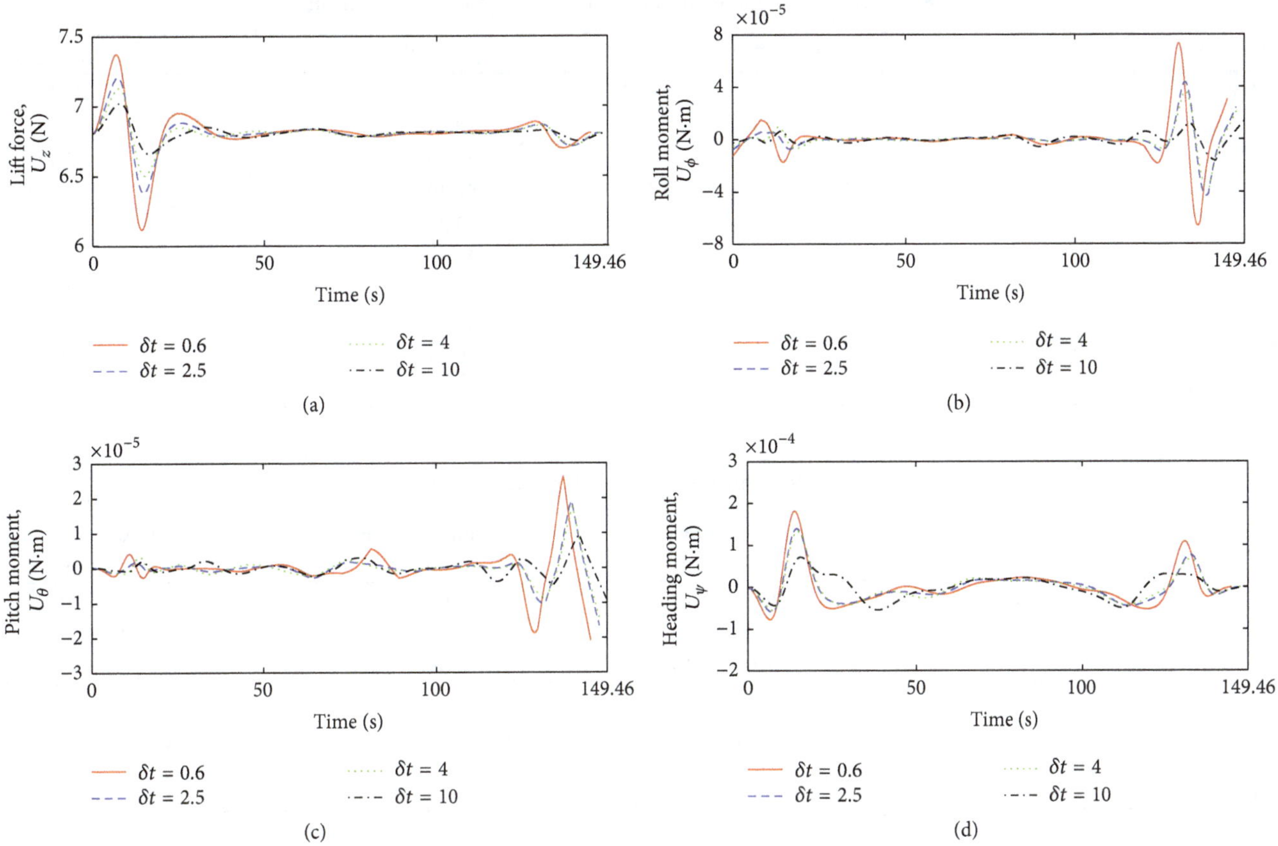

FIGURE 14: Induced efforts: (a) lift force and aerodynamical (b) roll and (c) pitch and (d) heading moments, respectively.

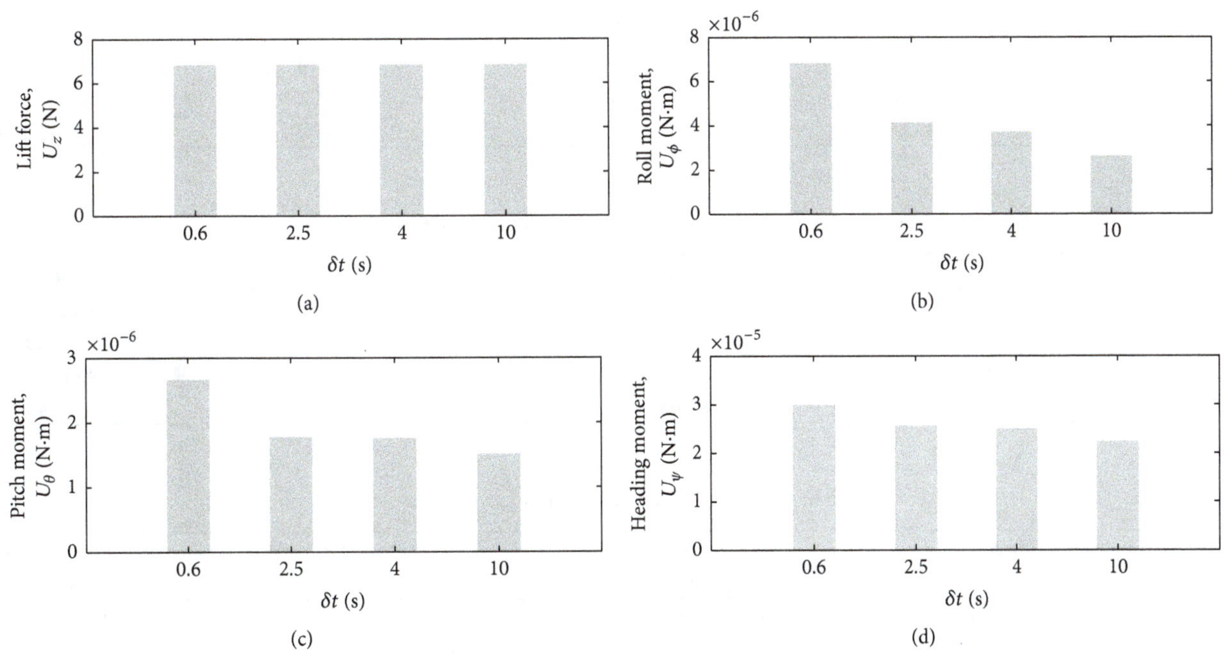

FIGURE 15: Mean values of (a) lift force and aerodynamical (b) roll and (c) pitch and (d) heading moments, respectively.

TABLE 8: Waypoints configuration with their optimal overfly-times schedule.

Waypoint	t_{scenario_1} (s)	t_{scenario_2} (s)	t_{scenario_3} (s)
Initial	0	0	0
1	10	18.44	15
2	25	34.55	28.46
3	43	47.26	45.42
4	68	60.24	63.21
5	80	71.67	75.37
6	95	89.44	94.81
7	118	105.39	113
8	130	120.76	125
Final	140	140	140

TABLE 9: Comparison results of induced efforts.

Scenario	Lift force (N)	Roll moment (N·m)	Pitch moment (N·m)	Heading moment (N·m)
1	6.8166	1.5055×10^{-5}	1.0145×10^{-5}	3.111×10^{-5}
2	6.8092	2.5521×10^{-6}	1.3739×10^{-6}	2.4893×10^{-5}
3	6.8101	6.2292×10^{-6}	2.631×10^{-6}	2.8259×10^{-5}

TABLE 10: Waypoints configuration with their optimal overfly-times schedule.

Waypoint	$t_{\delta t=0.6}$ (s)	$t_{\delta t=2.5}$ (s)	$t_{\delta t=4}$ (s)	$t_{\delta t=10}$ (s)
Initial	0	0	0	0
1	10.6	12.5	14	17.87
2	25.6	27.5	27.39	33.67
3	43.22	45.5	47	45.48
4	67.4	65.5	64	63.46
5	79.4	77.77	76	75.48
6	95.6	97.17	95.61	92.11
7	117.4	115.5	114	110.17
8	129.4	130.45	130.22	127.27
Final	144.68	146.22	147.54	149.46

TABLE 11: Comparison results of trajectory length and fitness value.

δt (s)	Trajectory length (m)	Fitness value
0.6	216.29	8.1647×10^{3}
2.5	206.21	5.5334×10^{3}
4	203.97	4.8964×10^{3}
10	203.77	4.7394×10^{3}

TABLE 12: Comparison results of induced efforts.

δt (s)	Lift force (N)	Roll moment (N·m)	Pitch moment (N·m)	Heading moment (N·m)
0.6	6.8101	6.7975×10^{-6}	2.6577×10^{-6}	2.9807×10^{-5}
2.5	6.8094	4.1188×10^{-6}	1.7692×10^{-6}	2.5535×10^{-5}
4	6.8093	3.6939×10^{-6}	1.7541×10^{-6}	2.4895×10^{-5}
10	6.8089	2.5992×10^{-6}	1.5055×10^{-6}	2.2336×10^{-5}

Acknowledgments

This paper is partly supported by the National Natural Science Foundation of China (Grants nos. 61473183 and U1509211), National Postdoctoral Program for Innovative Talents (no. BX201600103), and the China Postdoctoral Science Foundation (no. 2016M601600). The authors would like to thank Bahij Eliker, Zahia Oueldkherroubi, and Nafissa Zouad for their support and helpful feedback for drafting the paper.

References

[1] M. Huerta, *Integration of civil Unmanned Aircraft Systems (UAS) in the National Airspace System (NAS) Roadmap*, Federal Aviation Administration, 2013.

[2] L. Lai, C. Yang, and C. Wu, "Time-optimal control of a hovering quad-rotor helicopter," *Journal of Intelligent & Robotic Systems*, vol. 45, no. 2, pp. 115–135, 2006.

[3] Y. Bouktir, M. Haddad, and T. Chettibi, "Trajectory planning for a quadrotor helicopter," in *Proceedings of the 16th Mediterranean Conference on Control and Automation*, pp. 1258–1263, Ajaccio, France, June 2008.

[4] A. Chamseddine, Y. Zhang, and C.-A. Rabbath, "Adaptive trajectory planning for a quad-rotor unmanned aerial vehicle," in *Proceedings of the AIAA Guidance, Navigation, and Control Conference*, pp. 2–5, Toronto, Canada, August 2010.

[5] A. Chamseddine, T. Li, Y. Zhang, C. A. Rabbath, and D. Theilliol, "Flatness-based trajectory planning for a quadrotor unmanned aerial vehicle test-bed considering actuator and system constraints," in *Proceedings of the American Control Conference*, pp. 920–925, Montreal, Canada, June 2012.

[6] A. P. Schoellig, F. L. Mueller, and R. D'Andrea, "Optimization-based iterative learning for precise quadrocopter trajectory tracking," *Autonomous Robots*, vol. 33, no. 1-2, pp. 103–127, 2012.

[7] E. Kahale, P. Castillo, and Y. Bestaoui, "Minimum time reference trajectory generation for an autonomous quadrotor," in *Proceedings of the International Conference on Unmanned Aircraft Systems*, pp. 126–133, Orlando, Fla, USA, March 2014.

[8] M. Hehn and R. D'Andrea, "Quadrocopter trajectory generation and control," in *Proceedings of the IFAC World Congress*, vol. 44, pp. 1485–1491, 2011.

[9] M. Hehn, R. Ritz, and R. D'Andrea, "Performance benchmarking of quadrotor systems using time-optimal control," *Autonomous Robots*, vol. 33, no. 1-2, pp. 69–88, 2012.

[10] M. Hehn and R. Dandrea, "Real-Time Trajectory Generation for Quadrocopters," *IEEE Transactions on Robotics*, vol. 31, no. 4, pp. 877–892, 2015.

[11] J. Jamieson and J. Biggs, "Path planning using concatenated analytically-defined trajectories for quadrotor UAVs," *Aerospace*, vol. 2, no. 2, pp. 155–170, 2015.

[12] D. Mellinger and V. Kumar, "Minimum snap trajectory generation and control for quadrotors," in *Proceedings of the IEEE International Conference on Robotics and Automation*, pp. 2520–2525, IEEE, Shanghai, China, March 2011.

[13] D. Mellinger, N. Michael, and V. Kumar, "Trajectory generation and control for precise aggressive maneuvers with quadrotors," *International Journal of Robotics Research*, vol. 31, no. 5, pp. 664–674, 2012.

[14] J. Yu, Z. Cai, and Y. Wang, "Optimal trajectory generation of a quadrotor based on the differential flatness," in *Proceedings of the 28th Chinese Control and Decision Conference*, pp. 678–683, Yinchuan, China, March 2016.

[15] K. Eliker, H. Bouadi, and M. Haddad, "Flight planning and guidance features for an UAV Flight Management Computer," in *Proceedings of the 21st IEEE International Conference on Emerging Technologies and Factory Automation*, 6, 1 pages, Berlin, Germany, September 2016.

[16] I. D. Cowling, O. A. Yakimenko, J. F. Whidborne, and A. K. Cooke, "Direct method based control system for an autonomous quadrotor," *Journal of Intelligent & Robotic Systems*, vol. 60, no. 2, pp. 285–316, 2010.

[17] A. Chamseddine, Y. Zhang, C. A. Rabbath, C. Join, and D. Theilliol, "Flatness-based trajectory planning/replanning for a quadrotor unmanned aerial vehicle," *IEEE Transactions on Aerospace and Electronic Systems*, vol. 48, no. 4, pp. 2832–2848, 2012.

[18] S. Taamallah, X. Bombois, and P. M. J. Van den Hof, "Trajectory planning and trajectory tracking for a small-scale helicopter in autorotation," *Control Engineering Practice*, vol. 58, pp. 88–106, 2017.

[19] F. Morbidi, R. Cano, and D. Lara, "Minimum-energy path generation for a quadrotor UAV," in *Proceedings of the 2016 IEEE International Conference on Robotics and Automation*, pp. 1492–1498, Stockholm, Sweden, March 2016.

[20] M. De Lellis Costa de Oliveira, *Modeling, identification and control of a quadrotor aircraft [M.S. thesis]*, Faculty of Electrical Engineering, Prague, Czech Republic, 2011.

[21] L. Derafa, T. Madani, and A. Benallegue, "Dynamic modelling and experimental identification of four rotors helicopter parameters," in *Proceedings of the IEEE International Conference on Industrial Technology (ICIT '06)*, pp. 1834–1839, Mumbai, India, December 2006.

[22] H. Bouadi and F. Mora-Camino, "Modeling and adaptive flight control for quadrotor trajectory tracking," *Journal of Aircraft*, vol. 55, no. 2, pp. 666–681, 2018.

[23] G. G. Rigatos, *Nonlinear Control and Filtering Using Differential Flatness Approaches: Applications to Electromechanical Systems*, Springer, Cham, Switzerland, 2015.

[24] L. Biagiotti and C. Melchiorri, *Trajectory Planning for Automatic Machines and Robots*, Springer, Berlin, Germany, 2009.

[25] T. Coleman, M. A. Branch, and A. Grace, *Optimization Toolbox: for Use with Matlab*, Matlab The Mathworks Inc., 1999.

10

Minimizing the Average Delay at Intersections via Presignals and Speed Control

Mina Ghanbarikarekani,[1] **Xiaobo Qu**,[2] **Michelle Zeibots**,[3] **and Weiwei Qi**[4]

[1]*School of Civil and Environmental Engineering, University of Technology Sydney, Sydney, NSW 2007, Australia*
[2]*Department of Architecture and Civil Engineering, Chalmers University of Technology, Gothenburg, Sweden*
[3]*Institute for Sustainable Futures, University of Technology Sydney, Sydney, NSW 2007, Australia*
[4]*School of Civil Engineering and Transportation, South China University of Technology, Guangzhou 510641, China*

Correspondence should be addressed to Weiwei Qi; qwwhit@163.com

Academic Editor: David Z. W. Wang

Urban intersections have been well recognized as bottlenecks of urban transport systems. It is thus important to propose and implement strategies for increasing the efficiency of public and private transportation systems as a whole. In order to achieve this goal, an additional signal could be set up near the intersection to give priority to buses through stopping vehicles in advance of the main intersection as a presignal. It has been increasingly popular in urban cities. While presignals indeed reduce the average delay per traveler, they cause extra stops of private vehicles, which might compromise the overall efficiency, safety, and sustainability. This paper aims to propose a model to improve presignals by reducing the vehicles' number of stops behind the presignals. By applying the method, vehicles would be able to adjust their speed based on traffic conditions as well as buses' speed and approach. Numerical analyses have been conducted to determine the conditions required for implementing this method.

1. Introduction

Traffic congestion is one of the most significant issues in big cities that dissatisfy inhabitants. This matter has remarkably inevitable impacts on people's daily life in terms of travel time and health that resulted from increasing delay, traffic congestion, and air pollution [1–4]. Therefore, promising methodologies, which mitigate traffic congestion, play an important role in improving the life situation and transportation systems in main cities, specifically in their central business district [5]. Public transportation plays a vital role in decreasing traffic congestion as it has higher passenger factor than private vehicles, while it occupies less road space [6]. Hence, one of the most efficient ways that results in calming traffic congestion is developing public transit and encouraging people to use it rather than their private vehicles [7]. To serve this purpose, it is necessary to enhance the performance of public transportation systems in terms of delay and travel time.

Traffic control strategies often prioritize public transportation to motivate people not to use their private vehicles [8–10]. Presignaling for buses has been proven one of the effective strategies to reduce the average delay per passenger [11]. Presignal is a kind of traffic signal that is implemented in advance of main signalized intersections and provides priority for buses in choosing their approach behind the main intersections by giving red signal to private vehicles. Although this method has remarkably efficient effects on reducing buses' delay and travel time and increasing their speed, it has several disadvantages in traffic flow. More specifically, presignals impose additional stops to private cars that lead to the increase of their delay and travel time. In other words, using presignals enhances the functionality of buses through prioritizing them at intersections, while it does not consider private vehicles.

This study aims to give priority to buses as well as improve the performance of the private vehicles through minimizing their number of stops while reaching presignals. For the sake of alleviating the cars' number of stops, it is suggested to amend their speed based on traffic signal timing, buses approach, and traffic condition ahead. Consequently, the traffic flow of the private cars will be continuous and balanced rather than being discrete.

The rest of the paper is structured as follows. Section 2 explains background studies carried out related to the topic. Section 3 describes the methodology, the essential assumptions, and its procedure. Numerical analysis of the model, the results, and discussion are presented in Section 4. Section 5 shows the conclusion.

2. Literature Review

Presignalizing is one of the most recent innovations that have been paid attention over the last two decades. The bus presignal was defined by Peake [12] as the traffic signals implemented behind intersections for managing traffic flow and giving buses priority. The first time using presignals for prioritizing buses was proposed by Wu and Hounsell [13]. In their research, three different categories were presented including (1) implementing presignals with buses uncontrolled, (2) presignals with controlling both buses and cars, and finally (3) giving the red signal to private vehicles during arriving buses to presignal and then giving the red signal to the bus lane. It needs to be mentioned that, according to studies conducted by Wu and Hounsell [13], Kumara and Hounsell [14], He et al. [15], Xuan [16], and Xuan et al. [17], using and implementing presignals upstream of the urban intersections lead to the decrease of the discharge rate of the intersections and the waste of the intersection's green time. In other words, fewer vehicles can be discharged from presignals than the capacity and green time of the main intersection.

Kumara and Hounsell [14] proposed queue relocation to avoid wasting the main intersections' green time. Two priority methods in presignalized intersections, queue relocation and bus priority, were suggested in their research. Queue relocation keeps private vehicles at presignal stop line due to saving the green time of the main intersection. Moreover, bus priority is supplied by detectors embedded in vehicles. They indicated that presignals play a remarkable role in prioritizing buses in oversaturated intersections throughout queue relocation and bus priority.

Another solution offered to solve the problem of intersections' discharge rate is using mid-block presignals posed by Xuan et al. [17]. They demonstrated that this solution causes traffic flow to be stored efficiently between presignals and main intersections. Moreover, for increasing the intersections' discharge rate, an adaptive algorithm was proposed by He et al. [15]. This algorithm can control presignals regarding real-time demand for private and public transportation. It was achieved by accurate statistics and real-time detection. They indicated that presignals with adaptive control algorithm lead to the stimulation of the buses' use and reduction of person delay more than continuous and interrupted bus lane strategies, respectively. Increasing the intersections flow capacity is one of the significant parameters that affects the intersections efficiency fundamentally. Another considerable study carried out for increasing the intersections' capacity is Xuan's research [16]. Xuan has proposed a method in order to raise the signalized intersections flow capacity using tandem design. In this study, left-turning and through-moving vehicles were sorted throughout a mid-block presignal. Xuan conducted studies on the requirements of the blocks' length for reaching the optimal capacity and the effects of the tandem design on minimizing these requirements. It was shown that this tandem design had increased the flow capacity of cars and buses in intersections equipped with presignals.

In addition to the low discharge rate of the intersections, Wu and Hounsell [13] have explained another issue of presignals in their research and that is an extension of queue length to the upstream intersection. They have suggested vital assumptions for estimating traffic signal timing to avoid these mentioned issues.

Kejun [18] has conducted a study about prioritizing buses at a single intersection using presignal and passive priority through introducing Bus Advance Area between stop lines of the presignalized and main intersections. In addition, Kejun investigated the efficiency of presignal by simulating the studied intersection in VISSIM; consequently, it was demonstrated that although presignals increase the efficiency of the buses, they have a negative impact on the private vehicles' performance because of their additional stop behind presignals. For eliminating this impact at intersections equipped with presignal, He et al. [19] have suggested a control algorithm with an online performance. This algorithm has been modelled in microsimulation software, VISSIM. In their research, they implemented the algorithm with the bus lane, mixed lane, and presignal strategies. In the end, by comparing these methods, it was demonstrated that using presignals with the proposed algorithm not only prioritizes buses but also maintains proficiency of the private vehicles. In addition to He et al.'s studies, Guler and Menendez have played a significant role in improving the functionality of private vehicles as well as bus priority. Guler and Menendez [20] have estimated delays of cars and buses in presignalized intersections analytically using queuing theory. In their research, they have computed traveler delays in implemented presignals and allocated a lane to buses. By comparing the commuters' delay, it was concluded that presignal systems minimize the delay more than dedicated bus lanes. Consequently, implementing presignals has fundamental effects on buses due to their high capacities. In addition, Guler and Menendez [21] have presented a practical instruction about using presignals upstream of the intersections. The influences of implementing presignal on intersections comparing to other bus priority strategies pave the way to determine the conditions of applying presignals in arterials. It should be mentioned that they have proposed the presignals instruction to improve transit services and private transportation systems simultaneously. Furthermore, Guler et al. [22] have suggested an innovative strategy to prioritize buses as well as improving the functionality of cars. This strategy provides dynamic timing, for presignals, which can be activated or deactivated depending on the traffic situation. In their research, it has been considered that implementing presignals leads to increasing cars' delay in undersaturated intersections and reducing their discharging rate in oversaturated intersections. However, using the single-lane strategy can mitigate the mentioned undesirability.

3. Methodology

In this section, we first present some essential assumptions for the model. We further describe the model and its procedures in Section 3.2. In addition, the model's objective function based on the initial queue delay and the required parameters for the model are elaborated.

3.1. The Model Assumptions. In this study, there are some assumptions notified for proposing and presenting a model:

(i) Without loss of generality, we assume there are two lanes.

(ii) The intersection is controlled by a fixed-time traffic signal and equipped with presignal.

(iii) It is assumed that the studied intersection is isolated; that is, the performance of the adjacent intersections has no influence on it.

(iv) A Variable Message Sign (VMS) is installed for alarming the private cars.

(v) The Automatic Vehicle Location system (AVL) is utilized in each bus for the sake of determining the buses' location, their approach, and selected lane.

(vi) Detectors are needed throughout the area behind the stop line where buses can change their lane. These detectors also declare the existence of other cars and the bus's selected lane.

(vii) In the point VMS is installed, another detector is needed to count the private cars that occupied the distance between the VMS and the area behind the stop line.

(viii) The movement of the buses and cars is assumed static, and their speed value is constant, which is based on road rules.

3.2. The Model Procedure. The study aims to minimize the number of stops for optimizing the performance of the buses and private vehicles in urban arterials with signalized intersections equipped with presignals. Therefore, it is essential that

(i) buses be located in the first line of the queue behind the stop line; presignals are the appropriate methods for meeting this requirement by prioritizing buses behind intersections;

(ii) cars be at a moderate speed, without any stop or with the minimum stops behind presignals. To this end, it is required to minimize the initial queue behind presignals. More specifically, it would be necessary to minimize the initial queue delay [23] at presignals through the objective function proposed in (1).

Objective function:

$$F(x) = \min [d_3] \quad (1)$$

subject to

$$d_3 = \frac{1800 \times Q_b \times (1+u) \times t}{CT}$$
$$\text{If } v_{\text{car,suggested}} \geq 5\,\text{km/h} \quad (2)$$
$$\text{then } Q_b = 0,$$

where

T is the duration of analysis period (h),

Q_b is the initial queue at the start of period T (veh),

C is the adjusted lane group capacity (veh/h),

t is the duration of unmet demand in T (h), ($t = 0$ if $Q_b = 0$, otherwise $t = \min[T, Q_b/C[1-\min(1, V/C)]]$ if $Q_b \neq 0$),

u is the delay parameter ($u = 0$ if $t < T$, otherwise $u = 1 - CT/Q_b[1 - \min(1, V/C)]$ if $t \geq T$),

$v_{\text{car,suggested}}$ is the suggested speed of the car in arterials using the model (km/h).

It is required to identify some parameters in order to estimate the suggested speed of cars ($v_{\text{car,suggested}}$). These parameters would be gained through the following three steps.

Step 1 (estimate the optimal distance for implementing presignals). At this distance, buses could select their desirable lane at the main intersection. Additionally, the AVL applied in buses declare their real-time position to private vehicles in order to reduce their speeds due to prioritizing the buses and their selected lane (shown in Figure 1 as d_{bus}).

There are several factors that need to be considered for the sake of estimating this distance. These factors are based on buses movements for changing their lane and reaching the stop line of the main intersection. In this paper, the distance of implementing presignals (d_{bus}) is obtained by modifying the distance proposed by Guler and Menendez [21]. They assumed that bus stopped at the presignal so its initial speed to move towards the main intersection was zero. However, in this research, we suppose that buses do not stop at presignals and they reach the main intersection with an initial speed. Moreover, it is assumed that buses change their lane and arrive at the main intersection when traffic signal starts at red. Therefore, d_{bus} can be calculated by the following equation:

$$d_{\text{bus}} = \left[(c - r_{ms})_1 \times \left(\frac{C}{k_{\text{jam}}} - v_1 \right) + (c - r_{ms})_2 \times \left(0 - \frac{C}{k_{\text{jam}}} \right) \right] \times 1000, \quad (3)$$

where

d_{bus} is the distance between the presignal and the main intersection (m),

c is the common cycle length of the traffic signal (h),

r_{ms} is the red duration at the traffic signal (h),

C is the total capacity across all lanes at the main signal (veh/h),

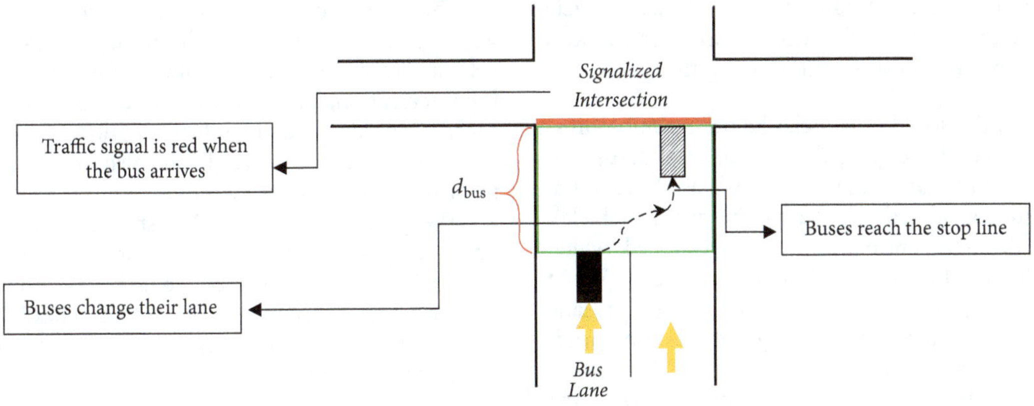

FIGURE 1: The buses' approach for changing lane and reaching the stop line.

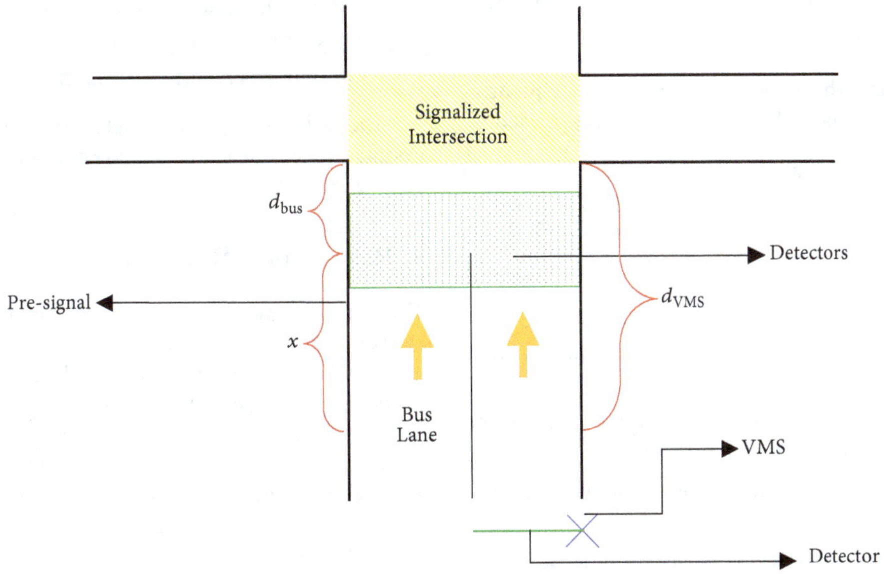

FIGURE 2: The schematic intersection components.

k_{jam} is jam density (density at zero speed) (veh/km),

C/k_{jam} is the bus moving speed for changing lane and reaching the intersection,

v_1 is the bus initial speed, which is not zero,

$(c - r_{ms})$ is the green duration of the traffic signal,

$(c - r_{ms})_1$ is the required time for buses to change the lane with initial speed of v_1,

$(c - r_{ms})_2$ is the required time for buses to reach the stop line with speed of C/k_{jam}.

Step 2 (determine the distance between cars and intersection for installing a VMS in order to give the car real-time information about its appropriate speed). The required distance lets the lead vehicle accelerate or decelerate its speed. In other words, it would be an essential distance for a car to moderate and justify its speed in the critical traffic situation. This distance (4) is related to the distance between bus and intersection, and the drivers' reaction time and their current speed. It should be considered that the distance of presignal (d_{bus}) estimated in (3) has to be added to this distance. Its details are depicted in Figure 2.

To calculate the distance:

$$x = \frac{v_{current} t_{reaction}}{3.6} \quad (4)$$

$$d_{VMS} = x + d_{bus}, \quad (5)$$

where

$v_{current}$ is the speed of the lead vehicle, which is assumed constant and based on the road rules (km/h),

$t_{reaction}$ is the reaction time of the lead vehicle once deciding to change its speed (s), which is assumed 2 s,

x is the distance between the lead vehicle and the presignal,

d_{VMS} is the essential distance between the lead vehicle and the intersection which is required for the lead vehicle in the critical situation to stop the car (m).

Step 3 (optimize the speed of vehicles before reaching presignals for the sake of reducing their number of stops). In order to minimize the cars' number of stops, speed and initial queue delay are considered. Alleviating the private vehicles' stops is done through controlling the vehicles speed, which leads to eliminating the vehicles' initial queue delay. When the bus arrives at d_{bus}, cars would be declared about its lane changing. As mentioned above, cars get the alarm earlier in d_{VMS}. Thus, cars get the chance to reduce their speed pertaining to several factors: buses travel time, speed, selected lane, d_{bus}, traffic signal timing, whether it is red or green, and the performance of the buses while crossing the intersection. Using the mentioned factors, the desired arrival time for cars is estimated. Then its speed can be calculated in accordance with the distance, d_{VMS}.

In order to estimate the vehicle's speed and consequently the following vehicles' speed, the below equations are used:

$$t_B = \max\left[t_{bus}, r_{ms}\right]$$
$$t_{bus} = \frac{3.6 d_{bus}}{v_{bus}}, \tag{6}$$

where

t_B is the bus's travel time according to traffic signal timing (s),

t_{bus} is the bus's travel time (for changing its lane) to pass the area behind the stop line (s),

d_{bus} is the distance between the presignal and the main intersection (m),

v_{bus} is the bus's mean speed of near the intersection (km/h),

r_{ms} is the duration of red time in traffic signal (s).

$$t_{car} = \frac{3.6 d_{VMS}}{v_{car,current}}, \tag{7}$$

t_{car} is the car's travel time to reach the intersection (s),

d_{VMS} is the proposed distance for installing the VMS (m),

$v_{car,current}$ is the car's mean speed in arterials (km/h).

$$t_C = t_B + t_{car}, \tag{8}$$

t_C is the car's travel time according to the leading traffic condition (s).

$$v_{car,suggested} = \frac{d_{VMS}}{t_C} \times 3.6 = v_{lead}, \tag{9}$$

$v_{car,suggested}$ is the suggested speed of the car in arterials according to the leading traffic condition, in terms of the existence of the buses changing their lanes and traffic signal timing (km/h).

The suggested speed for private vehicles estimated in (9), $v_{car,suggested}$, is the lead vehicle's speed. As a result, the private vehicles are informed of the traffic condition prior to reaching the intersection, and they are notified about the appropriate speed. Consequently, not only the number of stops for private vehicles would be reduced, but also the buses would be prioritized to select their approach.

The point is that the VMS should be installed in the optimum place to inform the private vehicles in an appropriate distance. This distance should be efficient in informing private vehicles even in the critical traffic condition. It should be considered that traffic condition is influenced by several factors which are as follows:

(i) Occupying the mentioned distance by private vehicles travel in front of the lead vehicle.

(ii) The bus, which travels in the bus lane, changes its lane in d_{bus} distance, or keeps moving in its lane.

(iii) Considering time and type of traffic signal in terms of red or green and its duration.

The value of speed depicted in the VMS needs to be updated when the traffic signal changes from red to green or vice versa.

4. Results and Discussion

The proposed model would be analyzed numerically through an example of signalized intersection equipped with a presignal system for both before and after implementing the model. In addition, vehicles' delay could be estimated in the mentioned cases, and the desirability of the model would be demonstrated by comparing the results. For the sake of analyzing this model, some assumptions have been assumed which are presented in Table 1. More specifically, the signal cycle length is 60 s, and the ratio of green time to cycle length (g/c) is 0.5. The density and capacity of the arterial are 100 veh/km and 800 veh/h, respectively. In addition, the proportion of traffic flow to capacity (V/C) is 1. The speed of buses and cars is hypothetically 20 km/h and 40 km/h before implementing the model.

The essential distance for buses to change their lane before the intersection (d_{bus}) and the required distance between the lead vehicle and the intersection for installing VMS (d_{VMS}) could be calculated using the mentioned parameters (assumed in Table 1) and the equations (proposed in the previous section). On the other hand, suggested speed and travel time of the cars would be estimated to be shown on VMS.

The most important difference between the current situation and applying the proposed model is forming a queue of vehicles behind presignals due to the red signal imposed on cars. This issue would cause initial queue delay in the case of implementing presignal without any modification. Consequently, there would be a remarkable increase in delay. In other words, implementing the model would eliminate the cars' stops behind the presignal. Therefore, the cars' delay and per person's delay related to additional stop would be alleviated. As it has been shown in Table 1, the delay has been reduced 20% through implementing the model in $V/C = 1$

TABLE 1: Assumptions for parameters used in the model.

Parameters	Values (current situation)	Values (after implementing the model)
Duration	1 h	1 h
Number of buses	12	12
Number of cars (V)	800 veh	800 veh
Bus headway	300 s	300 s
Car headway	4.5 s	4.5 s
Cycle length (c)	60 s	60 s
Red duration (r_{ms})	30 s	30 s
Capacity (C)	800 veh/h	800 veh/h
Jam density (K_{jam})	100 veh/km	100 veh/km
Initial speed of bus (V_{1b})	20 km/h	20 km/h
Initial speed of car (V_{1c})	40 km/h	40 km/h
Green duration (G)	30 s	30 s
Time for bus to change lane	15 s	15 s
Time for bus to reach the stop line	15 s	15 s
Capacity/density	8	8
d_{bus}	83.33 m	83.33 m
Distance of VMS for car (x)	- -	22.22 m
Reaction time (t)	2 s	2 s
d_{VMS}	- -	106 m
Bus travel time (t_{bus})	15 s	15 s
Suggested travel time of bus (t_B)	30 s	30 s
Car travel time (t_{car})	- -	9.5 s
Suggested travel time of car (t_C)	- -	39.5 s
Suggested speed of car ($V_{\text{car, suggested}}$)	- -	9.62 km/h
Traffic flow (V)	800 veh/h	800 veh/h
X (V/C)	1.00	1.00
Initial queue ($Q_{(b)}$)	10 veh	0 veh
Uniform delay ($d_{(1)}$)	15 s	15 s
Incremental delay ($d_{(2)}$)	64 s	64 s
Progression adjustment factor (PF)	1.667	1.667
Initial queue delay ($d_{(3)}$)	22.5 s	0 s
Delay	111.14 s	88.64 s
Travel time bus	126.14 s	103.64 s
Travel time car	150.64 s	128.14 s
g/c	0.5	0.5
Average car occupancy	2 p	2 p
Delay/person	55.57 s/pp	44.32 s/pp

and $g/c = 0.5$, and it would be an illustration for the efficient effect of the model on both cars' delay and per person's delay.

In order to conduct numerical analyses of the model, the fluctuation of g/c and V/C would be considered, and its influence on cars' delay would be determined. Figure 3 compares the amount of delay in the current situation and after implementing the model at $g/c = 0.5$ and different V/C ratios.

As is shown in Figure 3, implementing the proposed model has declined delay at all V/C ratios. It has to be mentioned that the percentage of delay improvement is decreased as the ratio of V/C increased. In other words, the reduction of delay is 63% at undersaturated condition, while it is 20% at oversaturated one.

5. Conclusion

One of the negatively influential issues regarding urban life is the traffic jam. It has negative impacts on people's life and well-being due to increasing their travel time and delay. It has to be mentioned that it would not only disturb the dwellers' planning and timetable, but also threaten individuals' health because of air pollution.

The suggested strategy to decrease traffic jam is to encourage people to use public transportation through improving its systems. One of these motivations is to give priority to this mode by implementing presignals behind the signalized intersections. Presignal is a signal that is installed in advance of the intersections and gives private cars red signal to

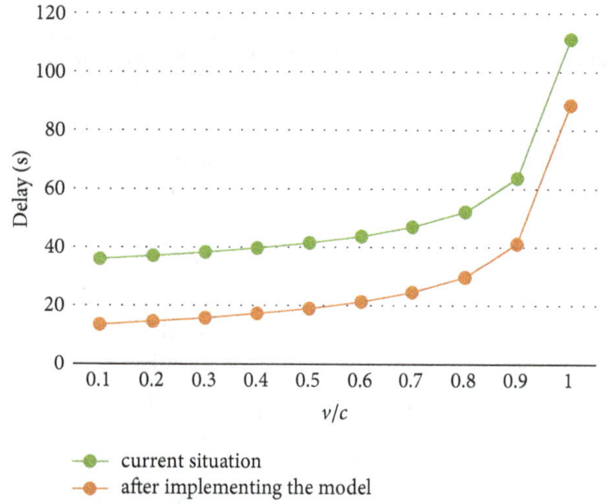

FIGURE 3: Comparison of private vehicles' delay for the current situation and the proposed presignal's model with different V/C ratios at $g/c = 0.5$.

prioritize buses in selecting their lane and approach behind the main intersection. By this way, conflicts between cars and buses while discharging the intersection are eliminated and the priority for locating in the first line of the queue and discharging the intersection is given to buses. It should be considered that as buses have much higher occupancies, the overall delay and travel time for passengers are decreased remarkably.

This strategy reduces buses' delay and travel time and increases their speed efficiently. However, cars are not taken into account in presignals, and they are given additional stops behind the presignal. More specifically, installing presignals enhance the functionality of buses by giving them priority and eliminating, or at least alleviating, the probable conflicts between them and cars. However, it worsens the performance of the private vehicles because of increasing their number of stops.

The study aims to develop this relatively modern method, presignals. In this research, it is considered to minimize the private vehicles' number of stops behind presignals. Thus, their travel time, delay, and speed are optimized. In other words, it is suggested to prioritize buses and develop cars' performance simultaneously by minimizing their number of stops behind presignals. For this purpose, it is proposed to balance the cars' speed according to buses' destination and traffic signal timing. Therefore, the delay of cars would be decreased remarkably.

Conflicts of Interest

The authors declare that they have no conflicts of interest.

Acknowledgments

The study is supported by the National Natural Science Foundation of China (no. 71701070) and the Natural Science Foundation of Guangdong Province (no. 2016A030310427).

References

[1] L. Han, D. Z. W. Wang, H. K. Lo, C. Zhu, and X. Cai, "Discrete-time day-to-day dynamic congestion pricing scheme considering multiple equilibria," *Transportation Research Part B: Methodological*, vol. 104, pp. 1–16, 2017.

[2] Y. Kuang, X. Qu, and S. Wang, "Propagation and dissipation of crash risk on saturated freeways," *Transportmetrica B*, vol. 2, no. 3, pp. 203–214, 2014.

[3] D. Z. W. Wang and B. Du, "Continuum modelling of spatial and dynamic equilibrium in a travel corridor with heterogeneous commuters-A partial differential complementarity system approach," *Transportation Research Part B: Methodological*, vol. 85, pp. 1–18, 2016.

[4] Z. Liu, S. Wang, B. Zhou, and Q. Cheng, "Robust optimization of distance-based tolls in a network considering stochastic day to day dynamics," *Transportation Research Part C: Emerging Technologies*, vol. 79, pp. 58–72, 2017.

[5] X. Qu, S. Wang, and J. Zhang, "On the fundamental diagram for freeway traffic: a novel calibration approach for single-regime models," *Transportation Research Part B: Methodological*, vol. 73, pp. 91–102, 2015.

[6] Z. Liu, Y. Yan, X. Qu, and Y. Zhang, "Bus stop-skipping scheme with random travel time," *Transportation Research Part C: Emerging Technologies*, vol. 35, pp. 46–56, 2013.

[7] A. Nayan and D. Z. W. Wang, "Optimal bus transit route packaging in a privatized contracting regime," *Transportation Research Part A: Policy and Practice*, vol. 97, pp. 146–157, 2017.

[8] Y. Bie, D. Wang, and X. Qu, "Modelling correlation degree between two adjacent signalised intersections for dynamic subarea partition," *IET Intelligent Transport Systems*, vol. 7, no. 1, pp. 28–35, 2013.

[9] Z. Liu, S. Wang, W. Chen, and Y. Zheng, "Willingness to board: a novel concept for modeling queuing up passengers," *Transportation Research Part B: Methodological*, vol. 90, pp. 70–82, 2016.

[10] Q. Meng and X. Qu, "Bus dwell time estimation at bus bays: a probabilistic approach," *Transportation Research Part C: Emerging Technologies*, vol. 36, pp. 61–71, 2013.

[11] Y. Bie, Z. Liu, and Y. Wang, "A real-time traffic control method for the intersection with pre-signals under the phase swap sorting strategy," *PLoS ONE*, vol. 12, no. 5, Article ID 0177637, 2017.

[12] M. Peake, "Getting buses to the front of the queue," in *Proceedings of the IPENZ Transportation Conference*, Queenstown , New Zealand, 2006.

[13] J. Wu and N. Hounsell, "Bus priority using pre-signals," *Transportation Research Part A: Policy and Practice*, vol. 32A, no. 8, pp. 563–583, 1998.

[14] S. S. P. Kumara and N. B. Hounsell, "Bus priority and queue relocation capabilities of pre-signals: an analytical evaluation," in *Proceedings of the 11th IFAC Symposium on Control in Transportation Systems*, pp. 391–396, Delft, The Netherlands, 2006.

[15] H. He, S. I. Guler, and M. Menendez, "Adaptive control algorithm to provide bus priority with a pre-signal," *Transportation Research Part C: Emerging Technologies*, vol. 64, pp. 28–44, 2016.

[16] Y. Xuan, *Increasing the flow capacity of signalized intersections with pre-signals: theory and case study [PhD thesis]*, University of California Berkeley, 2011.

[17] Y. Xuan, M. J. Cassidy, and C. F. Daganzo, "Using a pre-signal to increase bus- and car- carrying capacity at intersections: theory and experiment," in *TRB 2012 Annual Meeting*, 2011.

[18] L. Kejun, "Bus priority signal control at isolated intersection," in *International Conference on Intelligent Computation Technology and Automation*, vol. 2, pp. 234–237, IEEE Computer Society, Montpellier, France, 2008.

[19] H. He, S. I. Guler, and M. Menendez, "Providing bus priority using adaptive re-signals," *Online Pre-Signal Strategies TRB - Final after Reviews*, 2014, http://docs.trb.org/prp/15-2793.pdf.

[20] S. I. Guler and M. Menendez, "Analytical formulation and empirical evaluation of pre-signals for bus priority," *Transportation Research Part B: Methodological*, vol. 64, pp. 41–53, 2014.

[21] S. I. Guler and M. Menendez, "Pre-signals for bus priority: basic guidelines for implementation," *Public Transport*, vol. 7, no. 3, pp. 339–354, 2015.

[22] S. I. Guler, V. V. Gayash, and M. Menendez, "Bus priority at signalized intersections with single-lane approaches: A novel pre-signal strategy," *Transportation Research Part C: Emerging Technologies*, vol. 63, pp. 51–70, 2016.

[23] *Highway Capacity Manual: Metric Units*, Transportation Research Board, Washington, DC, USA, 2000.

11

Route Planning for Military Ground Vehicles in Road Networks under Uncertain Battlefield Environment

Tan Zhao,[1] Jincai Huang,[1] Jianmai Shi,[1,2] and Chao Chen[1]

[1]*Science and Technology on Information Systems Engineering Laboratory, National University of Defense Technology, Changsha, Hunan 410073, China*
[2]*School of Traffic and Transportation Engineering, Central South University, Changsha 410075, China*

Correspondence should be addressed to Jianmai Shi; jianmaishi@gmail.com

Academic Editor: Luigi Dell'Olio

Route planning for military ground vehicles in the uncertain battlefield is a special kind of route planning problem, as the military vehicles face a great of uncertain and unpredicted attacks. This paper models these uncertainties in the road network by a set of discrete scenarios. A kth shortest-path method is introduced to find intact routes from the origin to the destination for each vehicle. A binary integer programming is presented to formulate the problem. As the combination of the uncertainties results in a huge number of scenarios, we employed the sample average approximation method to obtain a robust solution for the problem. The solution approach is illustrated and tested through three road networks with different scales. The computational results show that, for networks of small scale, our method can provide a good solution with a sample of small size, while, for the large network, with sample of small size, this method usually leads to a suboptimal solution, but a good solution can still be obtained as the sample size grows bigger. In addition, variation trend of the deviation with different sample size indicates that a sample of larger size can bring more stability to the results.

1. Introduction

During wartime, Military Ground Vehicles (MGVs) usually have to maneuver out of their rear bases or concealed places (origins), pass through a complex road network, and arrive at the predetermined areas (destinations) to complete their operational missions, as in Figure 1. For example, mobile missile launchers depart from the military bases and move to the missile launch sites to complete the launch mission. Generally, origins are well fortified and disguised, which can ensure the safety of the MGVs. However, when the MGVs depart from the origins and move on the road network, they may be detected by the enemy's reconnaissance sensors. Because of the lack of fortification and camouflage resources, MGVs are exposed to man-made attacks during their movement in the road network. Such attacks, mainly arising from ground assault and air raid, are likely to occur frequently in battlefield. In fact, according to South Asia Terrorism Portal (SATP) [1], there were 309 attacks to the North Atlantic Treaty Organization (NATO) supply transport in Pakistan from 2008 to 2014. The number of deaths due to such attacks attains to 143. The problem of maneuvering in networks prone to attacks is prominently important due to the numerous zones that are currently at war around the world. According to the data given by the Global Security Organization (GSO) [2] there were 35 major ongoing wars and more than 25 minor conflicts around the world by the end of 2015. All these regions are highly prone to attacks where enemies perform attacks to military staff and vehicles travelling in the road networks. Thus, properly planning routes for MGVs in uncertain battlefield road networks, so as to reduce the risk of being attacked, is an important issue worth studying in the military field.

A classical problem related to MGVs' route planning is the Convoy Movement Problem (CMP) [3–5]. In the CMP, there is a set of convoys with a specific origin-and-destination (OD) pair. The objective is to find the routes for every convoy that minimize the total overall travel time while adhering to some strategic constraints [6]. Time is the most important factor in this problem since present-day military

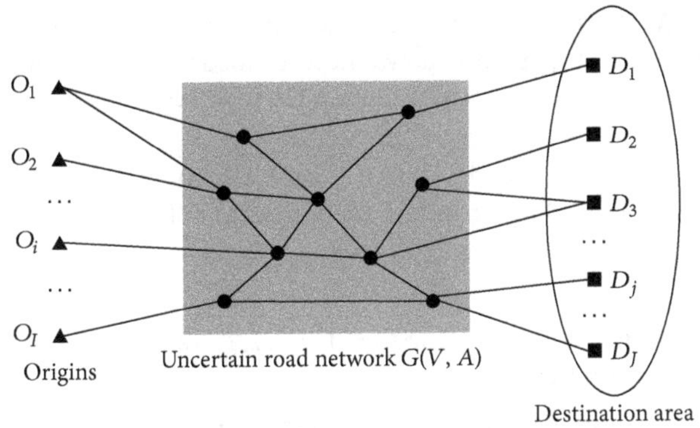

FIGURE 1: An illustration of MGVs' maneuvering.

engagements emphasize the need for mobility more than ever [3]. However, if we consider route planning associated with a mission, MGVs need not accomplish the maneuver mission as early as possible but must reach destinations before the mission deadline. Thus, in this paper we will consider the routing time of each MGV as a strong constraint rather than the objective for costs. Given the deadline of the mission, all MGVs involved in the maneuvering are requested to arrive at their destinations in a time no longer then this timeline.

Another important feature of the studied MGV routing problem is that the destination for each MGV is not deterministic and there are a number of candidate destinations for the MGVs to choose, which is quite different from the one-to-one OD pairs in CMP. Destination location is a critical decision for MGVs, which can be coordinately optimized with the routes. This is a practical requirement in military operations; for example, there usually are a set of potential launch areas for the mobile missile launchers to complete their mission and the final launch destination has to be determined according to the situation of the battlefield and the road network. Thus, the MGV routing problem includes two main parts: for each vehicle, select a destination among the candidate locations and a route among the kth-shortest paths between its origin and destination.

The emergency evacuation problem is concerned with shelter selection and route planning for evacuates, which displaces people from evacuation sources (areas suffered natural or human-made disasters) and transports them to evacuation destinations (safe shelters) [7]. The evacuation problem is related to our research in the location and routing decisions, but most of the work in this field needs not to consider the uncertain attacks during the movement in the road network. A comprehensive review of this topic can be seen in the work of Altay and Green III [8] and Galindo and Batta [9].

In the evacuation problem, there are usually a couple of candidate shelters and a given road network with deterministic delivery capacity in each road arc. The decision maker has to determine the suitable shelters and choose paths for all evacuees to optimize the evacuation cost and efficiency. Yamada [10] modeled the city road network as an undirected graph, and by solving a shortest-path problem on this graph, they obtained the optimal evacuation plan for assigning each resident in the city to one of the places for refuge. Cova and Johnson [11] focused on traffic delays occurring at intersections. To reduce these delays, they presented a network flow model for identifying optimal lane-based evacuation routing plans in a complex road network. Besides network-flow-based approach, the kth-shortest-path method is another method widely employed in the evacuation problem. Campos et al. [12] first presented this method for vehicle flow allocation in emergency transportation planning. Their goal is to send a greater number of vehicles in a minimum time outside a region under menace of some catastrophic event. Stepanov and Smith [13] applied the $M/G/c/c$ state dependent queuing models to cope with congestion and time delays on arcs. In their model, the set of feasible and potential egress routes is defined with the kth-shortest-path algorithm. He et al. [7] presented a k-shortest-path-based evacuation routing method with consideration of police resource allocated in transportation network. They proposed a nonlinear mixed integer programming to minimize overall evacuation clearance time. In general, the main purpose of the evacuation is to minimize the overall travelling time, and the uncertainty in the evacuation problem usually arises from the congestions and time delay occurring at intersections or arcs, which is different from the uncertain attacks considered in this paper.

To the best of our knowledge, existing studies rarely considered the arcs' uncertainty that arcs may be disrupted or travelling vehicles may be attacked during moving on the arc. In wartime, it is important for MGVs, such as missile launchers and movable radars, to quickly and safely arrive at their destination for completing the mission. In this work, we model the road uncertainty by a set of discrete scenarios. Each scenario, occurring with a specific probability, specifies a subset of arcs that become potentially insecure, on which MGVs travelling may be destroyed by intentional attacks. The optimization objective is to maximize the expectation that all MGVs accomplish the maneuver mission securely.

The main contributions of this paper are as follows.

We employ the k-shortest-path method to describe the scenario-based route planning problem. A binary integer programming (BIP) model is developed for the MGVs' route planning in battlefield road network, considering arcs' insecurity and the uncertainty in travelling time.

In each scenario, the instance of the problem is deterministic and can be solved by CPLEX solver [14]. To cope with the huge number of scenarios caused by uncertainties in the road network, we adapt the sample average approximation (SAA) method to address the computational complexity and propose a solution approach based on SAA.

Three road networks with different scales are presented to test the performance of the SAA-based algorithm. Sensitivity analysis on the sample size is conducted to search for a balance between efficiency and accuracy.

The rest of this paper is organized as follows. First we present the formulation of the problem in Section 2; next we introduce the SAA background and describe the solving process based on SAA in Section 3; then the computational experiments are presented and results are discussed in Section 4; the conclusion is given in the final section.

2. Model Formulation

In this section, we present the problem description in detail, the underlying assumptions, and the model formulation.

We define the problem on an undirected road network graph $G = (V, A)$, where V is the node set and A is the set of arcs. There are a number of I origins for the MGVs, and a set of J potential destinations. The total number of MGVs in all origins is W, which satisfies $I \leqslant W \leqslant J$. The security of arcs is assumed to be independent. Considering a route L, containing arcs $l_1, l_2, \ldots, l_n$, with secure probability (i.e., the probability a vehicle travels through the arc safely) $p_1, p_2, \ldots, p_n$, respectively, the secure probability of route L can be formulated as

$$P_L = \prod_{l=1}^{n} p_l. \tag{1}$$

We are supposed to select a route for each MGV so that the sum of all routes' P_L is maximized. We employ the kth-shortest-path method to help find the possible routes for each OD pair in the road network, based on which the routing problem of MGV is transformed into MGV allocation problem. Through allocation of MGVs to different kth-shortest paths of OD pairs, their route and destinations can be optimized to maximize the overall security.

2.1. Assumptions. In order to facilitate the formulation of the problem, several underlying assumptions are proposed.

There are two statuses for each arc in the road network: secure and insecure. For an arc in the secure status, MGVs can safely travel through the arc, while, for an arc in the insecure status, there is a probability that the MGV is attacked.

Each road arc is associated with a capacity limiting the maximum number of MGVs that can travel through. We consider the capacity of each arc is limited in the whole maneuvering process. Unlike the hypothesis in CMP [3] that no pair of convoys is allowed to occupy the same part of the network at the same time, we employ the vehicle hiding description presented by Yang et al. [15] as the hypothesis. In their work, they set $v_{ij}(P)$ as standardized value of the flow between node i and j in a given transport program P, and the information entropy between node i and j is expressed as

$$I\left(v_{ij}(P)\right) = -\sum_{i}\sum_{j\neq i} v_{ij}(P) \ln v_{ij}(P). \tag{2}$$

It is observed that the information entropy $I(v_{ij}(P))$ increases along with the $v_{ij}(P)$ decreasing. If all vehicles are distributed to different roads, we can obtain higher network information entropy, which also provides a better performance in elusiveness. Therefore, the capacity of each arc is assumed to be artificially limited in the entire maneuver process in our paper.

The MGVs considered are identical and are able to execute missions independently. To ensure the safety of vehicles and reduce losses during the moving process, sending vehicles to separate terminals is a bright choice. In this paper, we suppose that each destination can accommodate one vehicle at most.

As destinations are also assumed to be homogenous in this paper, the main issue we need to focus on is to select a route to an alternative destination for each vehicle.

2.2. Notations. To describe the problem more clearly, the notations used in the model are presented as follows.

Indices

i: index of origins, and $i = 1, 2, \ldots, I$, where I is the total number of origins;

j: index of alternative destination, and $j = 1, 2, \ldots, J$, where J is the total number of destinations;

k: index of kth-shortest path, and for each pair of OD $k = 1, 2, \ldots, K_{ij}$, where K_{ij} is the total number of shortest paths considered between origin i and destination j;

l: index of arc s, and $l = 1, 2, \ldots, L$, where L is the total number of arcs involved in the road network;

s: index of scenarios, and $s = 1, 2, \ldots, S$, where S is the total number of scenarios considered.

Parameters

w_i: the number of MGVs dispatched from origin i;

p_s: the probability that scenario s occurs;

q_l: the probability that arc l is in insecure state;

p_{sl}: the probability that a vehicle travels through arc l safely in scenario s, where the arc l can be either secure or insecure;

θ_l: the probability a vehicle travels through the insecure arc l safely;

a_{sl}: 1 if arc l is in secure state in scenario s, 0 otherwise;

P_{ijks}: the secure probability of the kth-shortest path from origin i to destination j in scenario s;

R_{ijk}: a set of arcs, composed of all arcs in the kth-shortest path from origin i to destination j;

t_{sl}: time consumed for travelling through arc l in scenario s, which is obtained considering distance of the arc and the vehicle velocity in the specific scenario;

T_e: time length of the mission, that is, the time allowed to finish the movement;

α_{lijk}: 1 if arc l is included in the kth-shortest path from origin i to destination j, 0 for otherwise;

λ_l: the capacity of arc l.

Decision Variable

x_{ijk}: 0-1 variable. 1 for dispatching a vehicle from origin i to destination j through the kth-shortest path, 0 for otherwise.

As each scenario specifies a subset of arcs which become insecure, given scenario s (in which a_{sl}, $\forall l$ are known), the probability that it occurs can be obtained through the following function:

$$p_s = \prod_{l\in L:a_{sl}=0} q_l \cdot \prod_{l\in L:a_{sl}=1} (1-q_l). \tag{3}$$

2.3. Formulation. We formulated the problem as follows:

$$\max \quad Z = \sum_s p_s \sum_i \sum_j \sum_k x_{ijk} P_{ijks} \tag{4}$$

$$\text{s.t.} \quad \sum_j \sum_k x_{ijk} = w_i, \quad \forall i, \tag{5}$$

$$\sum_i \sum_k x_{ijk} \le 1, \quad \forall j, \tag{6}$$

$$x_{ijk} \sum_{l\in R_{ijk}} t_{sl} \le T_e, \quad \forall i, j, k, \tag{7}$$

$$p_{sl} = \begin{cases} 1 & \text{if } a_{sl} = 1 \\ \theta_l & \text{if } a_{sl} = 0 \end{cases} \tag{8}$$

$$P_{ijks} = \prod_{l\in R_{ijk}} p_{sl} \quad \forall i, j, k \tag{9}$$

$$\sum_i \sum_j \sum_k \alpha_{lijk} x_{ijk} \le \lambda_l \quad \forall l \tag{10}$$

$$x_{ijk} \in \{0, 1\} \quad \forall i, j, k. \tag{11}$$

The objective function of the problem is to maximum the sum of expectations of all vehicles travelling securely to their destinations. Constraints (5) ensure that all vehicles at origin i are dispatched. Constraints (6) restrict that the number of vehicles in each destination should be no more than one. Constraints (7) present that in any scenario, each vehicle's moving time should be less than the time length restricted by the mission. In our paper, the time consumed in each arc, t_{sl}, is assumed to different under different scenario. Constraints (8) show that if an arc is in secure status, the MGV can pass through it safely with probability 1, and the safely travelling probability turns to be θ_l if the arc is insecure. Constraints (9) are the variants of formula (1) in an appropriate form. Constraints (10) show that the flow in each arc is limited. Constraints (11) state that decision variables, x_{ijk}, are all 0-1 variables.

It is observed that in a specified scenario, the problem is a deterministic BIP, which can be solved by CPLEX solver. If there are r arcs involved in a problem and each arc can either be in secure status or insecure status, 2^r scenarios need to be considered. Thus, the number of scenarios grows exponentially with the number of arcs. It would be a tough work to get the optimal solution with all scenarios considered. To cope with this problem, we introduce the SAA method to solve the problem in the next section.

3. Solving Methodology

In this section, we present the solution approach based on the SAA method.

3.1. Problem Description Based on SAA. The SAA method is a Monte Carlo simulation based approach to stochastic discrete optimization problems [16]. In this method, the expected value of the objective function is approximated by solving the problem for a sample of scenarios. The sample size of the SAA problem is much smaller than the number of scenarios in the *true* problem and as the sample size increases, SAA converges exponentially fast to the true problem [16]. The SAA method has been applied in many stochastic programs lacking exact algorithms, for example, the network design [17–19] and empty container repositioning [20, 21].

The SAA method repeats the sampling and solving process for several times to select a best solution. In each sample, N scenarios are randomly generated, and thus the objective function (4) of the true problem is approximated with

$$\widehat{Z}_N(x) := \max \frac{1}{N} \sum_{n=1}^{N} \sum_i \sum_j \sum_k x_{ijk} P^n_{ijk}. \tag{12}$$

For each scenario in a sample, both the time, t_l, and the secure probability, p_l, of each arc are fixed. The true problem (4)–(11) in each scenario can be rewritten as

$$\max \quad Z = \sum_i \sum_j \sum_k x_{ijk} P_{ijk} \tag{13}$$

ST (5)-(6), (10)-(11), and

$$x_{ijk} \sum_{l\in R_{ijk}} t_l \le T_0, \quad \forall i, j, k, \tag{14}$$

$$P_{ijk} = \prod_{l\in R_{ijk}} p_l, \quad \forall i, j, k. \tag{15}$$

This problem is a deterministic BIP, which can be solved by CPLEX solver. However, since the time, t_{sl}, varies from scenario to scenario, a solution suiting a scenario well may be an infeasible solution to another scenario. The solution we selected should be universally applicable. A solving process to this matter is given in Procedure 1.

First, we solve the problem of each scenario and get the optimal solutions separately. The solutions recurring for many times are thought to be of better performances than these hardly recurring, which are optimal in just some extreme scenarios. With all scenarios considered, we keep several solutions with the most recurring times. (In fact, in the experiment it is observed that the first three or four solutions consist of best solutions to more than a half of all scenarios.) From these solutions, the solution which is feasible in all scenarios and has the largest value of formula (12) is selected as the solution to the sample, denoted by x.

The process of sampling N scenarios and solving to obtain $\widehat{Z}_N$ and x_N is repeated M times. Calculate the average of the

Procedure: Selecting a solution for a sample
(1) Input the N randomly-generated scenarios, and initialize the solution $x = \emptyset$
(2) Solving the problem of each scenario with CPLEX, record the best solution to each scenario
(3) Counting the solutions, sort the diverse solutions in non-increasing order of counts
(4) Choose the first R diverse solutions (R is the number of solutions to check)
(5) For solution x_r $(r = 1, 2, \ldots, R)$, calculate the $\widehat{Z}_N(x_r)$ in formula (12)
(6) Check solution x_r in all scenarios, if x_r is not feasible overall, update $\widehat{Z}_N(x_r) = 0$
(7) Choose x_{r^*} with the biggest $\widehat{Z}_N(x_{r^*})$, and let $x = x_{r^*}$
(8) Return x

PROCEDURE 1: The solving process in a sample.

(1) Choose initial sample sizes N and N', a decision rule for determining repeating times M and a rule for increasing the sample sizes N and N'.
(2) For repetition m $(m = 1, 2, \ldots, M)$, generate a sample of size N, and solve the SAA problem to get the objective value $\widehat{Z}_N^m(x)$, and the corresponding solution x_m.
(3) Calculate the sample average $\overline{Z}_N^M$ and the corresponding deviation $\widehat{\sigma}_{\overline{Z}_N^M}$.
(4) For repetition m $(m = 1, 2, \ldots, M)$, generate a sample of size N' and calculate $\widehat{Z}_{N'}(x_m)$.
(5) Select a solution with the maximum objective value $x^* = \arg\max\{\widehat{Z}_{N'}(x) : x \in x_1, x_2, \ldots, x_M\}$, $\widehat{Z}_{N'} = \widehat{Z}_{N'}(x^*)$, and calculate the corresponding deviation $\widehat{\sigma}_{\widehat{Z}_{N'}}$.
(6) Calculate the two gaps, *gap* and $\widehat{\sigma}_{\overline{Z}_N^M + \widehat{Z}_{N'}}$.
(7) If convergence criterions are met then stop; if not, change N and N', and repeat from step (2).

ALGORITHM 1: SAA algorithm.

optimal objective function values of the M samples, namely, the value of the sample average, as

$$\overline{Z}_N^M := \frac{1}{M} \sum_{m=1}^{M} \widehat{Z}_N^m(x). \tag{16}$$

It is noteworthy that the sample size N is a tradeoff between computational complexity and accuracy. A larger value of N brings a more accurate estimate of the true problem, but an increase in computing complexity as well. The size N can be selected dynamically. The SAA solving process starts with a smaller size, and then we can regulate the sample size based on the acquired accuracy and the computational complexity.

The selected solution, given a candidate solution $\overline{x}$, for example, a feasible solution from the optimal solutions of M samples, needs to be evaluated. The evaluation of objective function is conducted by taking sample size of N' $(N' \gg N)$. That is, a number of N' scenarios are randomly generated for the evaluation. Let

$$\widehat{Z}_{N'}(\overline{x}) := \frac{1}{N'} \sum_{n=1}^{N'} \sum_{i} \sum_{j} \sum_{k} \overline{x}_{ijk} P_{ijk}^{n}. \tag{17}$$

Since N' is much larger than N, the objective value obtained $\widehat{Z}_{N'}(\overline{x})$ is considered as an estimator of the true objective function (3) [19]. In a common SAA problem, the quality of the solution is evaluated by calculating the optimality gap and the variance of optimality gap. In our problem, as the involved expectations are all numbers around 1, to highlight the dispersion of the result, we replace the variance of optimality gap with the deviation of optimality gap. Here, we define the optimality gap as

$$\text{gap} := \overline{Z}_N^M - \widehat{Z}_{N'}(\overline{x}). \tag{18}$$

And we define the deviation of optimality gap as

$$\widehat{\sigma}_{\overline{Z}_N^M + \widehat{Z}_{N'}} := \widehat{\sigma}_{\overline{Z}_N^M} + \widehat{\sigma}_{\widehat{Z}_{N'}}, \tag{19}$$

where

$$\begin{aligned} \widehat{\sigma}_{\overline{Z}_N^M} &:= \sqrt{\frac{1}{M(M-1)} \sum_{m=1}^{M} \left(\widehat{Z}_N^m(x) - \overline{Z}_N^M\right)^2}, \\ \widehat{\sigma}_{\widehat{Z}_{N'}} &:= \sqrt{\frac{1}{N'(N'-1)} \sum_{n=1}^{N'} \left(Z_n(\overline{x}) - \widehat{Z}_{N'}(\overline{x})\right)^2}. \end{aligned} \tag{20}$$

3.2. SAA Algorithm. Based on description above, the SAA algorithm process is presented in Algorithm 1.

4. Experiment Design and Computational Results Analysis

4.1. Experiment Design. In this part, we examine the performance of the algorithm on different dataset. We selected three road networks: a small-scale designed network with 13 nodes

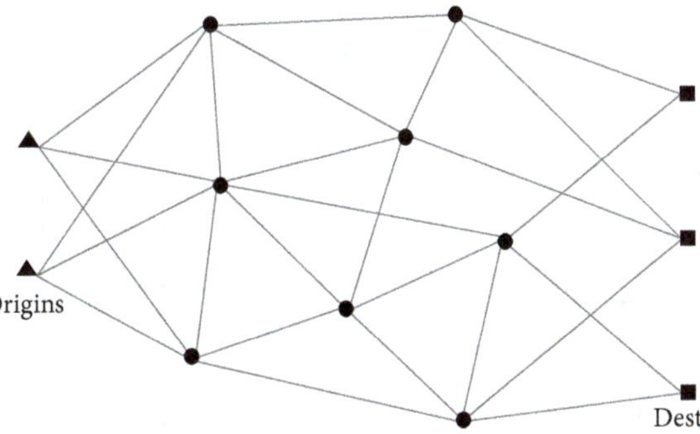

FIGURE 2: Network 1: the small-scale random network.

TABLE 1: The problems to be solved in the 3 networks.

Number	Network	Amount of nodes	Involved nodes	Involved arcs	Number of origins (I)	Number of destinations (J)	Number of MGVs dispatched (w_i)	Road capacity (λ_l)
(1)	Random	13	13	27	2	3	$[1, 1]^a$	2
(2)	Changsha City	46	37	57	3	5	$[1, 2, 1]$	2
(3)	Fujian Province	226	112	147	3	15	$[2, 3, 2]$	2

[a]Values in square brackets stand for the number of vehicles dispatched from the origins separately.

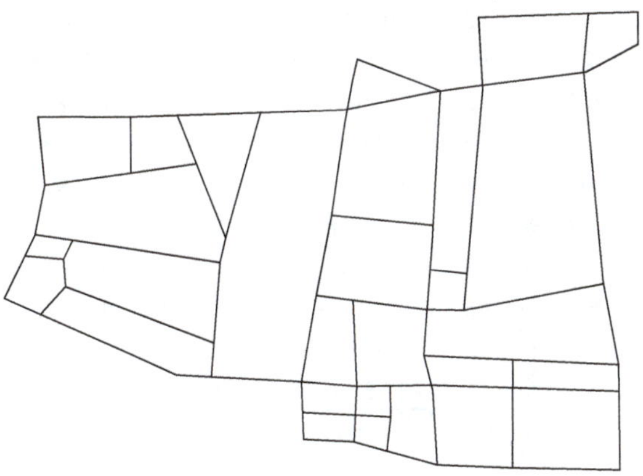

FIGURE 3: Network 2: a network from Changsha, China.

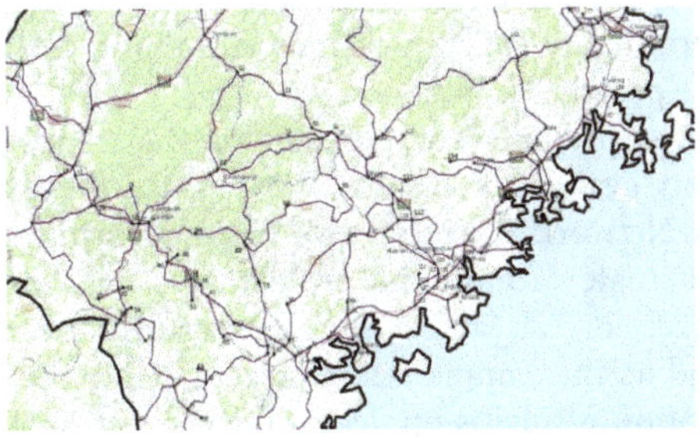

FIGURE 4: Network 3: the arterial traffic network of Fujian Province of China.

and 27 arcs (Figure 2), a middle scale practical network with 46 nodes and 77 arcs in Hunan Province of China (Figure 3), and a large scale practical network with 226 node and 239 arcs in southeast of China (Figure 4). The lengths of the arcs in Network 1 were randomly generated, while in Network 2 and Network 3 the lengths were calculated by the commercial software, ArcGIS, based on the practical data of the two regions.

The kth-shortest-path datasets were derived by the methods in [22–24]. Other key parameters are given as follows:

> θ is a constant and is set to be 0.2 in all experiments;
>
> the value of t_{sl} is set to be the road length divided by the velocity, which varies in a range around ±10% of a given velocity as the scenario varies;
>
> q_l is randomly generated.

The data for all the experiments can be found in the supplementary material noted as dataset. Table 1 shows a brief look of the problems considered in 3 networks, respectively. For the fulfillment of SAA, the value of M is set to be 10 and keeps constant for all sample sizes. The sample size N ranges from 100 through 1000, with an interval of 100. And the value of N' is set to be 10,000. The algorithms were coded in MATLAB by invoking CPLEX v.12.6. And all the programs were executed on a 64-bit computer with a 3.2 GHz CPU and 8.0 GB of physical RAM.

4.2. Computational Results and Analysis. The computational results for the three networks are presented in Tables 2, 3, and 4, respectively. We give the minimum, maximum, average, and corresponding deviation of each sample size, as the results of the SAA process. The calculating time of each SAA process is also listed as the CPU time. It can be seen that the calculating time increases as the sample size grows, where the growth rate keeps almost constant for each network separately. This is an acceptable result, since that we viewed the working time of each module of the program and found the scenario generating and CPLEX solving parts, both of

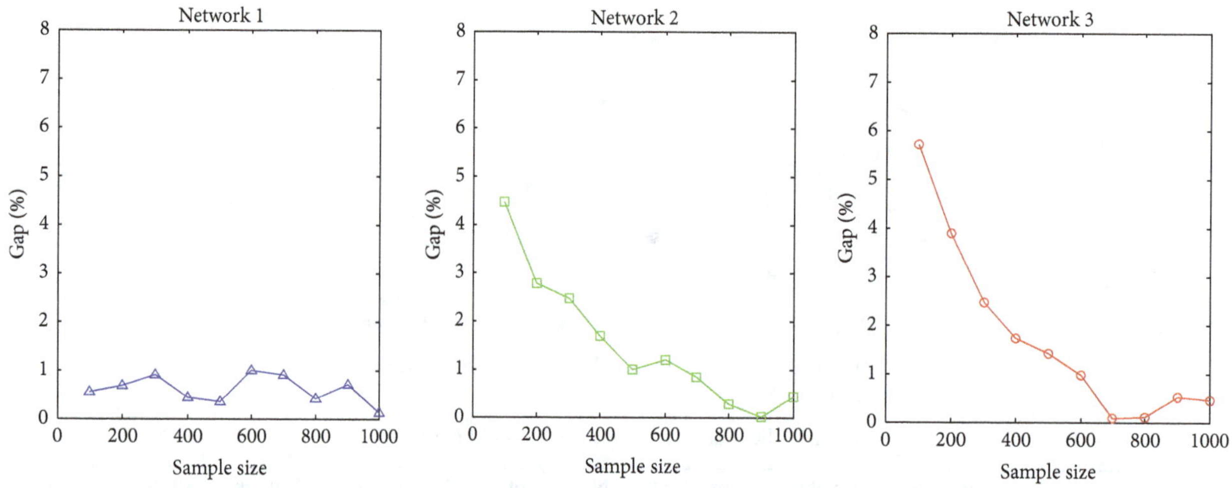

FIGURE 5: The line graph of sample size and *gap* (%).

TABLE 2: SAA results for test Network 1.

Sample Size (N)	SAA process					Evaluation		Optimality gap		
	$Z_{\min}$	$Z_{\max}$	$\overline{Z}_N^M$	$\widehat{\sigma}_{\overline{Z}_N^M}$	CPU time[a] (s)	$\widehat{Z}_{N'}(\overline{x})$	$\widehat{\sigma}_{\widehat{Z}_{N'}}$	*gap*	*gap* (%)[b]	$\widehat{\sigma}_{\overline{Z}_N^M+\widehat{Z}_{N'}}$
100	0.532	0.5864	0.5579	$5.3275e-03$	0.83	0.5548	$6.7234e-03$	$3.0553e-03$	0.5477	$1.2051e-02$
200	0.5456	0.5731	0.5586	$3.8670e-03$	1.54	0.5548	$6.7919e-03$	$3.7395e-03$	0.6695	$1.0659e-02$
300	0.5305	0.5783	0.5498	$5.4558e-03$	2.28	0.5548	$6.7253e-03$	$-4.983e-03$	0.9063	$1.2181e-02$
400	0.5438	0.5589	0.5525	$1.7122e-03$	3.01	0.5550	$6.7581e-03$	$-2.4441e-03$	0.4423	$8.4703e-03$
500	0.5450	0.5684	0.5567	$3.0237e-03$	3.76	0.5548	$6.836e-03$	$1.967e-03$	0.3533	$9.8597e-03$
600	0.5397	0.5595	0.5495	$2.0991e-03$	4.49	0.5549	$6.8159e-03$	$-5.4032e-03$	0.9833	$8.9149e-03$
700	0.5493	0.5695	0.5600	$2.6198e-03$	5.20	0.5549	$6.8249e-03$	$5.0636e-03$	0.9043	$9.4447e-03$
800	0.5501	0.5656	0.5565	$1.7425e-03$	5.94	0.5542	$6.7327e-03$	$2.2589e-03$	0.4059	$8.4752e-03$
900	0.5497	0.5651	0.5588	$1.3669e-03$	6.73	0.5550	$6.8224e-03$	$3.9135e-03$	0.7003	$8.1893e-03$
1000	0.5417	0.5630	0.5542	$2.5303e-03$	7.39	0.5549	$6.6988e-03$	$-6.5813e-03$	0.1187	$9.2291e-03$

[a]The CPU time presents the total time cost in the SAA sampling and solving process. [b]gap (%) = $(|\text{gap}|/\overline{Z}_N^M) * 100\%$.

which require repeating N (the value of sample size) times and take the most of the time (more than 95% of the whole). The evaluation part shows the results of employing the selected solution of the SAA process to a much bigger sample of size N'. And the gaps between these results are given in the optimality gap part. Several discussions about the optimality gap and the deviations are given as in Tables 2, 3, and 4.

Figure 5 shows the trends of *gap* (%) for the 3 networks while the sample size increases. Here the index, *gap* (%), means the closeness between the SAA results and evaluation results, which can be viewed as the estimator of the true problem. For Network 1, the *gaps* (%) are all below 1% and fluctuate irregularly, while, for Network 2 and Network 3, the values start from 4.46% and 5.69%, respectively, as the sample size is 100 and decrease distinctly as the sample size increases. Although there are several fluctuations in both of them, the downtrends are obvious. These differences between Network 1 and Network 2/3 arise from the solutions selected in each network. In fact, the solution selected for Network 1 stays constant from sample size 100 through 1000. The fluctuation is caused by the uncertainty of sampling. However, the solutions selected for Network 2 and Network 3 vary as the sample size increases. Better solution brings smaller optimality gap, which indicates that better solution to the true problem is obtained.

Figure 6 presents comparison for the trends of SAA results and evaluation results. We can see the evaluation results of Network 1 stay relative constant, while the SAA results fluctuate around the evaluation results in a quite small range. For Network 2, the SAA results have a distinct downtrend as the sample size increases, while the trend of the evaluation results is opposite. The gap between the two values is first below 1% when sample size is 500, after which the SAA results fluctuate around the evaluation results. The smallest gap between the two values reaches 0.013% when sample size is 900. The discussion for Network 2 can be applied to Network 3 as well, in which these trends are more obvious. The SAA results rise and evaluation results fall distinctly as the sample size increases from 100 through 600. From 600 through 1000, the SAA results fluctuate. And the

TABLE 3: SAA results for test Network 2.

Sample size (N)	SAA process					Evaluation		Optimality gap		
	$Z_{\min}$	$Z_{\max}$	$\overline{Z}_N^M$	$\hat{\sigma}_{\overline{Z}_N^M}$	CPU time (s)	$\hat{Z}_{N'}(\overline{x})$	$\hat{\sigma}_{\hat{Z}_{N'}}$	*gap*	*gap* (%)	$\hat{\sigma}_{\overline{Z}_N^M+\hat{Z}_{N'}}$
100	0.6988	0.7319	0.7175	3.7930e − 03	3.2	0.6855	1.5881e − 02	3.1969e − 02	4.4557	1.9674e − 02
200	0.6934	0.7197	0.7058	2.9367e − 03	5.97	0.6862	1.5822e − 02	1.9587e − 02	2.7752	1.8760e − 02
300	0.6943	0.7108	0.7035	1.6460e − 03	8.65	0.6862	1.5815e − 02	1.7336e − 02	2.4642	1.7461e − 02
400	0.6923	0.7169	0.7036	2.8679e − 03	11.45	0.6918	1.2582e − 02	1.1795e − 02	1.6764	1.5449e − 02
500	0.6872	0.712	0.6996	2.5119e − 03	14.17	0.6927	1.2499e − 02	6.9041e − 03	0.9869	1.5011e − 02
600	0.6929	0.7099	0.7013	1.6643e − 03	17.03	0.6929	1.2475e − 02	8.3798e − 03	1.195	1.4140e − 02
700	0.6922	0.7033	0.6998	1.2746e − 03	19.75	0.694	1.2445e − 02	5.8449e − 03	0.8352	1.3719e − 02
800	0.7004	0.7077	0.7035	7.1520e − 04	22.41	0.7015	1.0357e − 02	2.0178e − 03	0.2868	1.1072e − 02
900	0.6947	0.7098	0.7016	1.9050e − 03	25.22	0.7015	1.0322e − 02	9.2599e − 05	0.0132	1.2227e − 02
1000	0.6959	0.7039	0.6995	9.1399e − 04	27.99	0.7025	1.0217e − 02	−2.9942e − 03	0.4281	1.1131e − 02

TABLE 4: SAA results for test Network 3.

Sample size (N)	SAA process					Evaluation		Optimality gap		
	$Z_{\min}$	$Z_{\max}$	$\overline{Z}_N^M$	$\hat{\sigma}_{\overline{Z}_N^M}$	CPU time (s)	$\hat{Z}_{N'}(\overline{x})$	$\hat{\sigma}_{\hat{Z}_{N'}}$	*gap*	*gap* (%)	$\hat{\sigma}_{\overline{Z}_N^M+\hat{Z}_{N'}}$
100	1.0532	1.1621	1.0922	1.0115e − 02	36.02	1.0301	2.2434e − 02	6.2161e − 02	5.6911	3.255e − 02
200	1.0519	1.1191	1.0744	7.0514e − 03	64.02	1.0328	2.2288e − 02	4.1606e − 02	3.8724	2.934e − 02
300	1.034	1.1179	1.0641	7.8205e − 03	92.09	1.038	1.8193e − 02	2.6097e − 02	2.4525	2.6013e − 02
400	1.0535	1.1072	1.0683	4.8197e − 03	120.2	1.0499	1.6410e − 02	1.8438e − 02	1.7259	2.1229e − 02
500	1.0517	1.07	1.0626	2.2199e − 03	148.23	1.0476	1.6670e − 02	1.5009e − 02	1.4125	1.889e − 02
600	1.0344	1.0796	1.047	4.2979e − 03	176.4	1.0571	1.3658e − 02	−1.009e − 02	0.9638	1.7956e − 02
700	1.0381	1.0881	1.0628	5.6404e − 03	204.6	1.0619	1.3634e − 02	8.4620e − 04	0.0796	1.9275e − 02
800	1.0412	1.0829	1.0546	4.0003e − 03	232.48	1.0557	1.3209e − 02	−1.1158e − 03	0.1058	1.7209e − 02
900	1.0439	1.0745	1.054	2.7739e − 03	260.69	1.0595	1.3410e − 02	−5.4381e − 03	0.5159	1.6184e − 02
1000	1.0432	1.0601	1.0535	1.7549e − 03	288.78	1.0584	1.3530e − 02	−4.8823e − 03	0.4634	1.5284e − 02

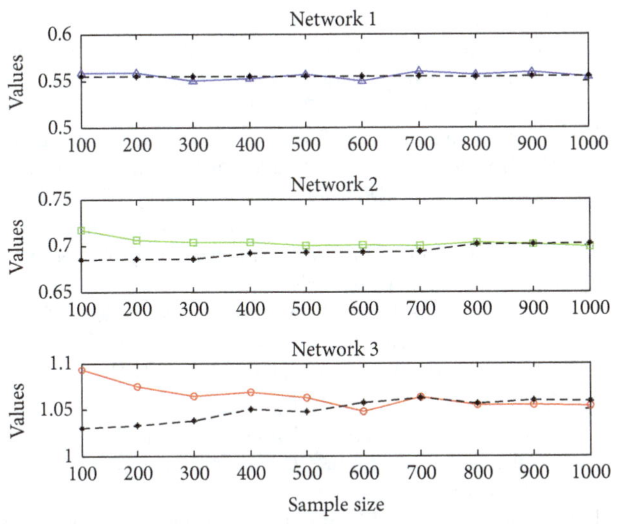

FIGURE 6: The comparison between SAA results (the colored solid lines) and evaluation results (the black dashed lines) with sample size varying.

smallest gap between the two is 0.08%, when the sample size is 700.

The results of Network 1 indicate that, for a network of small scale, a good solution can be obtained with a considerably small sample size. But for networks of bigger scales, such as Network 2 and Network 3, as the number of arcs considered in the problem increases, the number of possible scenarios increases. A sample of small size cannot reflect the panorama of the problem. In our algorithm, 10 samples are generated and solved, and the sample solution with the biggest expectation value is selected, while, for a sample of small size, a suboptimal solution may have a better performance than the "good" solution (as we cannot tell whether it is the optimal one) does. This explains why for networks of bigger scales, suboptimal solutions are usually selected when the sample size is small. But if the sample size is big enough, the good solutions can still be obtained constantly.

Figure 7 presents the deviation trends with the sample size increasing. All the deviations of the three networks have downward trends, which are especially evident in Network 2

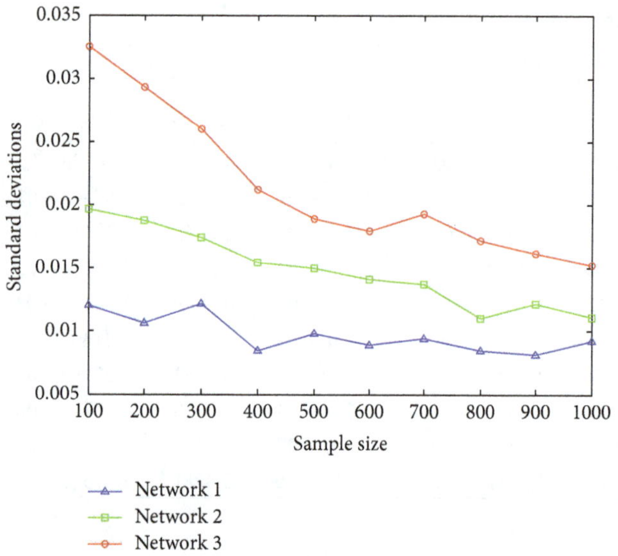

FIGURE 7: The line graph of sample size and deviations of 3 networks.

and Network 3. The deviation consists of the sampling deviation and the evaluation deviation. The evaluation deviation reflects the quality of a given solution. As all solutions selected in sample sizes are identical in Network 1, the evaluation deviations vary a little. The variation of the deviation is mainly caused by the sampling deviations, which manifests volatility. For networks 2 and 3, of each sample size, the evaluation deviation is much bigger than the sampling deviation. And as the selected solutions vary, the evaluation deviations have evident downtrends. In fact, the sampling deviations of Network 2 and 3 are fluctuating, of which the overall trends are drops as well. As the longest time spent on sampling and calculating is just 288.78 s (for Network 3, of sample size 1000), the time spent on calculating is relatively short and grows linearly with the growth of the sample size, we can solve the problem with bigger sample size to get a good solution of higher accuracy to reflect the true problem better.

5. Conclusion

In this paper, we addressed the problem MGVs maneuver in unsecure road networks in battlefield environment. We describe the uncertainty of the network based on discrete scenarios and introduce the kth-shortest-path method to calculate the possible paths between origins and destinations. The problem is formulated into a binary integer programming model. Due to the huge number of scenarios generated by the combination of uncertain arcs, we proposed a SAA method to solve the model. Experiments based on three networks with different scales were conducted to test the efficiency of the solution approach. Computational results show that bigger sample brings relevant progress of the approximation and an acceptable increasing in computational complexity.

In our paper, we did not consider the fortification or camouflage resource for the maneuver, and we only need to decide routes for all vehicles. In some cases, there are protective resources available in the MGV moving process, which can be allocated to some road arcs and reduce the risk for being attacked. Therefore, it is a valuable extension for integrating the protective resources in the MGV routing problem in future works.

Conflicts of Interest

The authors declare that there are no conflicts of interest regarding the publication of this paper.

Acknowledgments

The research is supported by the National Natural Science Foundation of China (no. 71771215 and no. 71471174).

References

[1] SATP, NATO related attacks in Pakistan, http://www.satp.org/satporgtp/countries/pakistan/database/natoattack.htm.

[2] GSO, The World at War, http://www.globalsecurity.org/military/world/war/index.html.

[3] A. L. Tuson and S. A. Harrison, "Problem difficulty of real instances of convoy planning," *Journal of the Operational Research Society*, vol. 56, no. 7, pp. 763–775, 2005.

[4] P. N. Ram Kumar and T. T. Narendran, "Integer programming formulation for convoy movement problem," *International Journal of Intelligent Defence Support Systems*, vol. 1, no. 3, pp. 177–188, 2008.

[5] K. S. Krishna and P. N. Ram Kumar, "On the amenability and suitability of ant colony algorithms for convoy movement problem," *Procedia—Social and Behavioral Sciences*, vol. 189, pp. 3–16, 2015.

[6] Y. Pinedo and D. Ruben, *Route Optimization While Improving Safety Using Escort Vehicles*, State University of New York at Buffalo, ProQuest Dissertations Publishing, 2013.

[7] Y. He, Z. Liu, J. Shi, Y. Wang, J. Zhang, and J. Liu, "K-shortest-path-based evacuation routing with police resource allocation in city transportation networks," *PLoS ONE*, vol. 10, no. 7, Article ID 131962, 2015.

[8] N. Altay and W. G. I. Green III, "OR/MS research in disaster operations management," *European Journal of Operational Research*, vol. 175, no. 1, pp. 475–493, 2006.

[9] G. Galindo and R. Batta, "Review of recent developments in OR/MS research in disaster operations management," *European Journal of Operational Research*, vol. 230, no. 2, pp. 201–211, 2013.

[10] T. Yamada, "A network flow approach to a city emergency evacuation planning," *International Journal of Systems Science*, vol. 27, no. 10, pp. 931–936, 1996.

[11] T. J. Cova and J. P. Johnson, "A network flow model for lane-based evacuation routing," *Transportation Research Part A: Policy and Practice*, vol. 37, no. 7, pp. 579–604, 2003.

[12] V. B. G. Campos, P. A. L. Da Silva, and P. O. B. Netto, "Evacuation transportation planning: a method of identify optimal independent routes," *Urban Transport V*, vol. 44, pp. 555–564, 1999.

[13] A. Stepanov and J. M. Smith, "Multi-objective evacuation routing in transportation networks," *European Journal of Operational Research*, vol. 198, no. 2, pp. 435–446, 2009.

[14] CPLEX Optimizer, https://www.ibm.com/analytics/data-science/prescriptive-analytics/cplex-optimizer.

[15] P. Yang, W. D. Liu, and M. Li, "An optimization model of transportation scheme in preoperational preparations of conventional missile," *Fire Control & Command Control*, vol. 32, no. 2, pp. 41–43, 2007.

[16] S. Ahmed and A. Shapiro, "The sample average approximation method for stochastic programs with integer recourse," *SIAM Journal on Optimization*, vol. 12, no. 2, pp. 479–502, 2002.

[17] T. Santoso, S. Ahmed, M. Goetschalckx, and A. Shapiro, "A stochastic programming approach for supply chain network design under uncertainty," *European Journal of Operational Research*, vol. 167, no. 1, pp. 96–115, 2005.

[18] G. R. Patil and S. V. Ukkusuri, "Sample average approximation technique for flexible network design problem," *Journal of Computing in Civil Engineering*, vol. 25, no. 3, pp. 254–262, 2011.

[19] B. Ayvaz, B. Bolat, and N. Aydin, "Stochastic reverse logistics network design for waste of electrical and electronic equipment," *Resources, Conservation & Recycling*, vol. 104, pp. 391–404, 2015.

[20] Y. Long, L. H. Lee, and E. P. Chew, "The sample average approximation method for empty container repositioning with uncertainties," *European Journal of Operational Research*, vol. 222, no. 1, pp. 65–75, 2012.

[21] Y. Long, E. P. Chew, and L. H. Lee, "Sample average approximation under non-i.i.d. sampling for stochastic empty container repositioning problem," *OR Spectrum*, vol. 37, no. 2, pp. 389–405, 2015.

[22] J. Y. Yen, "Finding the K shortest loopless paths in a network," *Management Science*, vol. 17, no. 11, pp. 712–716, 1971.

[23] D. Eppstein, "Finding the k shortest paths," *SIAM Journal on Computing*, vol. 28, no. 2, pp. 652–673, 1998.

[24] H. Aljazzar and S. Leue, "K*: a heuristic search algorithm for finding the *k* shortest paths," *Artificial Intelligence*, vol. 175, no. 18, pp. 2129–2154, 2011.

12

Monitoring the Number of Pedestrians in an Area: The Applicability of Counting Systems for Density State Estimation

D. C. Duives, W. Daamen, and S. P. Hoogendoorn

Civil Engineering and Geosciences, Department of Transport & Planning, Delft University of Technology, Stevinweg 1, 2628 CN, Delft, Netherlands

Correspondence should be addressed to D. C. Duives; d.c.duives@tudelft.nl

Academic Editor: Juan C. Cano

Crowd monitoring systems are more and more used to support crowd management organizations. Currently, counting systems are often used to provide quantitative insights into the pedestrian traffic state, since they are fairly easy to install and the accuracy is reasonably good under normal conditions. However, there are no sensor systems that are 100% accurate. Detection errors might have severe consequences for the density state estimation at large squares. The consequences of these errors for pedestrian state estimation have not yet been determined. This paper studies the impact of one specific type of detection error on the functionality of counting camera systems for density state estimation, namely, a randomly occurring "false negative" detection error. The impact is determined via two tracks, a theoretical track and a simulation track. The latter track studies the distribution of the cumulative number of pedestrians after 24 hours for three stylized cases by means of Monte Carlo simulations. This paper finds that counting camera systems, which have a detection error that is not correlated with the flow rate, provide a reasonably good estimation of the density within an area. At the same time, if the detection error is correlated with the flow rate, counting camera systems should only be used in the situation where symmetric demand patterns are expected.

1. Introduction

Large-scale events are organized frequently. Worldwide, a sense of community and the urge for joint entertainment draws people together during sports, music, and religious events. All these types of events have in common the fact that large crowds move through pedestrian infrastructures that are not specifically designed to handle large crowds. As a result of overcrowding, accidents still occur, for example, during the Love Parade [1, 2], a Brazil Discotheque party [3], and the Hajj [4].

All over the world crowd management organizations are trying to keep events free from accidents. Safety and security officials generally rely on sight and video surveillance to manage the crowds' movements during large-scale events. Due to the large amount of information that these officials have to sift through, their reliance on video surveillance often results in a lack of oversight, a subjective estimation of the traffic state, and an inefficient allocation of resources.

More and more, crowd monitoring systems are introduced to support the crowd management organization sensor systems in combination with real-time state estimation which can help identify when the infrastructure is reaching its capacity and when action has to be taken to alleviate the pending conditions. Many distinct crowd monitoring alternatives have been presented over the last two years. During SAIL 2015, Kingsday 2016, and the Europride 2016 in Amsterdam a combination of counting systems and Wi-Fi sensors (and GPS trackers during SAIL) was used to determine the traffic state [5, 6]. Next to that, [7, 8] used Wi-Fi sensors and camera feeds to determine the movement behaviour in transport hubs. During the 4Daagse and the Hajj, RFID-chips are used to track the progress of aggregated groups of individuals [9, 10]. Besides that, social media (Twitter, Instagram) [11] and mobile phone signals [12] are scoured for the location of visitors to large-scale events.

Currently, counting camera systems are used to provide quantitative insights into the flows in an area, since they are

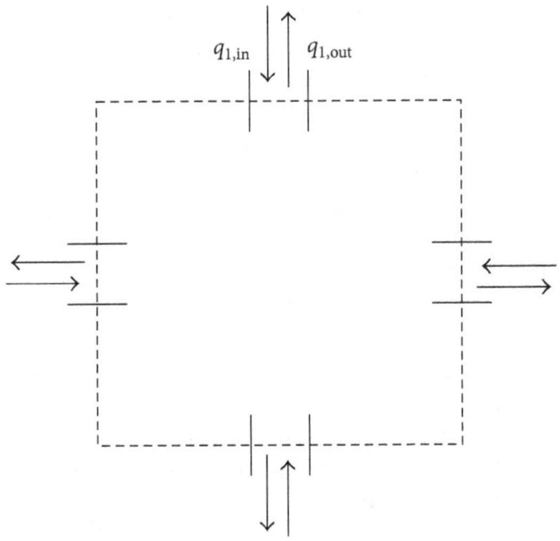

FIGURE 1: Visualisation of the theoretical framework.

fairly easy to install and the accuracy is reasonably good under "normal" traffic conditions. In some cases, similar systems are used to determine the density in an area. To the authors knowledge, there are no counting systems that are 100% accurate. While this does pose minor problems to determine the current flow rate in corridors, these counting errors might have severe consequences for the density estimation in areas because these minor errors add up over time. Yet, how severe these consequences are has not yet been determined.

Contemporary implementations of the sensor networks, like the one described above, have been piloted during the Europride in Amsterdam, The Netherlands. During the weekend of 23-24 June 2016 the entrances of a square were monitored by 12 distinct counting cameras. The cumulative curves of the different days (Figure 11) illustrate that the result of a minor detection error can lead to an extensive estimation error at the end of the day. However, before we can design a solution to this problem, we first need to understand the type of detection error that these cameras make and the extent to which the error impacts the density state estimation.

Given that there is very limited information on the mathematical shape of the detection errors of counting camera systems, this research will study one error type and assume the mathematical shape of this detection error based on the experience of the authors. In the following paper the impact is determined by a randomly occurring "false negative" detection error on the quality of the density state estimation of counting systems during large-scale pedestrian events. This paper focuses on density state estimation since this property of the crowd is often used by crowd managers to gauge the safety of the visitors. The impact is determined via two approaches, a theoretical approach and a simulation approach. While the first approach provides insights into the general influence of detection errors on the density state estimation, the second approach provides insights into the extent of this impact for several stylized situations that often occur during large-scale events.

Section 2 will elaborate on the theoretical framework that is used to determine the generic influence of detection errors. Moreover, the mathematical properties of the influence of the detection errors are discussed. Section 4 discusses the results of three stylized cases in which the dynamics of the system become more and more complex. Beforehand, the particulars of the stylized cases are introduced in Section 3. The last section (Section 5) discusses the results of the cases and draws some general conclusions regarding the influence of a randomly occurring "false negative" detection error on density state estimation.

2. The Expected Quality Reduction Based on Theory

This section studies the impact of a detection error on the density state estimation by means of a theoretical framework. Section 2.1 presents the mathematical description of the system. Sections 2.2 and 2.3 describe the impact of the detection error on the mean and standard deviation of the variance of the density.

2.1. Theoretical Framework. For traffic state estimation, any part of a pedestrian infrastructure can be simplified as a closed area with a collection of entrances n where pedestrians can enter and/or exit the area (Figure 1). The flow q at these entrances has a direction d, that is, in and out. Furthermore, these flows change size dynamically and are thus depending on time t. Moreover, the sizes of in- and outflows of the area are also dependent on the entrance n. As such, the flow is represented as $q_{n,d}(t)$. In this theoretical framework, the use of multiple cameras with dissimilar detection errors to detect the flow at one entrance has not been accounted for.

The characteristics of the problem above are similar to the classical traffic management problems of queue length estimation upstream of traffic signals and density estimation on highway links, which traffic engineers have been studying for quite some years (e.g., [13–19]). Many distinct types of filters, such as Kalman Filters, Particle filters, and Hidden Markov Models, have been proposed to improve the estimation of the number of vehicles in a queue and/or on a link in the highway network. These filters make use of three properties of vehicular traffic flows, being (A) flow conservation, (B) directionality of traffic, and (C) travel time optimization. Here, the first property, that is, flow conservation, assumes that all vehicles that enter a link will exit the link. The second property, that is, directionality, assumes that traffic that enters a link at one end will at some point exit the other end of the area. The third property, that is, the optimization of travel times, describes the process in which vehicles will try to optimize their travel time at all moments in time and will not randomly stop or drive slower than allowed and/or practically possible.

Yet, the situation at hand is more complex, since the three properties, which describe the orderly flow of vehicular traffic in networks, do not necessarily hold for pedestrian traffic flows in pedestrian infrastructures. That is, shops, stores, and houses act as dynamic sources and sinks on the link, directionality cannot be assumed on any link, travel times are not necessarily dependent on the density of the traffic flow, and, most importantly, pedestrians can move from each source to each sink of a link without restrictions. So, even though the queue estimation problem has been under consideration for quite some time and sophisticated filtering techniques to improve the state estimation in vehicular networks have been developed, it is, due to the differences in the behaviour between pedestrian and vehicular traffic flows, not trivial to apply the developed methods and theories to the specific case of traffic flow estimation in pedestrian infrastructure.

This paper studies the influence of a randomly occurring "false negative" detection error on density state estimation in order to provide insights into the nature of the issues that future pedestrian state estimation algorithms will need to solve. In general, the aim of a state estimation algorithm is to determine the state of the traffic flow by means of any of the three macroscopic flow characteristics, velocity v, density ρ, and flow q. As mentioned before, in the management of pedestrian traffic flows the density is predominantly used as state estimator, since this macroscopic flow characteristics can most directly be related to the safety of the visitors through the level of service of [20]. In general the following computation is used to determine the density $\rho_l(t)$ of link l based on a camera counting system, which detects pedestrians when they enter and exit the link:

$$\rho_l(t) = \frac{\sum_{t=0}^{t_{\text{now}}} q_l^*(t)}{A_l} + \rho_0,$$
$$q_l^*(t) = \sum^{N} q_{l,n}^*(t), \tag{1}$$
$$q_{l,n}^*(t) = \sum^{D} q_{l,n,d}^*(t) = q_{l,n,\text{in}}^*(t) - q_{l,n,\text{out}}^*(t),$$

where $q_l^*(t)$ represents the aggregated flow at all entrances, $q_{l,n,d}^*(t)$ the flow measured at entrance n in direction d, t the time at which the flow as measured, and A_l the walkable area. Here, flows into the area are assumed to be positive and flows out of the area negative. In the remainder of this paper, the dynamics of one area will be studied; as such $q_{l,n,d}^*(t)$ is simplified to $q_{n,d}^*(t)$. Here, it is important to note that $q_{n,d}^*(t)$ is an approximation of the flow at one of the entrances, because the reported flow at any entrance at any point in time includes detection errors. A quantification of the effect of the detection error on the density results can be determined if one realizes that $\rho_l^*(t)$ represents a distribution of density realizations (see (2)-(3)) that can be described by its statistical properties. In this study the effect of the detection error on both the mean and the variance of the distribution of density state estimation at the end of the day is studied.

$$q_{n,d}^*(t) \approx q_{n,d}(t) - \varepsilon_{q_{n,d}(t)}, \tag{2}$$

where

$$\varepsilon_{q_{n,d}(t)} = B(q_{n,d}(t), p_{\text{detect}}). \tag{3}$$

This process, in which there is a given chance of not-missing a number of pedestrians i, can be approximated by means of a binomial distribution $B(q_{n,d}(t), p_{\text{detect}})$; see equation (3). In this process the number of trials represents the total flow at location n with directionality d and the chance of success (p_{detect}) represents the chance that a single pedestrian is detected by the counting camera at time t. The output of a counting camera is the number of pedestrians that crossed a certain cross-section in one minute.

In order to compute this flow, counting camera systems make use of intricate detection and tracking algorithms. In recent years a multitude of approaches to automated pedestrian detection and tracking have been presented. The detection approaches vary widely from detection using global models [21–23] to detection based on more local features and/or parts of pedestrians [24–26]. More and more, machine learning strategies are used to improve the detection of pedestrians under a multitude of conditions (e.g., [27, 28]). Also a multitude of tracking algorithms have been presented, some of which use intermediate models to predict the future movement of pedestrians (e.g., [29]), while others use the information of the image to determine the motion between images (e.g., [30, 31]).

No detection and/or tracking algorithm is flawless. Pedestrians can be missed by the counting camera system for numerous reasons. The reasons are, amongst others, changing light conditions and abrupt changes in speed and direction of a pedestrian or because pedestrians are occluded by other pedestrians. Depending on the applied detection and tracking algorithms and the situation on-site, one or more types of errors are prone to occur. As a result of the above-mentioned issues, a lack of detection of pedestrians ("false negatives") and double registrations of pedestrians ("false positives") can arise. Depending on the situation, these types of errors can be uncorrelated or correlated in time and space.

A comprehensive study on the effect of distinct error types is out of the scope of this paper. Given the complexity of an analysis of the impact with a large number of degrees of freedom, the authors feel that an in-depth study regarding one type of detection error will provide better insights in the applicability of counting camera systems when this specific type of detection error is encountered. In the remainder of this paper only the effect of one certain specific error type will be studied, namely, that of a randomly occurring "false negative" error. Examples of this type of detection error are the occlusion of a child behind its parents or the lack of contrast between the colour of the pedestrian's clothes and the background. A "false negative" detection error always results in a lower amount of registered detection than the pedestrians that passed the counting camera and each pedestrian has the same chance to not be detected given a certain flow rate.

The effect of detection errors on the density estimation in an area can be determined if one knows what the chance is that a pedestrian is detected. There is limited knowledge on the average detection errors of counting cameras. Moreover, the details on the relation of the detection error and the flow rate are currently lacking. Therefore, some assumptions on the mathematical properties of the detection errors are assumed, namely, that the detection errors are random, independent, and not spatially and temporally correlated. Even though many detection errors might not adhere to these assumptions, preliminary insights into the impacts that can be expected under these relatively "nice" conditions are

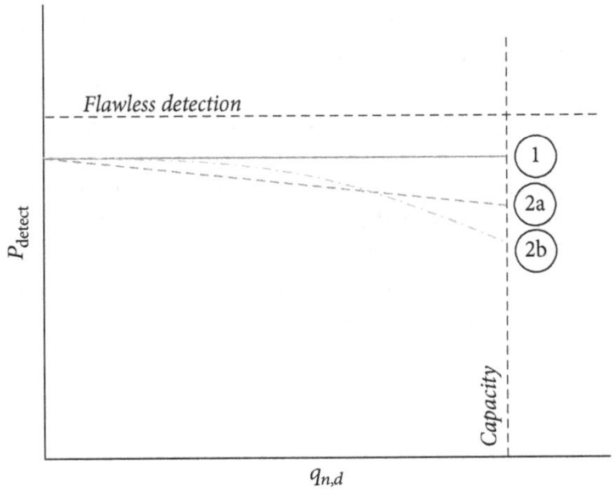

FIGURE 2: Visualisation of the possible mathematical descriptions of the counting error, where 1: uncorrelated with the flow, 2: linear, and 3: exponential.

deemed to be valuable to practitioners, especially given the fact that more complex types of detection errors will only deteriorate the density estimation results.

Logically speaking, three distinct situations can be defined for a randomly occurring "false negative" detection error with these mathematical properties, namely,

(1) The detection error is not correlated with the flow.

(2) The detection error is correlated with the total flow at each entrance.

(3) The detection error is correlated with the total flow and the direction of flow.

Figure 2 illustrates a few possible shapes of the function of the detection probability p_{detect} in situations 2 and 3 (i.e., detection error correlated with flow). This function defines the relation between the flow rate at a cross-section and the probability of detection of each individual that is part of that flow. As one can see, any detection probability function is limited by two boundaries, namely, (1) at $p_{\text{detect}} = 1.0$ the detection is completely accurate and (2) at the capacity of the entrance ($q_n = q_{\text{capacity}}$ in P/s) as no more pedestrians can move through an entrance within a time instance. Besides that, the function can be concave (1), linear (2a) or nonlinear (2b). In this paper, a linear shape of the probability function is assumed. Even though a nonlinear function might be more realistic in this respect, the linear approximation, which has computational advantages, will still illustrate the main implications of a concave relation. As such, the shapes of the detection error are approximated by $p_{\text{detect}}(t) = 1 - \alpha_n * q_n(t)$ in case 2 and $p_{\text{detect}}(t) = 1 - \alpha_{n,d} * q_n(t)$ in case 3, where the realizations of p_{detect} are dependent on area l and entrance n. Moreover, the detection error is assumed to be correlated with the total absolute flow rate at each entrance $q_n(t)$, which represents the total number of pedestrians who passed a cross-section (i.e., $\sum_d q_{n,d}(t)$). In the remainder of this section the effect of detection errors as a result of case 3 is determined, since this formulation can easily be simplified to the other two cases.

2.2. Impact of Detection Errors on the Mean Density. In the paragraphs above a mathematical description was presented. In the following section the mathematical description is used to determine the mean of the maximum impact of the accumulation of this type of detection error, namely, the impact of the cumulative detection error at the end of a day of counting. This total error is computed step by step, in which the level of aggregation increases over the number of entrances and over time. The mean number of counted individuals for one location is given by (4), for a set of detectors, and a certain point in time is given by (5) and for an entire day it is given by (6).

$$\begin{aligned} E\left[q^*_{n,d}(t)\right] &= q_{n,d} p_{\text{detect}} = q_{n,d}\left(1 - \alpha_{n,d} q_n\right) \\ &= q_{n,d} - \alpha_{n,d} q_n q_{n,d}, \end{aligned} \tag{4}$$

$$\begin{aligned} E\left[\sum^N\sum^D q_{n,d}\right] &= E\left[q_{1,1}\right] + E\left[q_{1,2}\right] + E\left[q_{2,1}\right] + \cdots \\ &\quad + E\left[q_{n,d}\right] \\ &= \sum^N\sum^D q_{n,d} - \sum^N\left(q_n^2 * \sum^D \alpha_{n,d}\right), \end{aligned} \tag{5}$$

$$\begin{aligned} E\left[\sum^T\sum^N\sum^D q_{n,d}\right] &= E\left[\sum^N\sum^D q_{n,d}(t_1)\right] \\ &\quad + E\left[\sum^N\sum^D q_{n,d}(t_1)\right] + \cdots \\ &\quad + E\left[\sum^N\sum^D q_{n,d}(t_T)\right] \\ &= \sum^T\sum^N\sum^D q_{n,d}(t) \\ &\quad - \sum^T\sum^N\left(q_n(t)^2 * \sum^D \alpha_{n,d}\right). \end{aligned} \tag{6}$$

The derivation illustrates that the expected amount of detected individuals during the day depends quadratically on the flow rate at each corridor. As a result, the mean error in the cumulative sum of the amount of pedestrians in an area, and as such the density, increases very quickly at the moments that large numbers of pedestrians enter and/or exit the area simultaneously. Additionally, the mean of the estimation of the cumulative number of pedestrians can be more or less than zero. The sign of the cumulative number of pedestrians depends on the exact realizations of the flows and the detection errors $\alpha_{n,d}$. Consequentially, without knowing the dynamics that will occur in a pedestrian infrastructure, it is impossible to determine to what extent a counting camera system will underestimate the amount of individuals in an area. Moreover, based on theory, especially in pedestrian infrastructures with a very high demand, it might be difficult to use a counting camera system to determine the total amount of pedestrians in the infrastructure, as high flow rates induce relatively high detection errors.

2.3. The Impact of Detection Errors on the Variance of the Density. The total amount of detected individuals at the end of the day also depends on the variance of the amount of detection. The variance in the amount of pedestrians that is detected by a detector at a given location at time instance t is determined by

$$\begin{aligned} \operatorname{var}\left(q_{n,d}(t)\right) &= q_{n,d}(t)\left(1-p_{\text{detect}}\right)p_{\text{detect}} \\ &= q_{n,d}(t)\left(1-\left(1-\alpha_{n,d}q_{n}, d(t)\right)\right)\left(1-\alpha q_{n}, d(t)\right) \\ &= q_{n,d}(t)\left(\alpha_{n,d}q_{n}(t)\right)\left(1-\alpha_{n,d}q_{n}(t)\right) \\ &= \alpha_{n,d}q_{n,d}(t)\,q_{n}(t)-\alpha_{n,d}^{2}q_{n,d}(t)\,q_{n}(t)^{2}. \end{aligned} \tag{7}$$

If we assume that the variance of the detection in one minute is independent of the variance of the detection error in the following minute, the variance in the amount of detected pedestrians at a certain moment in time by all detectors in the system is defined by (8) and the variance in the cumulative amount of pedestrians that are present in the area, and as such the density, is defined by (9).

$$\begin{aligned} \operatorname{var}\left(\sum^{N}\sum^{D}q_{n,d}(t)\right) &= \sum^{N}\sum^{D}\operatorname{var}\left(q_{n,d}(t)\right) \\ &= \alpha_{1,1}q_{1,1}q_{1}(t)-\alpha_{1,1}^{2}q_{1,1}q_{1}(t)^{2}+\alpha_{1,2}q_{1,2}q_{1}(t) \\ &\quad -\alpha_{1,2}^{2}q_{1,2}q_{1}(t)^{2}+\alpha_{2,1}q_{2,1}q_{2}(t)-\alpha_{2,1}^{2}q_{2,1}q_{2}(t)^{2} \\ &\quad +\cdots+\alpha_{N,D}q_{N,D}(t)\,q_{N}(t)-\alpha_{N,D}^{2} \\ &\quad * q_{N,D}(t)\,q_{N}(t)^{2} \\ &= \sum^{N}\left(q_{n}(t)\sum^{D}\left(\alpha_{n,d}q_{n,d}(t)\right)\right) \\ &\quad -\sum^{N}\left(q_{n}(t)^{2}\sum^{D}\left(\alpha_{n,d}^{2}q_{n,d}(t)\right)\right), \end{aligned} \tag{8}$$

$$\begin{aligned} \operatorname{var}\left(\sum^{T}\sum^{N}\sum^{D}q_{n,d}(t)\right) &= \sum^{T}\operatorname{var}\left(\sum^{N}\sum^{D}q_{n,d}(t)\right) \\ &= \sum^{T}\sum^{N}\left(q_{n}(t)\sum^{D}\left(\alpha_{n,d}q_{n,d}(t)\right)\right) \\ &\quad -\sum^{T}\sum^{N}\left(q_{n}(t)^{2}\sum^{D}\left(\alpha_{n,d}^{2}q_{n,d}(t)\right)\right). \end{aligned} \tag{9}$$

From (9) four things can be deduced about the variance of the total number of detected pedestrians at the end of a day given the presence of randomly occurring "false negative" detection errors. Firstly, this number is quadratically dependent on the total flow and can, therefore, become dominant very quickly in the estimation of the cumulative number of pedestrians in the area. Secondly, the variance is highly dependent on the distribution of the flows over the two exits ($\alpha_{n,d}$) and the size of the flows, and, as a result, the variance can vary severely between cases and flow situations. Thirdly, the behaviour of the variance becomes more complex with an increase of the number of cameras that are needed to detect all in- and outflows of the area. Last of all, theoretically, the total variance of the system can become negative as a result of the third power of the flow rate in (9). However, in practice this is impossible, as the variance can never go below 0.

The results of this theoretical framework illustrate that systematic detection errors in counting camera systems can have a severe influence on the density state estimation in the area. Yet, this framework does not illustrate the size of the effect of these influences in a practical application of the system. For example, if the flows in the system remain small and a limited number of cameras is used, the theoretical insights above show that the influence of the detection errors on the density state estimation is limited. Whether this is the case is dependent on the flow rate, the exact counting camera system used, and the layout of the system. In order to get some feeling for the impact of a "false negative" systematic detection error in real-life cases, a case study is performed in Section 4.

3. Case Study Description

Section 4 studies the effect of distinct realizations of $\alpha_{n,d}$ by means of a case study containing three cases. The cases internalize the two different types of correlations between p_{detect} and $q_{n,d}(t)$. Before going into detail about the results of these case studies, this section first introduces the main parameters of the case study. In the following paragraphs the case study layout, the demand levels, and the description of the probability functions of the detection errors are detailed.

Case Study Research Aims and Control Variables. The theoretical framework identified the fact that the type of relation between the detection error and the flow rate at the entrances severely influences the mean and variance of the density state estimation. Moreover, this framework illustrated the large impact of spatial and temporal differences in the flow rate on the mean and variance of the density state estimation. By means of a case study, this paper studies attempts to quantify this impact by means of the following two questions:

(i) How large is the impact of the type of relation between a "false negative" detection error and the flow rate at an entrance on the statistics of the cumulative sum of the amount of pedestrians within the area at the end of the day?

(ii) How large is the impact of the demand pattern of the entrances on the statistics of the cumulative sum of the amount of pedestrians within the area at the end of the day?

In order to answer these two questions, the types of correlation and distinct types of demand patterns are adopted as

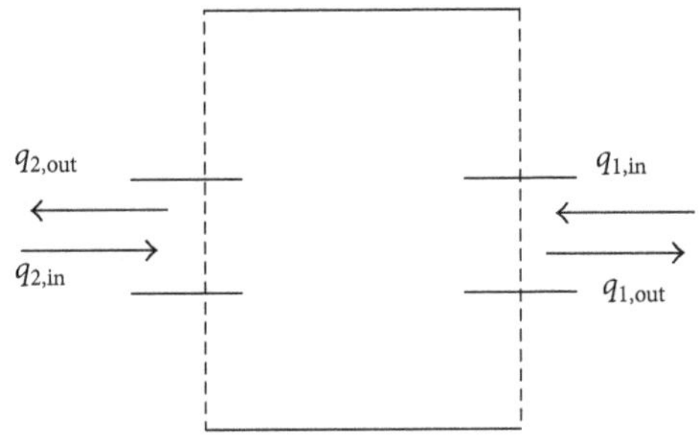

FIGURE 3: Visualisation of the case study layout.

control variables. Below, both control variables are elaborated upon in more detail.

Layout. The layout of the case study has been kept as simple as possible, while still ensuring that all degrees of freedom that are in the description of the function of the detection probability are also accounted for in the layout. A square with two entrances provides the possibility of studying the effect of counting errors that are dependent on the total flow at the entrances of an area with two entrances, which will hereafter be identified as a corridor, and are dependent on the direction of flow in the area. In Figure 3 a visualisation is provided of the case study area that is used in the remainder of this paper.

Cases. As mentioned in the theoretical framework, there are three possible scenarios with respect to the dependence of the probability of detection on the flow at an entrance, namely, (1) the detection error is not correlated with the flow, (2) the detection error is correlated with the total flow at an entrance but independent of the direction that is being counted, and (3) the detection error is correlated with the total flow and dependent on the direction of flow that is being counted. These three cases can be formalized by the three probability density functions, (10)–(13). Here, (10) represents situation 1, (11) situation 2, and (12)-(13) situation 3. Given that equations (12) and (13) are special cases of (11), these cases will not be treated in this study. The exact realizations of $\alpha_{n,d}$'s are specified in the next section.

$$P\left(X_{n,d,i}\right) = \alpha_1, \tag{10}$$

$$P\left(X_{n,d,i}\right) = \alpha_1 - \alpha_2 * q_n(t), \tag{11}$$

$$P\left(X_{n,\text{in},i}\right) = \alpha_1 - \alpha_{3,\text{in}} * q_{n,\text{in}}(t), \tag{12}$$

$$P\left(X_{n,\text{out},i}\right) = \alpha_1 - \alpha_{3,\text{out}} * q_{n,\text{out}}(t). \tag{13}$$

Demand Levels. The type of demand pattern is used to study the influence of the shape and size of the in- and outflow on the estimation of the total number of people in the area after 24 hours. Two distinct demand profiles have been adopted (see Figure 4). The first demand profile mimics a symmetrically loading and unloading demand profile. The in- and outflow at the entrances follow a similar pattern. The second demand profile mimics a nonsymmetric demand profile in which the inflow is gradual, but the outflow is abrupt. Especially, in situations where the detection error is dependent on the flow rate, these two demand profiles are expected to produce different density estimations at the end of the day.

Since there are two entrances in the case study, the parameters w_d are used to specify the distribution of demand over entrances, where $w_1 = q_1/\sum^D q_d$. For simplicity, it is assumed that w_d remains stable throughout the day.

Simulation of the Probability of Detection. When computing the cumulative number of pedestrians within an area, the authors are especially interested in the distribution of the cumulative number of pedestrians at the end of the day. Here, it is assumed that an accurate estimation of the number of pedestrians will result in a distribution with a mean near 0 and a small variance. In this study, a Monte Carlo simulation has been used to determine the properties of this distribution after 24 hours. Whether or not a pedestrian is detected is determined using a random drawn number $X_{i,\text{sim}}$ between 0 and 1. The function of the detection probability ((10)–(13)) is used as the threshold on the chance of detection given the demand $q_{n,d}(t)$ in the minute when the pedestrian passes the entrance or exit. A pedestrian for which $X_{i,\text{sim}} \leq p_{\text{detect}}$ is detected; otherwise it is not. In total, 10,000 simulations of the total demand curve (Figure 4) have been run.

4. Case Study Results and Synthesis

This section presents the results of the case studies. The first subsection describes the simplest scenario, in which the counting error is uncorrelated with the flow. Subsequently, the two more complex scenarios are discussed. That is, Sections 4.2 and 4.3 describe a correlation of the counting error with the total flow through the corridor, given symmetric and nonsymmetric demand profiles. Afterwards, Section 4.4 synthesizes the results.

4.1. Case 1: Detection Error Uncorrelated with Flow. If the detection error is uncorrelated with the flow, all individuals have the same chance of being detected. As a result, the distribution of the cumulative number of pedestrians in the area after 24 hours is only dependent on $\alpha_{n,d}$. In this case study $0.5 \leq \alpha_{n,d} \leq 1$ has been adopted.

Figure 5 shows the solution space of the Monte Carlo simulation for $\alpha_{n,d} = 0.95$. As expected, the peak of the cumulative curve is underestimated and shifted backwards in time. Additionally, the increase in the number of pedestrians during the morning and the decrease in the number of pedestrians in the evening are underestimated. As a result, in the morning, a lower number of pedestrians is estimated than is actually within the area, while during the afternoon the number of pedestrians is overestimated. Moreover, during the day, the variance of the distribution increases.

The manner of the distribution of the cumulative number of pedestrians in the area after 24 hours depends on the distribution of flows over the two exits and the detection errors are displayed in Figure 6. In case of a "false negative" detection error that is not correlated with the total flow in the corridor, the mean of the distribution of the cumulative number of pedestrians in the area is not influenced by the distribution of the flows over the two exits, nor by the

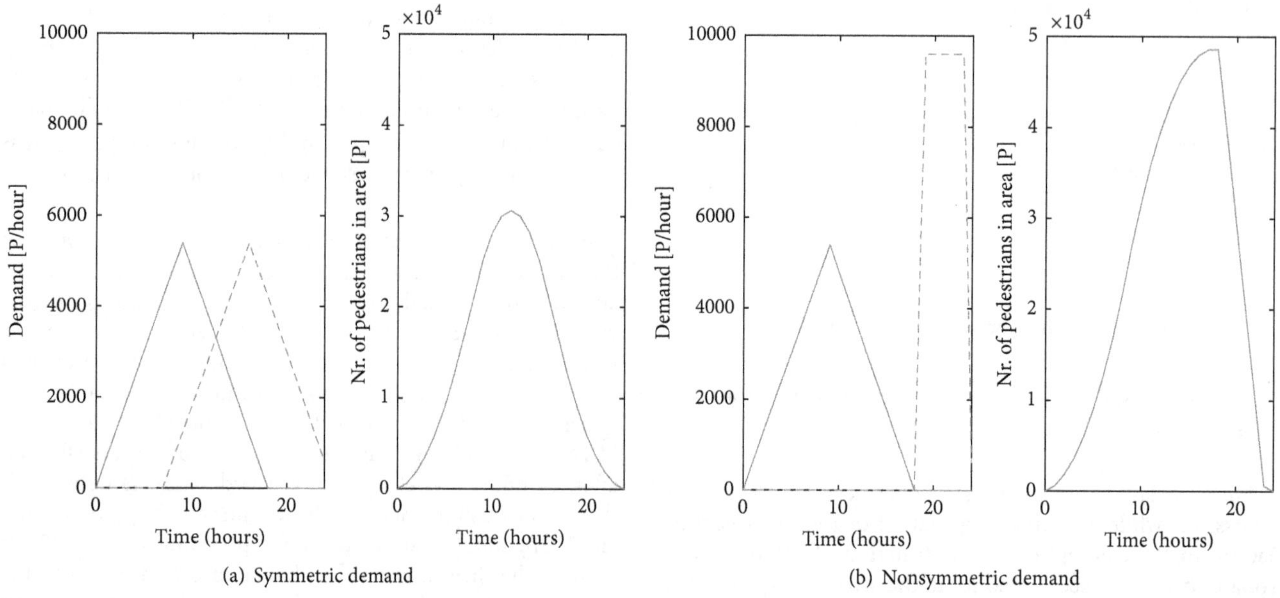

FIGURE 4: Visualisation of the demand at both entrances (left) and cumulative number of pedestrians in the region (right), where (a) illustrates the symmetric demand case and (b) the nonsymmetric demand case. The solid line represents the inflow, while the dashed line indicates the outflow.

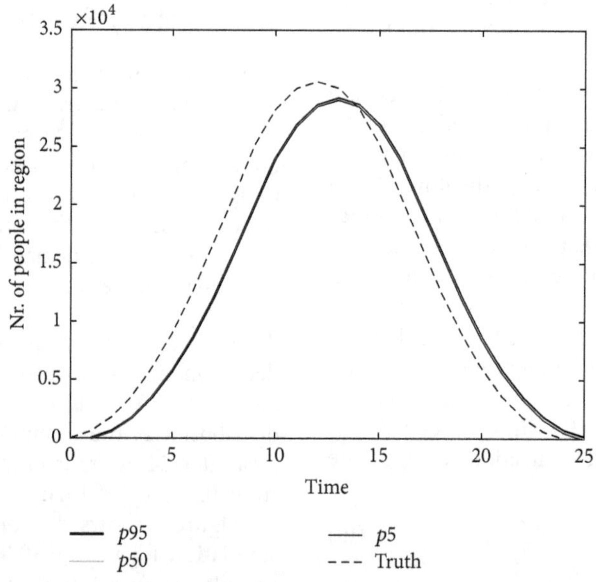

FIGURE 5: Case 1: visualisation of the solution space provided by the Monte Carlo simulation for $\alpha_n = 0.95$, where the dashed line of the total demand represents the ground truth, and $p5$, $p50$, and $p95$ represent the xth percentile of the distribution of realizations of the amount of pedestrians in the area at time t.

extent of the detection error. The variance of the distribution, however, is influenced by the detection error. If the detection error increases, the standard deviation of the distribution increases.

4.2. Case 2: Counting Error Correlated with Flow—Symmetric Demand. If the detection error is correlated with the flow, then the chance of being detected is governed by two types of detection errors. α_1 governs the base detection error that exists independent of the flow rate, which is set to $\alpha_1 = 0.01$. This value is typically found in our work with counting cameras. α_2 governs the increase in the detection error as a result of the increase in the flow rate $q_{n,d}(t)$ at the entrance. Literature does not provide information about the shape of the detection error. Therefore, the impact of several distinct α_2 will be tested, which are decided upon based on logic. In this case study $0.0000 \leq \alpha_2 \leq 0.0005$ has been adopted. These values might seem low. However, when taking into account a

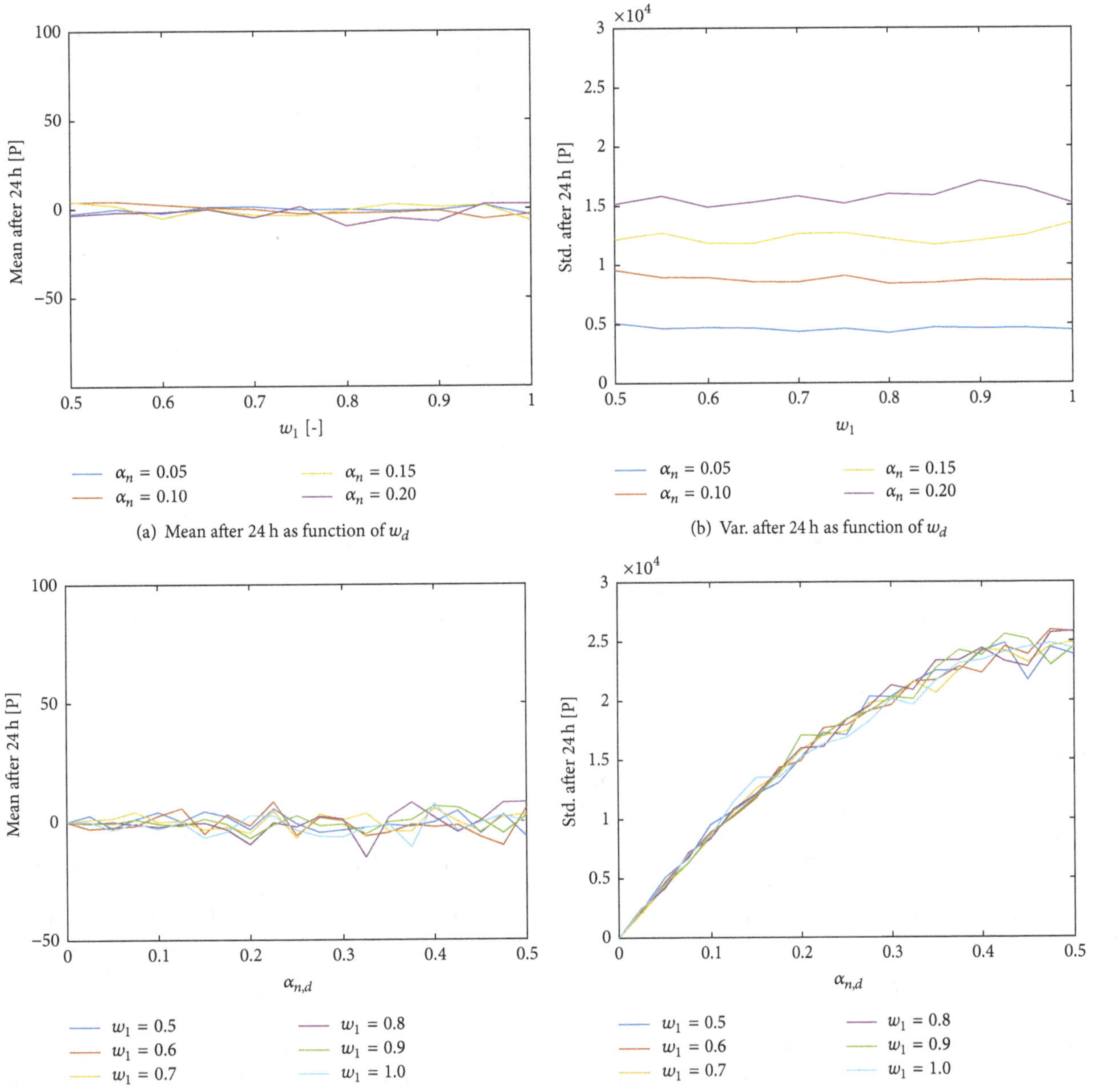

(a) Mean after 24 h as function of w_d

(b) Var. after 24 h as function of w_d

(c) Mean after 24 h as function of $\alpha_{n,d}$

(d) Var. after 24 h as function of $\alpha_{n,d}$

FIGURE 6: Case 1: visualisation of mean and standard deviation of the number of pedestrians in the region given a certain distribution over the entrances as a function of w_d and the detection error $\alpha_{n,d}$.

flow rate of 10,000 people per hour, the additional detection error is still substantial.

The solution space of the Monte Carlo simulation has been visualised for $\alpha_2 = 0.00005$ for two distinct distributions of the flow over the two entrances (see Figure 7). In the left situation flow is equally distributed over the two entrances, while in the right situation the flow is completely assigned to one of the two entrances. The less the flow is distributed over the entrances (i.e., $w_1 \rightarrow 1$), the more the solution differs with respect to the ground truth. Surprisingly, however, also in this case, at the end of the day, the mean error in the cumulative curves approximates 0 for all α_n's. This might, however, be the result of the choice for a symmetric demand curve. Even though the variance of the distribution increases throughout the day, the variance remains limited given the total flow through the area.

Figure 8 shows that the mean of the distribution is not affected by the distribution over the exits nor by the size of the detection error. The standard deviation is, however, dependent on the distribution over the exits. The less the flow is distributed over the two exits, the lower the standard deviation becomes. This was also found in the theoretical framework, which describes that for very high flow levels the variance will decrease. Additionally, a nonlinear trend is found regarding the influence of the detection error on the standard deviation. For detection errors $\alpha_{n,d} \leq 0.0001$ the

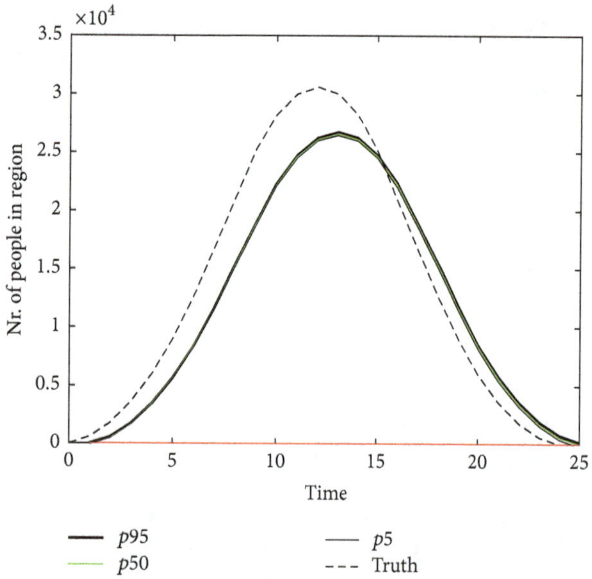

(a) $\alpha_{n,d} = 0.00005$, $w_1 = 0.6$

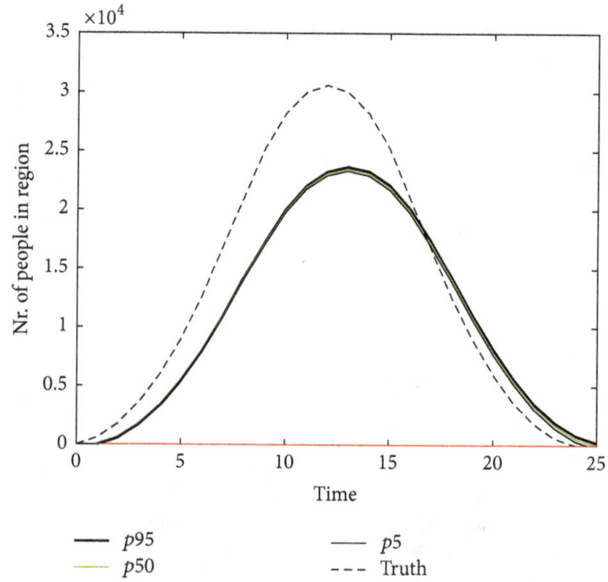

(b) $\alpha_{n,d} = 0.00005$, $w_1 = 1.0$

FIGURE 7: Case 2: visualisation of the solution space provided by the Monte Carlo simulation for given combinations of $\alpha_{n,d}$ and w_1, where the dashed line of the total demand represents the ground truth, and $p5$, $p50$, and $p95$ represent the xth percentile of the distribution of realizations of the amount of pedestrians in the area at time t.

standard deviation of the number of pedestrians in the area after 24 hours increases, while for detection errors for which $\alpha_{n,d} > 0.0001$ the standard deviation decreases again. This nonlinear trend in the simulation is, most likely, a result of the balance between the two components in the equation of the variance (see (9)). When the first part is dominant, the variance increases; when the second part is dominant the variance decreases.

4.3. Case 3: Detection Error Correlated with Flow—Non-symmetric Demand. Similar to case 2, the detection error is governed by two parameters, namely, α_1 and α_2. Here, α_1 represents the static detection error, while α_2 represents the increase of the detection error as a result of an increase in the flow rate. Also in this case study $0.0000 \leq \alpha_2 \leq 0.0005$ has been adopted. The main distinction between case 2 and case 3 is the demand curve. Where a symmetric demand curve has been adopted in case 2, in this case a smooth inflow and an abrupt outflow are simulated.

Figure 9 depicts the visualisation of the solution space provided by the Monte Carlo simulation for two different distributions of the flow over the two entrances, namely, $w_1 = 0.5$ (a) and $w_1 = 1.0$ (b). The shape of the two graphs differs from Figures 5 and 7 in two respects. Firstly, a sharp drop in the number of pedestrians in the area occurs at the end of the day, around hour 20. This is the result of the abrupt outflow that is being simulated and not a direct consequence of the detection error. Secondly, in both graphs the mean number of pedestrians is not zero in the area at the end of the day. This is due to the fact that the detection error during the inflow phase is limited, while the detection error during the outflow phase is quite substantial as a result of the high flow rates. The estimation error at the end of the day increases if a larger portion of the demand is directed towards one entrance. This is in contrast with the previous two cases, which found a limited estimation error at the end of the day because the detection error during the inflow phase was in balance with the detection error during the outflow phase.

A summary of the results for different distributions over the two entrances and different detection errors illustrates that the mean estimation error at the end of the day can be very large (i.e., approximately 30,000 pedestrians if $w_1 = 0.5$ or $\alpha_2 = 0.00025$) (see Figure 10). The mean of the estimation error is highly dependent on both w_1 and α_2. Unexpectedly, the mean estimation error is less for $\alpha_2 > 0.0001$. It is hypothesized that this is due the limiting effect of increased flow rates and increased detection errors on the estimation error. That is, if the detection error during the inflow phase is larger, the effect of a large detection error during the outflow phase is relatively smaller. When $\alpha_2 \leq 0$ the first part of the equation becomes dominant, which causes a quadratic dependence on the mean estimation error on the flow rate, which is larger if higher flow rates are encountered (i.e., $w_1 \rightarrow 1$).

Also the variance of the distribution of the cumulative number of pedestrians in the area is found to be dependent on the distribution over the entrances and the detection error. As can be expected based on the theoretical framework, the standard deviation decreases with an increase of the detection error.

4.4. Synthesis Case Studies. The case studies presented in Sections 4.1–4.3 illustrated the impact of the type of detection error on the distribution of the cumulative number of pedestrians in the area at the end of a day. Table 1 provides

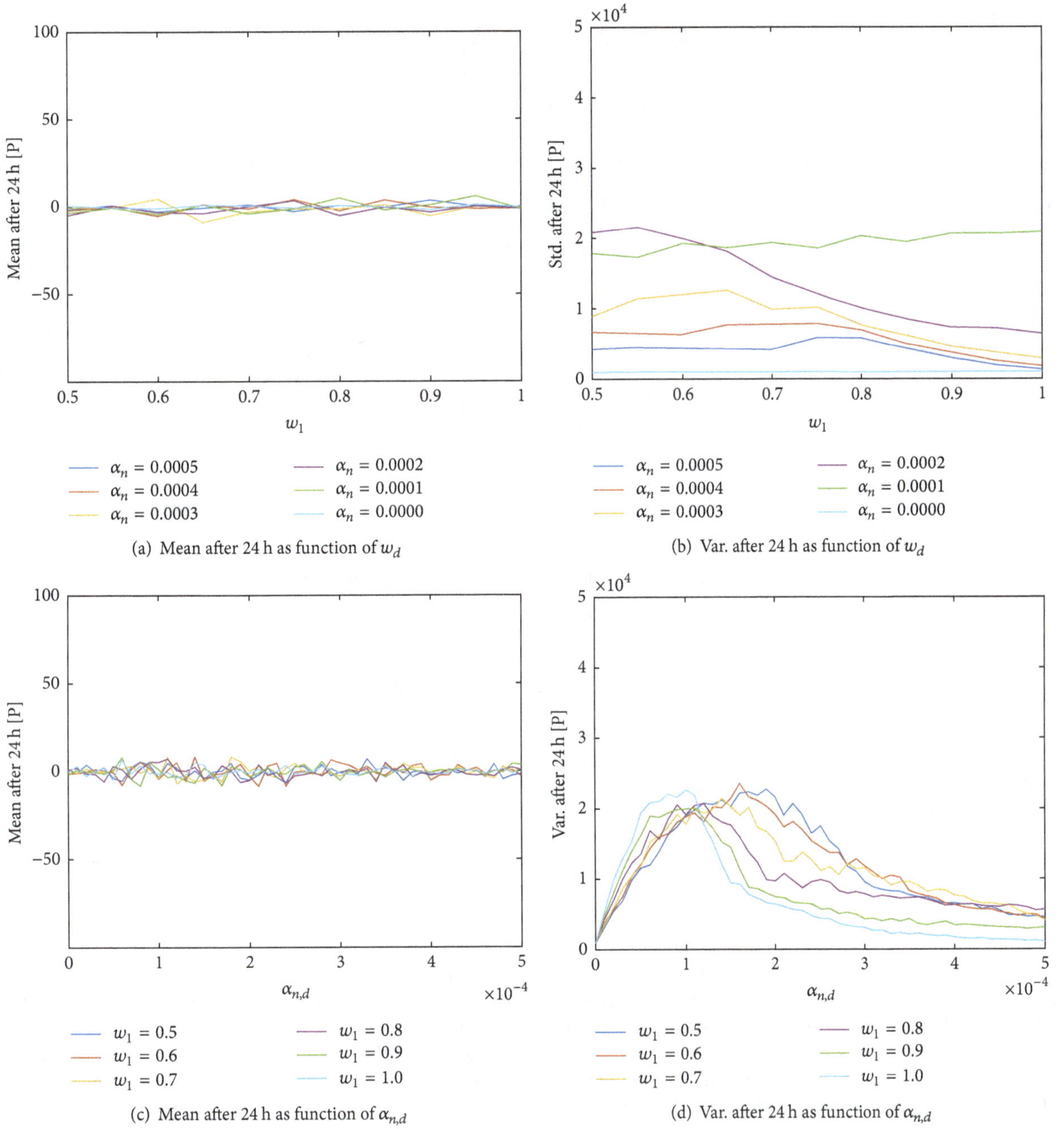

FIGURE 8: Case 2: visualisation of mean and standard deviation of the number of pedestrians in the region given a certain distribution over the entrances as a function of w_d and $\alpha_{n,d}$.

a summary of the impacts for each case. Based on this table several general conclusions can be drawn.

Table 1 shows that impact of a randomly occurring "false negative" detection error is limited if the detection error is not correlated to the flow rate at the entrances. In that case, the cumulative number of pedestrians is underestimated in the morning and overestimated in the afternoon. Yet, as long as the static detection error is limited ($\alpha_{n,d} \leq 0.05$), these impacts are limited too.

However, if the detection error is correlated to the flow rate at the entrances, the estimation of the cumulative number of pedestrians in an area, and as such the density, deviates more from the ground truth. Cases 2 and 3 illustrate that the demand profile at the entrances has a large impact on the density estimation in this case. For symmetric demand patterns, this effect is limited to an increase of the variance of the estimated density. However, for nonsymmetric demand patterns, the impact on both the mean and standard deviation of the density estimation after 24 hours is severe. The theoretical framework illustrates that these impacts especially exacerbate when high flow rates are encountered. Consequently, in situations where high flow rates and nonsymmetric demand

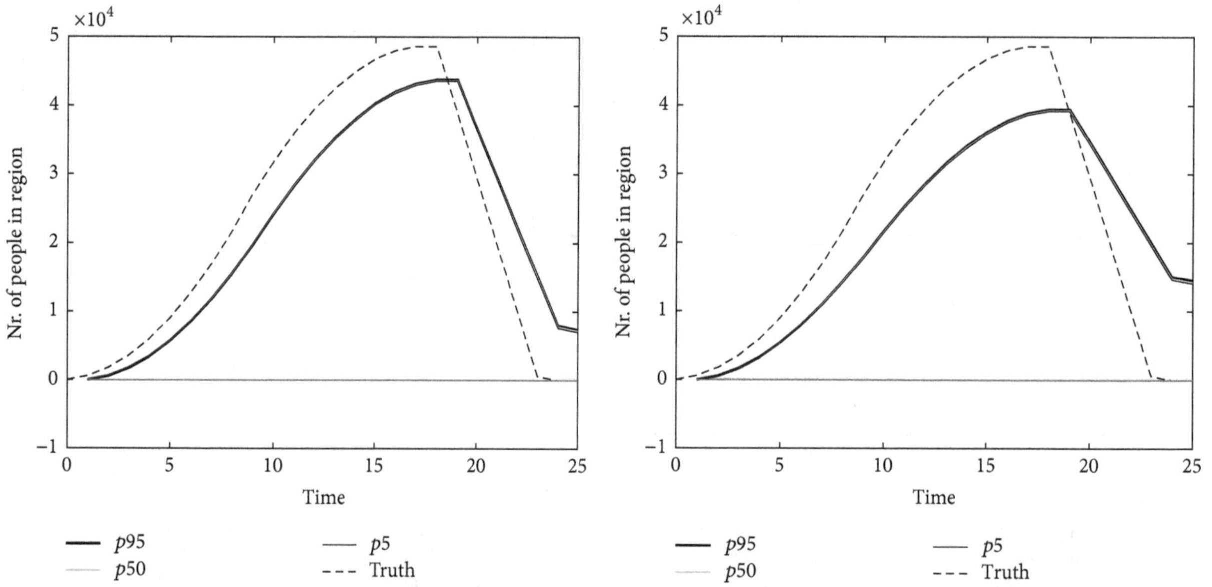

(a) $\alpha_{n,d} = 0.00005$, $w_1 = 0.5$

(b) $\alpha_{n,d} = 0.00005$, $w_1 = 1.0$

FIGURE 9: Case 3: visualisation of the solution space provided by the Monte Carlo simulation for given combinations of $\alpha_{n,d}$ and w_1.

TABLE 1: Synthesis of the case studies.

Cases	Symmetric in- and outflow	Error dependent on flow	Error dependent on w_1	Error dependent on $\alpha_{n,d}$	Max. avg. error	Max. std. error
Case 1: uncorrelated flow	Yes	No	No	Yes	Limited	Limited
Case 2: symmetric demand	Yes	Yes	Yes	Yes	Limited	Large
Case 3: nonsymmetric demand	Yes	Yes	Yes	Yes	Large	Large

profiles are encountered, such as during most large-scale events, detection systems for which the detection error is correlated with the flow rate will produce unrealistic density estimations.

Based on the cumulative curve presented in Figure 11 and the nonsymmetric demand pattern of the Europride 2016 at the Dam in Amsterdam, it can be concluded that it is very likely that the detection error of counting cameras during this event was flow dependent. As the authors have seen similar results during other large-scale events, this leads the authors to conclude that, given the insights presented in this paper, one should be very hesitant to perform density state estimation by means of counting cameras at large-scale events.

5. Conclusion and Future Research

This paper has studied the impact of one specific type of detection error on the functionality of counting camera systems for density state estimation during large-scale pedestrian events, namely, a randomly occurring "false negative". First, via a theoretical track, the theoretical impact of this type of detection error on the distribution of the density estimation (i.e., mean and variance) has been determined. Accordingly, the dynamics of the impact have been studied in more detail by means of a simulation case study.

Based on the theoretical framework and the results, this study concludes that counting camera systems, which have a randomly occurring "false negative" detection error that is not correlated with the flow rate, which makes use of cumulative curves, provide a reasonably good estimation of the density in an area. However, in real-life situations the demand generally changes dynamically during the day. This study illustrates that even under relatively "nice" conditions, it is already highly questionable whether the results from counting camera systems can be used in a real-time monitoring operation to determine the amount of pedestrians in an area. For offline analysis this type of system might be beneficial, as at least a part of the effect of the counting error can be rectified through the application of the mass-conservation law.

Moreover, this study shows that counting camera systems that have a detection error that is correlated with the flow rate should only be used in the situation where symmetric demand patterns are expected. In case of nonsymmetric demand patterns, the estimation of the number of pedestrians

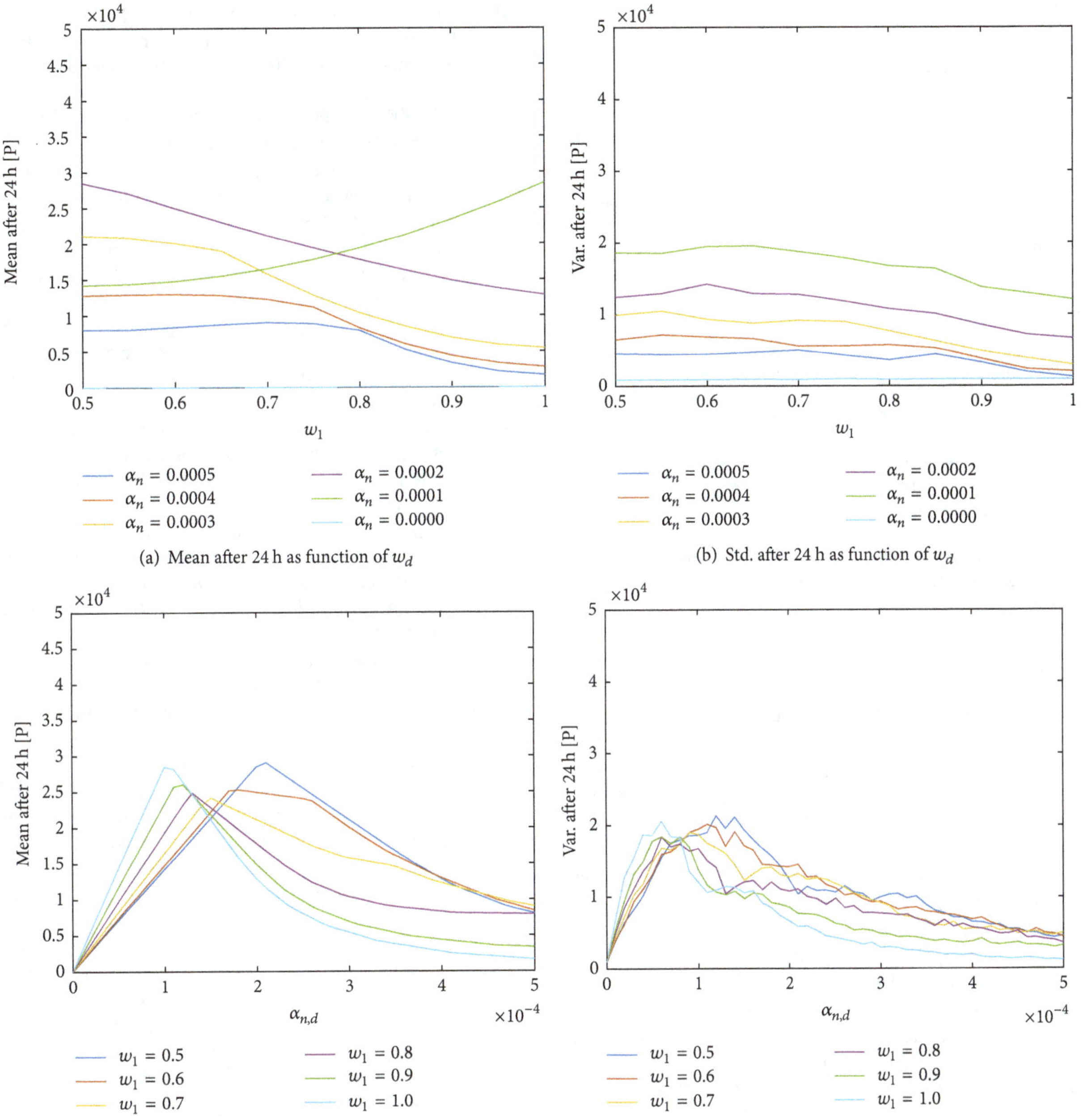

(a) Mean after 24 h as function of w_d

(b) Std. after 24 h as function of w_d

(c) Mean after 24 h as function of $\alpha_{n,d}$

(d) Std. after 24 h as function of $\alpha_{n,d}$

FIGURE 10: Case 3: visualisation of the mean and standard deviation of the number of pedestrians in the region given a certain distribution over the entrances as a function of w_d and $\alpha_{n,d}$.

in an area is unrealistic as the detection error during the inflow phase is in disbalance with the detection error during the outflow phase. Here, it needs to be noted that these conclusions only hold for the estimation methods featuring cumulative curves explored in this paper.

Above all, this study illustrates that knowledge on the types of detection error in counting cameras is essential to enabling density state estimation by means of counting camera systems. Given that only the effect of a type of detection error has been studied, more research is needed into the effect of more complicated detection errors, for instance, "false positives", detection errors that are correlated in time and space, and combinations thereof. Once the error is known, algorithms can be developed that account for the missing detection. More importantly, if the detection errors of these systems are understood better, it becomes easier to identify in which situations these systems should or should not be used for state estimation purposes.

Conflicts of Interest

The authors declare that they have no conflicts of interest.

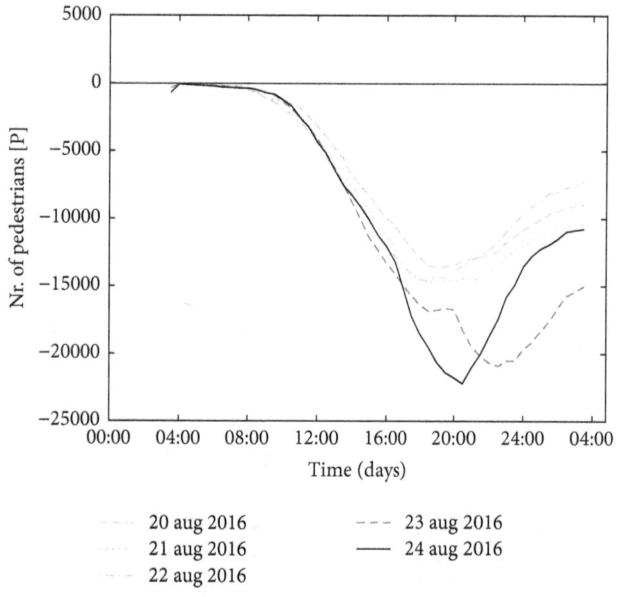

FIGURE 11: Cumulative number of pedestrians on the Dam during the first weekend of the Europride 2016.

References

[1] B. Krausz and C. Bauckhage, "Loveparade 2010: Automatic video analysis of a crowd disaster," *Computer Vision and Image Understanding*, vol. 116, no. 3, pp. 307–319, 2012.

[2] D. Helbing and P. Mukerji, "Crowd disasters as systemic failures: Analysis of the love parade disaster," *EPJ Data Science*, vol. 1, article 1, 2012.

[3] S. Darlington, M. Brocchetto, and D. Ford, "Fire rips through crowded brazil nightclub, killing 233," January 2013.

[4] S. Algadhi and H. S. Mahmassani, "Modelling crowd behavior and movement: application to makkah pilgrimage," in *Proceedings of the 11th Int'l Symposium on Transportation and Traffic Theory*, pp. 59–78, Yokohama, Japan, 1990.

[5] W. Song, J. Ma, and L. Fu, "Comparing three types of real-time data collection techniques: counting cameras, wi-fi sensors and gps trackers," in *Proceedings of the Pedestrian and Evacuation Dynamics 2016 (PED '16)*, W., J. Song, Ma., L., Fu., and L. Ma, Eds., pp. 568–574, 2016.

[6] W. Daamen, E. Kinkel, D. Duives, and S. P. Hoogendoorn, "Monitoring visitor flow and behaviour during a festival: the mysteryland case study," in *Proceedings of the 96th Annual meeting of the Transportation Research Board*, 2017.

[7] A. Danalet, B. Farooq, and M. Bierlaire, "A Bayesian approach to detect pedestrian destination-sequences from WiFi signatures," *Transportation Research Part C: Emerging Technologies*, vol. 44, pp. 146–170, 2014.

[8] M. Nikolic, B. Farooq, and M. Bierlaire, "Exploratory analysis of pedestrian flow characteristics in mobility hubs using trajectory data," in *Proceedings of the Swiss Transportation Research Conference (STRC)*, Ascona, Switzerland, April 2013.

[9] M. Yamin, "A framework for improved hajj management and research," in *Proceedings of the International Conference on Wireless Communications And Sensor Networks*, pp. 1–5, 2006.

[10] L. Hogenkamp, Festivak 2013: Voorspelling 4daagse: vanaf 2016 crowdmanagement via gps, November 2013.

[11] S. Bocconi, A. Bozzon, A. Psyllidis, C. T. Bolivar, and G.-J. Houben, "Social glass: A platform for urban analytics and decision-making through heterogeneous social data," in *Proceedings of the 24th International Conference on World Wide Web, WWW 2015*, pp. 175–178, Florence, Italy, May 2015.

[12] A. de Vries, App: Druktemeter, September 2014.

[13] D. Heidemann, "Queue length and delay distributions at traffic signals," *Transportation Research Part B: Methodological*, vol. 28, no. 5, pp. 377–389, 1994.

[14] Y. Wang, M. Papageorgiou, and A. Messmer, "A real-time freeway network traffic surveillance tool," *IEEE Transactions on Control Systems Technology*, vol. 14, no. 1, pp. 18–32, 2006.

[15] G. Vigos and M. Papageorgiou, "A simplified estimation scheme for the number of vehicles in signalized links," *IEEE Transactions on Intelligent Transportation Systems*, vol. 11, no. 2, pp. 312–321, 2010.

[16] C. Antoniou, H. N. Koutsopoulos, and G. Yannis, "Dynamic data-driven local traffic state estimation and prediction," *Transportation Research Part C: Emerging Technologies*, vol. 34, pp. 89–107, 2013.

[17] Z. Amini, R. Pedarsani, A. Skabardonis, and P. Varaiya, "Queue-length estimation using real-Time traffic data," in *Proceedings of the 19th IEEE International Conference on Intelligent Transportation Systems, ITSC 2016*, pp. 1476–1481, Rio de Janeiro, Brazil, November 2016.

[18] A. Nantes, D. Ngoduy, A. Bhaskar, M. Miska, and E. Chung, "Real-time traffic state estimation in urban corridors from heterogeneous data," *Transportation Research Part C: Emerging Technologies*, vol. 66, pp. 99–118, 2016.

[19] G. Zhu, K. Song, P. Zhang, and L. Wang, "A traffic flow state transition model for urban road network based on Hidden Markov Model," *Neurocomputing*, vol. 214, pp. 567–574, 2016.

[20] J. J. Fruin, *Pedestrian Planning and Design*, 1971.

[21] C. Papageorgiou and T. Poggio, "Trainable system for object detection," *International Journal of Computer Vision*, vol. 38, no. 1, pp. 15–33, 2000.

[22] D. M. Gavrila, "Pedestrian detection from a moving vehicle," in *Computer Vision — ECCV 2000*, vol. 1843 of *Lecture Notes in Computer Science*, pp. 37–49, Springer, Berlin, Germany, 2000.

[23] P. Felzenszwalb, "Learning models for object recognition," in *Proceedings of the 2001 IEEE Computer Society Conference on Computer Vision and Pattern Recognition. CVPR 2001*, pp. I-1056–I-1062, Kauai, HI, USA, 2001.

[24] A. Mohan, C. Papageorgiou, and T. Poggio, "Example-based object detection in images by components," *IEEE Transactions on Pattern Analysis and Machine Intelligence*, vol. 23, no. 4, pp. 349–361, 2001.

[25] P. Viola, M. J. Jones, and D. Snow, "Detecting pedestrians using patterns of motion and appearance," *International Journal of Computer Vision*, vol. 63, no. 2, pp. 153–161, 2005.

[26] K. Mikolajczyk, C. Schmid, and A. Zisserman, "Human detection based on a probabilistic assembly of robust part detectors," in *Computer Vision - ECCV 2004*, vol. 3021 of *Lecture Notes in Computer Science*, pp. 69–82, Springer, Berlin, Germany, 2004.

[27] X. Du, M. El-Khamy, J. Lee, and L. Davis, "Fused DNN: A deep neural network fusion approach to fast and robust pedestrian detection," in *Proceedings of the 17th IEEE Winter Conference on Applications of Computer Vision, WACV 2017*, pp. 953–961, Santa Rosa, Calif, USA, March 2017.

[28] J. Li, X. Liang, S. Shen, T. Xu, J. Feng, and S. Yan, "Scale-aware Fast R-CNN for pedestrian detection," *IEEE Transactions on Multimedia*, 2017.

[29] A. Bera, D. Wolinski, J. Pettré, and D. Manocha, "Realtime pedestrian tracking and prediction in dense crowds," *Group and Crowd Behavior for Computer Vision*, pp. 391–415, 2017.

[30] A. Leykin and R. Hammoud, "Robust multi-pedestrian tracking in thermal-visible surveillance videos," in *Proceedings of the 2006 Conference on Computer Vision and Pattern Recognition Workshops*, New York, NY, USA, June 2006.

[31] L. Fu, S. Yu, and Y. Jia, "A pedestrian tracking algorithm based on mean shift using color histogram equalization method," in *Proceedings of the 13th IEEE International Symposium on Autonomous Decentralized Systems, ISADS 2017*, pp. 235–240, Bangkok, Thailand, March 2017.

13

Dynamic Route Network Planning Problem for Emergency Evacuation in Restricted-Space Scenarios

Yi Hong, [1] **Deying Li**, [2] **Qiang Wu**, [3] **and Hua Xu** [1]

[1] *Information Engineering College, Beijing Institute of Petrochemical Technology, Beijing 102617, China*
[2] *School of Information, Renmin University of China, Beijing 100872, China*
[3] *National Engineering Research Center of Coal Mine Water Hazard Controlling, China University of Mining and Technology, Beijing, Beijing 100083, China*

Correspondence should be addressed to Yi Hong; hongyi@bipt.edu.cn

Academic Editor: Zhi-Chun Li

We investigate a dynamic route planning problem in restricted-space evacuation, namely, the *Multiobjective Dynamic Route Network Planning (MODRNP) problem*. It models the multisource to multidestination evacuation in restricted-space scenarios, with the objectives of minimizing the whole evacuation delay and maximizing the evacuation efficiency. We study the problem in 3D scenarios, which can provide intuition vision for the geographic space and contribute to the evacuation plan and implementation. Based on the auxiliary graph transformation, we propose a heuristic algorithm referred to the classical problem, Minimum Weighted Set Cover. We finally conduct extensive experiments to evaluate the performance of the proposed algorithm and give an application instance on a typical kind of restricted-space scenarios. The results indicate that the proposed algorithm outperforms the existing alternatives in terms of the utilization as well as timeliness.

1. Introduction

As a significant research problem, disaster prevention and relief became one of the new requirements for social development in recent years, and emergency evacuation planning is an important part of rescue and relief work. Emergency evacuation generally contains the evacuee evacuation [1] and the relief resource deployment [2], and the evacuee evacuation is the most urgent work of disaster relief. An effective evacuation path planning strategy can not only effectively reduce the evacuation time consumption, but also prevent unnecessary congestion in the evacuation process, even secondary damage. Thus, the optimization design and theoretical analysis for evacuation path planning are of great practical significance [3].

According to the disaster scenarios, evacuation path planning can be classified into free-space (e.g., urban streets and squares) planning and restricted-space (e.g., underground pipeline, tunnel, and mine laneway) planning [4, 5]. The former one has most applications in city-security and antiterrorist, which has been well studied based on one-source to one-destination model or one-source to multidestination model.

The later one involves in a large range of industrial applications, i.e., the evacuation planning in fire disaster, gas explosion, or water inrush in underground mine. The restricted-space has several complex characteristics which are different from the free space: **(a) the space and time consuming limitations on entrance and evacuation. The density of personnel is sparser; (b) the unsuitability of long operating time. The reliability of temporary shelters is weaker; (c) the trapped-vulnerability of internal structure in disasters.** When a disaster occurs, the number of available evacuation paths is much less. Thus, the restricted-space has higher accident rate and brings more difficulties and challenges to rescue and relief work, which has been urgent to be studied. To improve the quality of emergency response in restricted-space planning, we focus on dynamic path planning based on the multisource to multidestination model.

In practical, restricted-space emergency evacuations have different evacuation requirements according to the accident type, nature, and distribution. On the one hand, considering the limitations and unsuitability of restricted spaces, it may cause local or global congestions when evacuation traffic load is too large, e.g., the evacuees intensively rushing to a certain exit. Thus, evacuation requirements do not only include the shortest evacuation time consumption, but also the unobstructed connectivity of route network. The goals of the restricted-space emergency evacuation are minimizing the global evacuation time consumption and maximizing the utilization rate of each exit (balancing among the exits' traffic loads). **To minimize the global evacuation time consumption**, each evacuee tends to choose the shortest path and nearest exit, which may cause the imbalance of exit utilization, i.e., most evacuees intensively rushing to a certain exit via their shortest paths. **To maximize the utilization rate of each exit**, the evacuees should dispersedly choose the exits, which may cause that some evacuee chooses the suboptimal path rather than the shortest one and its evacuation time consumption will be lengthened. Thus, the two objectives are conflicting and they need a tradeoff to balance their realization, i.e., each evacuee should choose the global optimum rather than local optimum. On the other hand, due to the trapped-vulnerability of restricted spaces, the static planning strategies cannot adapt to the dynamic spreading of disasters or accidents and other environment changes. Thus, it is necessary to design dynamic evacuation planning based on environment changes. Based on the above considerations, we propose new evacuation routes planning problem, the Multiobjective Dynamic Route Network Planning (MODRNP) problem, and aim to design dynamic route network planning strategies and give theoretical analysis.

In this paper, firstly we introduce the restricted-space model and formulate the Multiobjective Dynamic Route Network Planning (MODRNP) problem in restricted-space evacuation. Secondly, we propose dynamic and global-optimized strategies to construct the evacuation route network, whose performance will be evaluated by the simulations and instance experiments at last. The rest of the paper is organized as follows. We summarize the related work in Section 2. Section 3 introduces the restricted-space model and problem definitions. Section 4 introduces the strategy framework for MODRNP Problem and describes the algorithms. Simulation results and corresponding discussions are given in Section 5. Section 6 concludes this paper.

2. Related Works

As a crucial kind of combinational optimization problem, the path planning problem has been widely studied such like recent research works on traveling salesman problem (TSP)[6], traffic assignment/vehicle routing problem [7–9], and evolutionary optimization problems, e.g., general transportation planning problems, facility location problem, and roadway repair problem [10]. Based on the existing theoretical achievements, the research on evacuation path planning can be classified as static planning and dynamic planning.

Research on the **static path planning problem provides theoretical guideline for the emergency analysis and escaping exercise preparation.** Most algorithms for it are based on Dijkstra algorithm, improved Dijkstra algorithm, Floyd algorithm, the first k shortest path algorithm, A^* algorithm, and dynamic programming. Under the static problem model, the solutions generally simplified the disaster type, evacuee psychological pattern, and other factors by introducing the path equivalent length, with the assumption that the disaster would be spreading steadily.

The studies on the **dynamic path planning problem** are the basis of real-time decision making in emergency response and management in disaster/accident. The solutions can be classified into two kinds, the approximate algorithms with high theoretical precision and the heuristic ones with high efficiency.

To design approximate algorithms for the dynamic problem model, most researchers firstly transformed the problem into the classical network flow problem model [11–14] and then utilized the classical polynomial time algorithms to design the path selection strategy. **By applying the maximum flow algorithm**, Dunn and Newton [15] constructed a path network with the goal of guiding the most survivors in accordance with the constructed path network under the condition of permission capacity. **By applying the minimum cost flow algorithm**, Yamada [16] assigned evacuation traffic aimed at minimizing the overall length of the chosen evacuation paths, and Cova and Johnson[17] focused on complex path network with the multiple objectives of minimizing the intersections between the evacuation paths and minimizing the overall evacuation length. **By applying the fast flow algorithm**, the authors in [18, 19] aimed at the application scenarios modeled as directed graph. But the above algorithms could be only applied to the case of grid structure path network and unified path capacity. **By applying the contraflow algorithm**, the authors in [20] considered the setup time of the contraflow operation and proposed a two-layer algorithm to minimize the evacuation time consumption and the setup time for flow patterns. Since the algorithms for the flow problems have high time or space complexity, most of these strategies lack high efficiency in practice.

To design heuristic algorithms for the dynamic problem model, researchers in [21, 22] designed simulated evolutionary algorithms. They took full advantages of the positive feedback information mechanism of ant colony algorithm and the fast convergence of the genetic algorithm [23]. For the airport emergency evacuation scenarios, the authors in [24] performed a series of simulations based on an agent-based model (ABM) to determine the collective behaviors and the overall evacuation time, which are affected by the environment and complex structures. And [25] was motivated by the improvement of the marine evacuation system plan in a practical case, Dalian offshore airport of China, and developed a modeling framework to evaluate and optimize the emergency evacuation capability. For the crowded building emergency evacuation scenarios, the authors in [26] studied the human behavior of crowd evacuation to guide the design of buildings and the reinforcement of security management. The proposed algorithms can solve the path planning problem in densely populated scenarios. Due to the sparsity of personnel in restricted spaces, these strategies cannot be applied directly in the evacuation planning.

The above related work had limitations at three aspects. Firstly, most research works considered application scenarios for two-dimensional (2D) planes, in which the proposed algorithms cannot adapt to three-dimensional (3D) application scenarios. Secondly, most existing solutions for the path network construction problem were with high time or space complexity, which may bring low efficiency in restrict space evacuation. Thirdly, the existing researches on dynamic planning of emergency evacuation paths were still limited; e.g., the authors in [27] conducted research on the vehicle routing problem for postdisaster scenarios. And their application effect in practical is not ideal in terms of dynamic adjustment and timely feedback. Therefore, we intend to introduce a new dynamic planning problem model and utilized new technology in solution for restricted-space scenarios.

3. Problem Formulation

3.1. Restricted-Space Model. We focus on the emergency evacuation in confined space works and model such a densely populated restricted-space, as a 3D connected graph $G = (V, E)$, where $V = \{v_1, \cdots, v_n\}$ is the set of predetermined observation vertices in the space, E is the set of bidirectional edges, and for each $v_i \in V$, $v_i = (ID_i, x_i, y_i, z_i)$ and for each $e_k \in E$, $e_k =< ID_i, ID_j > (1 \le i < j \le n)$. For any vertex subset $V' \subseteq V$, $G[V']$ is the subgraph of G induced by V'. Similarly, $G[E']$ is the subgraph of G induced by an edge subset $E' \subseteq E$. The source set S and the destination set D are important subsets of the vertex set in G, which are composed of the evacuees' initial positions and the feasible exits, respectively. Here we assume S and D are predefined.

When a disaster or an accident happens, the evacuees need to escape via the edge/segment in the restricted-space in the shortest possible time. And the traffic capacity of each edge is affected by several influence factors, which need to be considered in the decision of the most reliable evacuation paths. We list the main influence factors of the restricted-space as follows.

(a) *Edge types:* $\beta_1(e_k)$.The properties like the shape and slope of the edge/section decide the different types of the edge. Different edge types have different level effects on evacuee evacuation. Here we consider the scope of the edge, and for any pair of edges $< ID_i, ID_j >$ and $< ID_j, ID_i >$, $\beta_1(< ID_i, ID_j >) = -\beta_1(< ID_j, ID_i >) (1 \le i < j \le n)$.

(b) *Passibility:* $\beta_2(e_k)$. The passibility of the edge is greatly influenced by the obstacles and the vehicle of the evacuees. We assume that the evacuees adopt a uniform activity manner and have the same traveling speed. For any pair of edges $< ID_i, ID_j >$ and $< ID_j, ID_i >$, $\beta_2(< ID_i, ID_j >) = \beta_2(< ID_j, ID_i >) (1 \le i < j \le n)$.

(c) *Exit escaping priority:* $priority(d_i)$. Due to the exits' security levels, structural characteristics, or functional characteristics, different exits have different escaping priorities $\{priority(d_i) \mid \forall d_i \in D\}$ which have influence on the edge's traffic.

(d) *Evacuation priority:* $\alpha(e_k)$. The factor is generated from the incident relationship with the destinations in D, i.e., if edge e_k is an incident edge with more than one d_i in D, $\alpha(e_k) = \max_{d_i \in D \text{ and } d_i \in e_k} priority(d_i)$. For any pair of edges $< ID_i, ID_j >$ and $< ID_j, ID_i >$, $\alpha(< ID_i, ID_j >) = \alpha(< ID_j, ID_i >) (1 \le i < j \le n)$.

(e) *Travel length:* $length(e_k)$. The geometric length of the edge. For any pair of edges $< ID_i, ID_j >$ and $< ID_j, ID_i >$, $length(< ID_i, ID_j >) = length(< ID_j, ID_i >) (1 \le i < j \le n)$.

Based on these influence factors, the traffic capacity on an edge can be expressed in terms of the equivalent length/edge weight: $weight(e_k) = length(e_k) \cdot \beta_1(e_k) \cdot \beta_2(e_k) \cdot \alpha(e_k), 1 \le k \le |E|$. Thus, the constrained space can be further modeled as an edge-weighted graph $G = (V, E, W)$. Here we adopt travel length instead of traveling time to be the basis of traffic capacity for the reason that the traveling speed of each evacuee is regarded as a uniform value.

3.2. Disaster Spreading Model. When the disaster or accident happens, the spreading speed and mode have great influence on the decision of evacuation route. Generally, the disaster spreading pattern is determined by several influence elements, e.g., the fire spreading pattern is effected by carbon monoxide in the volume flow of smoke and fire gases CO, oxygen in the fire place O_2, heat flow, and so on. And the spreading pattern is directly reflected in the speed of disaster on each edge. Generally the spreading pattern can be modeled as a statistical model depending on the disaster itself.

To this end, we value the maximum spreading time duration of the disaster as T_{dis}; i.e., it takes T from the disasters happening to its whole coverage. Note that we assume that all the evacuees can find the feasible evacuation path and their evacuation time consumption is not exceeding T_{dis}. We also assume that T_{dis} can be splitted as several timeslots according to the spreading rule and the evacuation time can be measured as the same timeslot. Thus, the disaster spreading pattern can be modeled as a function of $\mathcal{F}(G_t, t)$, where $G_t = (V_t, E_t, W_t)$ is the temporal space subgraph of G determined by disaster spreading speed in the t-th timeslot. And the influence factor of the disaster is as follows.

(f) *Disaster spreading status:* $\beta_3(e_k, t)$. In each timeslot, we value $\beta_3(e_k, t)$ according to $\mathcal{F}(G_t, t)$. For any pair of edges $< ID_i, ID_j >$ and $< ID_j, ID_i >$, $\beta_3(< ID_i, ID_j >, t) = \beta_3(< ID_j, ID_i >, t) (1 \le i < j \le n)$.

Based on the influence factor, the edge weight in G_t can be further expressed as $weight(e_k, t) = weight(e_k) \cdot \beta_3(e_k, t), 1 \le k \le |E|$. Note that the edge weight is a time-related measure only in G_t depending on disaster spreading status.

3.3. Problem Definitions. We consider a Multiobjective Dynamic Route Network Planning problem in a restricted-space evacuation, which is to design a staged and adjusted construction scheme of evacuation path network from all the evacuees in S to the emergency exit set D in a 3D space graph $G = (V, E, W)$.

Firstly, the problem is motivated from two points of view: the one is for each evacuee and the goal is to decide its globally ideal escape exit based on evacuation priority and find the escaping path to the ideal exit; the other one is for each escape exit. In most of practical applications, the feasible escape exits have different evacuation priorities; e.g., in underground coal mine evacuation, the auxiliary and main shafts have higher priority than chambers or mobile capsules. Thus, to enhance the utilization rate of each exit, it is needed to assign relatively balanced count of evacuees to the exit such that the case of congesting in some an overdrawn exit can be avoided.

Secondly, the problem has two objectives; one is in time dimension, i.e., minimizing the whole evacuation travel length. Here *the whole evacuation travel length* is defined as the travel length consumed by the latest evacuee, which can be referred to as the length of the longest escaping path, $\max_{\forall d_i \in D, \forall s_j \in S} length(d_i, s_j)$. And the other one is in space dimension, i.e., maximizing the priority-oriented evacuation efficiency. *The priority-oriented evacuation efficiency* is a global measure representing the balance performance among each exit's utilization efficiency; i.e., a smaller efficiency stands for a higher balance performance. And the priority-oriented evacuation efficiency is defined as the maximum value among the utilization efficiency differences between any pairs of exits, $\max_{d_p, d_q \in D} |priority(d_p)/|d_p| - priority(d_q)/|d_q||$, where d_i represents two meanings, the exit d_i itself and the set of the sources in S escaping from d_i finally.

Thirdly, the problem has dynamic constrains, i.e., disaster spreading status, which is the most influential one among the factors for traffic capacity and is varying along with the evacuee evacuation. In each timeslot, the disaster spreading pattern is regarded as an input item of the path planning process and it can be modeled as a function of $\mathscr{F}(G_t, t)$, where G_t is the temporal version of G.

Our problem in restricted-space evacuation is defined as follows, which is a staged problem model.

Definition 1 (Multiobjective Dynamic Route Network Planning (MODRNP) problem).

Input: Given a 3D restricted-space model, graph $G = (V, E, W)$, source set S, and destination set D with its priority set, $priority(D)$, the disaster spreading pattern $\mathscr{F}(G_t, t)$.

Find: Find the route network in each timeslot t from all the sources to their optimal destinations, and the whole route network is constructed with two goals, minimizing the whole evacuation travel length $\max_{\forall d_i \in D, \forall s_j \in S} length(d_i, s_j)$ and maximizing the priority-oriented evacuation efficiency $\max_{d_p, d_q \in D} |priority(d_p)/|d_p| - priority(d_q)/|d_q||$.

Theorem 2. *MODRNP Problem in restricted-space evacuation is NP-hard.*

Proof. In order to proof the hardness of MODRNP Problem, we firstly decompose the problem in each timeslot as a subproblem and rewrite the subproblem into a mathematical formulation as below: In a 3D connected and edge-weighted graph $G_t = (V_t, E_t, W_t)$, given two vertex subsets of V_t, the source set S_t and the destination set D, and D's priority set of $priority(D)$, the problem is to find a path network from S_t to D such that

(i) the maximum difference among D's utilization efficiencies, $\max_{d_p, d_q \in D} |priority(d_p)/|d_p| - priority(d_q)/|d_q||$, can be minimized, where d_p is not only a destination in D, but also a set of nodes in S_t and connected to d_p by the constructed paths,

(ii) the maximum path length among all sources and destinations, $\max_{\forall d_i \in D, \forall s_j \in S_t} length(d_i, s_j)$, can be minimized.

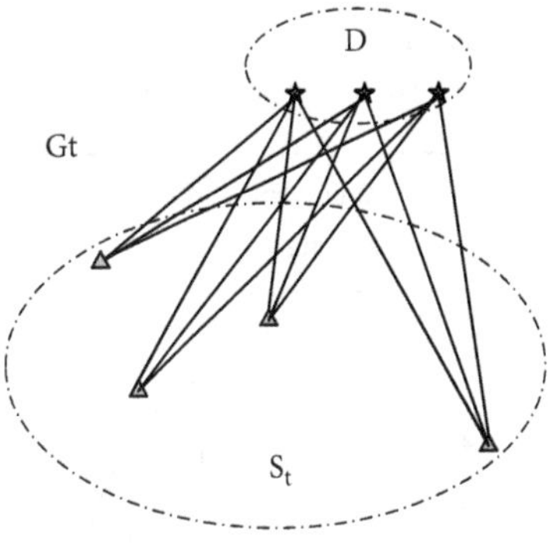

FIGURE 1: Bipartite graph transformation.

Then we consider a special case of it: Case (a). $\forall d_i \in D, \forall s_i \in S_t$, $length(d_i, s_j)$s are a uniform value; i.e., the maximum value in (ii) is unified as a constant. Thus, the minimization of the maximum value can be regarded a default implement. Based on Case (a), we transform these isometric paths into the corresponding one-hop edges; i.e., G_t is transformed into a bipartite graph with the two vertex sets S_t and D, $G_t[D, S_t]$ as shown in Figure 1.

Case (b). $\exists d_0 \in D$, $priority(d_0) = 0$, i.e., the objective measure in (i) $\min \max_{d_p, d_q \in D} |priority(d_p)/|d_p| - priority(d_q)/|d_q|| = \min_{d_p \in D} |priority(d_p)/|d_p| - 0/|d_0|| = \min_{d_p \in D}(priority(d_p)/|d_p|)$. For each node $d_i \in D$, d_i can be used to represent its accessible nodes in S_t, i.e., $d_i \subseteq S_t$. When each d_p is assigned the weight of $(priority(d_p) \cdot |S_t|)/(|d_p| \cdot |D|)$, the problem becomes to find a conditional set cover in the bipartite graph $G_t[D, S_t]$.

Based on the above problem variant, the subproblem in Cases (a) and (b) is to find a subcollection $\mathscr{C} \subseteq \mathscr{D}$ (here $\mathscr{D} = \{d_1, d_2, \ldots, d_{|D|}\}$) such that $\bigcup_{d_i \in \mathscr{C}} d_i = S$ and the total weight of d_i in $\mathscr{C}$ is minimized. Here we review an important and classical NP-hard problem, **Minimum Weighted Set Cover (MWSC) Problem**: Given a set $\mathscr{X}$ composed of n elements, a collection $\mathscr{D}$ of m subsets of $\mathscr{X}$ ($\mathscr{D} = \{\mathscr{X}_1, \mathscr{X}_2, \ldots, \mathscr{X}_m\}$), and each $\mathscr{X}_i \subseteq \mathscr{X}$ $(1 \le i \le m)$ has a weight $w(\mathscr{X}_i)$, the problem is to find a subcollection $\mathscr{C} \subseteq \mathscr{D}$ such that $\bigcup_{\mathscr{X}_i \in \mathscr{C}} \mathscr{X}_i = \mathscr{X}$ and $\sum_{\mathscr{X}_i \in \mathscr{C}} w(\mathscr{X}_i)$ is minimized. It is easy to discover that this special version of each subproblem is equivalent to MWSC problem, which was proved to be NP-hard [28]. Therefore, MODRNP Problem is NP-hard in general. □

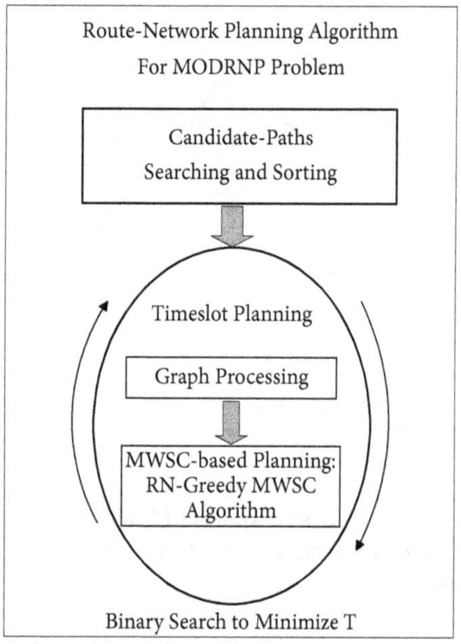

FIGURE 2: The framework of Route Network Planning Algorithm.

4. Multiobjective Dynamic Route Network Planning Algorithm

To solve the NP-hard problem MODRNP, we propose a partition strategy (the framework is shown in Figure 2), which has a uniform algorithm and progressive inputs in each timeslot. The division of timeslot is determined by the disaster spreading function $\mathcal{F}(G_t, t)$ and it generates the temporal graph G_t in timeslot t. In each timeslot, we construct the route network according to the following iterations.

In each timeslot t,

(i) Phase 1: update the space graph according to the spreading pattern, $G_t \leftarrow \mathcal{F}(G_t, t)$, where G_t can be regarded as a progressive input for the algorithm.

(ii) Phase 2: search for the current route network $\mathcal{P}_t$ in G_t, which connects the current source set S_t to the set D. Since the evacuees from the current sources may not reach the destinations D in this timeslot limited by the evacuee speed, it is also needed to extract the path set $\mathcal{P}_t^*$ from S_t to their destinations in the current timeslot, collected by D_t.

(iii) Phase 3: update S_t to D_t as the new source set in the next timeslot and enter timeslot $t + 1$.

In Phase 2, the searching strategy consists of three steps.

Firstly, we find all the shortest paths from each node of S_t to each node of D in G_t based on Dijkstra Algorithm and record all the obtained paths as a subgraph G_t'.

Secondly, we sort all the paths of G_t' in the nondecreasing order of their lengths to obtain an ordered path set $G_t' = \{path_1, path_2, \cdots, path_q\}$ as shown in step (3) in Algorithm 1.

Thirdly, we construct a route network from the current source set S_t to D, whose details are as follows: (1) construct a subgraph G_t' composed of all the shortest paths from each node of S_t to the whole destinations of D via Dijkstra Algorithm; (2) design a greedy strategy to select the optimal whole destination for each node of S_t in G_t' based on Minimum Weighted Set Cover and record the selected shortest paths as RN; (3) for each node in S_t, decide its current destination on its path in G_t' and store the corresponding paths into RN. The construction is under the goals of minimizing the longest path and maximizing the priority-oriented evacuation efficiency.

The former objective is realized by a binary search as shown by the while loop in steps (5)-(17). In each iteration of the binary search, for each $path_i \in G_t'$, starting from $Mid = \lfloor (1 + q)/2 \rfloor$, we **temporarily** remove all the $path \in G_t'$ whose length satisfies $length(path) \geq length(path_{Mid})$ and check if the subgraph of G_t' can still construct a route network without those edges, which is verified by a searching algorithm. **If a route network can be constructed, we obtain a route network RN on the subgraph of G_t' and proceed with the increased Mid. Otherwise, we keep those edges in G_t' and proceed with the decreased Mid.**

The latter objective, balancing the evacuation efficiency among all the nodes of D, is implemented by the route network construction in each iteration of the binary search: we propose a path planning algorithm, **RN-Greedy-MWSC**(S_t, D) (Algorithm 2). To compute evacuation paths from multiple sources to their most suitable destinations, we referred to the classical approximation algorithm with the ratio of $\ln(n + 1)$ for MWSC problem: select a subset $\mathcal{T} \in \mathcal{F}$ that minimizes the whole redundant weight $(w(\mathcal{T})/|\mathcal{T} \bigcap \mathcal{U}|)$ in each loop, where $\mathcal{U}$ stands for the uncovered elements in $\mathcal{X}$; repeat the greedy selection until the set of $\mathcal{T}$s cover $\mathcal{X}$. Thus, the criteria of greedy selection here are the redundant utilization, which is defined in step (8) of Algorithm 2; i.e., we repeat selecting a best d_i with the minimum $|redutil(d_i)/|d_i \bigcap \mathcal{S}||$ value in each iteration until each node S_t can be reached by the selected d_is.

Theorem 3. *The time complexity of Route Network Planning Algorithm for MODRNP Problem is* $\mathbf{O(n^2)}$.

Proof. For the four phases for Algorithm 1, the first one (Step (2)) is constructing all the candidate paths between S_t and D which has $\mathbf{O(n^2)}$ time complexity. In the second phase (Step (3)), sorting the candidate paths based on their lengths takes $\mathbf{O((|S| * |D|)^2)}$ time. Thus, the time cost in this phase is $O(n \log n)$. For Phase 3 (Steps (5)-(17)), the binary search for the current route network construction of the auxiliary graph has the time complexity of $\mathbf{O(|D| \cdot \log(|S| \cdot |D|))}$. In the final phase (Steps (18)-(21)), the worst-case time complexity is $\mathbf{O(|S|)}$. To conclude, the whole algorithm has the time complexity of $\mathbf{O(n^2)}$. □

5. Performance Evaluation

5.1. Simulation Plan. We perform simulation experiments based on the random generation of 3D networks $G = (V, E)$. The vertex set V stands for the observation nodes which are randomly deployed for a constrained space, and the deployment of it is randomly assigning n nodes' 3D

(1) **Construct a route network from S_t to D_t in timeslot t.** *in:* G_t, S_t, D *out:* $\mathcal{P}_t^*, D_t$
(2) **Candidate-paths searching:** Find all the shortest path from each node of S_t to each node of D in G_t based on Dijkstra Algorithm, and record all the paths into subgraph G_t'. Note that for each $d_i \in D$, d_i is also represented as the set of the sources in S_t which can reach it in G_t.
(3) **Candidate-paths sorting:** Sort the paths in G_t' according to their weights in the non-decreasing order and store the order in $G_t' = \{path_1, path_2, \cdots, path_q\}$.
(4) Set $\mathcal{P}_t^* \longleftarrow \emptyset, D_t \longleftarrow \emptyset, Min = 1, Max = q, Mid = 0$
(5) **while** $(Min \le Max)$ **do**
(6) $Mid = \lfloor \frac{Min + Max}{2} \rfloor, G_{temp} = \{path \mid path \in G_t' \text{ and } length(path) \ge length(path_{Mid})\}$;
(7) $\forall d_i \in G_t', d_i \longleftarrow d_i \setminus \{s_j \in G_{temp}\}$;
(8) **Graph processing:** $G_t' \longleftarrow G_t' \setminus G_{temp}, D \longleftarrow D \setminus \{d_i \mid \forall d_i \in G_{temp} \text{ and } |d_i| = 0\}$;
(9) **RN-Greedy-MWSC(S_t, D):** Find whether a path set can be constructed in G_t';
(10) **if** (RN-Greedy-MWSC(S_t, D) returns **False**) **then**
(11) $Max = Mid$;
(12) $\forall d_i \in G_{temp}, d_i \longleftarrow d_i \bigcup \{s_j \in G_{temp}\}$;
(13) $G_t' \longleftarrow G_t' \bigcup G_{temp}, D \longleftarrow D \bigcup \{d_i \mid \forall d_i \in D \cap G_{temp}\}$;
(14) **else**
(15) $Min = Mid, \mathcal{P}_t$ =RN-Greedy-MWSC(S_t, D);
(16) **end if**
(17) **end while**
(18) **for** each node $s_j \in S_t$ **do**
(19) Find s_j's path in $\mathcal{P}_t$ and calculate s_j's actual destination d_j^t on the path based on the evacuee speeding and t's duration;
(20) $D_t \longleftarrow D_t \bigcup \{d_j^t\}, \mathcal{P}_t^* \longleftarrow \mathcal{P}_t^* \bigcup \{path(s_j, d_j^t)\}$;
(21) **end for**
(22) **return** $\mathcal{P}_t^*, D_t$.

ALGORITHM 1: Route-Network Planning Algorithm.

(1) **Construct a route network from S_t to D.** *in:* S_t, D *out:* RN
(2) Set $avr = \sum_{d_i \in D} |d_i|/|D|$, $sumpri = \sum_{d_i \in D} priority(d_i)$.
(3) **for** each $d_i \in D$ **do**
(4) Use its redundant utilization as its weight, $redutil(d_i) = |d_i| - TH$, where $TH = avr \cdot (sumpri + priority(d_i))/sumpri$ is the threshold of the traveling-person number for an exit.
(5) **end for**
(6) Set $\mathcal{S} \longleftarrow S_t, \mathcal{D} \longleftarrow D, RN \longleftarrow \emptyset$.
(7) **while** $(\mathcal{D} \ne \emptyset)$ **do**
(8) Select the $d_i \in \mathcal{D}$ with the minimum average redundant utilization ratio, $|redutil(d_i)/|d_i \bigcap \mathcal{S}||$;
(9) **if** $(|d_i| > TH)$ **then**
(10) $d_i \longleftarrow$ {The first TH nearest nodes in d_i}
(11) **end if**
(12) $d_i \longleftarrow d_i \bigcap \mathcal{S}, \mathcal{D} \longleftarrow \mathcal{D} - \{d_i\}, \mathcal{S} \longleftarrow \mathcal{S} - d_i, RN \longleftarrow RN \bigcup \{d_i\}$.
(13) **end while**
(14) **if** $(\mathcal{S} \ne \emptyset)$ **then**
(15) **return** False;
(16) **else**
(17) **return** RN;
(18) **end if**

ALGORITHM 2: Function: RN-Greedy-MWSC Algorithm.

coordinate values. And the edge set E is randomly generated based on the deployment of V to guarantee that the 3D graph is connected: firstly allocate the connecting relationship to each pair of nodes in V; secondly determine whether the graph is connected. If being connected, the edge generation is successful; otherwise, repeat the edge generation process. We consider the following parameters and their variation modes are described as follows:

(i) $n = |V|$: the number of observation nodes in the constrained space;

(ii) $s = |S|$: the number of source nodes;

(iii) $d = |D|$: the number of destination nodes;

(iv) α: an edge's priority is decided by the relevance between the edge and the destinations with different

Table 1: Measurement index of algorithm performance.

Factor	Calculation	Measurement method
Priority	$\max \left\lvert \frac{priority(d_p)}{\lvert d_p \rvert} - \frac{priority(d_q)}{\lvert d_q \rvert} \right\rvert \ (\forall d_p, d_q \in D)$	A lower priority factor stands for the better performance.
Average length	$\sum(length(d_i, s_j) - SP(d_i, s_j)) / \sum SP(d_i, s_j)$ $(\forall d_i \in D, \forall s_j \in S)$, where $length(d_i, s_j)$ is the path length between d_i and s_j and $SP(d_i, s_j)$ is the length of shortest path between them.	A lower average length factor stands for the better performance.
Global length	$\frac{\lvert \max length(d_i, s_j) - \max SP(d_i, s_j) \rvert}{\max SP(d_i, s_j)} \ (\forall d_i \in D, \forall s_j \in S)$	A lower global length factor stands for the better performance.

priorities; i.e., the priorities of destinations are randomly valued in the range of $[1, |D|]$;

(v) Influence factor: edge type (β_1) is decided based on the slope of the edge; passibility (β_2) is assumed to be uniform in walking mode, i.e., $\beta_2 = 1$; disaster spreading status ($\beta_3(e_k, t)$) is determined by laws and characteristics of disaster spreading, where we adopt a random value function; travel length ($length(e_k)$) is calculated as the Euclidean distance.

For the evaluation of algorithm performances, we compare our algorithm with the shortest path (SP) algorithm in dynamic and static patterns; i.e., we apply the algorithms in each timeslot under $\mathcal{F}(G_t, t)$ for dynamic pattern; meanwhile, we apply these algorithms in static pattern only under the latest status of disaster spreading $\mathcal{F}(G_T, T)$ and without the consideration of the disaster changes. We discuss about the average behavioral characteristics of these algorithms (denoted as **Dynamic RN-Greedy-MWSC, Dynamic SP, Static RN-Greedy-MWSC, and static SP**).

We will investigate the influence of all the relevant parameters on our algorithms and the parameter setting is based on the scale proportions between the pairs of the space size, escaping evacuee (source) and exits (destination) in practical:

(a) n varies from 200 to 550 by the step of 50 and $s = 30$, $d = 6$.

(b) s varies from 10 to 45 by the step of 5 and $n = 300$, $d = 6$.

(c) d varies from 2 to 16 by the step of 2 and $n = 300$, $s = 30$.

For each parameter setting, we run 100 instances and compute their average for evaluation. Based on the parameters and their variation modes, we study the average characteristic behavior of the proposed algorithm and evaluate their average performance in terms of three items in Table 1.

5.2. Simulation Results and Analysis

5.2.1. Performance on Priority Factor. We first observe the impact of n, s, and d on the *priority factor*. Overall in the three subfigures of Figure 3, the priority factors of the dynamic and static SP with the variation of parameters are all higher than those of the dynamic and static RN-Greedy-MWSC. And dynamic RN-Greedy-MWSC is obviously lower than the static one during the second half of the variation.

(a) In Figure 3(a), comparing our algorithms with SP algorithms, with the growth of the space scale, the factors in the three compared algorithms are shown in parabolic shapes and the gap between RN-Greedy-MWSCs and SPs becomes bigger. It can be explained as that enlarging the space scale supplies more feasible evacuation paths. Thus, RN-Greedy-MWSCs can perform their advantages on balancing the utilization efficiency among these paths.

Comparing our algorithm in dynamic and static patterns, we can observe that, with the increasing of n, the factor's fluctuation in static RN-Greedy-MWSC is more obvious than that in dynamic RN-Greedy-MWSC and the dynamic one is trending downward on the whole. When $n > 250$, the advantage of dynamic one becomes obvious. It is because when the density of observation vertices deployment becomes larger, the connectivity of the candidate route network is improved, which makes more influence of disaster spreading on each edge. Dynamic RN-Greedy-MWSC considers the influence changes and thus can select paths with better priority-oriented utilization.

(b) As shown in Figure 3(b), comparing our algorithms with SP algorithms, SP algorithms perform higher than ours on the priority factor and present state of rises and falls. It is because that SP algorithms can guarantee minimizing the route lengths but cannot provide the utilization balance of the routes. Thus, with the changes of source number, the maintaining of the utilization balance may have instability.

Comparing our algorithm in dynamic and static patterns, we also find that when s is larger than 15, the variation trends for the two algorithms become similar and dynamic one maintains the advantage. The number of evacuee s raises the fluctuation in the two algorithms with the peak values at $s = 30$ in dynamic RN-Greedy-MWSC and $s = 35$ in static RN-Greedy-MWSC.

(c) From Figure 3(c), comparing our algorithms with SP algorithms, under a small number of destinations, SP algorithms have undesirable priority factor values but show improvement when $d > 10$. It can be explained that the number of the available and shorter paths can be increased with the growth of destination number, which can raise the efficiency of path utilization. The increasing of the destination

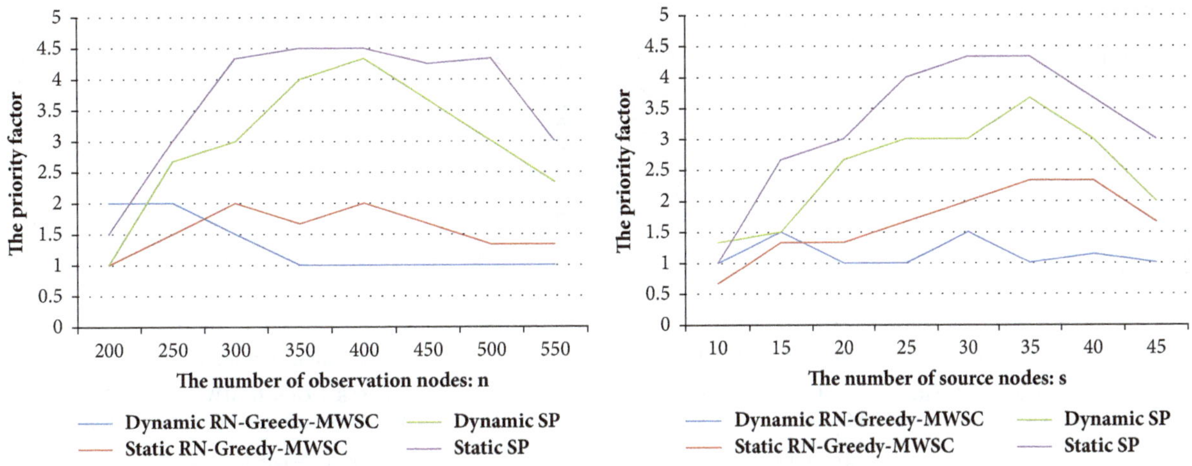

(a) The priority factor versus n

(b) The priority factor versus s

(c) The priority factor versus d

FIGURE 3: Performance on the priority factor.

number is beneficial to improve the utilization balance of the path planning algorithms.

Comparing our algorithm in dynamic and static patterns, we can learn that the priority factor gained from static RN-Greedy-MWSC can be guaranteed into range $[1, 2]$ and the dynamic one shows a trend of decline and becomes much closer to 0 along with the rise of the exit number. The reason is that with the adjustment of exit utilization based on disaster spreading, the priority-oriented utilization for each exit can be enhanced.

5.2.2. Performance on Average Length Factor. Secondly, we investigate the impact of n, s, and d on the *average length factor*. Note that, for the SP algorithm, the average length factors of dynamic and static SP are both 0 ($= length(d_i, s_j) - SP(d_i, s_j)$) which are the lower bounds of all the path-finding algorithms. These three subfigures show that dynamic and static RN-Greedy-MWSCs have gaps with SP on the average length performance. In Figure 4(a), with the growth of due to the consideration of the balanced utilization of each destination, the path assignment in RN-Greedy-MWSC can be adjusted under the change of the path network topological structure. Based on Figures 4(b) and 4(c), it is worth noting that, with the increasing of the number of sources or destinations, the gap between the average length of the escaping network generated by RN-Greedy-MWSC and that by SP tends to be gradually narrowing.

Comparing dynamic RN-Greedy-MWSC with static RN-Greedy-MWSC, the average length factors of the dynamic one with the variation of parameters are all higher than the static one. Based on Figure 4(a), the average length factor of dynamic RN-Greedy-MWSC is in volatility and that of static RN-Greedy-MWSC is gradually raising along with the rise of n. It is because, with the enlarging of the scale of restricted-space, the global lengths of paths between each pair of source and destination will be increased. From Figures 4(b) and 4(c), we can observe that the number of source nodes s has more influence on this factor of our solutions than the number of destination nodes d. With the increasing of s, the factor of two algorithms presents parabolic shape in the range $[10, 25]$ and the gap between the two algorithms reaches its maximum when $s = 20$. And with the growth of d, the factors of two algorithm both show a parabolic shape in a reasonable range $[0, 0.35]$.

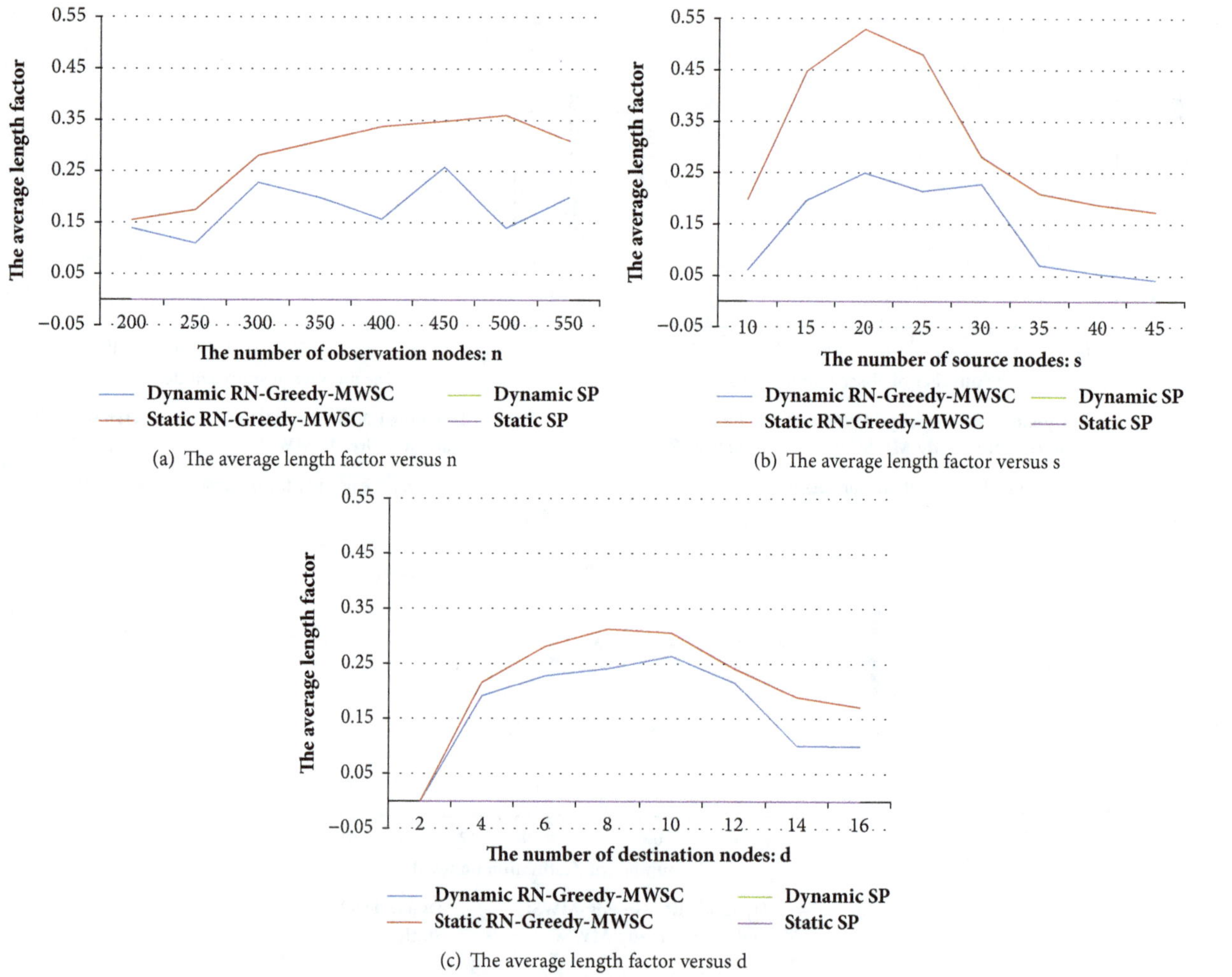

(a) The average length factor versus n

(b) The average length factor versus s

(c) The average length factor versus d

FIGURE 4: Performance on the average length factor.

5.2.3. Performance on Global Length Factor. Similar to the average length factor, the global length factors of dynamic and static SPs are both 0 (= $\max length(d_i, s_j) - \max SP(d_i, s_j)$) which are the lower bounds of our algorithms. Seen from Figure 5(a), the difference between the global length obtained from RN-Greedy-MWSCs and that from SPs is getting smaller with the rising of the space scale. And when n is larger than 450, the difference is getting 0 which indicates our strategy can enhance the utilization efficiency of the escaping exits with the minimization guarantee of the global length. On the contrary, as shown in Figures 5(b) and 5(c), the difference becomes larger with the increasing of the source number and could be controlled in a steady state when $s \geq 35$ or $d \geq 10$. This reason is that, with the rise of the source number or destination number, the utilization efficiency of shorter paths can be increased, which can narrow the gap of global path length between SP algorithms and RN-Greedy-MWSCs.

Comparing dynamic RN-Greedy-MWSC with static RN-Greedy-MWSC, seen from Figures 5(a) and 5(c), the difference between the global escaping time consumption obtained from dynamic solution and that from static scheme is getting smaller with the enlarging of space scale or the rising of the evacuee number. And when n is larger than 450 and $d \geq 14$, the difference is approaching 0 which indicates our strategies can enhance the priority utilization efficiency of multiple escaping exits with the complication of the space structure. On the contrary, as shown in Figure 5(b), the difference between the two algorithms is getting larger since the source count is larger than 25, which is also caused by the dynamic influence of the disaster.

Via the performance evaluations on the three measurement factors, we can draw the conclusions.

(A) The path planning algorithms: firstly, RN-Greedy-MWSC algorithms have more advantages on path utilization than SP algorithms. And the dynamic RN-Greedy-MWSC outperforms the static one in terms of priority factor when the scale of a restricted-space, the scale of sources, and destinations are relatively large; secondly, when evaluating on the average and global path length, the advantage of dynamic RN-Greedy-MWSC maintains signification.

(B) The influence factors: the influence of the scale of sources has more influence on the algorithm performance than that of destinations and space scale.

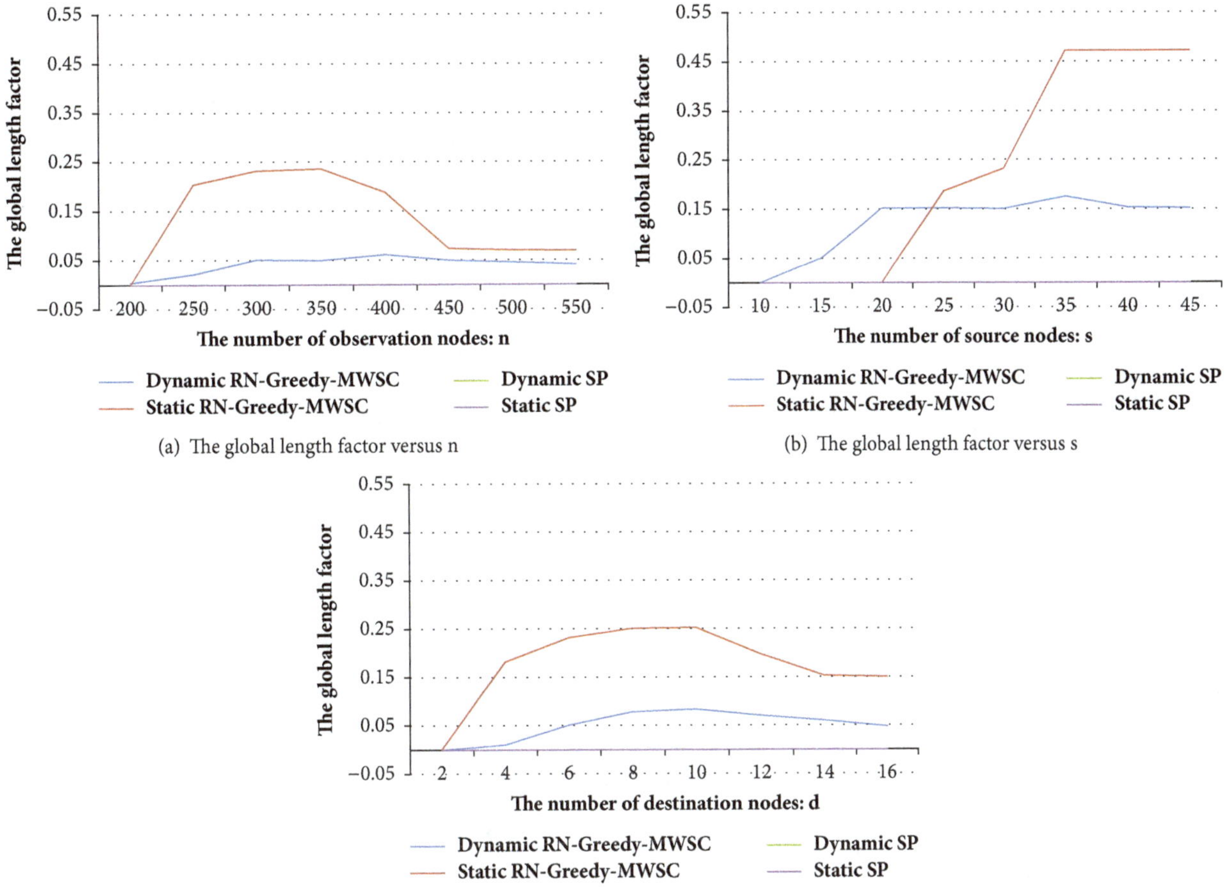

(a) The global length factor versus n

(b) The global length factor versus s

(c) The global length factor versus d

FIGURE 5: Performance on the global length factor.

TABLE 2: Instance experiments results.

	d1	d2	d3	PF	ALF	GLF
Dynamic Algorithm	2	1	7	1.28	0.1321	0.0251
Static Algorithm	3	0	7	∞	0	0

5.3. Application Instance Experiment and Discussion. We further perform an instance experiment on our algorithm in a representative kind of restricted-space application scenarios, underground coal mine water inrush. Here we adopt the 3D model data of a mine area in Zhangjiakou, Hebei Province of China from Kailuan Group Company. We preprocess and transform the original spatial data of the model; i.e., the vertex set in G is composed of all the traverse points, and the edge set of G maintains consistent direction with the original laneway. Water spreading status is obtained by the water spreading algorithm in [29]; i.e., in each timeslot, the prediction data of the water level on all observation points are the input of our algorithms. According to practical requirements, the priorities of three feasible exits, auxiliary shaft (d1), main shaft (d2), and air shaft(d3), are valued as follows: $priority(d1) = 5$, $priority(d2) = 2$, and $priority(d3) = 1.5$.

In the visualization results in Figure 6, we present the paths from the 10 evacuation source positions to their temporal destinations from timeslots 4 to 11, and finally all the evacuees successfully escape from the 3 red exits in the last subfigure. Note that the points in blue stand for the submerged part by water inrush, which are all avoided from by the escapers in our algorithms. In Table 2, the evaluation criteria, priority factor, average length factor, and global length factor are represented by PF, ALF, and GLF in short. The number of paths reaching these exits is denoted as d1, d2, and d3, respectively. The parameter results show that the dynamic algorithm outperforms the static one in terms of priority factor in a practical application scene with large scale.

To conclude, our algorithm has advantages on the performance of the escaping exit priority utilization and the evacuation time consumption. And when the scale of space, evacuee, and exits are relatively high, the advantage will be significant. Furthermore, the parameter, the number of evacuees, has more influence on our algorithm than the number of exits on the length factor. Therefore, our strategy can enhance the priority utilization efficiency of the escaping

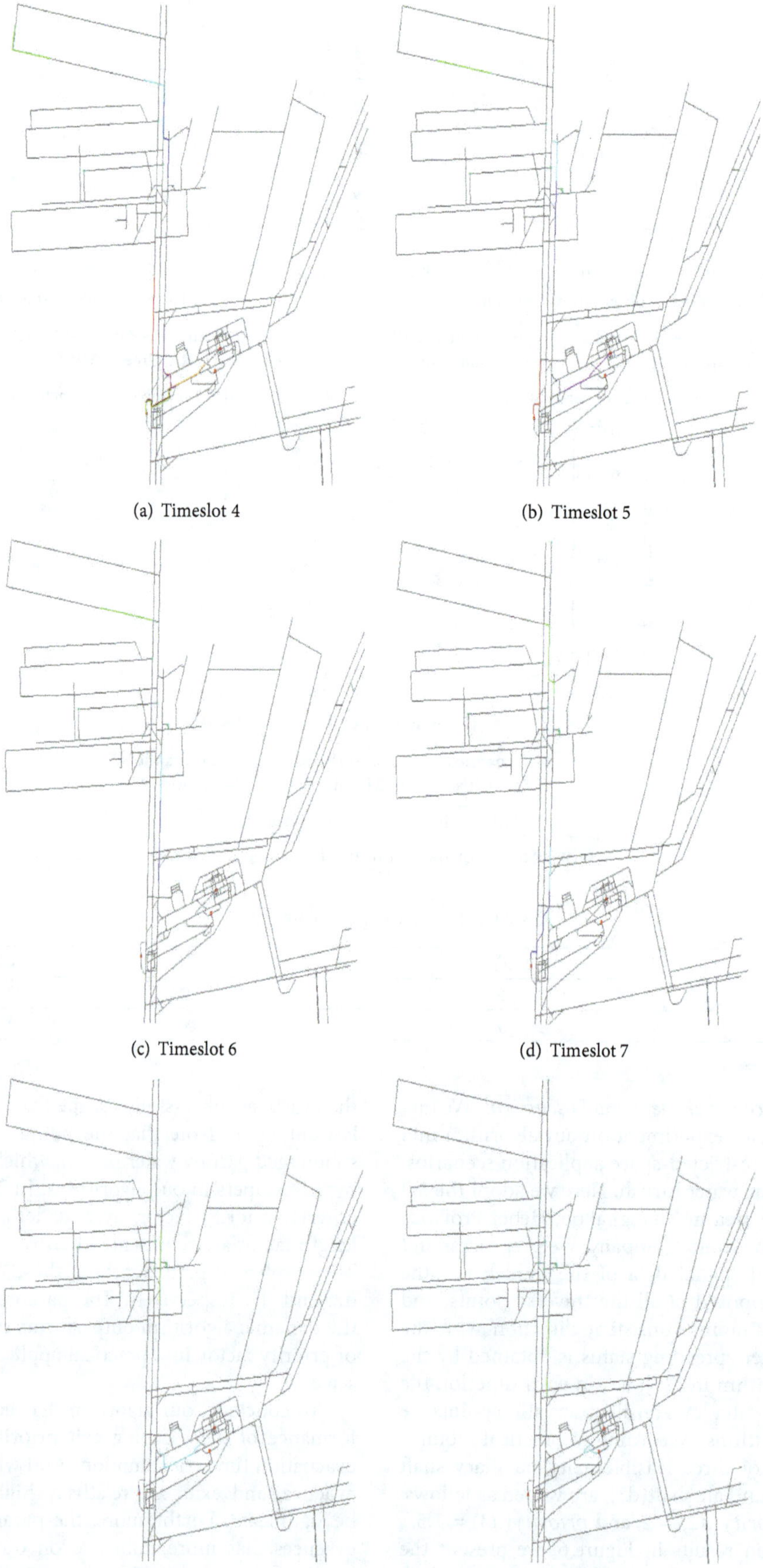

(a) Timeslot 4

(b) Timeslot 5

(c) Timeslot 6

(d) Timeslot 7

(e) Timeslot 8

(f) Timeslot 9

FIGURE 6: Continued.

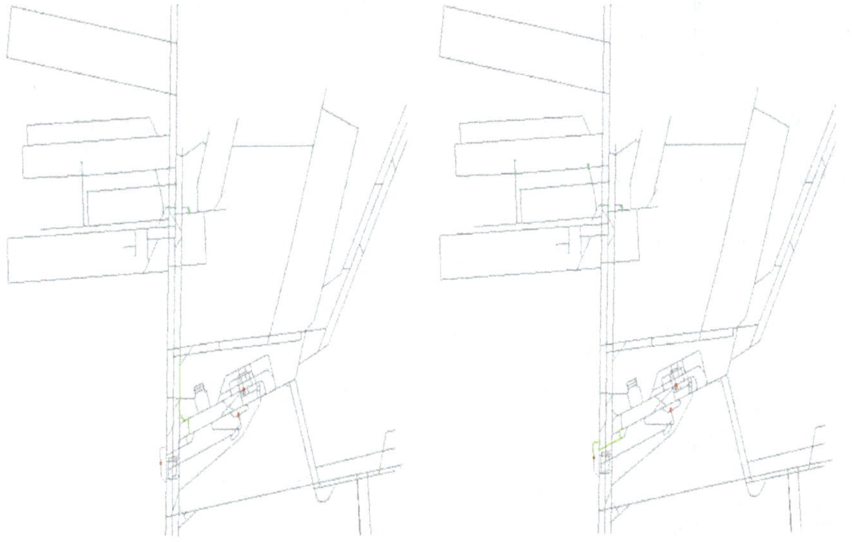

(g) Timeslot 10

(h) Timeslot 11

FIGURE 6: The visualization results of dynamic algorithm.

exits and guarantee a tolerable range of the global escaping time consumption.

6. Conclusions

In this paper, we propose an optimization problem for emergency evacuation planning in restricted-space scenarios, the Multiobjective Dynamic Route Network Planning (MOD-RNP) problem. To minimize the whole evacuation delay and maximize the priority-oriented evacuation efficiency, we design a dynamic and global-optimized strategy to construct the evacuation route network, based on Minimum Weighted Set Cover (MWSC). Based on the performance evaluation, our algorithm outperforms the shortest path algorithm in dynamic pattern on the priority utilization efficiency of the escaping exits, and it can guarantee that the deviation of the global travel length from the shortest one is in a reasonable range. Furthermore, we believe there are lots of room for further investigation and we plan to design distributed dynamic route network planning strategies to contribute or reorganize route network in real time.

Conflicts of Interest

The authors declare that they have no conflicts of interest.

Acknowledgments

This research was supported in part by Beijing Natural Science Foundation (4174090) and Program of Beijing Excellent Talents Training for Young Scholar (2016000020124G056).

Professor Wu and Professor Xu were supported in part by China National Scientific and Technical Support Program (2016YFC0801801), China National Natural Science Foundation (41430318, 41272276, 41572222, and 41602262), Beijing Natural Science Foundation (8162036), Fundamental Research Funds for the Central Universities (2010YD02), Innovation Research Team Program of Ministry of Education (IRT1085), and State Key Laboratory of Coal Resources and Safe Mining.

References

[1] Y.-J. Zheng, H.-F. Ling, X.-L. Xu, and S.-Y. Chen, "Emergency scheduling of engineering rescue tasks in disaster relief operations and its application in China," *International Transactions in Operational Research*, vol. 22, no. 3, pp. 503–518, 2015.

[2] L. Özdamar and O. Demir, "A hierarchical clustering and routing procedure for large scale disaster relief logistics planning," *Transportation Research Part E: Logistics and Transportation Review*, vol. 48, no. 3, pp. 591–602, 2012.

[3] M. Lahmar, T. Assavapokee, and S. A. Ardekani, "A dynamic transportation planning support system for hurricane evacuation," in *Proceedings of the Intelligent Transportation Systems Conference, ITSC '06*, pp. 612–617, 2006.

[4] Q. Lu, G. B. George, and S. Shekhar, "Capacity constrained routing algorithms for evacuation planning: a summary of results," in *Proceedings of the International Symposium on Advances in Spatial & Temporal Databases*, vol. 3633, 8, pp. 291–307, 2005.

[5] S. Kim, B. George, and S. Shekhar, "Evacuation route planning: Scalable heuristics," in *Proceedings of the 15th ACM International Symposium on Advances in Geographic Information Systems, GIS '07*, vol. 20, 8, pp. 1–8, November 2007.

[6] A. H. Sampaio and S. Urrutia, "New formulation and branch-and-cut algorithm for the pickup and delivery traveling salesman problem with multiple stacks," *International Transactions in Operational Research*, vol. 24, no. 1-2, pp. 77–98, 2017.

[7] F. Hernandez, M. Gendreau, and J.-Y. Potvin, "Heuristics for tactical time slot management: a periodic vehicle routing problem view," *International Transactions in Operational Research*, vol. 24, no. 6, pp. 1233–1252, 2017.

[8] M. Kammoun, H. Derbel, M. Ratli, and B. Jarboui, "An integration of mixed VND and VNS: the case of the multivehicle covering tour problem," *International Transactions in Operational Research*, vol. 24, no. 3, pp. 663–679, 2017.

[9] P. Chen, B. Golden, X. Wang, and E. Wasil, "A novel approach to solve the split delivery vehicle routing problem," *International Transactions in Operational Research*, vol. 24, no. 1-2, pp. 27–41, 2017.

[10] Y.-J. Zheng, S.-Y. Chen, and H.-F. Ling, "Evolutionary optimization for disaster relief operations: a survey," *Applied Soft Computing*, vol. 27, pp. 553–566, 2015.

[11] S. Kim and S. Shekhar, "Contraflow network reconfiguration for evacuation planning: a summary of results," in *Proceedings of the 13rd ACM International Symposium on Geographic Information Systems, ACM-GIS '05*, pp. 250–259, 2005.

[12] G. J. Lim, S. Zangeneh, M. Reza Baharnemati, and T. Assavapokee, "A capacitated network flow optimization approach for short notice evacuation planning," *European Journal of Operational Research*, vol. 223, no. 1, pp. 234–245, 2012.

[13] A. Takizawa, M. Inoue, and N. Katoh, "An emergency evacuation planning model using the universally quickest flow," *The Review of Socionetwork Strategies*, vol. 6, no. 1, pp. 15–28, 2012.

[14] N. Baumann and M. Skutella, "Solving evacuation problems efficiently Earliest arrival flows with multiple sources," in *Proceedings of the 47th Annual IEEE Symposium on Foundations of Computer Science, FOCS '6*, pp. 399–408, 2006.

[15] C. E. Dunn and D. Newton, "Optimal routes in GIS and emergency planning applications," *Area*, vol. 24, no. 3, pp. 259–267, 1992.

[16] T. Yamada, "A network flow approach to a city emergency evacuation planning," *International Journal of Systems Science*, vol. 27, no. 10, pp. 931–936, 1996.

[17] T. J. Cova and J. P. Johnson, "A network flow model for lane-based evacuation routing," *Transportation Research Part A: Policy and Practice*, vol. 37, no. 7, pp. 579–604, 2003.

[18] N. Kamiyama, N. Katoh, and A. Takizawa, "An efficient algorithm for evacuation problems in dynamic network flows with uniform arc capacity," *Lecture Notes in Computer Science*, vol. 4041, pp. 231–242, 2006.

[19] N. Kamiyama, N. Katoh, and A. Takizawa, "An efficient algorithm for the evacuation problem in a certain class of networks with uniform path-lengths," *Discrete Applied Mathematics: The Journal of Combinatorial Algorithms, Informatics and Computational Sciences*, vol. 157, no. 17, pp. 3665–3677, 2009.

[20] J. W. Wang, H. F. Wang, W. J. Zhang, W. H. Ip, and K. Furuta, "Evacuation planning based on the contraflow technique with consideration of evacuation priorities and traffic setup time," *IEEE Transactions on Intelligent Transportation Systems*, vol. 14, no. 1, pp. 480–485, 2013.

[21] O. L. Huibregtse, S. P. Hoogendoorn, A. Hegyi, and M. C. Bliemer, "A method to optimize evacuation instructions," *OR Spectrum*, vol. 33, no. 3, pp. 595–627, 2011.

[22] M. Goodwin, O.-C. Granmo, and J. Radianti, "Escape planning in realistic fire scenarios with Ant Colony Optimisation," *Applied Intelligence*, vol. 42, no. 1, pp. 24–35, 2014.

[23] M. Goerigk, K. Deghdak, and P. Heßler, "A comprehensive evacuation planning model and genetic solution algorithm," *Transportation Research Part E: Logistics and Transportation Review*, vol. 71, pp. 82–97, 2014.

[24] M. Manley, Y. S. Kim, K. Christensen, and A. Chen, "Airport Emergency Evacuation Planning: An Agent-Based Simulation Study of Dirty Bomb Scenarios," *IEEE Transactions on Systems, Man, and Cybernetics: Systems*, vol. 46, no. 10, pp. 1390–1403, 2016.

[25] S.-Q. Tong, N. Wang, and N.-Q. Song, "Emergency evacuation capability evaluation and optimization for an offshore airport: The case of Dalian Offshore Airport, Dalian, China," *Safety Science*, vol. 92, pp. 128–137, 2017.

[26] H. Liu, B. Xu, D. Lu, and G. Zhang, "A path planning approach for crowd evacuation in buildings based on improved artificial bee colony algorithm," *Applied Soft Computing*, vol. 68, pp. 360–376, 2018.

[27] N. Al Theeb and C. Murray, "Vehicle routing and resource distribution in postdisaster humanitarian relief operations," *International Transactions in Operational Research*, vol. 24, no. 6, pp. 1253–1284, 2017.

[28] M. R. Garey and D. S. Johnson, "'Strong' NP-completeness results: motivation, examples, and implications," *Journal of the ACM*, vol. 25, no. 3, pp. 499–508, 1978.

[29] C.-P. Li, Z.-X. Li, Y.-X. Zheng, Li. Xin, D.-Y. Hou, and Y.-D. Zhou, "Three-dimensional dynamic simulation modeling of water inrushes in underground mines," *Journal of University of Science and Technology Beijing*, vol. 35, no. 2, pp. 140–147, 2013.

14

Random Regret-Minimization Model for Emergency Resource Preallocation at Freeway Accident Black Spots

Siliang Luan,[1,2,3] Qingfang Yang,[1,2,3] Wei Wang,[1,2,3] Zhongtai Jiang,[1,2,3] Ruru Xing,[1,2,3] and Ruijuan Chu[1,2,3]

[1]*School of Transportation, Jilin University, Changchun 130022, Jilin, China*
[2]*Jilin Research Center for Intelligent Transportation System, Jilin University, Changchun 130022, Jilin, China*
[3]*Jilin Province Key Laboratory of Road Traffic, Jilin University, Changchun 130022, Jilin, China*

Correspondence should be addressed to Wei Wang; wwei@jlu.edu.cn and Zhongtai Jiang; 543605081@qq.com

Academic Editor: Richard S. Tay

The preallocation of emergency resources is a mechanism increasing preparedness for uncertain traffic accidents under different weather conditions. This paper introduces the concept of accident probability of black spots and an improved accident frequency method to identify accident black spots and obtain the accident probability. At the same time, we propose a three-stage random regret-minimization (RRM) model to minimize the regret value of the attribute of overall response time, cost, and demand, which allocates limited emergency resources to more likely to happen accident spots. Due to the computational complexity of our model, a genetic algorithm is developed to solve a large-scale instance of the problem. A case study focuses on three-year rainy accidents' data in Weifang, Linyi, and Rizhao of China to test the correctness and validity of the application of the model.

1. Introduction

Freeways play a very important role in the transport system, which accounted for 2.8% of national highways and carried 1/2 of business passengers turnover and 40% of freight and cargo turnover in China in 2016. However, because of the high speed of vehicles, once the incident happens on the freeway, it will lead to relatively heavy casualties. Generally, in response to these unexpected situations, freeway operational companies often configure certain amount of emergency resources around the accident sites to provide assistance to victims conveniently. Hence, the preallocation of emergency resources, as the first step to rescue, directly impacts the efficiency of the rescue service. The reliable decision support models are heavily needed to improve the freeway service before incidents happen.

Increased attention to literature has been focused on stochastic programming (SP) model in order to address the problem of emergency recourse allocation. Zografos et al. constructed an incident-management program to minimize the incident delay and proved that the deployment of traffic-flow restoration units is essential to finish this goal [1]. Later, Zografos et al. developed decision support systems (DSSs) that can be used for incident response logistics and improved the quality of the decision. He also proposed a mathematical model to decrease incident response time considering users' requirements [2]. Baker et al. introduced criteria for budgets and workload and developed an integer, nonlinear mathematical programming model to allocate emergency medical services, especially for ambulances to meet the government's criterion of response time [3]. Ozbay et al. put forward a flexible optimal dispatching model to the operator with the probabilistic constraint for the potential accident locations, that is, the concept of the quality of service [4]. In order to satisfy the demands with a given flood occurrence probability, Garrido et al. addressed an issue of the delivery supplies and formulated a mathematical optimization model considering the level of inventory of emergency resources and the availability of vehicles [5]. Rawls and Turnquist focused on the special disasters like hurricanes and so forth, which had uncertainty in demand for the stocked supplies. He presented a two-stage stochastic mixed integer program before the disasters [6]. Zhang developed a multiagent-based decentralized resource allocation approach for different emergency events based on the domain transportation theory [7].

The above literatures have studied a variety of approaches with respect to the allocation problem of different types of emergency resources in specific disasters. Overall, there are four obvious features and problems. Firstly, many efforts in emergency response researches focus on natural disasters in the cities, such as hurricane, earthquake, and flood [5, 6, 8], but relatively fewer scholars pay attention to applying approaches on the freeway. In terms of problems of emergency resources allocation in the cities, due to the complexity and high density of the urban road networks, the researchers mostly select the facility locations and then allocate the resources [9]. However, the freeway network has lower road coverage and density than urban road network and supply spots are relatively fixed and less. Obviously, the existing approaches cannot be adapted to the freeway network. In addition, most scholars have studied how to distribute resources to deal with natural disasters, such as floods and earthquakes, but they ignored the fact that accidents happen more easily on rainy, foggy, and snowy days. It is the lack of researches in this field, which needs more adequate study. Secondly, the goal of emergency resource preallocation is to prepare for the future incidents to promise to give assistance in time. There were some works on the potential accidents or secondary locations, such as introducing the quality of service to evaluate the probability for the future incidents [5, 6], and given probabilities of secondary disasters to allocate resources [9, 10]. These methods of representation can, to some extent, indicate the probability of an accident occurring on a freeway, but they are not precise and specific and have a large number of errors. In particular, the locations of accidents are unfixed so that the accident data is mass and disorderly. Under this condition, allocating emergency resources is really difficult. Hence, it is very necessary to investigate the accident-prone location, black spots, prior to the deployment of emergency resources. Thirdly, in the research of this issue, current researches are basically a single-objective or multiobjective SP model to analyze and solve the problem of the allocation of emergency resources. The ultimate goal of optimization is the least cost or the shortest response time. However, these models have a serious problem: a well-performed attribute can compensate for the performance of other poorer attributes. Therefore, the final decision is not globally optimal. In our paper, a random regret-minimization (RRM) function is introduced, which is a convex function that can make up for the previous deficiency. When the attributes are equally important, the better performance of an attribute is only a half-compensating effect on the overall regret value.

In the study of the identification of accident black spots and the calculation of accident probability, the research of these issues is relatively mature. Deacon et al. focused on the intersection and nonintersection's black spots and distinguished between short highway segments and large segments based on accident statistics [11]. Saccomanno et al. established a multivariate Poisson regression and empirical Bayesian models for the potential accidents and applied them to the highway [12]. Gregoriades and Mouskos illustrated an integration of Bayesian Networks model with the simulator to assess the accident risk index, which was used to identify accident black spots on road networks [13]. Washington et al. proposed a combination of equivalency calculation only considering property damage and quantile regression technique to identify hot spots in a transportation network [14]. Debrabant et al. addressed issues of discrete distribution and overdispersion data from hospital records from Funen and Denmark to present an autoregressive Poisson-Tweedie model [15]. The above studies used different methods to identify accident black spots. Especially for our paper, identifying the black spots objectively and accurately will help decision maker allocate limited resources to accident-prone sites in the event of insufficient funds and improve traffic safety management.

Regret is one of the common and widespread negative emotions which usually expresses dissatisfaction and disappointment in people's feelings about what has been chosen. Initially, the regret theory (RT) was proposed by Bell et al. in 1982 and applied to the field of economics. The RT represents a status that a decision maker, who engages in commodity trading, tend to avoid the risk of loss as much as possible, in other words, to avoid regret [16]. Over the years, Daskin et al. applied P-minimax regret method to solve the facility location problems and presented a new model that optimizes the worst-case performance over a set of scenarios in 1997 [17].

In 2008, Chorus et al. presented random regret-minimization (RRM) model rooted in RT at the first time and provided several useful features for travel demand analysis [18]. Later, the RRM model was gradually improved through in-depth studies of many scholars and applied to more and more fields. We introduce the latest publications and give an overview of them. In 2010, Chorus developed a new discrete choice model to improve the RRM 2008 model with respect to foregone alternatives and revealed a promising performance of the new RRM-model through the application of travel mode choices, travel information acquisition choices, parking lot choices, and shopping location choices [19]. Dekker et al. introduced the concept of need-satisfaction into the hybrid choice model and better understood the behavioral processes underlying leisure activity participation based on the RRM [20]. In the same year, Hess et al. stated the contrast between RRM model and the Random Utility Maximization (RUM) model and found out that the differences between two models, apart from the factor of alternatives, were driven by the datasets [21]. Van Cranenburgh et al. proposed new methodological insights on RRM models, which are the μRRM model and the Pure-RRM model illustrated by reanalyzing ten datasets [22]. Guevara et al. developed an approach to achieve consistency, asymptotic normality, and relative efficiency of the estimators while sampling of alternatives and testing their approaches were more practical than a truncated model through real data experiments [23]. Jang et al. defined the regret as a function of physical attributes of choice alternatives and proposed a nonlinear psychophysical representations of the perception of attributes levels, which greatly enriched the RRM [24]. Rasouli and Timmermans compared original specification of RRM with the logarithmic specification of RRM and found that although logarithmic specification became theoretically inferior to original specification, the results of case study illustrated that the original specification outperformed the new specification for the collected data [25]. In 2018, Rasouli and Timmermans summed up the issues of recent researches in this field and developed a new sight of Chorus and Van Cranenburgh models [26]. Dekker and Chorus interpreted the choice probability as a well-behaved approximation and developed a measure of consumer surplus for RRM [27].

We give a brief overview of these studies on the RRM models and summarize three characteristics of these studies. First, incremental changes have been suggested and new insights of the RRM model have been added. An original RRM model just considered the choice alternatives. Later, a logarithmic specification was expressed. The scholars developed different specifications from various angles to provide a good deal of insight into the fundamental theory involved. Second, the empirical comparisons with RUM constitute above all publications, because some peers have questioned the performance of the RRM model. Therefore, their compromise effects and the differences have been argued and observed. Third, the applications of the RRM model have involved various fields including marketing decisions, travel route choices, travel information acquisition choices, parking lot choices, and shopping location choices. With its application enlarged, its utility gets more and more important. Thus, in this paper, we aim to broaden the range of the application of the RRM model continually and use it in the emergency management.

In this paper, we put forward a three-stage RRM model to solve the emergency resource preallocation problem on the freeway from the perspective of regret based on the above research basis. Before the decision is made, we first identify the black spots on the freeway using the improved accident frequency method and then calculate the probability of an accident. In particular, because of the different external conditions such as the geographic location of the freeway and the condition of the road surface, the black spots in the discussed area are not completely consistent under different weather conditions. Thus, only the accident data under similar weather conditions can be collected.

We attempt to address the emergency resource allocation problem with different approaches from other researches, which is a three-stage RRM model. Since the response time is the most vital factor in the rescue, a minimum response time model is established in the first stage with the specified coverage constraint about time. We use a genetic algorithm to generate the initial plan set with regard to the number of resources dispatching from supply locations to black spots. In the second stage, a new solution set is generated based on the original set of plans, combining the probability of accidents occurring in black spots under specific circumstances. The third stage is comparison of plans to select the final decision. RRM model is formulated comparing three attributes of every plan, i.e., response time, storage and procurement costs, and the demand. The plan with the minimal regret value is regarded as our final decision.

The rest of this paper is organized as follows. In Section 2, we describe the mathematical formulation including two parts, which is the identification of black spots and a three-stage RRM model. In Section 3, we illustrate the use of the model through a case study for a three-stage RRM model for emergency resources preallocation in anticipation of rainy day in three cities in China. Section 4 provides conclusions and directions for further work.

2. Methodology

This section introduces the identification of freeway black spots using the approach of improved accident frequency at first and then describes the model development, that is, a three-stage RRM model, which promises the minimization of regret value, considering the probability of black spots. Our goal is to find the most optimized plan to allocate resources from supply locations to black spots on the freeway.

2.1. The Identification of Black Spots. The diagnosis of accident black spots does not only solve the existing traffic safety problems, but also provide first-hand information for the prediction and prevention of traffic accidents. The black spots seriously reduce the service quality of the road network, since the cumulative number of incidents at each black spot accounts for a high proportion of the total number of accidents, which has a great impact on the overall safety of the transportation system. For emergency management, the identification of black spots on freeways also plays a vital role as it provides guidance about preallocation emergency resources to nearby areas prone to accidents.

To begin, we divide the road section into units so that it is easier to process accident data statistically. And then the improved accident frequency method is used to identify black spots. We select a critical number of accidents as the criteria for identification. If the number of accidents on a certain road section is greater than the critical value, it will be marked as an accident-prone point. The advantage of this method is easy to calculate and select and clear at a glance. It is especially suitable for the processing of massive accident data [28].

The improved accident frequency method is described as follows. Firstly, calculate the average number of accidents λ on a unit.

$$\lambda = \frac{\sum m_i}{n} \tag{1}$$

where m_i is the number of accidents on the road section i and n is the total number of units.

If the confidence level is α, the critical number of accidents R is

$$R = \lambda + u_{(1-\alpha)/2} \cdot \sqrt{\lambda}, \quad i = 1, 2, \ldots, n \tag{2}$$

We compare the number of actual accidents with the critical value R. If it is greater than the critical value R, we can determine that the road section is a black spot.

The "spot" in the black spots can be a point, a road section, an entire road, or an area, so the spot in this paper refers to a road segment. However, the above method will miss some real black spots resulting from concentrated accident locations which were divided into two by a fixed segment; hence, after the completion of the above identification, it is necessary to use the section to cut surface technology in order to correct the results. The principle of cut surface technology is to move one or more adjacent unit sections without exceeding the critical value R to a suitable location in light of their distribution. This supplementary calculation, the nonfixed description of black spots, has greatly improved the accuracy of the accident probability calculation.

2.2. A Three-Stage RRM Model for the Emergency Resources Allocation. The problem of freeway emergency resource preallocation can be considered as a risky decision issue. Existing studies have been focusing on a variety of factors in the SP model. If certain attributes perform well, good-performing attributes will fully compensate for other worse performing attributes. Obviously, this is not acceptable for an emergency rescue decision, because the result of emergency decision does not depend on the longest board, but the shortest board. However, the RRM model can exactly make up for the shortages caused by the ordinary SP model.

RRM was proposed by Chorus based on classical regret theory in 2008. RRM is a general choice model that deals with multiple alternatives with multiattributes and it can weigh the performance of multiple attributes to avoid the regret. The regret results from the unselected schemes that perform better than the selected scheme [18]. Consider the traveler who faces a choice between alternatives i, j, and k. The alternatives i, j, and k have the attributes of x, y, and z; that is, $i = \{x_i, y_i\}$, $j = \{x_j, y_j\}$. The regret is obtained by the comparison of that alternative with the best of other two alternatives:

$$R_i = \max\{R_{ij}, R_{ik}\} \tag{3}$$

$$R_j = \max\{R_{ji}, R_{jk}\} \tag{4}$$

$$R_k = \max\{R_{ki}, R_{kj}\} \tag{5}$$

Take an example of the calculation of R_{ij}:

$$R_{ij} = \varphi_x(x_i, x_j) + \varphi_y(y_i, y_j) + \varphi_z(z_i, z_j) \tag{6}$$

where $\varphi_x, \varphi_y, \varphi_z$ is an attribute-regret function using the following formulas to obtain. Let β be the estimated parameter of the attribute.

$$\varphi_x(x_i, x_j) = \max\{0, \beta_x \cdot (x_j - x_i)\} \tag{7}$$

$$\varphi_y(y_i, y_j) = \max\{0, \beta_y \cdot (y_j - y_i)\} \tag{8}$$

$$\varphi_z(z_i, z_j) = \max\{0, \beta_z \cdot (z_j - z_i)\} \tag{9}$$

We finally select the minimum regret value from R_i, R_j, R_k: $\min\{R_i, R_j, R_k\}$.

Based on the above theory, we propose a three-stage RRM emergency resource preallocation at freeway black spots. The first stage is that the SP model generates a preliminary plan based on the response time. The second stage is to form a new set of solutions by combining the probability of accidents occurring at black spots. The third stage is to compare the schemes obtained in the second stage by the calculation of the RRM model and finally choose the scheme with the smallest regret value.

Stage 1: Generate a Preliminary Plan Based on Response Time. In emergency rescue, the response time is the most important decisive factor. Certain supply locations do not cover many black spots, so that rescue personnel cannot reach the black spots within the response time range specified by the system. Therefore, we need to make a preliminary match between supply locations and black spots. Due to the vast area, i.e., the large number of supply locations and black spots, preliminary screening can simplify calculation and increase the speed of operations.

In the first step, we propose a stochastic programming model with the smallest response time. We assume that the road from supply locations to the black spots is travelable and the shortest distance. Supply can be prepositioned at the location i, and black spots can be defined as j. We use t_{ij} to denote emergency resource transportation time from supply location i to black spot j, t_0 to denote the rescue system specified response time, and $\overline{t_i}$ to denote the average processing time after receiving the alarm at the depot i. Let l_{ij} be the shortest distance from a supply location i to the black spot j, and let v^k be the average transportation speed under special scenario k ($k \in K$): $t_{ij} = l_{ij}/v^k$. In addition, the parameter θ_{ij} represents whether accident black spots j are within the coverage of the depot i and a_i represents the maximum stock capacity of the supply location i. We use x_{ij} ($x_{ij} \in X_n$) to denote the number of emergency resources dispatched from depot i to black spot j. Finally, an initial scheme set S_1 is calculated by a genetic algorithm and X_n is the element of the initial scheme set S_1.

The first stage of the stochastic programming model is given as

$$\min \sum_{i=1}^{n_i}\sum_{j=1}^{n_j}\theta_{ij}t_{ij}x_{ij} \tag{10}$$

subject to

$$\sum_{j=1}^{n_j} x_{ij} \le a_i \tag{11}$$

$$\theta = \begin{cases} 0 & t_{ij} > t_0 - \overline{t_i} \\ 1 & t_{ij} \le t_0 - \overline{t_i} \end{cases} \tag{12}$$

$$x_{ij} \geq 0,$$
$$x_{ij} \in X_n, \tag{13}$$
$$x_{ij} \text{ is integer}$$

The objective function of the first stage (4) incorporates the total response time in order to provide an incentive to execute the resources preparedness at the lowest time within the constraint of response time coverage. Constraint (5) states that the number of emergency resources in depot i must be within the maximum inventory capacity a_i at that point. Constraint (6) states that if the black spot j is within the coverage of supply location i, $\theta_{ij} = 1$; otherwise, $\theta_{ij} = 0$. Constraint (7) requires x_{ij} must be a positive integer. After the calculation of the genetic algorithm, several different solution sets will be generated. The solution set will be stored in the form of a matrix X_n, which is a set of solutions, and finally, form the initial solution set S_1 ($X_n \in S_1$).

Stage 2: A New Solution Set S_2 Based on the Accident Probability of Black Spots. Through formulas (1) and (2), we can obtain the results of identifying black spots, so that we can calculate the relative accident probability p_j^k of black spots under the circumstance k and form a diagonal matrix P. Each element X_n in the set S_1 is weighted by this diagonal matrix P to form a new element X_n'. Formula (8) is as follows:

$$X_n' = X_n \cdot P \tag{14}$$

Through the above formula, we can obtain a new set of solutions S_2 ($X_n' \in S_2$).

Stage 3: The Best Solution Based on RRM Model. After the calculation of the above two steps, the initial set S_1 is generated by statistic programming model on the basis of geographical location and response time. In addition, a new solution set S_2 is generated based on the accident probability of black spots. However, the two stages only consider the response time and accident probability but do not comprehensively consider the costs of storage and procurement of emergency resources, and the demand for black spots. Some emergency resources cannot be stored permanently and need to be replaced regularly; therefore, decision makers hope to use the lowest cost to maximize benefits. We define that there are m plans obtained by stage 2. Let C_s, C_b, respectively, be the inventory costs and the procurement costs of emergency resources, and let C be the attribute of the total cost. The supply locations in the rescue network have different resource allocation standards and construction costs resulting in different rescue service [29]. In this paper, we divide the supply locations into three grades. Grade 1 is the top of them, which means that it has the greatest rescue ability. Otherwise, Grade 3 is the bottom of them, which has the worst rescue service. We denote $G(i)$ as the service level of the depot i,

$$G(i) = \begin{cases} 3 & \text{Grade 1} \\ 2 & \text{Grade 2} \\ 1 & \text{Grade 3.} \end{cases} \tag{15}$$

The purpose of the emergency resources preallocation is that, in the event of a traffic accident, the supply locations can distribute enough resources to meet the needs of the accident site in the shortest possible time. Hence, in addition to considering the response time and cost, we also consider whether the scheme can meet the requirements of black spots in the RRM model. We denote W as the shortage of commodity at black spots and denote d_j as the demand of black spot j.

According to formula (3), we put forward to RRM model for emergency resource preallocation problem.

$$\min \quad R_n(k) \tag{16}$$

$$R_1 = \max\{R_{12}, R_{13}, \cdots, R_{1m}\} \tag{17}$$

$$R_2 = \max\{R_{21}, R_{23}, \cdots, R_{2m}\} \tag{18}$$

$$\vdots \tag{19}$$

$$R_m = \max\{R_{m1}, R_{m2}, \cdots, R_{m(m-1)}\} \tag{20}$$

$$R_{n_1n_2} = \varphi_T(T_{n_1}, T_{n_2}) + \varphi_C(C_{n_1}, C_{n_2}) + \varphi_W(W_{n_1}, W_{n_2}) \tag{21}$$

$$\varphi_T(T_{n_1}, T_{n_2}) = \max\{0, \beta_T \cdot (T_{n_2} - T_{n_1})\} \tag{22}$$

$$\varphi_C(C_{n_1}, C_{n_2}) = \max\{0, \beta_C \cdot (C_{n_2} - C_{n_1})\} \tag{23}$$

$$\varphi_W(W_{n_1}, W_{n_2}) = \max\{0, \beta_W \cdot (W_{n_2} - W_{n_1})\} \tag{24}$$

$$T_n = \sum_{i=1}^{n_i}\sum_{j=1}^{n_j} t_{ij}x_{ij} \tag{25}$$

$$C_n = \sum_{i=1}^{n_i}\sum_{j=1}^{n_j} (C_s + C_b) G_i x_{ij} \tag{26}$$

$$W_n = \sum_{j=1}^{n_j} (d_j - x_j) \tag{27}$$

subject to

$$\sum_{j=1}^{n_j} x_{ij} \leq a_i \tag{28}$$

$$\sum_{i=1}^{n_i}\sum_{j=1}^{n_j} x_{ij} \leq A \tag{29}$$

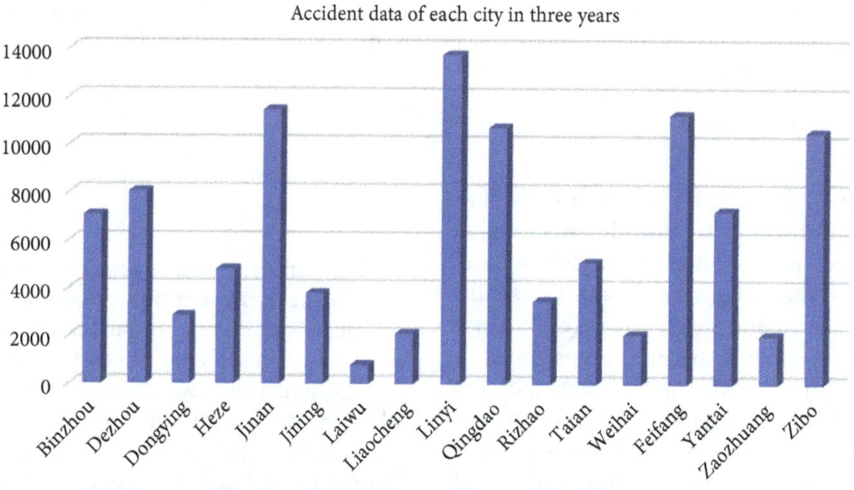

FIGURE 1: The accident data for each city in three years.

$$x_{ij} \geq 0,$$

$$x_{ij} \in X'_{n}, \qquad (30)$$

$$x_{ij} \text{ is integer}$$

where $R_n(k)$ is the regret value of plan n under the scenario k, and $R_1, R_2, \cdots, R_m$, respectively, is the regret value of the plan $1, 2, \cdots, m$. Let $\varphi_T, \varphi_C, \varphi_W$ be the attribute-regret function. We use β_T to denote the estimated parameters of response time, β_C to denote the estimated parameters of the total cost, and β_W to denote the estimated parameters of unfilled demand.

The goal of the objective function (16) is to select the minimal regret value from all schemes. Equations (17), (18), and (20), respectively, indicate that the regret value of scheme $1, 2, \cdots, m$ equals the regret associated with the comparison of that alternative with the best of the other alternatives. Equation (21) represents the binary regret associated with alternative n_1 when compared to alternative n_2. Equations (22), (23), and (24) state that either alternative n_1 performs better than n_2 in terms of the response time attribute, the total cost attribute, and the demand attribute, in case there is no attribute-regret, or alternative n_1 performs worse than n_2, in case the regret value associated with these attributes is a linear function of the difference in attribute values. Equation (25) indicates the meaning of the attribute of the response time. The total cost attributes including storage cost and acquisition cost, as well as considering the rescue service level of the depot i, are represented by formula (26). The difference between the resources dispatched to the black spots and its demand is stated by formula (27). Constraint (28) requires that the number of emergency resources in depot i must be within the maximum inventory capacity at that point. Constraint (29) states that the total number of resources in the study area cannot exceed the total number of emergency resources provided by the government. Constraint (30) requires that x_{ij} must be a positive integer and belong to X'_n.

Because the parameters belong to different dimensions in $\varphi_T, \varphi_C, \varphi_W$ and the numerical differences are relatively large, it is not appropriate to use them directly in the RRM model. Therefore, it is necessary to carry out the nondimensionalization of $(T_{n_2} - T_{n_1})$, $(C_{n_2} - C_{n_1})$, and $(W_{n_2} - W_{n_1})$, called the normalization of parameter data, that is, convert the parameters into relative dimensionless numbers. In this paper, a standardized method is used to perform a dimensionless treatment of $(T_{n_2} - T_{n_1})$, $(C_{n_2} - C_{n_1})$, and $(W_{n_2} - W_{n_1})$, and they are separately converted into dimensionless standardized data to eliminate the impact of the data dimension.

Standardized formula is as follows:

$$y_i = \frac{x_i - \overline{x}}{s} \qquad (31)$$

where $\overline{x} = (1/n)\sum_{i=1}^{n} x_i$ and $s = \sqrt{(1/(n-1))\sum_{i=1}^{n}(x_i - \overline{x})^2}$.

3. Case Study

We present a case study to demonstrate our approach. Three-year accidents' data of three big cities in Shandong province is analyzed and processed to verify the effectiveness of the three-stage RRM model. Although our model can cover several types of emergency resources, in this case study, we only consider single type resources (wreckers) for the sake of clarity in the representation.

3.1. Identify Black Spots and Calculate the Accident Probability of Black Spots. In this paper, we analyze the freeway accident data on rainy days in Shandong province in years from 2014 to 2016 and count the accident distributions of each city. The results are as in Figure 1.

As shown in Figure 1, we choose Linyi with the largest number of accidents as our study area. Meanwhile, we select its adjacent cities Weifang and Rizhao to identify three cities' accident black spots on the freeway.

TABLE 1: The number of accidents on the unit road section.

No.	Original Pile No.	Final Pile No.	m_i
1	K69+909	K71+000	3
2	K71+000	K72+000	0
3	K72+000	K73+000	3
⋮	⋮	⋮	⋮
52	K120+000	K121+000	2
53	K121+000	K122+000	4
54	K122+000	K123+000	5
⋮	⋮	⋮	⋮
101	K188+000	K189+000	21
102	K189+000	K191+000	13
103	K191+000	K192+000	27
⋮	⋮	⋮	⋮
121	K232+000	K233+000	30
122	K233+000	K234+000	27
123	K234+000	K236+414	39
	Total		1109

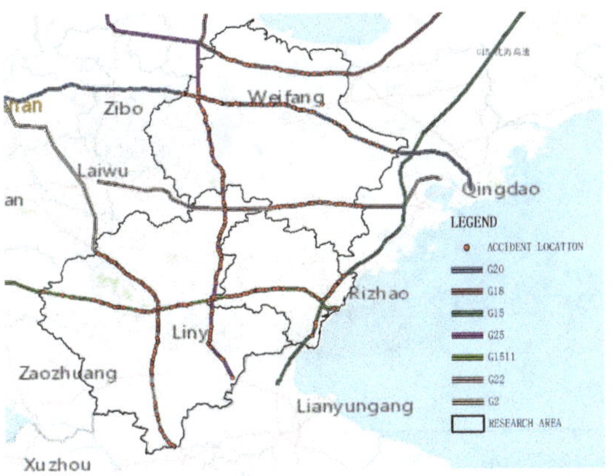

FIGURE 2: Accident locations marked on the map of ArcGIS.

Weifang contains four freeways, which are G18, G25, G20, and S16. Rizhao includes four freeways, which are G15, G1511, G25, and S7801. Linyi has four freeways, which are G2, G20, G22, and G25.

We calibrate the latitude and longitude positions of accidents in three years on the map of ArcGIS. As shown in Figure 2, we can see that the actual accident points are marked on the map. However, due to the large volume of accident data and its inconsistent locations, their distributions are very fragmented. If we do not classify them, it is very difficult to allocate emergency resources in such a disorganized situation. Therefore, we need to process the three-year accidents' data on the freeway and identify the black spots from them.

Take the G20 as an example to describe the identification process. G20 has a total length of 419 kilometers, which starts from east of Qingdao and passes through Weifang, Zibo, and Jinan. In this paper, we only identify the black spots of G20 in the Weifang section.

(1) Determination of the Critical Number of Accidents. We decide 1 km as a unit length and divide G20 into several unit lengths, 123 unit road sections in total. The number of accidents m_i occurring on each unit i is shown in Table 1.

The average number of accidents λ on a unit is

$$\lambda = \frac{\sum m_i}{n} = \frac{1109}{123} \approx 9 \tag{32}$$

If the confidence level is 95%, the critical number of accidents R is

$$R = \lambda + u_{(1-\alpha)/2} \cdot \sqrt{\lambda} = \lambda + 1.96\sqrt{\lambda} \approx 15 \tag{33}$$

(2) Obtain Accident Black Spots through the Primary Election. We compare the actual number of accidents on the G20 with the critical value R and then obtain the results of black spots through primary selection. There are 31 accident black spots in total as shown in Table 2.

(3) The Modification of Black Spots. According to the accidents occurring on the adjacent road sections, we use "section interception technology" to modify the original units and merge adjacent black spots, and then we obtain the results that include 14 black spots in Table 3.

We also calculate the critical value R and identify black spots for other freeways in the same method. The results are in Table 4.

We totally identified 88 accident black spots on the freeway in the three cities using the above method.

(4) The Calculation of Accident Probability of Black Spots. After screening out the black spots, we need to calculate their accident probabilities. In the study area, there are seven freeways with different traffic volumes and number of accidents. We have mentioned before that the accident

TABLE 2: Preliminarily select black spots.

Original Pile No.	Final Pile No.	The amount of accidents	Original Pile No.	Final Pile No.	The amount of accidents
K99	K100	16	K206	K207	20
K135	K136	20	K207	K208	18
K136	K137	16	K208	K209	16
K144	K145	18	K209	K210	16
K157	K158	21	K210	K211	17
K158	K159	23	K211	K212	42
K159	K160	16	K212	K213	22
K165	K166	17	K213	K214	34
K168	K169	26	K216	K217	47
K177	K178	24	K220	K221	40
K188	K189	21	K225	K226	38
K191	K192	27	K230	K231	37
K194	K195	18	K232	K233	30
K197	K198	16	K233	K234	27
K200	K201	18	K234	K235	39

TABLE 3: The results of the modification.

Black spots j		The number of accidents
Original Pile No.	Final Pile No.	
K101	K102	36
K107	K108	36
K131	K133	36
K136	K141	96
K142	K144	36
K147	K151	60
K157	K158	36
K169	K172	60
K177	K186	156
K189	K191	84
K194	K196	36
K197	K199	36
K200	K206	144
K213	K216	132

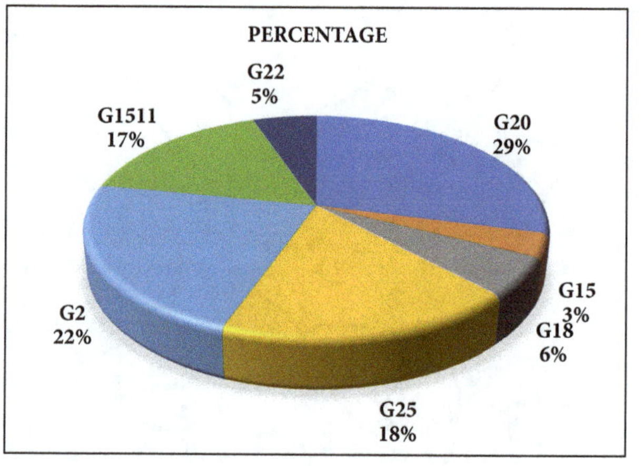

FIGURE 3: The percentage of accidents on every freeway.

probability of black spots is just relative value based on the proportion of three-year accidents in this region. As Figure 3 shows, the percentage of accidents on every freeway is counted as in Figure 3.

We can get the results of the accident probability on the basis of the above percentage in Table 5.

According to the accident probability, the summary graph is shown in Figure 4.

We mark the black spots with different proportions that depend on the accident probability on the map of ArcGIS in Figure 5. The results are more obvious, clear, and orderly than before.

3.2. The Three-Stage RRM Model of Allocating Emergency Resource. We apply the three-stage RRM model to the freeway in Linyi, Weifang, and Rizhao based on the accident data in 2013-2016. We take the wrecker as an example under the natural environment where the rainfall intensity is less than $2.5\ mm \cdot h^{-1}$, and the average speed v is $82.4\ km/h$ to verify the RRM emergency resource preallocation model.

(1) The First Stage: Generate a Preliminary Plan. The goal of this stage is to generate a preliminary plan based on the minimum of response time through the genetic algorithm. We change the initial scale of the population, the number of iterations, and so on, in order to obtain several schemes.

Freeway operation company requires that the maximum response time in the system t_0 is 30 minutes and the average process time $\overline{t_l}$ is 5 minutes. According to the judging conditions $t_0 - \overline{t_l} = 25$min, we can get the value of θ_{ij} whether supply location i can cover the black spot j. Finally, we get 15 schemes, which compose the plan set S_1. It is stipulated that government provides 75 wreckers for three cities that are stored in road administration brigades and road administration squadrons and their rescue grade, respectively, are 1 and 2. The road administration brigades

TABLE 4: The result of the identification of black spots in the study area.

Name	G15	G18	G2	G20	G22	G25	G1511
The average number of accidents λ	3	4	8	9	3	5	5
The critical value R	7	8	14	15	7	10	10
The number of black spots	4	7	21	14	7	22	13

TABLE 5: The results of accident probability.

No.	Name	Original Pile No.	Final Pile No.	Accident Probability
1	G15	K704	K707	0.012234
2		K712	K713	0.009176
3		K714	K715	0.006117
4		K742	K744	0.006117
5	G18	K363	K364	0.00534
⋮		⋮	⋮	⋮
11		K443	K445	0.00534
12	G2	K542	K543	0.006662
⋮		⋮	⋮	⋮
32		K707	K709	0.006662
33	G20	K101	K102	0.010531
⋮		⋮	⋮	⋮
46		K213	K216	0.038614
47	G22	K101	K103	0.005706
⋮		⋮	⋮	⋮
53		K159	K160	0.005706
54	G25	K1406	K1408	0.005464
⋮		⋮	⋮	⋮
75		K1604	K1606	0.010927
76	G1511	K16	K18	0.009786
⋮		⋮	⋮	⋮
88		K156	K158	0.009786

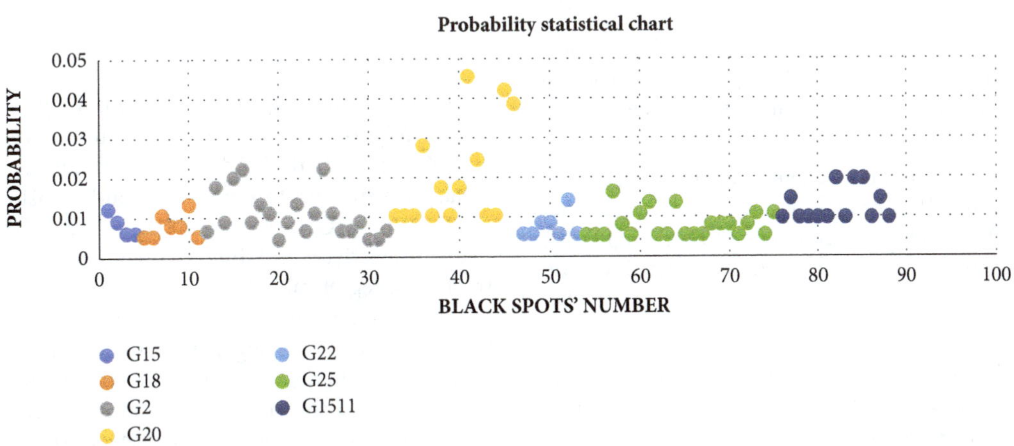

FIGURE 4: Probability statistical chart.

TABLE 6: The preliminary schemes.

Black spot	Scheme No.														
	1	2	3	4	5	6	7	8	9	10	11	12	13	14	15
1	2	2	2	2	2	2	2	2	2	2	2	0	2	2	2
2	0	0	2	0	0	0	0	0	0	2	0	0	2	0	1
3	0	2	0	0	0	0	0	0	0	0	0	0	0	0	0
4	0	0	0	0	0	0	0	0	0	0	0	0	0	0	1
⋮	⋮	⋮	⋮	⋮	⋮	⋮	⋮	⋮	⋮	⋮	⋮	⋮	⋮	⋮	⋮
80	2	2	2	0	0	0	1	0	1	0	1	0	0	0	1
81	2	2	2	2	2	2	2	2	2	2	2	2	2	2	2
82	1	2	0	1	0	1	1	0	0	0	2	0	2	2	1
83	2	2	2	2	2	2	2	2	2	2	2	2	2	2	2
84	2	2	2	2	2	2	2	2	2	2	2	2	2	2	2
85	0	0	2	0	1	2	1	2	0	0	0	0	0	0	0
86	2	0	0	2	2	2	2	2	2	2	2	2	2	2	2
87	2	2	2	0	0	0	1	1	0	0	0	2	2	2	2
88	2	2	2	0	0	0	1	0	1	0	1	0	0	0	1

FIGURE 5: The map with final black spots.

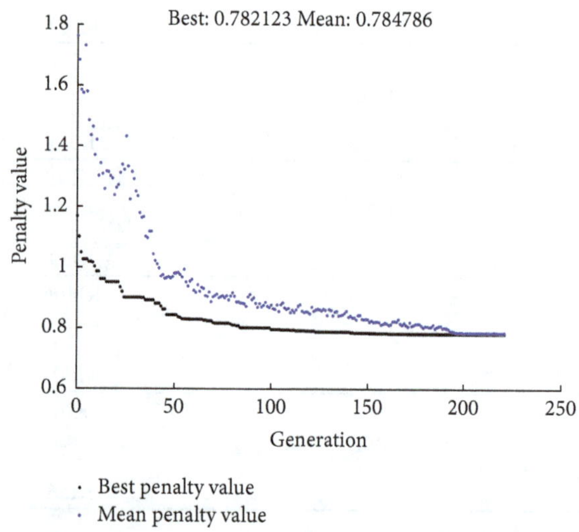

FIGURE 6: The result of genetic algorithm operation.

and road administration squadrons make up the supply location i, and they are 18 in all. We get 15 schemes that can represent how many wreckers allocate in every supply location, which is shown in Table 6.

Figure 6 shows the result of Scheme No. 1 through genetic algorithm operation. The optimal local value is 0.782123 after 221 iterations. Other schemes are obtained by changing the size of the initial population and the methods of hybridization or selection, and so forth.

(2) Generate Scheme Set S_2 on the Basis of Accident Probability. According to the accident probability of black spots in Table 4, we generate a diagonal matrix: $P = \begin{bmatrix} 0.012234 & \cdots & 0 \\ \vdots & \ddots & \vdots \\ 0 & \cdots & 0.009786 \end{bmatrix}$. Then, we apply formula (8) in Section 2.2 and finally obtain the new scheme set S_2.

(3) Generate the Most Optimal Plan through the Calculation of the RRM Model. In this part, we put every element in the scheme set S_2 into the RRM model to calculate. At first, we use a standard method to perform a nondimensional treatment of $(T_{n_2} - T_{n_1})$, $(C_{n_2} - C_{n_1})$, and $(W_{n_2} - W_{n_1})$, then compare 15 plans, respectively, 210 times to obtain $\varphi_T, \varphi_C, \varphi_W$, and select the lowest regret value $R_n(k)$.

According to the degree of importance of each attribute, we use the expert scoring method to determine the estimated parameter values, $\beta_T = -0.45$, $\beta_C = -0.3$, $\beta_W = -0.25$. The results are shown in Table 7 and Figure 7. In addition to the analysis of the results, except for scheme 15, the maximum regret value of other alternatives is generated from the compassion with $R_{n_1,15}$. Comparing all regret values of each plan, scheme 15 has the lowest total regret value and

TABLE 7: The regret value of each plan.

R_1	$R_{1,2}$	$R_{1,3}$	$R_{1,4}$	$R_{1,5}$	$R_{1,6}$	$R_{1,7}$	$R_{1,8}$	$R_{1,9}$	$R_{1,10}$	$R_{1,11}$	$R_{1,12}$	$R_{1,13}$	$R_{1,14}$	$R_{1,15}$
1.065	0.297	0.143	0.143	0.306	0.143	0.462	0.445	0.429	0.569	0.419	0.286	0.000	0.000	1.065
R_2	$R_{2,1}$	$R_{2,3}$	$R_{2,4}$	$R_{2,5}$	$R_{2,6}$	$R_{2,7}$	$R_{2,8}$	$R_{2,9}$	$R_{2,10}$	$R_{2,11}$	$R_{2,12}$	$R_{2,13}$	$R_{2,14}$	$R_{2,15}$
0.908	0.000	0.071	0.071	0.080	0.071	0.462	0.374	0.357	0.303	0.214	0.214	0.000	0.000	0.908
R_3	$R_{3,1}$	$R_{3,2}$	$R_{3,4}$	$R_{3,5}$	$R_{3,6}$	$R_{3,7}$	$R_{3,8}$	$R_{3,9}$	$R_{3,10}$	$R_{3,11}$	$R_{3,12}$	$R_{3,13}$	$R_{3,14}$	$R_{3,15}$
1.734	0.811	1.036	0.594	1.117	0.704	1.109	0.870	1.074	1.237	1.087	0.242	0.611	0.551	1.734
R_4	$R_{4,1}$	$R_{4,2}$	$R_{4,3}$	$R_{4,5}$	$R_{4,6}$	$R_{4,7}$	$R_{4,8}$	$R_{4,9}$	$R_{4,10}$	$R_{4,11}$	$R_{4,12}$	$R_{4,13}$	$R_{4,14}$	$R_{4,15}$
1.139	0.217	0.442	0.000	0.523	0.110	0.515	0.303	0.480	0.643	0.493	0.143	0.017	0.000	1.139
R_5	$R_{5,1}$	$R_{5,2}$	$R_{5,3}$	$R_{5,4}$	$R_{5,6}$	$R_{5,7}$	$R_{5,8}$	$R_{5,9}$	$R_{5,10}$	$R_{5,11}$	$R_{5,12}$	$R_{5,13}$	$R_{5,14}$	$R_{5,15}$
1.051	0.071	0.143	0.214	0.214	0.214	0.534	0.517	0.500	0.445	0.357	0.357	0.000	0.071	1.051
R_6	$R_{6,1}$	$R_{6,2}$	$R_{6,3}$	$R_{6,4}$	$R_{6,5}$	$R_{6,7}$	$R_{6,8}$	$R_{6,9}$	$R_{6,10}$	$R_{6,11}$	$R_{6,12}$	$R_{6,13}$	$R_{6,14}$	$R_{6,15}$
1.029	0.107	0.332	0.000	0.000	0.413	0.462	0.303	0.370	0.533	0.383	0.143	0.000	0.000	1.029
R_7	$R_{7,1}$	$R_{7,2}$	$R_{7,3}$	$R_{7,4}$	$R_{7,5}$	$R_{7,6}$	$R_{7,8}$	$R_{7,9}$	$R_{7,10}$	$R_{7,11}$	$R_{7,12}$	$R_{7,13}$	$R_{7,14}$	$R_{7,15}$
0.767	0.164	0.461	0.143	0.143	0.470	0.200	0.214	0.570	0.502	0.583	0.286	0.000	0.000	0.767
R_8	$R_{8,1}$	$R_{8,2}$	$R_{8,3}$	$R_{8,4}$	$R_{8,5}$	$R_{8,6}$	$R_{8,7}$	$R_{8,9}$	$R_{8,10}$	$R_{8,11}$	$R_{8,12}$	$R_{8,13}$	$R_{8,14}$	$R_{8,15}$
0.863	0.244	0.469	0.000	0.026	0.549	0.136	0.310	0.434	0.438	0.448	0.071	0.043	0.000	0.863
R_9	$R_{9,1}$	$R_{9,2}$	$R_{9,3}$	$R_{9,4}$	$R_{9,5}$	$R_{9,6}$	$R_{9,7}$	$R_{9,8}$	$R_{9,10}$	$R_{9,11}$	$R_{9,12}$	$R_{9,13}$	$R_{9,14}$	$R_{9,15}$
0.803	0.023	0.248	0.000	0.000	0.329	0.000	0.462	0.231	0.449	0.156	0.000	0.000	0.000	0.803
R_{10}	$R_{10,1}$	$R_{10,2}$	$R_{10,3}$	$R_{10,4}$	$R_{10,5}$	$R_{10,6}$	$R_{10,7}$	$R_{10,8}$	$R_{10,9}$	$R_{10,11}$	$R_{10,12}$	$R_{10,13}$	$R_{10,14}$	$R_{10,15}$
0.605	0.000	0.030	0.000	0.000	0.111	0.000	0.231	0.071	0.286	0.143	0.143	0.000	0.000	0.605
R_{11}	$R_{11,1}$	$R_{11,2}$	$R_{11,3}$	$R_{11,4}$	$R_{11,5}$	$R_{11,6}$	$R_{11,7}$	$R_{11,8}$	$R_{11,9}$	$R_{11,10}$	$R_{11,12}$	$R_{11,13}$	$R_{11,14}$	$R_{11,15}$
0.693	0.000	0.092	0.000	0.000	0.172	0.000	0.462	0.231	0.143	0.293	0.000	0.000	0.000	0.693
R_{12}	$R_{12,1}$	$R_{12,2}$	$R_{12,3}$	$R_{12,4}$	$R_{12,5}$	$R_{12,6}$	$R_{12,7}$	$R_{12,8}$	$R_{12,9}$	$R_{12,10}$	$R_{12,11}$	$R_{12,13}$	$R_{12,14}$	$R_{12,15}$
1.419	0.712	0.937	0.000	0.495	1.018	0.605	1.010	0.700	0.832	1.138	0.845	0.512	0.451	1.491
R_{13}	$R_{13,1}$	$R_{13,2}$	$R_{13,3}$	$R_{13,4}$	$R_{13,5}$	$R_{13,6}$	$R_{13,7}$	$R_{13,8}$	$R_{13,9}$	$R_{13,10}$	$R_{13,11}$	$R_{13,12}$	$R_{13,14}$	$R_{13,15}$
1.551	0.486	0.783	0.429	0.429	0.720	0.522	0.784	0.731	0.891	1.055	0.905	0.571	0.286	1.551
R_{14}	$R_{14,1}$	$R_{14,2}$	$R_{14,3}$	$R_{14,4}$	$R_{14,5}$	$R_{14,6}$	$R_{14,7}$	$R_{14,8}$	$R_{14,9}$	$R_{14,10}$	$R_{14,11}$	$R_{14,12}$	$R_{14,13}$	$R_{14,15}$
1.326	0.261	0.557	0.143	0.186	0.566	0.296	0.559	0.462	0.666	0.829	0.679	0.286	0.060	1.326
R_{15}	$R_{15,1}$	$R_{15,2}$	$R_{15,3}$	$R_{15,4}$	$R_{15,5}$	$R_{15,6}$	$R_{15,7}$	$R_{15,8}$	$R_{15,9}$	$R_{15,10}$	$R_{15,11}$	$R_{15,12}$	$R_{15,13}$	$R_{15,14}$
0.220	0.000	0.139	0.000	0.000	0.220	0.000	0.000	0.000	0.143	0.109	0.047	0.000	0.000	0.000

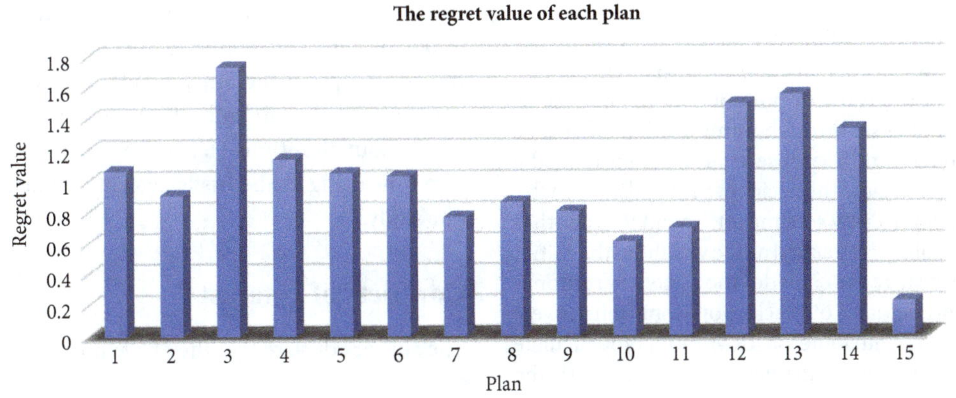

FIGURE 7: The regret value of each plan.

is the best option for the decision maker. Since scheme 15 is the best plan in all solutions, it can generate the lowest regret value among them. Therefore, if the decision makers choose other schemes rather than scheme 15, the regret will be larger than the regret of scheme 15.

In order to prove the superiority of our proposed method, we compare it with the SP model considering response time, total cost, and the demand as well and solve it by a genetic algorithm. The final SP solution and the final RRM solution are considered as two alternatives. Then we compare them

TABLE 8: The comparing result.

Selected plan	SP model				RRM model			
	T	C	W	Final Result	φ_T	φ_C	φ_W	R
SP plan	36.98	55.18	45.22	137.38	2.21	2.64	0.69	5.54
RRM plan	55.93	48.79	49.19	153.91	0.00	0.00	0.00	0.00

separately by the SP model and RRM model. The results are shown in Table 8.

As shown in Table 8, although the result of SP plan is smaller than the RRM plan through the calculation of the SP model, the attribute of response time extremely compromises the final result in SP plan. In addition, we can see that the regret of the SP plan is larger than the regret of the RRM model when we choose the SP plan. The regret value of every attribute of the SP plan performs worse than the RRM plan, but we cannot find this result from the calculation of SP model. Therefore, the results of the comparison can confirm that the RRM approach we proposed is better than SP model in emergency resource preallocation problem.

4. Conclusions

A three-stage RRM model considering the accident probability of black spots is established to study the emergency resource preallocation. Firstly, the improved accident frequency method is used to identify the black spots, in order to calculate the accident probability. Although some scholars considered the probabilistic constraints in the model for the resource allocation problem, they assumed a given probability rather than identifying the accident black spots. With no doubt, the identification of the black spot can be beneficial to summarize the mass accident data, which makes the occurring accident probability more authentic and believable.

Secondly, we come up with a three-stage RRM model for the emergency resources preallocation in this paper. Because response time plays the most important role in the rescue, we select the initial plan set through minimization stochastic programming model with response time constraint in the first stage. Then, we combine the accident probability with the initial plan set to obtain a new scheme set. To put it in another way, accident probability is attached as the weight to the initial plan set. In the third stage, we consider the factors of response time, total cost, and demand of black spots comprehensively and regard them as the attribute of RRM model to calculate every scheme's regret value and select the scheme with the smallest regret value.

In the case study, in order to verify the correctness and validity of our approach, we analyze three-year rainy day accidents' data of Linyi, Weifang, and Rizhao, identify 88 black spots, and calculate their relative accident probability. In addition, we calculate the regret value of 15 plans through the RRM model and compare the final selection with the result of the SP model. The case study shows the following results:

(1) The improved accident frequency method is effective for limited accident data. The method extremely simplifies the complexity of data and highly summarizes where accidents occur more. The identification of black spots lays the foundation for the following emergency resource preallocation issue.

(2) The three-stage RRM model for emergency resource preallocation is more advantageous for the emergency decision than common SP model, since the characteristics of semi-compensation in RRM model can consider every attribute comprehensively, while the better performance of attribute can completely compromise the total value in SP model. Therefore, our approach has good applicability to the preallocation of emergency resources on the freeway.

Several avenues present themselves as direction for further work. At first, due to the limitation of data acquisition and the confidentiality of certain data by related departments, there is no specific number of casualties. Hence, we cannot put this factor into the identification of black spots. Actually, the number of casualties cannot be ignored in the issue of identifying black spots. Although the improved accident frequency has relatively errors, this approach has already made the most of available data. In the future, the number of casualties per accident can be fully taken into consideration as one of the identification factors for black spots to improve the accuracy. On the other hand, the RRM model needs to compare every plan with others, which obviously increases the computation time. Hence, how to decrease the computation time using the heuristic algorithm is also the next step of our study.

Conflicts of Interest

The authors declare that they have no conflicts of interest.

Acknowledgments

This research has been jointly supported by National Key Technology Support Program (Grant no. 2014BAG03B03).

References

[1] K. G. Zografos, T. Nathanail, and P. Michalopoulos, "Analytical framework for minimizing freeway-incident response time,"

Journal of Transportation Engineering, vol. 119, no. 4, pp. 535–549, 1993.

[2] K. G. Zografos, K. N. Androutsopoulos, and G. M. Vasilakis, "A real-time decision support system for roadway network incident response logistics," *Transportation Research Part C: Emerging Technologies*, vol. 10, no. 1, pp. 1–18, 2002.

[3] J. R. Baker, E. R. Clayton, and B. W. Taylor, "A non-linear multi-criteria programming approach for determining county emergency medical service ambulance allocations," *Journal of the Operational Research Society*, vol. 40, no. 5, pp. 423–432, 1989.

[4] K. Ozbay, C. Iyigun, M. Baykal-Gursoy, and W. Xiao, "Probabilistic programming models for traffic incident management operations planning," *Annals of Operations Research*, vol. 203, no. 1, pp. 389–406, 2012.

[5] R. A. Garrido, P. Lamas, and F. J. Pino, "A stochastic programming approach for floods emergency logistics," *Transportation Research Part E: Logistics and Transportation Review*, vol. 75, pp. 18–31, 2015.

[6] C. G. Rawls and M. A. Turnquist, "Pre-positioning of emergency supplies for disaster response," *Transportation Research Part B: Methodological*, vol. 44, no. 4, pp. 521–534, 2010.

[7] J. Zhang, M. Zhang, F. Ren, and J. Liu, "An innovation approach for optimal resource allocation in emergency management," *IEEE Transactions on Computers*, p. 1-1, 2016.

[8] N. Ahmed, "An analytical decision model for resource allocation in highway maintenance management," *Transportation Research Part A: General*, vol. 17, no. 2, pp. 133–138, 1982.

[9] H. O. Mete and Z. B. Zabinsky, "Stochastic optimization of medical supply location and distribution in disaster management," *International Journal of Production Economics*, vol. 126, no. 1, pp. 76–84, 2010.

[10] F. Fiedrich, F. Gehbauer, and U. Rickers, "Optimized resource allocation for emergency response after earthquake disasters," *Safety Science*, vol. 35, no. 1–3, pp. 41–57, 2000.

[11] J. A. Deacon, C. V. Zegeer, and R. C. Deen, "Identification of hazardous rural highway locations," *Transportation Research Record*, vol. 543, pp. 16–33, 1975.

[12] F. F. Saccomanno, R. Grossi, D. Greco, and A. Mehmood, "Identifying black spots along highway SS107 in Southern Italy using two models," *Journal of Transportation Engineering*, vol. 127, no. 6, pp. 515–522, 2001.

[13] A. Gregoriades and K. C. Mouskos, "Black spots identification through a Bayesian Networks quantification of accident risk index," *Transportation Research Part C: Emerging Technologies*, vol. 28, pp. 28–43, 2013.

[14] S. Washington, M. M. Haque, J. Oh, and D. Lee, "Applying quantile regression for modeling equivalent property damage only crashes to identify accident blackspots," *Accident Analysis & Prevention*, vol. 66, pp. 136–146, 2014.

[15] B. Debrabant, U. Halekoh, W. H. Bonat, D. L. Hansen, J. Hjelmborg, and J. Lauritsen, "Identifying traffic accident black spots with Poisson-Tweedie models," *Accident Analysis & Prevention*, vol. 111, pp. 147–154, 2018.

[16] D. E. Bell, "Regret in Decision Making under Uncertainty," *Operations Research*, vol. 30, no. 5, pp. 961–981, 1982.

[17] M. S. Daskin, S. M. Hesse, and C. S. Revelle, "α-reliable p-minimax regret: a new model for strategic facility location modeling," *Location Science*, vol. 5, no. 4, pp. 227–246, 1997.

[18] C. G. Chorus, T. A. Arentze, and H. J. P. Timmermans, "A Random Regret-Minimization model of travel choice," *Transportation Research Part B: Methodological*, vol. 42, no. 1, pp. 1–18, 2008.

[19] C. G. Chorus, "A new model of random regret minimization," *European Journal of Transport and Infrastructure Research*, vol. 10, no. 2, pp. 181–196, 2010.

[20] T. Dekker, S. Hess, T. Arentze, and C. Chorus, "Incorporating needs-satisfaction in a discrete choice model of leisure activities," *Journal of Transport Geography*, vol. 38, pp. 66–74, 2014.

[21] S. Hess, M. J. Beck, and C. G. Chorus, "Contrasts between utility maximisation and regret minimisation in the presence of opt out alternatives," *Transportation Research Part A: Policy and Practice*, vol. 66, no. 1, pp. 1–12, 2014.

[22] S. van Cranenburgh, C. A. Guevara, and C. G. Chorus, "New insights on random regret minimization models," *Transportation Research Part A: Policy and Practice*, vol. 74, pp. 91–109, 2015.

[23] C. Guevara, C. G. Chorus, and M. Ben-Akiva, "Sampling of alternatives in random regret minimization models," *Transportation Science*, vol. 50, no. 1, pp. 306–321, 2016.

[24] S. Jang, S. Rasouli, and H. Timmermans, "Incorporating psycho-physical mapping into random regret choice models: model specifications and empirical performance assessments," *Transportation*, vol. 44, no. 5, pp. 999–1019, 2017.

[25] S. Rasouli and H. Timmermans, "Specification of regret-based models of choice behaviour: formal analyses and experimental design based evidence," *Transportation*, vol. 44, no. 6, pp. 1555–1576, 2017.

[26] S. Rasouli and H. Timmermans, "Issues in the specification of regret-only choice models: a rejoinder to Chorus and Van Cranenburgh," *Transportation*, vol. 45, no. 1, pp. 257–263, 2018.

[27] T. Dekker and C. G. Chorus, "Consumer surplus for random regret minimisation models," *Journal of Environmental Economics and Policy*, vol. 7, no. 3, pp. 269–286, 2018.

[28] M. Ziying, *Emergency Facilities Layout Optimal Method of Highway Based on Resource Sharing*, HeFei University of Technology, Hefei, China, 2013.

[29] X. Hongyan, *A Study on Expressway Traffic Incident Emergency Rescue System*, Southwest Jiaotong University, Cnki, Chengdu, China, 2011.

15

Causation Analysis of Hazardous Material Road Transportation Accidents by Bayesian Network using Genie

Xiaoli Ma, Yingying Xing, and Jian Lu

The Key Laboratory of Road and Traffic Engineering, Ministry of Education, Tongji University, 4800 Cao'an Road, Shanghai 201804, China

Correspondence should be addressed to Jian Lu; jianjohnlu@tongji.edu.cn

Academic Editor: Xiaobo Qu

With the increase of hazardous materials (Hazmat) demand and transportation, frequent Hazmat road transportation accidents had arisen the widespread concern in the community. Thus, it is necessary to analyze the risk factors' implications, which would make the safety of Hazmat transportation evolve from "passive type" to "active type". In order to explore the influence of risk factors resulting in accidents and predict the occurrence of accidents under the combination of risk factors, 839 accidents that have occurred for the period 2015–2016 were collected and examined. The Bayesian network structure was established by experts' knowledge using Dempster-Shafer evidence theory. Parameter learning was conducted by the Expectation-Maximization (EM) algorithm in Genie 2.0. The two main results could be likely to obtain the following. (1) The Bayesian network model can explore the most probable factor or combination leading to the accident, which calculated the posterior probability of each risk factor. For example, the importance of three or more vehicles in an accident leading to the severe accident is higher than less vehicles, and in the absence of other evidences, the most probable reasons for "explosion accident" are vehicles carrying flammable liquids, larger quantity Hazmat, vehicle failure, and transporting in autumn. (2) The model can predict the occurrence of accident by setting the influence degrees of specific factor. Such that the probability of rear-end accidents caused by "speeding" is 0.42, and the probability could reach up to 0.97 when the driver is speeding at the low-class roads. Moreover, the complex logical relationship in Hazmat road transportation accidents could be obtained, and the uncertain relation among various risk factors could be expressed. These findings could provide theoretical support for transportation corporations and government department on taking effective measures to reduce the risk of Hazmat road transportation.

1. Introduction

In recent years, the demand for hazardous materials (Hazmat) has increased, resulting in increasing transportation requirement. More than 95% Hazmat require off-site transportation in China, and 63% are transported by road in Brazil, as well as 90% in the United States [1, 2]. However, Hazmat could provide the great convenience for people's life, but also significant risks to environment and human health exist. For instance, a total of 3744 heavy trucks were involved in severe accidents in the United States, of which 3% were carrying Hazmat [3]. On January 11, 2015, a tanker truck carrying gasoline collided with a bus in Pakistan, causing 57 deaths. And a tanker truck carrying liquid ammonia collided with a van, resulting in a large-scale spill of liquid ammonia, leading to 28 deaths and the number of poisons was up to 350 on March 29, 2005, which arose on negative social impacts in China [4].

Hazmat transportation accidents would be able to produce catastrophic influence on human health, public safety, environment, and property due to the special characteristic of Hazmat, attracting more attention from general public and government on the management of Hazmat road transportation. Thus, how to improve the transportation condition and reduce the risk of transportation have become important and urgent problems for the industrial development. A growing amount studies about Hazmat transportation and production have been conducted [5–7]. Therefore, the need for investigating risk factors that contribute to Hazmat accident and the relationship of risk factors are highlighted to reduce the risk of Hazmat transportation. To that end, the effective method to describe and evaluate the accident process is causation analysis, which could be used to determine government priorities related to the implementation of prevention measures [8]. And causation analysis also could provide the theoretical support for actionable information of controlling over the risk factors for the transportation corporations. In addition, exploring the most probable factor or combination leading to accidents and predicting accidents are the important research topics in the field of Hazmat safety, reducing the frequency and severity of accidents.

2. Literature Review

The purpose of this study is to explore risk factors to reduce the risk of Hazmat road transportation. Many studies have been conducted by using statistical methods. Haastrup and Brockhoff [9] statistically analyzed the cases of Hazmat accidents in Western Europe, and 39% of accidents occurred during transportation; in 682 accidents the consequence included fatality. A study about Hazmat transportation accidents divided risk factors into human, vehicle, packing, transportation facilities, road conditions, and environmental conditions [5]. Shen [10] studied 708 accidents with Hazmat in China from 2004 to 2011 and found that accidents easily occurred at expressways, and the higher probability of spill accident is associated with accident type. Fang et al. [11] concluded that speeding was the main reason for Hazmat transportation accidents through the analysis of accident data between 1999 and 2013.

Although statistical methods could analyze the relationships between accidents and the risk factors, they cannot account for the interplay among different factors and fail to reflect the fact that an accident is not usually the result of a single factor [12]. The use of causation analysis theory for accidents could extract the accident mechanism and accident models from a large number of typical accidents. For instance, Jason et al. [13] conducted the study about the influence of vehicle, occupant, driver, and environmental characteristics on accident injuries involved with heavy-duty trucks, and the conclusion was obtained by using the heteroskedastic ordered probit models, which showed that the likelihood of severe accident is estimated to rise with the more vehicles involved in accident. Uddin and Huynh [14] used an ordered probit model to explore the relationship among drivers, vehicles, roadways, environment, temporal characteristics, and the severity of accident. There was a study by using logit model to study the driver's behaviors effect on accidents, and the results indicated that the more significant risk factors were speeding, not using seatbelt, drivers' age, and drivers with no valid license [15]. In addition, the Bayesian network and tree-based methods were considered to explore deeper accident mechanisms, which is increasingly utilized in traffic accidents analysis. For instance, Oña et al. [16] classified traffic accidents based on the severity of injuries by using the Bayesian networks; the factors associated with fatal or severe accidents were identified by inference, such as accident type, the driver's age, and lighting. In order to simplify the model, Mujalli et al. [17] used Bayesian networks to reduce the number of variables in the study of analyzing the accidents severity on rural roads, and the result showed that the number of variables could reduce up to 60% (the variables considered are accident type, age, atmospheric factors, gender, lighting, number of injured, and occupants involved), maintaining the good performance of models. Zhao et al. [18] pointed that the three most significant factors influencing Hazmat transportation by applying Bayesian networks were human factors, the transport vehicles and facilities, and the packaging and loading of Hazmat. Chen et al. [19] analyzed the between-accident variance and within-accident correlations by using Bayesian network and explored the risk factors influencing accidents and their heterogeneous impacts on accident severity in rural roads. And in order to improve the efficiency of emergency rescue of Hazmat transportation road accidents, a study was conducted to evaluate the time of accidents dealing based on the Bayesian network [20]. In addition, the Bayesian network model could also be used to describe the probability and risk of accidents [21–24].

However, despite many studies on the traffic accidents and Hazmat accidents, most of them are studied based on the analysis of specific, isolated, and single factor [25, 26]. Moreover, the characteristic of Hazmat was not taken into consideration during the analysis of accidents, limiting the studies of risk factors in Hazmat road transportation. In addition, the statistical methods could reveal the inherent rules on the occurrence of accidents, but the relationship of risk factors was not observed, which cannot reflect the accident mechanism. The application of causal analysis model (such as Bayesian network) can explain the correlation between risk factors and further explain the accident mechanism, but the Bayesian network structure may exist subjectivity due to the experts' knowledge, leading to incorrect description of relationships between nodes in the Bayesian network structure. Therefore, Hazmat road transportation accidents in China from 2015 to 2016 are considered as the research object to explore the potential risk factors of accidents based on experts' knowledge. The Bayesian network is used to explore the most probable factor or combination leading to accident and determine the correlation between the risk factors, providing the decision-making basis for Hazmat transportation corporations and government departments to reduce the risk of Hazmat transportation.

3. Database

The Hazmat transportation accident data was obtained from State Work Accident Briefing System, and Chemical Accidents Information Network for two years (2015-2016) in China, and the weather data was obtained from the China Meteorological Administration. The regional distribution of Hazmat transportation accidents is shown in Figure 1. The database considered in the study contains 839 records, and each record contains detailed information including the date, time, location, type of accidents, type and number of vehicles involved in accident, driver characteristic, the quantity and categories of Hazmat, accident consequence, causes of accident, and a detailed description of the accident. Sixteen variables extracted from the database were considered as the significant factors, which are shown in Table 1.

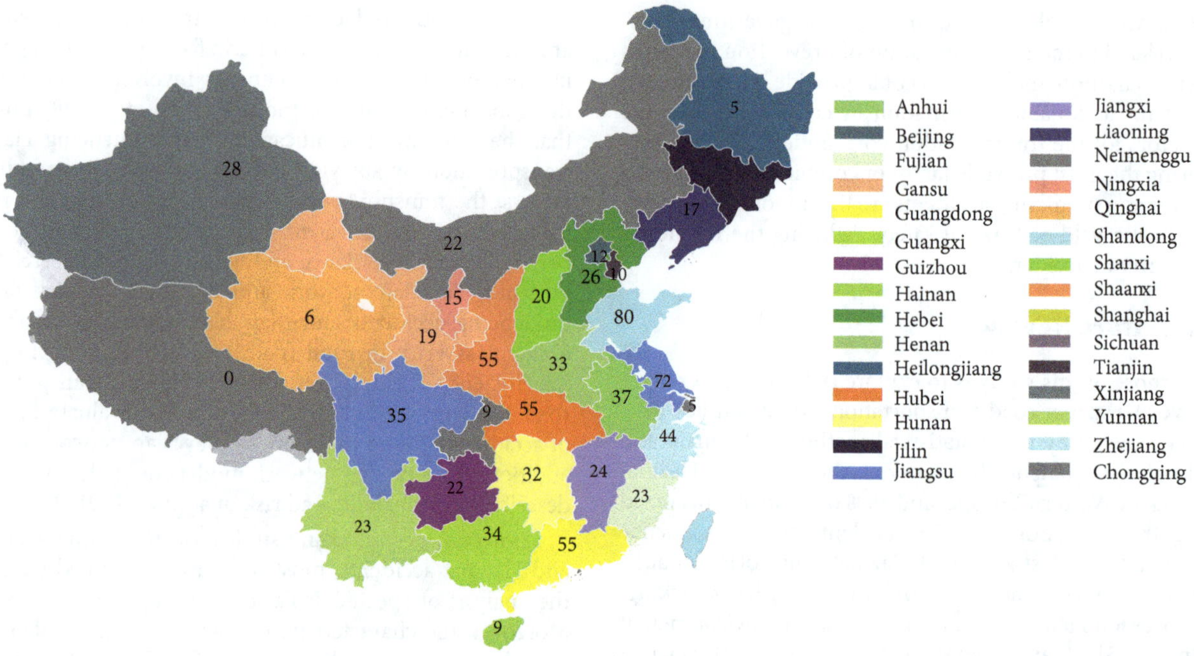

FIGURE 1: Regional distribution of accidents.

Accident information is accident type (rear-end, sideswipe, rollover, collision, and vehicle failure) and accident consequence (explosion, fire, spill, and nonspill). Previous studies [14, 27] divided injury severity into five categories; the accident severity in the paper is considered as no injury, severe injury, and fatality. Simplified classification of accident severity could ease the issue of potential relationship of related consequences of an accident and ensure the sufficient sample size for the Bayesian network model [28, 29]. In the paper, the simplified classification of accident severity would obtain the better results.

Hazmat information is Hazmat categories and quantity of Hazmat transportation.

Driver information is characteristics of the driver, such as age and behavior.

Location information is road surface condition and accident location (such as Group one, Group two, Group three, and Group four); the special road section including intersection, freeway service areas, toll stations, and gas stations are considered in the study.

Vehicle information is type and number of vehicles involved in accident.

Environment information is time distribution of accident (hour, day, and month), visibility (dawn: 5:00 to 6:59 am, day: 7:00 am to 4:59 pm, dusk: 5:00 to 6:59 pm, and dark: 7:00 pm to 4:59 am), and weather conditions (sunny, cloudy, rainy and snowy, and fog and haze) [18].

4. Methodology

4.1. Definition of Bayesian Network. Bayesian network is considered as the effective method to describe the causality between the risk factors and the output in the system, also referred to as the belief network. The Bayesian network is a Directed Acyclic Graph (DAG) and nodes represent variable status, while the directed edges represent dependencies between variables. The relationship or confidence coefficient between variables could be described by using Conditional Probability Table (CPT). The Bayesian formula is considered as the basis for the Bayesian network model, which could be expressed as

$$P(X \mid Y) = \frac{P(Y \mid X) \times P(X)}{P(Y)} \tag{1}$$

where $P(X \mid Y)$ is the probability of X under the condition of a known event Y. $P(Y \mid X)$ is the conditional probability of Y at the occurrence of X. And the joint distribution of two random variables X and Y can be expressed as

$$P(X,Y) = P(X)P(Y \mid X) \tag{2}$$

where $P(X)$ is called the prior probability and $P(Y \mid X)$ is the posterior probability. Combined with the chain rules, reducing the complexity of the probability model, the joint distribution of n variables is

$$\begin{aligned} &P(X_1, X_2, \cdots, X_n) \\ &= P(X_1)P(X_2 \mid X_1)\cdots P(X_n \mid X_1, X_2, \cdots, X_{n-1}) \end{aligned} \tag{3}$$

and the joint distribution also could be expressed as

$$P(X_1, X_2, \cdots, X_n) = \prod_{i=1}^{n} P(X_i \mid Parent(X_i)) \tag{4}$$

where $X = \{X_1, X_2, \cdots, X_n\}$, setting S is a network structure, P is a set of local probability distributions associated with each variable, X_i denotes the variable node, and $Parent(Xi)$ denotes the father node of Xi in S.

TABLE 1: Variables of Hazmat road transportation accidents.

Factors	Variables	Variables description	Discretization	Frequency	Percentage
Hazmat factors	Hazmat categories	Explosives	1	27	3.20%
		Toxic gases	2	158	18.90%
		Flammable liquids	3	429	51.10%
		Corrosives	4	121	14.40%
		others	5	104	12.40%
	Quantity of Hazmat	<10	1	127	15.10%
		10-24	2	284	33.80%
		25-39	3	358	42.70%
		≥40	4	70	8.40%
Driver factors	Age	24-35	1	144	17.20%
		36-45	2	644	76.70%
		46-60	3	51	6.10%
	Behaviors	Inappropriate driving	1	13	1.50%
		Speeding	2	36	4.30%
		Fatigue driving	3	20	2.40%
		Normal driving	4	770	91.80%
Location factors	Accident location	Group one	1	360	42.90%
		Group two	2	336	40.00%
		Group three	3	59	7.00%
		Group four	4	84	10.10%
	Special section	Intersection	1	18	2.10%
		Freeway service areas	2	50	6.00%
		Toll stations	3	78	9.30%
		Gas stations	4	23	2.70%
		Normal	5	670	79.90%
Environment factors	Road surface	Dry	1	794	94.60%
		Wet	2	45	5.40%
	Season	Spring	1	227	27.10%
		Summer	2	258	30.70%
		Autumn	3	186	22.20%
		Winter	4	168	20.00%
	Weekly distribution	Weekends	1	198	23.60%
		Weekdays	2	641	76.40%
	Weather	Sunny	1	202	24.10%
		Cloudy	2	347	41.40%
		Rainy & snow	3	268	31.90%
		Fog & haze	4	22	2.60%
	Visibility	dawn	1	94	11.20%
		day	2	409	48.70%
		dusk	3	60	7.20%
		dark	4	276	32.90%
Vehicle factors	Total vehicle involved in accident	1	1	503	59.90%
		2	2	276	32.90%
		3	3	31	3.70%
		≥4	4	29	3.50%
	Type of vehicle	Bus & Truck	1	13	1.55%
		Private cars & Truck	2	42	5.01%
		Non-motor & Truck	3	11	1.31%
		Bus&Private cars&Truck	4	10	1.19%
		Trucks	5	763	90.94%

TABLE 1: Continued.

Factors	Variables	Variables description	Discretization	Frequency	Percentage
Accidents factors	Accident type	Rear-end	1	189	22.50%
		Sideswipe	2	20	2.40%
		Rollover	3	340	40.50%
		Collision	4	145	17.30%
		Vehicle failure	5	145	17.30%
	Accident consequence	Explosion	1	25	3.00%
		Fire	2	96	11.40%
		Spill	3	682	81.30%
		Non-spill	4	36	4.30%
	Severity of accident	No injury	1	656	78.19%
		Severe injury	2	139	16.57%
		Fatality	3	44	5.24%

The construction of the Bayesian network model consists of following steps:

(1) Parameter determination: analyze the risk factors of Hazmat road transportation, and determine the variables needed for modeling (nodes of the Bayesian network), which could be shown in Table 1.

(2) Structure learning: determine the dependencies or independencies relationships between variables (nodes), so that a directed acyclic network structure was constructed.

(3) Parameter learning: based on the given Bayesian network structure, determine the CPT for each node, and the dependence relationship between random variables could be described quantitatively.

4.2. Structure Learning. The scientific network structure needs continuous iterations. At present, there are three methods to construct a Bayesian network structure [30]. **(1)** Construct the network structure subjectively through experts' knowledge. **(2)** Determine the network structure objectively via the analysis of data. **(3)** Construct the network structure based on experts' knowledge and data analysis. The method used in the paper for accident causation analysis is that establishing a preliminary Bayesian network structure based on the model assumption and then the network structure is adjusted with experts' knowledge and data analysis, avoiding the disadvantage of strong subjectivity and enormous amount of data computing. The Bayesian network structure is constructed as shown in Figure 2.

Steps for Building a Bayesian Network Structure

(1) Establish a preliminary Bayesian network structure based on the assumptions of model.

(2) Use Delphi method to determine the relationship between risk factors. In general, there are four possible relationships between variables:

(A) F_i directly lead to F_j, which could be represented as $F_i \longrightarrow F_j$.

(B) F_j directly lead to F_i, which could be represented as $F_i \longleftarrow F_j$.

(C) The relationship between variables cannot be determined, which could be represented as $F_i \longleftrightarrow F_j$.

(D) There is no relationship between variables, which could be represented as $F_i \mid F_j$.

(3) Synthesize results from multiple experts. D-S evidence theory is used to reduce the subjectivity of experts' knowledge, and the correlation between variables could be determined. The Dempster synthesis rule formula could be expressed as

$$M(A) = K \cdot \sum_{A_1 \cap A_2 \cap \cdots \cap A_n} m_1(A_1) m_2(A_2) \cdots m_n(A_n)$$

$$\forall A \subseteq \Theta,\ A \neq \varnothing,\ A_1, A_2, \ldots A_n \subset \Theta,$$

$$K = \left(\sum_{A_1 \cap A_2 \cap \cdots \cap A_n \neq \varnothing} m_1(A_1) m_2(A_2) \cdots m_n(A_n) \right)^{-1} = \left(1 - \sum_{A_1 \cap A_2 \cap \cdots \cap A_n = \varnothing} m_1(A_1) m_2(A_2) \cdots m_n(A_n) \right)^{-1} \quad (5)$$

where A represents the possible relationship between variables, m_i represents the mass function, equaling to the expert opinions, and n represents the number of experts.

(4) As the relationship of variables cannot be obtained by Delphi and D-S evidence theory, the mutual information value of variables should be calculated. And the entropy can be expressed as

$$H(F_i) = \sum_{F_i} P(F_i) \log \frac{1}{P(F_i)} = -\sum_{F_i} P(F_i) \log P(F_i) \quad (6)$$

Conditional entropy is a measure of the uncertainty of a random variable F_i under the condition of giving F_j, which can be expressed as

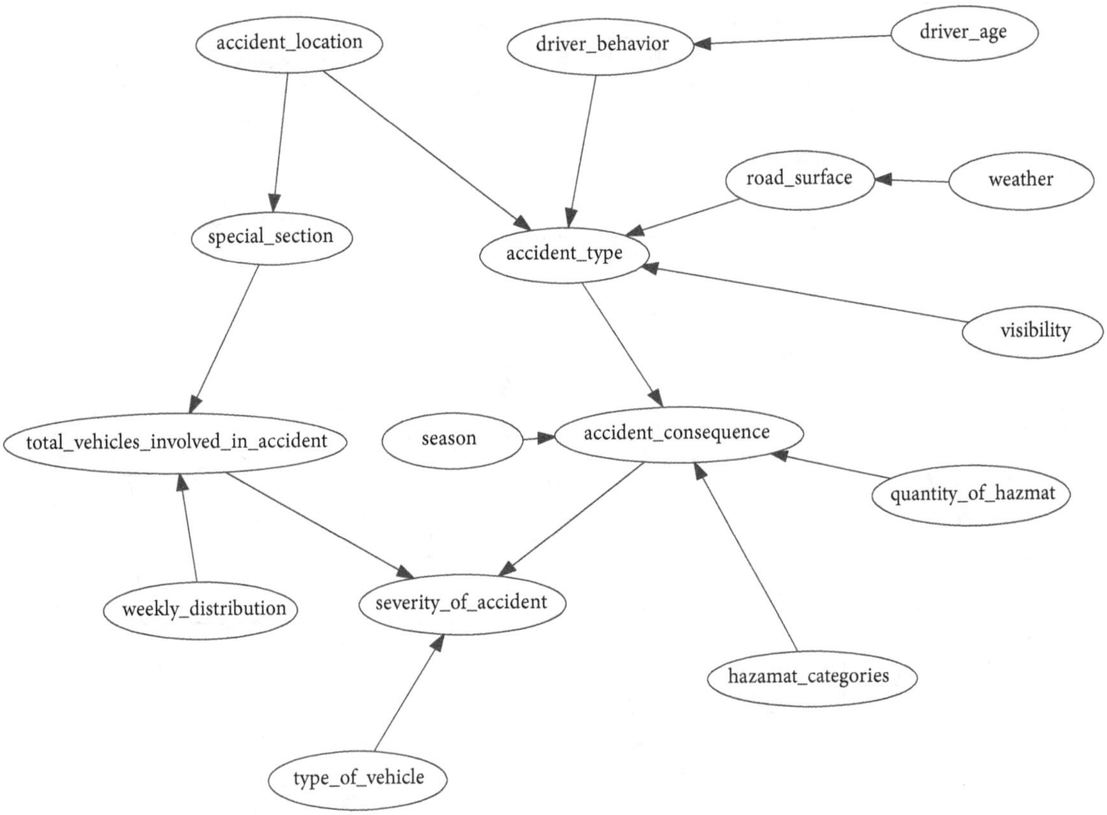

FIGURE 2: The Bayesian network structure for Hazmat road transportation accidents.

$$H(F_i \mid F_j) = \sum_{F_i} P(F_i \mid F_j) \log \frac{1}{P(F_i \mid F_j)} = -\sum_{F_i} P(F_i \mid F_j) \log P(F_i \mid F_j) \quad (7)$$

Before obtaining F_j, the uncertainty of F_i is $H(F_i)$, and after obtaining F_j, the uncertainty of F_i is $H(F_i \mid F_j)$, so that the difference of $H(F_i)$ and $H(F_i \mid F_j)$ is considered as the mutual information, which is expressed as

$$I(F_i, F_j) = H(F_i) - H(F_i \mid F_j) = \sum_{F_i}\sum_{F_j} P(F_i, F_j) \cdot \log_2 \frac{P(F_i, F_j)}{P(F_i) P(F_j)} \quad (8)$$

4.3. Parameter Learning. There are missing data on Hazmat road transportation accidents; the Expectation- Maximization (EM) algorithm is considered as the effective method to perform the maximum likelihood estimation for a set of parameters θ from the incomplete dataset [31–33]. The EM algorithm starts with randomly assigning a configuration θ^0 for θ by the system. Suppose that θ^t is the outcome after t iterations. The calculation process mainly involved two steps: Expectation Step (E-Step) and Maximization Step (M-Step).

Consider that D_m is missing sample, and X_m is the set of all variables with missing value in the sample D_m. Set $X_m = x_m$, and the complete dataset would be obtained by adding x_m to D_m. All of the possible result would be considered by EM algorithm due to that X_m may have more possibility, so the weight w_{x_m} is assigned for each possible result by EM algorithm, and the weighted sample could be given by

$$(D_m, X_m = x_m)\left[w_{x_m}\right] \quad (9)$$

where $w_{x_m} = P(X_m = x_m \mid D_m, \theta^t)$, and the weight ranges from 0 to 1.

E-Step: suppose the log-likelihood function of θ based on D^t.

$$m(\theta \mid D^t) = \sum_{t=1}^{m} \sum_{x_m \in X_m} P(X_m = x_m \mid D_m, \theta^t) \cdot \log P(D_m, X_m = x_m \mid \theta) \quad (10)$$

where $D = (D_1, D_2, \ldots, D_m)$, and $m(\theta \mid D, \theta^t) = m(\theta \mid D^t)$ is referred to as the expected log-likelihood function. In the iteration, due to the characteristic of D, which is invariant, the formula could be expressed as

$$M(\theta \mid \theta^t) = m(\theta \mid D, \theta^t) = \sum_{t=1}^{m} \sum_{x_m \in X_m} P(X_m = x_m \mid D_m, \theta^t) \cdot \log P(D_m, X_m = x_m \mid \theta) \quad (11)$$

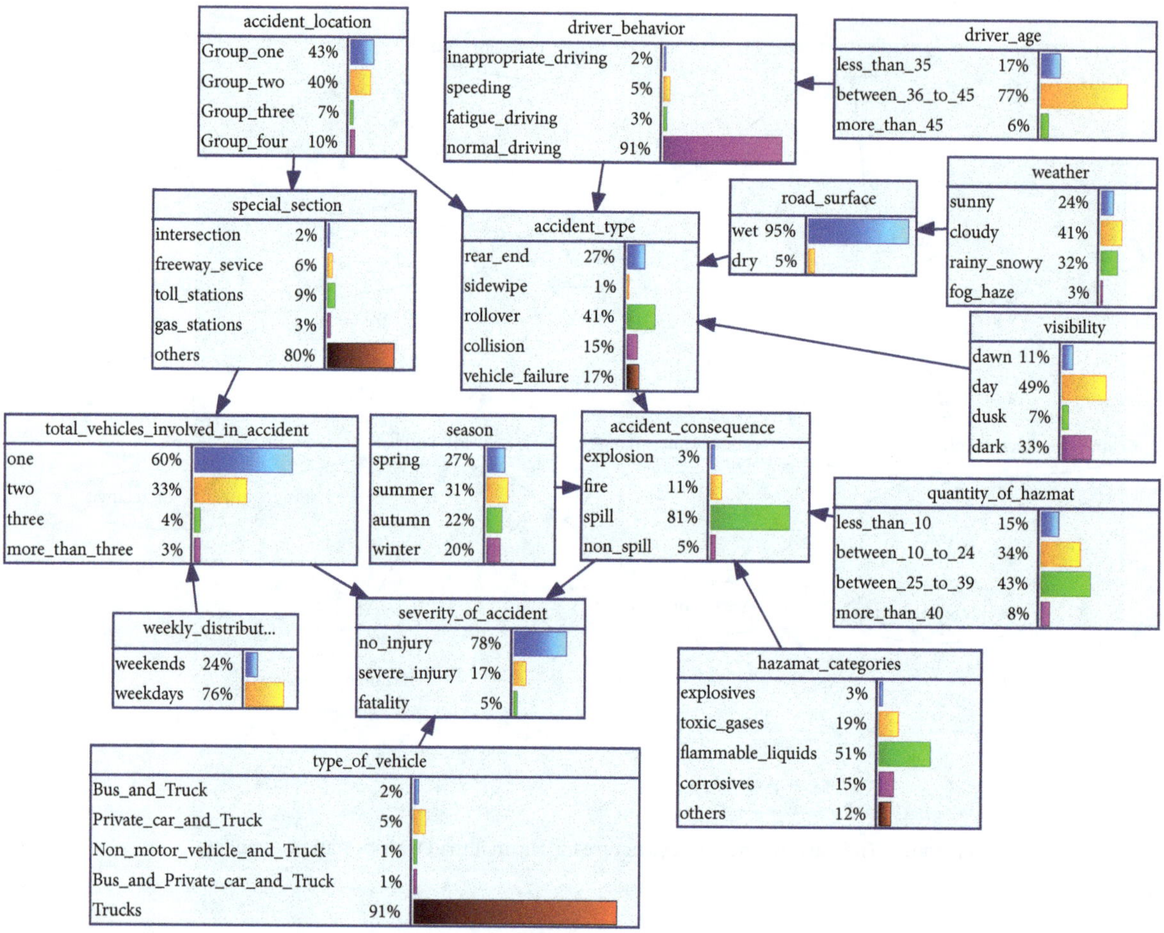

FIGURE 3: The Bayesian network model after parameter learning in Genie 2.0.

M-Step: calculate the value of θ when $M(\theta \mid \theta^t)$ have reached the maximum:

$$\theta_{ijk}^{t+1} = \begin{cases} \dfrac{m_{ijk}^t}{\sum_{k=1}^{r_i} m_{ijk}^t} & \sum_{k=1}^{r_i} m_{ijk}^t > 0 \\ \dfrac{1}{r_i} & \sum_{k=1}^{r_i} m_{ijk}^t \le 0 \end{cases} \quad (12)$$

where m_{ijk}^t is the sum of sample weights in the dataset D^t.

5. Results

The guidance for the variable selection and classification were followed by the analysis of accident data and previous studies [6, 34–36]. In the paper, sixteen variables are considered as the significant risk factors, as shown in Table 1. There are numerous types of software to establish the Bayesian network efficiently, such as Netica, Genie, Bayes Net Toolbox, and Analytica. In the paper, Genie2.0 (developed by the Decision Systems Laboratory, the University of Pittsburgh) was considered as the effective tool to finish the Bayesian network parameter learning by using EM algorithm, which would make the construction, analysis, and visualization of Bayesian network be performed efficiently, simplifying the calculation. And the network parameters are repeatedly iterated by using the accident data; the conditions for the termination of calculation are as follows: (1) the variation of the posterior probability for single risk factor is less than 1%; (2) the cumulative variation of posterior probability for the entire network is less than 15%. The results were shown in Figure 3.

5.1. Causal Inference. The Bayesian network could be used to calculate the posterior probability of risk factors under conditions of an accident and obtain the most likely factors or combinations that caused accidents. Set the "explosion" in "accident consequence" as the example to explore the causal inference, and the evidence variable is "explosion". As shown in Figure 4, the probabilities of risk factors are obtained through the update function of the Genie. And the probability of "autumn" in "season" increases from 22% to 35%; "vehicle failure" (referred as the tire blowout, spontaneous combustion, tanker damage) in "accident type" increases from 17% to 37%; the quantity of Hazmat increases from 8% to 20% for the category of more than 40 tons; "flammable liquids" in "Hazmat categories" increase from 51% to 65%; and the explosives increase from 3% to 8%. These findings mean that, in the absence of other evidences, the most probable reasons for "explosion" are vehicles carrying

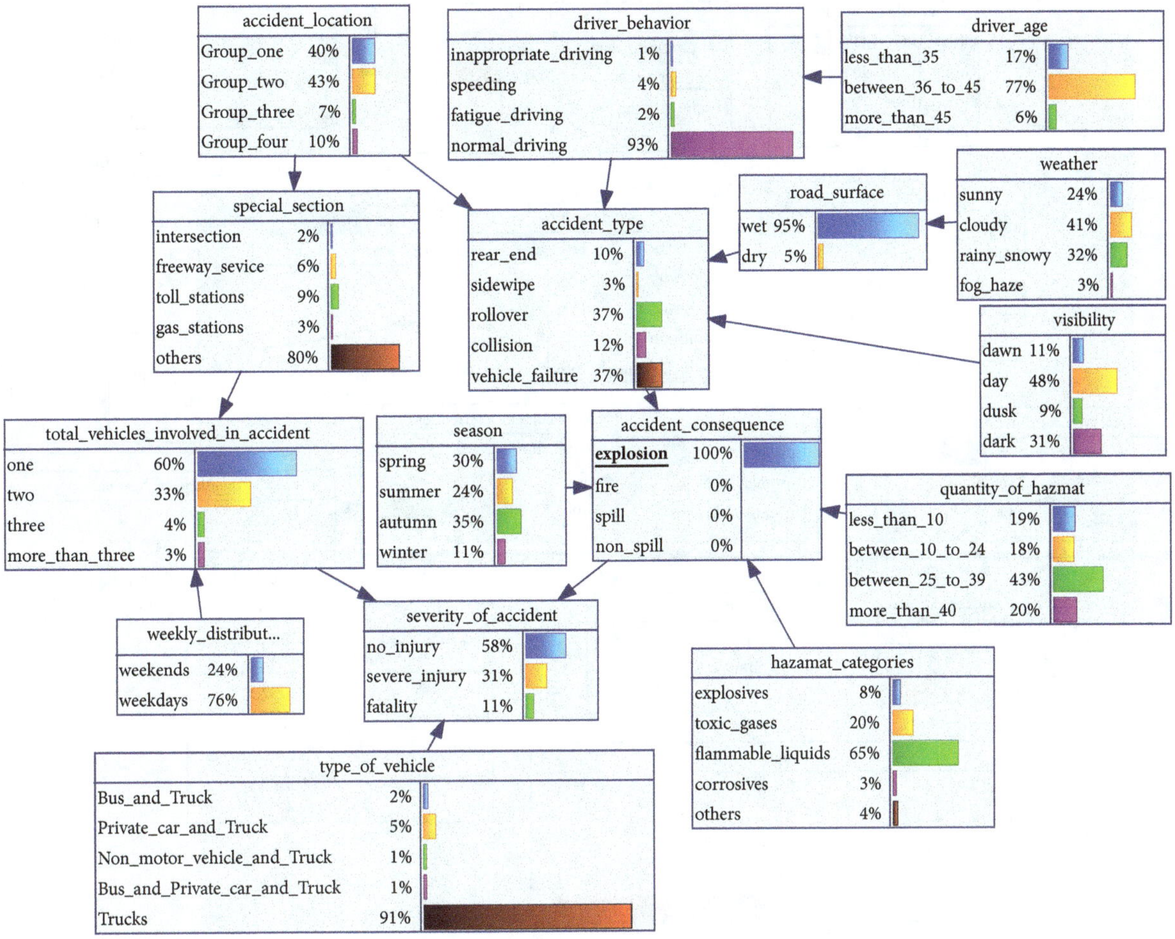

FIGURE 4: Posterior probability when the variable is "explosion".

flammable liquids, larger quantity of Hazmat, vehicle failure, and transporting in autumn.

In addition, if the "fatality" in the "severity of accident" is considered as the evidence variable, the probability change of "total vehicle involved accident" could be obtained. The probability of "three" increases from 4% to 11%, and "more than three" is increasing from 3% to 9%. This may be explained by the fact that the importance of 3 or more vehicles in an accident leads to the severe accident being higher than less vehicles. Moreover, as for the accident consequence, the probability of "spill" decreases; meanwhile the "explosion" (3% to 6%) and fire (11% to 18%) have increased. Due to the special characteristic of Hazmat, explosion and fire would cause a larger area affected and can easily result in casualties, especially in the urban road and higher population densities [26].

5.2. Accident Prediction. Based on the bidirectional reasoning, not only could the Bayesian network model obtain the risk factors or the combination caused accidents, but also the probability of accidents could be calculated under the risk factors or combination, for example, in Genie, setting the "speeding" in "driver behavior" as an evidence variable, meaning that the status of evidence variable is considered as 100%. As can be seen from Figure 5, the probability of "rear-end" in "accident type" is found to increase from 27% to 42%, indicating that the drivers' speeding could be more prone to lead to rear-end accidents. This is because the vehicle is difficult to control under the condition of speeding, and the braking time is longer. And previous studies have shown that driving behavior could significantly affect the severity of traffic accidents [37–39].

As shown in Figure 6, in addition to "speeding", it is assumed that the transportation route is on low-class roads; that is, "Group four" in the "accident location" is considered as the evidence variable, and the probability of the entire network is automatically updated. It can be found that the probability of "rollover" in "accident type" further increases from 42% to 97%. This finding shows that "driver behavior" and "accident location" would affect the probability of "rollover" accident on different degrees. Therefore, when the driver is speeding on low-class roads, the more attention should be paid on the rollover accident.

6. Discussion and Conclusions

6.1. Hazmat Factors. Flammable liquids have the highest posterior probability (0.51) and would easily result in explosion. This could be explained by that increasing demand for the flammable liquid and decreasing reliability of transporting

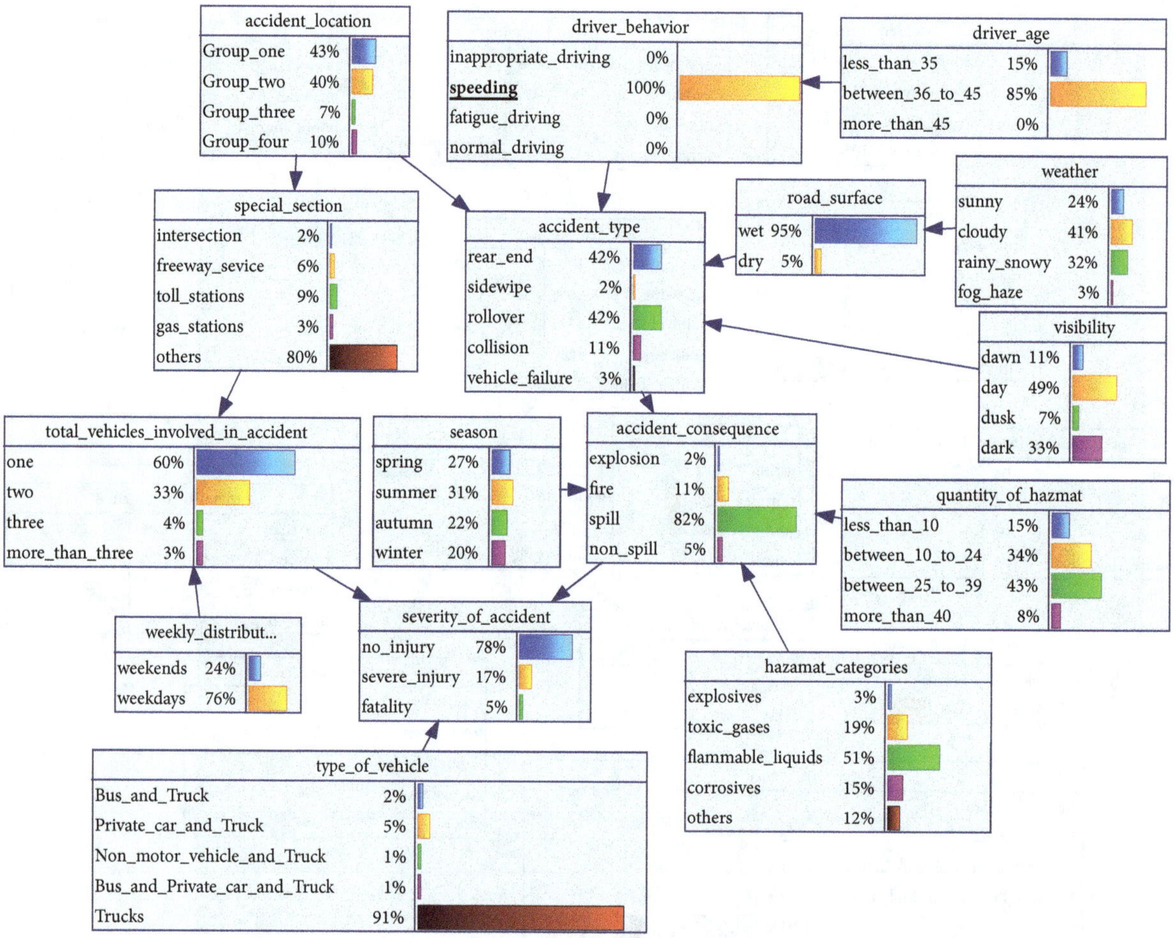

FIGURE 5: Accident prediction when the evidence variable is "speeding".

flammable liquids due to the single-mode packaging. The quantity of Hazmat transported would significantly affect the severity of accident. The larger the quantity of Hazmat transportation, the larger the inertia of the transportation vehicles, making it not easy to control the emergency [40]. Moreover, the larger quantity of Hazmat transportation is prone to the serious consequences, such as explosion and spill, threatening people's health and environment [10].

6.2. Driver Factors. Previous studies have shown the relationship between driver's age and the severity of accidents [27, 41, 42]. According to the model results, the younger driver (less than 35) would be more prone to inappropriate driving behavior, which indicates the need for carrying out education programs and training for younger drivers. Tavris et al. [43] also found that younger drivers were much more likely to be involved in severe and fatal accidents. As for the driving behavior, speeding is more likely to lead to rollover accident, especially on the low-class road. This could ascribe the small amount of lanes and the road condition defects on low-class roads, and the speeding would make Hazmat slosh or move around inside the tank, which can constantly shift the vehicle weight, leading to vehicle to rollover due to the off balance [44, 45].

6.3. Location Factors. The model results show that "Group one" (the posterior probability is 0.43) and "Group two" (the posterior probability is 0.40) in "accident location" are likely to be associated with severe accidents, which could be attributed by the combination of higher average speed and larger speed dispersion. More importantly, "Group one" and "Group two" roads are considered as the major transport corridors for Hazmat [10, 46]. In addition, some special sections would also considered as the significant risk factors; this could be explained by the fact that there are more interference factors (such as line of sight, pedestrians, and signal lights) at intersections and the greater potential explosion risk around the gas stations [47].

6.4. Environment Factors. Hazmat road transportation accidents would easily occur at summer (the posterior probability is 0.31), which is attributed to the characteristic of Hazmat, such as flammable and explosive. And the posterior probability of accidents occurring at weekdays is 0.76, which could be explained by that freeway could be toll-free on important holidays, resulting in significant increase of traffic volume, which could decrease the speed of vehicles. Moreover, Hazmat transportation vehicles were not allowed to drive on freeway (Pan, 2013). Weather is a significant factor for the Hazmat transportation, with cloudy having the

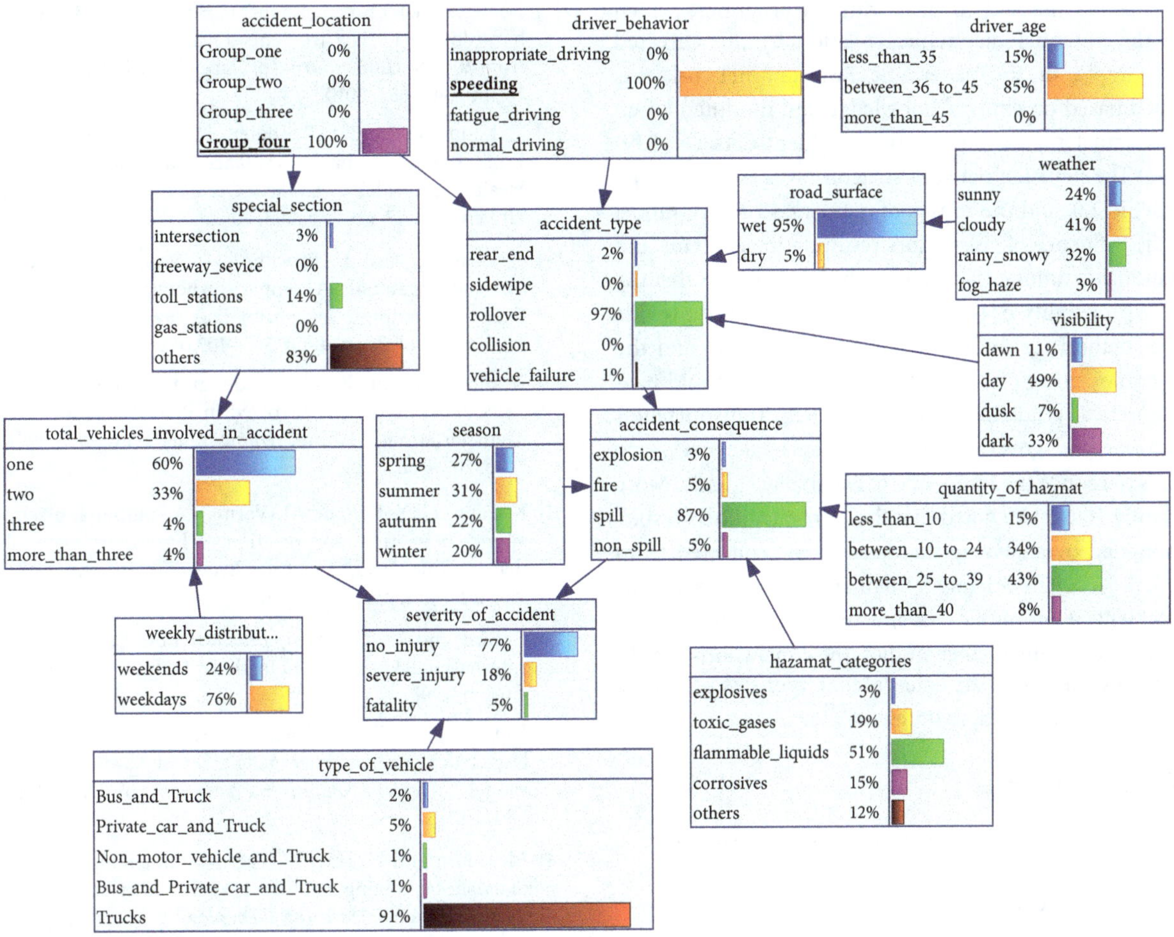

FIGURE 6: Accident prediction when the evidence variable are "speeding" and "Group four".

highest posterior probability (0.41) followed by rainy (0.32). This could be ascribed that the driver's mood and visual would be decreased in cloudy and rainy, and the rainy would lower the friction coefficient of roads due to the thin film of water existing between the road surface and tires, which could make the road slippery, increasing the braking distance effectively [48, 49]. Regarding visibility, daytime has the highest posterior probability (0.49), and the dark is 0.33. This is because most transportation corporations are more likely to transport Hazmat at daytime in China [50]. In addition, poor visibility at night would make drivers tired, resulting in driver fatigue, especially from 11:00 pm to 3:00 am [51]. In the sample of accident data, drivers are more prone to fatigue status accounting for 62% of total accidents from 7:00 pm to 4:59 am.

6.5. Vehicle Factors. As for the total vehicles involved in accident, "more than three" would easily result in higher severity of accidents. And the private car involved in accident would cause the severe accident. Two reasons could explain these findings: one is that more vehicles would cause more people involved in accidents, resulting in more people injured; another one is the disparity in mass and speed of trucks compared to other vehicles. In case of an accident, lighter vehicles (such as private cars) usually absorb the greatest part of the kinetic energy and suffer from more severe injury.

6.6. Accident Factors. Many studies have shown the significant relationship of accidents type and severity, indicating that the rollover accident is associated with the higher severity of accident [16, 44]. The Bayesian network results show that rollover accident has the highest posterior probability (0.41). The reason could be that Hazmat sloshing or moving around inside the tank can constantly shift the vehicle weight, making the vehicle off balance, causing the transportation vehicle to roll over, especially during abrupt evasive maneuvers or turning the vehicle [10]. In addition, as for the consequence of accident, the posterior probability of spill could reach up to 0.81, threatening human health and environment. The result could be explained by that Hazmat releasing could immediately result in poisoning and suffocation, which is difficult for people on-site to escape quickly, resulting in severe and fatality accident [50].

In summary, the occurrence of Hazmat road transportation accidents is unexpected, random, dangerous, and potential. Frequent accidents imply that it is necessary to explore risk factors by using accident mechanism. Bayesian network is the effective method to deal with uncertainties, which exhibit the potential hierarchical relation by the Directed Acyclic Graph. In the paper, the Bayesian network was developed based on experts' knowledge and modified based on the Hazmat road transportation accident data (N=839) in China. The Bayesian network structure was established by using Genie 2.0, and the results of network structure model reveal the influence of risk factors resulting in accidents and the relationship among risk factors. The study shows that the posterior probability of the Bayesian network could provide effective method for finding the important factors and the factors combination of accidents. These findings could provide theoretical guidance, which could help transportation corporations and government departments take necessary measures to reduce the frequency of Hazmat accidents. More importantly, it must be noted that the aforementioned results were obtained by analyzing the data sample collected from State Work Accident Briefing System and Hazardous Chemical Accidents Communications, which could be existing limitations. As for the further studies, the conclusions should be more generalizable if the dataset had larger size of sample and accidents from multiple states.

Conflicts of Interest

The authors declare that they have no conflicts of interest.

Acknowledgments

This study has been supported by projects of the National Natural Science Foundation of China (no. 71671127).

References

[1] K. G. Zografos and K. N. Androutsopoulos, "A decision support system for integrated hazardous materials routing and emergency response decisions," *Transportation Research Part C: Emerging Technologies*, vol. 16, no. 6, pp. 684–703, 2008.

[2] F. G. Cordeiro, B. S. Bezerra, A. S. P. Peixoto, and R. A. R. Ramos, "Methodological aspects for modeling the environmental risk of transporting hazardous materials by road," *Transportation Research Part D: Transport and Environment*, vol. 44, pp. 105–121, 2016.

[3] Federal Motor Carrier Safety Administration (FMCSA), "Large Truck and Bus Crash Facts 2014," https://www.fmcsa.dot.gov/sites/fmcsa.dot.gov/files/docs/Large-Truck-and-Bus-Crash-Facts-2014%28April%202016%29.pdf, 2016.

[4] J. Zhao L, "Risk Analysis of Dangerous Chemicals Transportation," *Systems Engineering-Theory Practice*, vol. 27, no. 12, pp. 117–122, 2007.

[5] W. Y. Hua and A. Tong P, "Risk Analysis on Road Transport System of Dangerous Chemicals," *China Safety Science Journal*, vol. 15, no. 2, pp. 8–12, 2005.

[6] L. Zhao J, P. Wu, and K. Xu, "Statistic analysis and countermeasures on dangerous chemical accidents in China," *China Safety Science Journal*, 2009.

[7] J. Yang, F. Li, J. Zhou, L. Zhang, L. Huang, and J. Bi, "A survey on hazardous materials accidents during road transport in China from 2000 to 2008," *Journal of Hazardous Materials*, vol. 184, no. 1-3, pp. 647–653, 2010.

[8] T. Kauppinen and J. Rantanen, "Work and Health Country Profiles and National Surveillance Indicators in Occupational Health and Safety," *Applied Occupational & Environmental Hygiene*, vol. 17, no. 9, p. 603, 2002.

[9] P. Haastrup and L. Brockhoff, "Severity of accidents with hazardous materials. A comparison between transportation and fixed installations," *Journal of Loss Prevention in the Process Industries*, vol. 3, no. 4, pp. 395–405, 1990.

[10] X. Shen, Y. Yan, X. Li, C. Xie, and L. Wang, "Analysis on Tank Truck Accidents Involved in Road Hazardous Materials Transportation in China," *Traffic Injury Prevention*, vol. 15, no. 7, pp. 762–768, 2014.

[11] K. Fang, G. Y. Ke, and M. Verma, "A routing and scheduling approach to rail transportation of hazardous materials with demand due dates," *European Journal of Operational Research*, vol. 261, no. 1, pp. 154–168, 2017.

[12] F. Bird and G. Germain, *Practical Loss Control Leadership*, International Loss Control Institute, Duluth, GA, USA, Revised edition, 1990.

[13] J. D. Lemp, K. M. Kockelman, and A. Unnikrishnan, "Analysis of large truck crash severity using heteroskedastic ordered probit models," *Accident Analysis & Prevention*, vol. 43, no. 1, pp. 370–380, 2011.

[14] M. Uddin and N. Huynh, "Factors influencing injury severity of crashes involving HAZMAT trucks," *International Journal of Transportation Science and Technology*, vol. 7, no. 1, pp. 1–9, 2018.

[15] E. K. Adanu and S. Jones, "Effects of Human-Centered Factors on Crash Injury Severities," *Journal of Advanced Transportation*, vol. 2017, no. 1528, pp. 1–11, 2017.

[16] J. De Oña, R. O. Mujalli, and F. J. Calvo, "Analysis of traffic accident injury severity on Spanish rural highways using Bayesian networks," *Accident Analysis & Prevention*, vol. 43, no. 1, pp. 402–411, 2011.

[17] R. O. Mujalli and J. De Oña, "A method for simplifying the analysis of traffic accidents injury severity on two-lane highways using Bayesian networks," *Journal of Safety Research*, vol. 42, no. 5, pp. 317–326, 2011.

[18] L. J. Zhao, X. L. Wang, and Y. Qian, "Analysis of factors that influence hazardous material transportation accidents based on Bayesian networks: a case study in China," *Safety Science*, vol. 50, no. 4, pp. 1049–1055, 2012.

[19] C. Chen, G. Zhang, X. C. Liu et al., "Driver injury severity outcome analysis in rural interstate highway crashes: a two-level Bayesian logistic regression interpretation," *Accident Analysis & Prevention*, vol. 97, pp. 69–78, 2016.

[20] J. Chen, M. Zhang, S. Yu, and J. Wang, "A Bayesian Network for the Transportation Accidents of Hazardous Materials Handling Time Assessment," *Procedia Engineering*, vol. 211, pp. 63–69, 2018.

[21] M. Deublein, M. Schubert, B. T. Adey, J. Köhler, and M. H. Faber, "Prediction of road accidents: a Bayesian hierarchical approach," *Accident Analysis & Prevention*, vol. 51, pp. 274–291, 2013.

[22] C. Xu, W. Wang, P. Liu, and Z. Li, "Calibration of crash risk models on freeways with limited real-time traffic data using Bayesian meta-analysis and Bayesian inference approach," *Accident Analysis & Prevention*, vol. 85, pp. 207–218, 2015.

[23] C. Tang, Y. Yi, Z. Yang, and J. Sun, "Risk analysis of emergent water pollution accidents based on a Bayesian Network," *Journal of Environmental Management*, vol. 165, pp. 199–205, 2016.

[24] X. Zou and W. L. Yue, "A Bayesian Network Approach to Causation Analysis of Road Accidents Using Netica," *Journal of Advanced Transportation*, 2017.

[25] R. Bubbico, S. Di Cave, B. Mazzarotta, and B. Silvetti, "Preliminary study on the transport of hazardous materials through tunnels," *Accident Analysis & Prevention*, vol. 41, no. 6, pp. 1199–1205, 2009.

[26] R. A. Garrido and A. C. Bronfman, "Equity and social acceptability in multiple hazardous materials routing through urban areas," *Transportation Research Part A: Policy and Practice*, vol. 102, pp. 244–260, 2016.

[27] T. Usman, L. Fu, and L. F. Miranda-Moreno, "Analysis of factors affecting winter collision severity," in *Meeting of the Transportation Research Board*, Washington, DC, USA, 2013.

[28] F. Chen and S. Chen, "Injury severities of truck drivers in single- and multi-vehicle accidents on rural highways," *Accident Analysis & Prevention*, vol. 43, no. 5, pp. 1677–1688, 2011.

[29] S. Islam, S. L. Jones, and D. Dye, "Comprehensive analysis of single- and multi-vehicle large truck at-fault crashes on rural and urban roadways in Alabama," *Accident Analysis & Prevention*, vol. 67, pp. 148–158, 2014.

[30] Q. Xiaohu, L. Li, and Z. Ying, "A traffic accident prediction method based on Bayesian network model," *Computer Simulation*, vol. 22, no. 11, pp. 230–232, 2005.

[31] S. L. Lauritzen, "The EM algorithm for graphical association models with missing data," *Computational Statistics & Data Analysis*, vol. 19, no. 2, pp. 191–201, 1995.

[32] F. V. Jensen and T. D. Nielsen, "Bayesian Networks and Decision Graphs," *Technometrics*, vol. 50, no. 1, p. 362, 2012.

[33] J. Zhou, W. Xu, X. Guo, and J. Ding, "A method for modeling and analysis of directed weighted accident causation network (DWACN)," *Physica A: Statistical Mechanics and its Applications*, vol. 437, pp. 263–277, 2015.

[34] J. Abellán, G. López, and J. de Oña, "Analysis of traffic accident severity using decision rules via decision trees," *Expert Systems with Applications*, vol. 40, no. 15, pp. 6047–6054, 2013.

[35] R. O. Mujalli, G. López, and L. Garach, "Bayes classifiers for imbalanced traffic accidents datasets," *Accident Analysis & Prevention*, vol. 88, pp. 37–51, 2016.

[36] A. Iranitalab and A. Khattak, "Comparison of four statistical and machine learning methods for crash severity prediction," *Accident Analysis & Prevention*, vol. 108, pp. 27–36, 2017.

[37] L. Fridstrøm and S. Ingebrigtsen, "An aggregate accident model based on pooled, regional time-series data*," *Accident Analysis & Prevention*, vol. 23, no. 5, pp. 363–378, 1991.

[38] G. F. Ulfarsson, S. Kim, and E. T. Lentz, "Factors affecting common vehicle-to-vehicle collision types: Road safety priorities in an aging society," *Transportation Research Board*, vol. 1980, no. 1, pp. 70–78, 2006.

[39] G. Zhang, K. K. W. Yau, X. Zhang, and Y. Li, "Traffic accidents involving fatigue driving and their extent of casualties," *Accident Analysis & Prevention*, vol. 87, pp. 34–42, 2016.

[40] A. Vorster, "Transporting dangerous goods worldwide: materials handling logistics," *South African Pharmaceutical Cosmetic Review*, 2015.

[41] A. P. Jones and S. H. Jørgensen, "The use of multilevel models for the prediction of road accident outcomes," *Accident Analysis & Prevention*, vol. 35, no. 1, pp. 59–69, 2003.

[42] S. Kaplan and C. G. Prato, "Risk factors associated with bus accident severity in the United States: a generalized ordered logit model," *Journal of Safety Research*, vol. 43, no. 3, pp. 171–180, 2012.

[43] D. R. Tavris, E. M. Kuhn, and P. M. Layde, "Age and gender patterns in motor vehicle crash injuries: Importance of type of crash and occupant role," *Accident Analysis & Prevention*, vol. 33, no. 2, pp. 167–172, 2001.

[44] C. S. Duncan, A. J. Khattak, and F. M. Council, "Applying the ordered probit model to injury severity in truck-passenger car rear-end collisions," *Transportation Research Record*, no. 1635, pp. 63–71, 1998.

[45] A. Montella, L. Imbriani, and F. Mauriello, "Factors Contributing to Run-off-the-Road Severe Crashes," in *Proceedings of the Transportation Research Board 94th Annual Meeting*, Washington, DC, USA, 2015.

[46] A. Oggero, R. M. Darbra, M. Muñoz, E. Planas, and J. Casal, "A survey of accidents occurring during the transport of hazardous substances by road and rail," *Journal of Hazardous Materials*, vol. 133, no. 1-3, pp. 1–7, 2006.

[47] R. Tay and S. M. Rifaat, "Factors contributing to the severity of intersection crashes," *Journal of Advanced Transportation*, vol. 41, no. 3, pp. 245–265, 2010.

[48] Q. Lin and W. A. Nixon, "Effects of Adverse Weather on Traffic Crashes: Systematic Review and Meta-Analysis," *Transportation Research Record Journal of the Transportation Research Board*, vol. 2055, no. 2055, pp. 139–146, 2008.

[49] P. Konstantopoulos, P. Chapman, and D. Crundall, "Driver's visual attention as a function of driving experience and visibility. Using a driving simulator to explore drivers' eye movements in day, night and rain driving," *Accident Analysis & Prevention*, vol. 42, no. 3, pp. 827–834, 2010.

[50] H.-D. Zhang and X.-P. Zheng, "Characteristics of hazardous chemical accidents in China: A statistical investigation," *Journal of Loss Prevention in the Process Industries*, vol. 25, no. 4, pp. 686–693, 2012.

[51] J. Pahukula, S. Hernandez, and A. Unnikrishnan, "A time of day analysis of crashes involving large trucks in urban areas," *Accident Analysis & Prevention*, vol. 75, pp. 155–163, 2015.

16

Modelling the Operating Speed in Segments of Two-Lane Highways from Probe Vehicle Data: A Stochastic Frontier Approach

António Lobo, Marco Amorim, Carlos Rodrigues, and António Couto

Research Centre for Territory, Transports and Environment, Faculty of Engineering, University of Porto, Rua Dr. Roberto Frias, s/n, 4200-465 Porto, Portugal

Correspondence should be addressed to António Lobo; lobo@fe.up.pt

Academic Editor: Abdelaziz Bensrhair

Most of the existing operating speed statistical models are applicable to individual design elements, particularly horizontal curves and tangents. A segment approach to operating speed has rarely been followed, with a few exceptions mainly related to the performance assessment of urban and freeway corridors, or design consistency studies using speed profiles built from successive design elements. This study introduces a new model to predict operating speeds in segments of two-lane highways. The maximum operating speed is given by a stochastic frontier function of the average daily traffic and road geometrics; the asymmetric disturbance accounts for the diversity in drivers' behaviour and vehicle characteristics, allowing estimating any percentile speed. The model was calibrated using probe vehicle data from noncongested roads. The accuracy of the average daily traffic in representing the actual driving conditions was further validated using simultaneous speed-traffic measurements. The new model aims to assist practitioners in the evaluation of design consistency from a macroscopic perspective since the early stages of road planning and design, as well as to support the definition of speed limits at new or existing infrastructures.

1. Introduction

Increasing efforts to understand and predict operating speeds in roadway facilities have been developed across the past decades, following the tendency of researchers and practitioners to use the operating speed as an instrument to define road geometrics and promote design consistency. The great majority of the existing models are focused on the purely geometric effects on driving speeds, exploring the relations between geometric parameters and the speeds practiced under free-flow conditions in individual design elements, especially in horizontal curves or tangents. Such models are commonly known as spot speed models, of which the *Transportation Research Circular E-C151* [1] provides a comprehensive review.

The operating speed in a road segment, defined as a continuous sequence of design elements composing a stretch of road, has been much less studied, despite the relevance that segment speed estimations based on roadway characteristics may present to the evaluation of road design and operations. The use of segment speed prediction models since the early stages of road design allows practitioners to verify, for each design iteration, if the differences between the selected design speed and the expected operating speed lie within acceptable limits. A sensitivity analysis on such differences could shed light on the variables most influencing design consistency, thus providing an insight into the most effective actions to control it in the following iterations of the design process, with the ultimate goal of obtaining a satisfactory final solution. This procedure is in line with the suggestions made by Fitzpatrick et al. [2] regarding the use of alignment indices to evaluate design consistency from a macroscopic perspective. Nevertheless, to improve safety performance, Fitzpatrick et al. [2] advocate that such analysis should be complemented by a conventional speed profile approach based on the evaluation of speed reductions between successive design elements.

Segment speed models may also play an important role in the definition of speed limits in planned roads or in the adjustment of the existing ones. Speed limits are legislated by road functional class, but road designers and managers may establish speed zones with different limits if the statutory limits do not comply with the specific road or traffic conditions. The most common approach consists in setting the speed limit near the 85th-percentile speed (V85) at representative locations of the proposed zone [3]. The use of segment speed models avoids a subjective choice of representative sites by considering the aggregated characteristics of entire zone. Additionally, the model can be used in alternative methods to define speed limits, such as the expert judgment [4–6], assisting experts in the proposal of adequate speed limits.

Finally, practitioners may use segment speed estimations in the evaluation and monitoring of existing roads when real-world speed data is not available. The evaluation of travel times, costs, and emissions, the identification of safety hazards, and the information of road users are examples of applications that frequently rely on the knowledge of operating speeds and that may lead practitioners to take measures aimed at improving the system's performance and efficiency.

To address the lack of operating speed research in road segments, this study presents a new model for two-lane highways. The main objective is to deliver a novel segment speed tool, capable of predicting any user-specified percentile, to assist practitioners in the design, safety analysis, and performance evaluation of new and existing roadway infrastructures, focusing on rural roads where speed monitoring equipment is rarely available. The segment speed model builds on the authors' previous research on spot speed modelling, retaining the operating speed frontier model (OSFM) formulation that allows percentile speed estimation [7, 8]. In the new model, the geometric and traffic characteristics are represented by a frontier function, while an asymmetric disturbance term accounts for nonquantified factors related to drivers' behaviour and vehicle technology. Segment speeds were assessed from probe vehicle data, geometric characteristics were assessed from both probe vehicle data and in situ measurements, and traffic effects were evaluated through the annual average daily traffic.

The remainder of the paper is structured as follows. First, a background on the most relevant achievements contained in the literature concerning the study of segment speeds is provided. The data description details the speed and road geometry data collection using probe vehicles and the additional data requirements in terms of geometric and traffic characteristics. Then, the model description presents the OSFM formulation and the methods used to derive the new segment speed model from a previous spot speed model presented by the authors [8]. The model estimation introduces the segment speed model for two-lane highways, calibrated for Portuguese conditions. The paper proceeds with the comparison between the OSFM and a speed profile method, and with the model validation using simultaneous speed and traffic data from road segments beyond the calibration sample. The main conclusions of this study are presented in the final section.

2. Background

The array of segment speed models presented in the literature is much smaller than the comprehensive set of spot speed models developed to date [1]. The main cause of the lack of development of segment speed studies is probably related to the need for more time and significantly more resources to perform data collection.

The traditional techniques to assess segment speeds are the stationary observer method and the moving observer method [9]. In the stationary observer method, an observer is placed at each end of the survey segment recording each vehicle's passing time and license plate, in order to calculate the time spent to traverse the segment. In the moving observer method, an observer in a test vehicle chases randomly chosen vehicles across the segment, recording the corresponding travel times. The underlying speed concept in both methods is the space mean speed, represented by the harmonic mean of speeds over a length of roadway, which can be derived from the average travel time of the observed vehicles.

In freeways and urban roads, researchers have taken advantage of available technology devices to reducing human intervention in data collection. Dion and Rakha [10], Li et al. [11], and Tam and Lam [12] applied the principle of the stationary observer by using automatic vehicle identification (AVI) data from toll collection points to compute travel time and speed. Jenelius and Koutsopoulos [13] and Wang et al. [14] applied the principle of the moving observer to measure travel times and speeds in urban links using floating car data (FCD) from probe systems installed in taxi fleets. Rakha and Zhang [15], Soriguera and Robusté [16], and Martínez-Díaz and Pérez [17] derived space mean speeds and corresponding confidence intervals from time mean speeds in freeways, taking advantage of existing loop detectors. While these studies have predominantly followed a statistical approach, authors such as Park et al. [18], van Lint et al. [19], and Tang et al. [20] applied neural networks to forecast speed and travel time using data from fixed devices installed in urban freeway corridors, highlighting the advantages of these methods to provide short-term predictions in periods of congestion.

In relation to two-lane highways, only a few design guides propose methods to define operating speeds over a length of roadway. In the UK, both operating and design speeds are established on the basis of geometric variables, particularly the bendiness and mean visibility observed on a minimum 2 km long segment [21]. In Germany, the operating speed is determined by the curvature change rate and pavement width [22]. Other official guidelines propose different design speeds according to the road functional classification and provide methods to ensure design consistency by limiting speed differences between successive design elements. This is the case of the US reference manual *A Policy on Geometric Design of Highways and Streets* [23] and the Portuguese guidelines [24].

The *Highway Capacity Manual* (HCM) [25], which is regarded by practitioners worldwide as a standard reference on capacity and level-of-service procedures, also uses a segment approach to estimate the free-flow speed (FFS). First, a base free-flow speed (BFFS) based on the design speed, posted speed limit, or speeds observed in similar facilities

(a) (b)

FIGURE 1: Examples of the considered roads: N 101 (a) and N 206 (b).

must be adopted. Then, the FFS is estimated by reducing the BFFS for the effects of the cross-section width and density of access points. The HCM justifies that no further guidance on estimating the BFFS is provided because of the diversity of local conditions that play a major role in drivers' speed choice. However, it is possible to conclude that the BFFS is the speed observed at roads with a similar functional classification and design standards, with no access points, and lane and shoulder widths equal to or greater than 3.6 m and 1.8 m, respectively, since the proposed corrections to the BFFS are only applicable for higher densities of access points and smaller cross-section widths. The HCM suggests the use of this method only when speed measurements are not possible, for example, at the planning and design stages of a new road, but speed estimation methods for other facilities are still required if the BFFS is established by comparison with similar roads.

In sum, the academic community has developed some methods to estimate segment speeds in freeways and small urban links, using data from existing monitoring equipment to simplify data collection. In turn, reference manuals provide some guidance to define design and/or operating speeds in roads of different functional classifications, including in two-lane highways, but are relatively coy about explaining the underlying methods and assumptions. Therefore, the segment speed model developed in this study aims at presenting a significant contribution to this branch of knowledge.

3. Data Description

The new segment speed model was calibrated using speed, traffic, and geometrics data from nine road segments contained in three two-lane highways in Portugal: N 14, N 101, and N 206. In these roads, design speeds range from 40 to 60 km/h, junctions are usually at-grade, and direct access to marginal properties is allowed. Additionally, roadside occupation has a great influence on the density of intersections and lateral clearance. The curve radius and lane width are sometimes smaller than the minimum values proposed by the guidelines [24]. The case-study roads were selected to reflect the diversity of characteristics among the National Roads, a category that represents the vast majority of the country's highway network. Roads of higher functional classifications, that is, Principal and Complementary Itineraries, were not included in the present study due to data availability issues.

Each calibration segment is between 2 and 4 km long. The total extension of analysed roadway is of 23.5 km. The segments are located outside the urban areas, crossing either completely rural settings or zones with scattered buildings (Figure 1). The pavement is well maintained and the markings are clearly visible. The topographic features are consistent with the classification of level or rolling terrain [25]. The traffic is characterized by a noncongested flow, except for the possible occurrence of isolated incidents. Segments containing signalized intersections or roundabouts were excluded from this analysis, ensuring that all the vehicles traversing the calibration segments have right-of-way.

Speed data was derived from probe vehicle data provided by a company specialized in digital mapping and applications. This company runs numerous probe information devices, installed on board of professional and private vehicles, which feed a real-time traffic information system available to media partners and GPS applications. A high-rate GPS system (>1 Hz) provides an accurate representation of each probe vehicle's trajectory, allowing identifying the vehicles that have traversed each calibration segment. The space mean speed of each vehicle, consisting of the model's dependent variable, was calculated from the travel time given by the difference between GPS timestamps corresponding to the beginning and the end of the segment. To ensure a homogenous modelling sample, 75 vehicles observed in each segment in both directions were considered, corresponding to the amount of probe vehicles at the calibration segment with less recorded observations. In total, 675 observations were used for model estimation.

The independent variables included in the segment speed model are based on both geometric and traffic characteristics. The geometric design is characterized by the horizontal alignment, cross-section width, and density of intersections, which is in line with previous research on the speed effects produced by variables measured over a length of roadway [8, 21, 22, 25, 26]. The horizontal alignment of the calibration segments was reproduced in a CAD software, using probe data collected by the instrumented vehicle of the Traffic Analysis Laboratory of the Faculty of Engineering of the University of

TABLE 1: General data on calibration segments.

Variable description	Mean	Standard deviation	Minimum	Maximum
Space mean speed (km/h)	49.2	9.2	21.0	80.0
Bendiness (*B*) [degrees/km]	306.7	207.0	39.0	682.3
One-direction paved width (PW) [m]	4.2	0.5	3.4	5.4
Extra lateral clearance (ELC) [m]	1.2	0.4	0.7	2.2
Density of intersections (DI) [No./km]	4.0	1.8	0.5	7.0
Annual average daily traffic (AADT) [vehicles/day]	8,736	5,545	1,750	18,135

Porto. A similar reproduction of the vertical alignment was not possible due to unreliable altimetry data. The *Highway Capacity Manual* [25] suggests that relevant speed reductions may occur in segments with grades exceeding 3% for more than 1 km. However, this situation does not occur in the calibration segments, although steeper grades may locally be present. Thus, the lack of variables characterizing the vertical alignment is not expected to represent a relevant omission in the model.

Contrary to spot speed models usually dealing with free-flow vehicles, segment speed models should account for the traffic effects. First, the evaluation of the operational performance of a roadway infrastructure using the latter models will be more realistic if traffic conditions are considered. Second, independently of the method used to collect segment speed data, it is difficult to guarantee that sample vehicles traverse the entire segment under free-flow conditions; hence speed observations may have been affected by the presence of other vehicles on the road.

Specifically, the geometric and traffic characteristics included in the model were the bendiness (*B*), the one-direction paved width (PW) and its standard deviation (SDPW), the extra lateral clearance (ELC), the density of intersections (DI), and the annual average daily traffic (AADT). *B* corresponds to the sum of the deflection angles of the horizontal alignment per kilometre. PW represents the mean value of the lane and right shoulder combined width. The mean was calculated from measurements taken at a set of locations within the segment. The spacing between measurement spots is approximately 500 m. PW experiments frequent variations according to the marginal land use and topography across Portuguese National Roads, especially concerning the shoulder width. Therefore, SDPW was included in the model to account for the effects of such diversity within a road segment. ELC represents the distance between the right shoulder external limit and any fixed object at the roadside. This variable was defined by the mean value corresponding to the same sites where PW was measured. DI corresponds to the number of intersections or interchanges per kilometre. An aggregated daily traffic measure, AADT, was considered because the time of day at which probe vehicles were recorded is unknown. Then, AADT was used as a proxy for the momentary traffic conditions faced by the test vehicles. Traffic data was either provided by a previous accident modelling study conducted at the same roads by Costa [27] or estimated from in situ measurements using automatic traffic counters (VIACOUNT II). The use of daily traffic measures as speed predictors is not new among the literature, being present in studies such as Lamm and Choueiri [28], Jessen et al. [29], and Schurr et al. [30].

Because all the variables encompass a spatial dimension, reflecting either a value per unit of distance or an average value representative of the entire segment, the segment length was not included in the analysis. The minimum length of 2 km required to select a calibration segment ensures a reasonable distance to assess space mean speeds, as well as the necessary differentiation from spot speed studies. Table 1 presents the general description of the data used to estimate the segment speed model.

4. Model Description

The OSFM used in this research was introduced in previous spot speed studies by the authors [7, 8] to address some limitations of the widespread percentile-specific models [1]. Traditionally, the calibration of speed models is made by applying a linear regression between speed factors and a specific percentile speed obtained by the aggregation of values collected in situ. Besides the obvious disadvantage of supporting the estimation of only one percentile speed, speed data aggregation in conventional models causes a loss of information that reduces the total variability and the nature of the variability associated with the regression function. Consequently, the estimated effects of the independent variables may be biased [31]. Therefore, modelling the entire speed distribution may contribute to rectifying this problem.

The literature contains only two other models supporting a speed distribution that can be used to estimate any percentile spot speed. Medina and Tarko [32] employed an ordinary least squares method to estimate a customized regression equation for any intended percentile. The resulting models represent a linear combinations of factors associated with the mean and the standard deviation of the speed distribution. Hewson [33] generalized this approach by using quantile regression.

The OSFM [7, 8], albeit using the entire speed distribution, follows a completely different approach based on stochastic frontier models from the econometric analysis [34, 35]. The OSFM is formed by a deterministic speed frontier function and two disturbance terms. The speed frontier is given by an exponential function of a set of independent variables characterizing the most relevant road conditions, being estimated by the maximum likelihood method using the total number of speed observations. The exponential form allows for the interaction between the order of magnitude of speeds and the impacts of the independent variables. The values upon the speed frontier represent the upper limit of the operating speed estimated for each combination of independent variables; hence, the speed frontier is dubbed the maximum operating speed ($V\max$). In other words, $V\max$ reflects the speed pattern of the fastest drivers and varies according to the conditions presented by the road. $V\max$ results from the statistical processing of operating speeds collected under good weather and pavement conditions; thus it should not be mistaken for speed concepts representing safety-critical situations for the vehicle's dynamic equilibrium.

The segment speed model is based on the authors' latest spot speed model for horizontal curves and tangents of Portuguese National Roads, introduced in Lobo et al. [8]. In this spot speed model, $V\max$ is composed of variables characterizing the local conditions, namely, on-site geometric and visibility parameters, and variables reflecting the recent driving experience, that is, geometric and roadside interference parameters measured over the preceding segment. The general formulation presented by Lobo et al. [8] that supports the prediction of $V\max$ at an individual design element is given by

$$V\max_e = \exp\left(\beta_0 + \sum_{k=1}^{n_k} \beta_k \ln X_k + \sum_{m=1}^{n_m} \beta_m \ln X_m\right), \quad (1)$$

where $V\max_e$ is the maximum operating speed at a given design element, X_k is on-site characteristic k, X_m is segment characteristic m, and β are regression coefficients.

The new segment speed model is developed from the previous spot speed model, not only because both models adopt an OSFM formulation, but also because the aggregated effect produced by the segment characteristics on the operating speed is assumed identical in both models except for a scale factor. In other words, the segment effects included in the spot speed model translate into the segment speed model through the calibration of a single coefficient affecting all those variables. The variable SC, used as a predictor in the segment speed model, combines the segment characteristics considered in the spot speed model, such as $\mathrm{SC} = \exp(\sum_{m=1}^{n_m} \beta_m \ln X_m)$. Additionally, the new model accounts for specific effects affecting segment speeds that were absent from the spot speed model. These effects are related to the presence of traffic and the dispersion of the cross-section width, as previously discussed in this paper. Variables characterizing individual design elements, that is, measured locally instead over a length of roadway, are excluded from the segment speed model. Equation (2) represents the general formulation of the model to estimate $V\max$ at a given road segment.

$$V\max_s = \exp\left(\alpha_0 + \alpha_1 \ln \mathrm{SC} + \sum_{q=2}^{n_q} \alpha_q \ln X_q\right), \quad (2)$$

where $V\max_s$ is the maximum operating speed at a given road segment, SC is the variable representing the aggregated segment characteristics from the spot speed model, X_q is segment characteristic q related to traffic or the dispersion of cross-section variables, α_0 is the regression constant, α_1 is the regression coefficient or scale factor of variable SC, and α_q are regression coefficients of the segment characteristics q.

The OSFM considers two disturbance terms: the noise term and the asymmetric disturbance term. The former term supports the stochastic nature of the speed frontier, representing random errors associated with model specification and data collection. It is frequently considered in regression models and assumes a normal distribution. The latter term is an asymmetrically distributed disturbance accounting for the deviations from $V\max$ that are attributed to nonquantified factors, that is, other factors than those represented by the deterministic frontier. Then, the asymmetric disturbance is suitable to represent the diversity of vehicle speeds observed at the same road element or segment, which, in turn, depends on the individual responses of each pair driver-vehicle to similar road conditions. In other words, the asymmetric disturbance in the OSFM accounts for the speed reductions from $V\max$ caused by the diversity in drivers' behaviour and vehicle technology. Therefore, the cumulative distribution function of this disturbance allows estimating any desired percentile speed. Similarly to the previous studies [7, 8], the asymmetric disturbance adopts an exponential distribution, such as $f(u) = \theta \times \exp(-\theta u)$, where θ is the rate parameter of the exponential function. The cumulative function of the asymmetric disturbance is given by $F(u) = 1 - \exp(-\theta u)$, and the inverse transform is $u = (-1/\theta) \times \ln(1 - F)$. Hence, the pth percentile speed at a given road segment can be estimated using

$$Vp_s = V\max_s \times \exp\left(\frac{1}{\theta} \ln p\right), \quad (3)$$

where Vp_s is the pth percentile speed a given road segment and p is the percentile value ($0 < p < 1$).

According to Greene [36], the maximum likelihood method is more efficient in processing asymmetric disturbances than the least squares method. Thus, the OSFM is estimated through the maximization of the log-likelihood function shown in

$$\begin{aligned} \ln L = {} & N \ln \theta + \frac{N}{2}\theta^2 \sigma_v^2 + \theta \sum_{j=1}^{n_j} \sum_{i=1}^{n_i} \left(v_{ij} - u_{ij}\right) \\ & + \sum_{j=1}^{n_j} \sum_{i=1}^{n_i} \ln \Phi\left(-\frac{v_{ij} - u_{ij}}{\sigma_v} - \theta \sigma_v\right), \end{aligned} \quad (4)$$

where L is the likelihood function, N is the total number of observations, σ_v is the standard deviation of the noise term distribution, v_{ij} is the noise term for vehicle i in segment j, u_{ij} is the asymmetric disturbance for vehicle i in segment j, and $\Phi(\cdot)$ is the standard normal distribution function.

5. Model Estimation

The calibration of the segment speed model for National Roads in Portugal was made through a stochastic frontier regression between the space mean speeds of all sample

TABLE 2: Results of the stochastic frontier regression.

Variable	Coefficient	Standard error
Constant	4.846	0.197[a]
ln SC	4.462	0.437[a]
ln SDPW	−0.125	0.019[a]
ln AADT	−0.064	0.019[a]

Note. Log-likelihood = 143.617; number of observations = 675; $\sigma_u = 0.168$; $\sigma_v = 0.124$; $\theta = 5.947$. [a]Significant at 1% level.

vehicles and the road geometrics and traffic. The regression is based on

$$\ln Vp_s = \alpha_0 + \alpha_1 \ln \mathrm{SC} + \alpha_2 \ln \mathrm{SDPW} + \alpha_3 \ln \mathrm{AADT} + \frac{1}{\theta} \ln p, \tag{5}$$

where $\mathrm{SC} = \mathrm{PW}^{0.079} \times \mathrm{ELC}^{0.008} \times B^{-0.027} \times \mathrm{DI}^{-0.036}$.

Equation (5) represents a linearized form of (3) and details the model formulation with the previously described variables. The exponents of the factors composing SC originate from the spot speed model for Portuguese National Roads presented by Lobo et al. [8], being recalibrated in the segment speed model by the scale factor α_1. To comply with the OSFM formulation, the speed predictors are transformed in log terms, which implies that the model is not applicable when ELC, B, DI, or SWPD is null.

The model was estimated through the maximum likelihood method, using the econometric software Limdep [37]. The results are shown in Table 2. These results translate into the final specification for the model to estimate Vmax and the pth percentile speed in segments of Portuguese National Roads represented by

$$V\max_s = \exp\left(4.846 + 4.462 \times \ln\left(\mathrm{PW}^{0.079} \times \mathrm{ELC}^{0.008} \times B^{-0.027} \times \mathrm{DI}^{-0.036}\right) - 0.125 \times \ln \mathrm{SDPW} - 0.064 \times \ln \mathrm{AADT}\right)$$
$$Vp_s = V\max_s \times \exp\left(\frac{1}{5.947} \times \ln p\right) \tag{6}$$

with ELC, B, DI, and SWPD > 0.

The coefficients of the individual variables can be interpreted as elasticities due to the model's functional form. The elasticities of PW, ELC, B, and DI are given by the original elasticities estimated in the spot speed model multiplied by the scale factor α_1. The results show that the geometric variables, particularly those related to the cross-section width, represent the most relevant effects affecting operating speeds. An increase in PW while keeping the remaining variables constant produces a positive impact on speed, materialized by an elasticity of 0.352. The impact of increasing ELC is also positive, albeit with a smaller elasticity of 0.036. Additionally, the negative coefficient of SDPW reflects that the variation of PW over a road segment has a negative impact on speed. Many Portuguese National Roads are characterized by an inconsistent paved width, mainly attributed to frequently changing shoulder widths according to the marginal land use, which introduces uncertainty in drivers' expectations about the road ahead, leading them to adopt a more cautious speed choice. The speed impacts caused by increases in the density of intersections and bendiness are negative, with elasticities of −0.161 and −0.120, respectively. The results also show that, as expected, increasing traffic leads to speed reductions. Particularly, the elasticity associated with AADT denotes that a traffic increase of 10% under a noncongested flow would produce a decrease in the space mean speed of 0.6%.

6. Model Validation

The model's capability of supporting accurate speed predictions was evaluated using two different procedures. First, the speed predictions returned by the model for each one of the calibration segments were compared against the results of a speed profile method. Then, the model was validated using simultaneous speed and traffic data collected in two additional road segments.

6.1. Comparison with a Speed Profile Approach. To the best of the authors' knowledge, there is no other segment speed model for two-lane rural highways that was developed outside the scope of road design and capacity manuals. In Portugal, the official guidelines propose a speed profile method to evaluate design consistency over a length of roadway from a safety perspective. Speed is computed element by element, taking into account the vehicle's dynamic equilibrium, to verify if the differences between successive design elements lie within the regulatory limits [24]. Based on this approach, a speed profile approach was used to estimate segment speeds at the nine calibration segments, CS1 to CS9, and compare the results with the OSFM estimations. To comply with the operational nature of this analysis, the dynamic equilibrium equations were replaced by a conventional V85 model previously developed for spot speed prediction in curves and tangents of Portuguese two-lane roads [26]. Then, an approximation of the travel time in each design element was obtained from the estimated V85; the model does not consider speed variations within the design elements. Finally, segment speeds were derived from the sum of travel times of all the design elements composing each segment. Figure 2 presents the observed 85th-percentile segment speeds (V85so) and the corresponding estimations using the OSFM (V85sm) and the speed profile method (V85ssp).

In seven out of nine cases, the differences between the OSFM estimations and the observed values are smaller than 4 km/h, outperforming the speed profile method. Although the latter method returns better predictions in the remaining two cases, the deviations from the observed values across the test cases are much more inconsistent, which may be related to the nonconsideration of traffic and acceleration/deceleration effects.

The mean absolute deviation (MAD) and the mean squared error (MSE) were calculated for both methods using the set of nine test cases. The OSFM estimations resulted in a MAD of 3.4 and a MSE of 32.1. The speed profile method produced a MAD of 8.4 and a MSE of 128.3. These

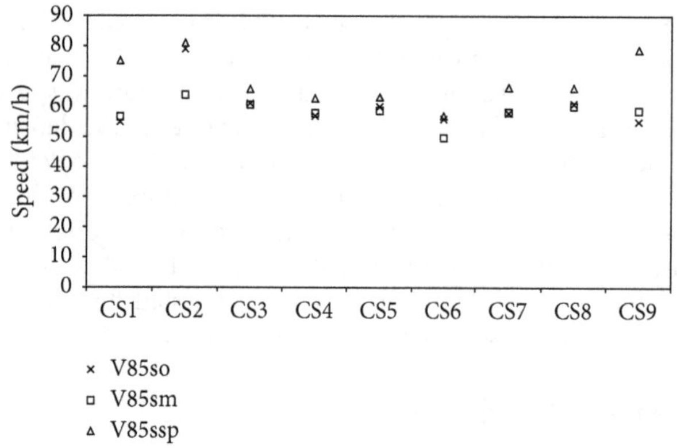

FIGURE 2: Comparison between the OSFM and a speed profile method.

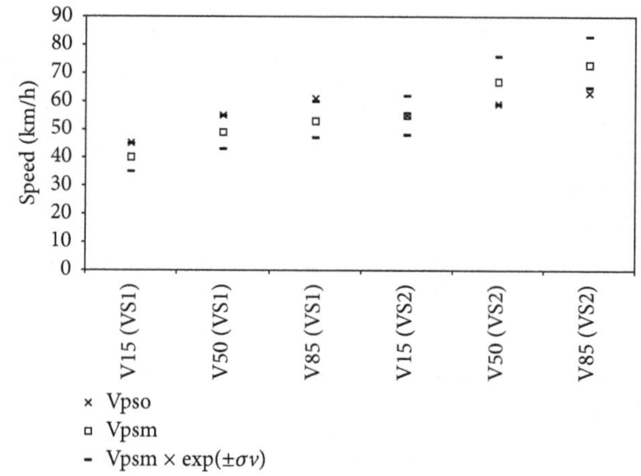

FIGURE 3: Validation results for conditions of high traffic volume.

TABLE 3: Geometric and traffic characteristics of the validation segments.

Variable	VS1	VS2
B (degrees/km)	199.3	63.5
PW (m)	4.4	6.6
ELC (m)	1.4	1.0
DI (number/km)	4.3	3.0
AADT (vehicles/day)	6,193	15,290

results show a much better fit of the OSFM compared to the speed profile approach, regarding the estimation of V85. Additionally, the OSFM was used to estimate the percentile speeds between the 1st and the 99th for all the calibration segments. A MAD of 4.5 and a MSE of 33.1 were obtained with this sample, denoting that the model's goodness-of-fit is reasonably stable across different percentiles.

6.2. Validation with Simultaneous Speed and Traffic Data. The segment speed model was validated in two additional road segments, VS1 and VS2, to test if the use of an aggregated traffic measure, AADT, does not affect the model's ability to cope with different traffic situations. In this sense, percentile speeds obtained from the model and from measurements conducted at the two validation segments were compared against each other for equivalent traffic flows. The comparison was made for high and low volume conditions, which were associated with peak and daytime off-peak periods, respectively. The nighttime off-peak period was not analysed because it is assumed that the traffic flow is low enough to ensure free-flow conditions for almost all the vehicles.

Both VS1 and VS2 are 3 km long and belong to two Portuguese National Roads not included in the modelling sample, respectively, N 108 and N 222. The former road is much older than the latter, which translates into lower standards in terms of horizontal curvature, cross-section width, and roadside interference. The traffic in VS1 is lower than in VS2 because the former is located farther away from a major urban centre. Table 3 contains the characteristics collected for both validation segments.

The procedure started with the implementation of 24-hour traffic measurements to characterize both high and low volume traffic conditions at each segment in a labour day. An automatic traffic counter (VIACOUNT II) was placed approximately at the midpoint of each segment. The highest half-hourly traffic volume was observed during the afternoon peak in both segments, specifically between 5:45 and 6:15 PM in VS1 and between 6:00 and 6:30 PM in VS2. The lowest half-hourly volume during the daytime was observed between the morning and the afternoon traffic peaks, from 4:45 to 5:15 PM in VS1 and from 4:15 to 4:45 PM in VS2. The half-hourly basis was selected to contain the use of resources in the following step, conducted at the equivalent time periods of another labour day. In that step, an observer recording the vehicles' license plates and passing times and an automatic traffic counter for simultaneous and reliable traffic measurements were placed at both ends of each validation segment. The license plate recognition allowed deriving segment percentile speeds from in situ observations (Vp_{so}) for high and low traffic volume conditions. In turn, the simultaneous traffic measurements allowed obtaining segment percentile speed estimations from the model (Vp_{sm}). Because the model uses AADT as an explanatory variable, the half-hourly traffic characterizing periods of high and low volume were converted to a daily basis by multiplying for 48. This transformation led to Vp_{sm} estimations corresponding to hypothetical situations of constant traffic flow during 24 hours. Although such situations are unrealistic, this procedure ensures that Vp_{sm} and Vp_{so} report similar traffic flows.

The comparison between Vp_{sm} and Vp_{so} was performed through the 15th, 50th, and 85th percentiles (V15, V50, and V85), which are the most commonly used in speed modelling applications. To better guide this comparison, the random error component was applied to speed estimations in the form of an interval defined around Vp_{sm} by the standard deviation of the noise term distribution, such that the limits of the interval are given by $Vp_{sm} \times \exp(\pm\sigma_v)$. The results are plotted in Figures 3 and 4, corresponding to the conditions of high and low traffic volumes, respectively.

For the condition of a high traffic volume, the observed percentile speeds lie within the interval defined by the

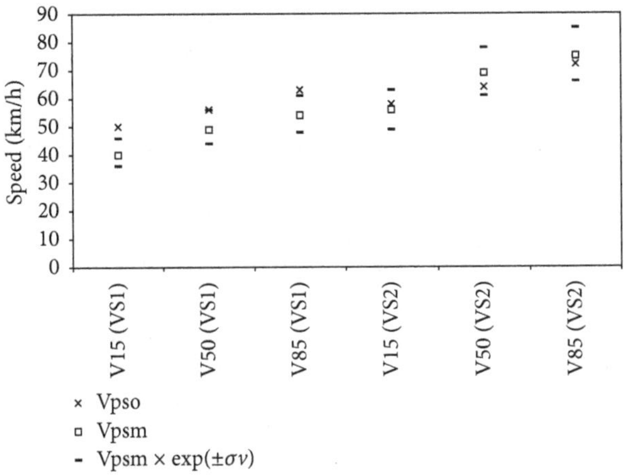

FIGURE 4: Validation results for conditions of low traffic volume.

standard deviation of the noise term in four out of six cases, with the V85 estimations sitting 1 km/h above and 2 km/h below the interval for VS1 and VS2, respectively. For a low traffic volume, the observed values are also contained in the interval in four out of six cases. The observed V15 and V85 in VS1 surpass the upper limit of the interval by 4 km/h and 2 km/h, respectively.

These results may denote a slight underestimation of the segment percentile speeds by the model in National Roads with lower design standards, with half of the observed values presented for VS1 being placed above the corresponding intervals. The model performs better in the case of VS2, representing a National Road with a higher design speed, with just one of the observed percentile speeds falling below the interval for a small margin. The segment speed model also seems to return more accurate predictions for medium-to-small percentiles.

However, from a global perspective, the observed percentile speeds are contained by the interval around the estimated values in eight out of twelve cases. In the remaining four cases, speed deviations from the corresponding intervals are, at most, of 4 km/h. Therefore, the selected speed predictors, including AADT, do not seem to affect the validity of the segment speed model, which has revealed the capability to support reliable speed estimations under different geometric and traffic conditions.

7. Conclusions

Speed is one of the most important factors affecting the convenience and efficiency of roadway infrastructures. Drivers' perception of the operating speeds of different routes and, consequently, of travel times and costs strongly influences their route choice. In turn, road designers and managers use the operating speed to evaluate and monitor the performance of infrastructures in terms of safety, traffic flow, and environmental efficiency.

Notwithstanding the notable contributions presented by numerous spot speed models existing in the literature, the evaluation of road performance can be significantly improved through the development of segment speed methods. The estimation of operating speeds over a length of roadway is usually associated with the measurement of the corresponding travel times, which can be very time and resource consuming, especially if no automatic data collection is involved. In practice, such difficulties have been hampering the development of new segment speed prediction tools; most research in this field has been limited to freeway and urban links, taking advantage of data collected by monitoring equipment owned by road authorities and concessionaires [17].

This study delivers the only operating speed model applicable to segments of two-lane highways that was developed to date outside the scope of official guidelines, with the major advantage of supporting any user-specified percentile speed prediction. The model can be particularly useful in road design and safety applications, such as the evaluation of design consistency from an early design stage and the definition of speed zones. Additionally, it can provide speed predictions to assist long-term network management actions in roads without permanent traffic monitoring equipment, a situation that is very common in the case of two-lane rural highways.

The model evolves from the authors' previous research on spot speed modelling, retaining the OSFM formulation presented in Lobo et al. [7, 8]. In this sense, road geometry and traffic characteristics are used to estimate a deterministic speed frontier, representing the maximum operating speed for each combination of explanatory variables. The asymmetric disturbance accounts for nonquantified factors associated with drivers' behaviour and vehicle technology, with the estimation of percentile speeds being made by means of the cumulative distribution function of this disturbance.

The segment speed model was calibrated using the entire speed distribution from a probe vehicle database, observed in nine segments of Portuguese National Roads. Because simultaneous speed and traffic data was not available for the calibration segments, AADT was used to represent traffic conditions. The model estimation results confirm the primary influence of geometric characteristics on the speeds practiced under noncongested traffic conditions; the impact of traffic is much smaller and predictably negative.

The OSFM was compared against a speed profile method with positive results. Additionally, it was validated using license plate recognition in two road segments beyond the calibration sample, each one presenting differentiated geometric and traffic characteristics. This procedure allowed obtaining simultaneous speed and traffic data, which was used to assess percentile speeds from both in situ measurements and model estimations. The model has proved reliable in its predictions, with the observed speed values lying within or sufficiently close to the interval around the estimated values defined by the standard deviation of the noise term distribution. Therefore, this study confirms the adequacy of using probe vehicles to collect speed and road geometry data for speed modelling applications.

It should be noted that the model was calibrated for noncongested National Roads in Portugal. Therefore, it is not valid for congested traffic flow. Additionally, its application to a different regional context should be carefully considered, especially in non-European countries where geometric standards may vary significantly. The OSFM formulation is, however, flexible enough to be replicated in other regions with different road conditions. The lack of tools for segment speed prediction in two-lane highways further highlights the contribution of the present study to road performance and safety applications, particularly to the control of design consistency from the early planning stages, but also to the definition of speed limits, the evaluation of travel times, and the recommendation of efficient routes.

Conflicts of Interest

The authors declare that there are no conflicts of interest regarding the publication of this article.

Acknowledgments

The authors would like to thank the Portuguese Foundation for Science and Technology for funding this research through the Ph.D. Studentship SFRH/BD/71322/2010 and InfoPortugal, S.A., for providing the probe vehicle data.

References

[1] Transportation Research Board (TRB), "Modeling operating speed," Synthesis Report The National Academies Press, Washington, DC, USA, 2011.

[2] K. Fitzpatrick, O. Wooldridge, J. M. Collins et al., *Alternative Design Consistency Rating Methods for Two-Lane Rural Highways*, Publication FHWA-RD-99-172, Federal Highway Administration, U.S. Deptartment of Transportation, McLean, VA, USA, 2000.

[3] Transportation Research Board (TRB), "Managing speed: review of current practice for setting and enforcing speed limits," Special Report 254, Austroads, Washington, DC, USA, 1998.

[4] B. Fildes, J. Langford, D. Andrea, and J. Scully, "Balance between harm reduction and mobility in setting speed limits: a feasibility study," Report AP-R272/05, Sydney, Australia, Austroads, 2005.

[5] National Cooperative Highway Research Program (NCHRP), "An Expert System for Recommending Speed Limits in Speed Zones," Project 3-67—Final Report, Transportation Research Board, Washington, DC, USA, 2006.

[6] G. H. D. A. Correia and A. B. Silva, "Setting speed limits on rural two-lane highways by modeling the relationship between expert judgment and measurable roadside characteristics," *Journal of Transportation Engineering*, vol. 137, no. 3, pp. 184–192, 2010.

[7] A. Lobo, C. Rodrigues, and A. Couto, "Estimating percentile speeds from maximum operating speed frontier," *Transportation Research Record*, vol. 2404, pp. 1–8, 2014.

[8] A. Lobo, A. Couto, and C. Rodrigues, "Flexible stochastic frontier approach to predict spot speed in two-lane highways," *Journal of Transportation Engineering*, vol. 142, no. 8, p. 04016032, 2016.

[9] K. M. Lum, H. S. L. Fan, S. H. Lam, and P. Olszewski, "Speed-flow modeling of arterial roads in Singapore," *Journal of Transportation Engineering*, vol. 124, no. 3, pp. 213–222, 1998.

[10] F. Dion and H. Rakha, "Estimating dynamic roadway travel times using automatic vehicle identification data for low sampling rates," *Transportation Research Part B: Methodological*, vol. 40, no. 9, pp. 745–766, 2006.

[11] R. Li, G. Rose, and M. Sarvi, "Evaluation of speed-based travel time estimation models," *Journal of Transportation Engineering*, vol. 132, no. 7, pp. 540–547, 2006.

[12] M. L. Tam and W. H. K. Lam, "Using automatic vehicle identification data for travel time estimation in Hong Kong," *Transportmetrica*, vol. 4, no. 3, pp. 179–194, 2008.

[13] E. Jenelius and H. N. Koutsopoulos, "Travel time estimation for urban road networks using low frequency probe vehicle data," *Transportation Research Part B: Methodological*, vol. 53, pp. 64–81, 2013.

[14] X. Wang, H. Liu, R. Yu, B. Deng, X. Chen, and B. Wu, "Exploring operating speeds on urban arterials using floating car data: Case study in Shanghai," *Journal of Transportation Engineering*, vol. 140, no. 9, Article ID 04014044, 2014.

[15] H. Rakha and W. Zhang, "Estimating traffic stream space mean speed and reliability from dual- and single-loop detectors," *Transportation Research Record*, no. 1925, pp. 38–47, 2005.

[16] F. Soriguera and F. Robusté, "Estimation of traffic stream space mean speed from time aggregations of double loop detector data," *Transportation Research Part C: Emerging Technologies*, vol. 19, no. 1, pp. 115–129, 2011.

[17] M. Martínez-Díaz and I. Pérez, "A simple algorithm for the estimation of road traffic space mean speeds from data available to most management centres," *Transportation Research Part B: Methodological*, vol. 75, pp. 19–35, 2015.

[18] D. Park, L. R. Rilett, and G. Han, "Spectral basis neural networks for real-time travel time forecasting," *Journal of Transportation Engineering*, vol. 125, no. 6, pp. 515–523, 1999.

[19] J. W. C. van Lint, S. P. Hoogendoorn, and H. J. van Zuylen, "Accurate freeway travel time prediction with state-space neural networks under missing data," *Transportation Research Part C: Emerging Technologies*, vol. 13, no. 5-6, pp. 347–369, 2005.

[20] J. J. Tang, F. Liu, Y. J. Zou, W. B. Zhang, and Y. H. Wang, "An improved fuzzy neural network for traffic speed prediction considering periodic characteristic," *IEEE Transactions on Intelligent Transportation Systems*, vol. 18, no. 9, pp. 2340–2350, 2017.

[21] Highways Agency (HA), *Design Manual for Roads and Bridges*, vol. 6, section 1, London, UK, 2002.

[22] Forschungsgesellschaft fuer Strassen-und Verkehrswesen (FGSV), *Richtlinien fuer die Anlage von Strassen Strassen-und Verkehrswesen*, Cologne, Germany, 2012.

[23] American Association of State Highway and Transportation Officials (AASHTO), *A Policy on Geometric Design of Highways and Streets*, Washington, DC, USA, 6th edition, 2011.

[24] Junta Autónoma de Estradas (JAE), *Norma de Traçado*, Almada, Portugal, 1994.

[25] Transportation Research Board (TRB), *Highway Capacity Manual 2010*, Washington, DC, USA, 2010.

[26] A. Lobo, C. Rodrigues, and A. Couto, "Free-flow speed model based on portuguese roadway design features for two-lane highways," *Transportation Research Record*, no. 2348, pp. 12–18, 2013.

[27] J. O. Costa, *Desenvolvimento dum modelo de previsпо de acidentes [Ph.D. Thesis]*, University of Minho, 2013 (Portuguese).

[28] R. Lamm and E. M. Choueiri, "Recommendation for evaluating horizontal design consistency based on investigations in the state of new york," *Transportation Research Record*, vol. 1122, pp. 68–78, 1987.

[29] D. R. Jessen, K. S. Schurr, P. T. McCoy, G. Pesti, and R. R. Huff, "Operating speed prediction on crest vertical curves of rural two-lane highways in Nebraska," *Transportation Research Record*, no. 1751, pp. 67–75, 2001.

[30] K. S. Schurr, P. T. McCoy, G. Pesti, and R. Huff, "Relationship of design, operating, and posted speeds on horizontal curves of rural two-lane highways in Nebraska," *Transportation Research Record*, no. 1796, pp. 60–71, 2002.

[31] J. P. Tarris, C. M. Poe, J. M. Mason Jr., and K. G. Goulias, "Predicting operating speeds on low-speed urban streets: regression and panel analysis approaches," *Transportation Research Record*, no. 1523, pp. 46–54, 1996.

[32] A. M. F. Medina and A. P. Tarko, "Speed factors on two-lane rural highways in free-flow conditions," *Transportation Research Record*, no. 1912, pp. 39–46, 2005.

[33] P. Hewson, "Quantile regression provides a fuller analysis of speed data," *Accident Analysis & Prevention*, vol. 40, no. 2, pp. 502–510, 2008.

[34] D. Aigner, C. A. Lovell, and P. Schmidt, "Formulation and estimation of stochastic frontier production function models," *Journal of Econometrics*, vol. 6, no. 1, pp. 21–37, 1977.

[35] W. Meeusen and J. van Den Broeck, "Efficiency estimation from cobb-douglas production functions with composed error," *International Economic Review*, vol. 18, no. 2, pp. 435–444, 1977.

[36] W. H. Greene, *Econometric Analysis*, Prentice Hall, Upper Saddle River, NJ, USA, 6th edition, 2008.

[37] W. H. Greene, *Limdep Version 9.0: Econometric Modeling Guide*, Econometric Software, Plainview, NY, USA, 2007.

17

An Overview of Nature-Inspired, Conventional, and Hybrid Methods of Autonomous Vehicle Path Planning

Ben Beklisi Kwame Ayawli,[1,2] Ryad Chellali,[3,4] Albert Yaw Appiah,[1,5] and Frimpong Kyeremeh[1,5]

[1] *College of Electrical Engineering and Control Science, Nanjing Tech. University, China*
[2] *Computer Science Department, Sunyani Technical University, Sunyani, Ghana*
[3] *Nanjing Forestry University, Nanjing, China*
[4] *Kita Technologies, Nanjing, China*
[5] *Electrical and Electronic Engineering Department, Sunyani Technical University, Sunyani, Ghana*

Correspondence should be addressed to Ben Beklisi Kwame Ayawli; bbkayawli@yahoo.com

Academic Editor: Jose E. Naranjo

Safe and smooth mobile robot navigation through cluttered environment from the initial position to goal with optimal path is required to achieve intelligent autonomous ground vehicles. There are countless research contributions from researchers aiming at finding solution to autonomous mobile robot path planning problems. This paper presents an overview of nature-inspired, conventional, and hybrid path planning strategies employed by researchers over the years for mobile robot path planning problem. The main strengths and challenges of path planning methods employed by researchers were identified and discussed. Future directions for path planning research is given. The results of this paper can significantly enhance how effective path planning methods could be employed and implemented to achieve real-time intelligent autonomous ground vehicles.

1. Introduction

In recent years, mobile robot path planning research aimed at achieving intelligent autonomous ground vehicles has received more attention. Path planning and motion control are very important but complex navigation tasks in robotics. Path planning in mobile robotics refers to strategies to determine how a mobile robot gets to its goal safely ensuring that obstacles are avoided [1]. To successfully achieve path planning and motion control, a vehicle is expected to possess the ability to perceive and detect obstacles to be avoided to enable it to reach its target safely. What is expected in robotics is to develop real-time intelligent autonomous navigation robots that can sense and interpret information collected from their environment to determine their position, direction to goal, avoid obstacles and perform smooth navigation in both structured and unstructured environments. The robots are expected to perform these tasks with safe and shortest path, and lowest time to their target and finally perform their assigned task without the intervention of humans [2–7]. There are many applications and benefits to be derived from intelligent autonomous robots. Some of which include: rescue operations during disaster, task performance in factories, homes, in areas of transportation, medicine, education, agriculture and many others.

Mobile robotics path planning is described as multi-objective optimization problem since it is required to generate the appropriate trajectories as well as obstacle avoidance in the environment [8]. Three tasks are required during mobile robot path planning. These include acquiring information from its environment, localization of its current position, and finally taking necessary decision based on the provided algorithm and the information acquired to execute its task [9, 10]. Lavalle [11] identified feasibility and optimality as path planning criterion with different concerns. The concern of feasibility criteria is to determine a plan to reach goal irrespective of the efficiency. Optimality is concerned with determining an optimized feasible plan to reach goal with efficient performance. Comparing the two, achieving a feasible path planning was described to be a problem but that of optimal path planning is more problematic. Path planning and obstacle avoidance methods could be classified into static or dynamic based on the environment and as global or local based on the path planning algorithm [12]. Environment composed of stationary objects is described as static while those with moving objects are termed as dynamic. When the path planning occurs in a static environment and all the information of the environment required by the robot to perform path planning and obstacle avoidance is available, it is described as global path planning. Local path planning on the other hand occurs when the robot is in motion and it reacts to changes in the environment to decide its movement [1, 10]. Path planning can also be described as either online or off-line and reactive or map based [13]. Off-line techniques compute the path before navigation while online techniques perform the computations and take decision incrementally during the navigation process. Data derived from cameras and sensors including ultrasonic sensors, infrared sensors and light sensors are interpreted in different ways by algorithms to embark on safe path planning. Control of navigation and obstacle avoidance for mobile robots can be described as a reactive approach [14–16] because the robot is expected to react against its changing environment by reacting immediately to new information received from sensors. Control of path planning is however described as deliberative approach because the robot is expected to provide the exact planning to achieve a goal relying on the geometric model of the environment and the appropriate theory. The most common control method of avoiding obstacles in cluttered environment by mobile robot is the reactive local navigation schemes where the robot's action depends on sensor information.

Ongoing research in mobile robotics is geared towards building autonomous and intelligent robots that perform robust motion planning and navigation in dynamic environments [17]. There are records of considerable number of research that aimed at addressing path planning and obstacle avoidance problem using diverse approaches and algorithms [17–29]. Despite the numerous efforts to address the problem of safe path planning of mobile robots, there still exists a challenge in achieving real safe and optimal path planning to achieve the use of intelligent autonomous vehicles [5, 30–35]. These outstanding challenges motivated this research to conduct an overview of the popular methods and strategies used by researchers to address the problem and to identify the strengths and challenges of these approaches and consider the way forward to achieving intelligent autonomous vehicles.

Mobile robot path planning methods could be categorized in different ways. In this paper, we classified them into nature-inspired computation, conventional, and hybrid methods. Methods and strategies which have something to do with the imitation of phenomena of nature are described as nature-inspired computation method. Those that are not related to imitation of nature phenomena are classified under conventional method. Methods that combine two or more strategies are described as hybrid methods in this paper.

The remainder of this paper is organized as follows: discussion of conventional mobile robot path planning methods is given in Section 2. The strengths and challenges of the conventional methods are given in Section 3. Discussion of nature-inspired mobile robot path planning methods with their strengths and challenges is covered in Sections 4 and 5. In Sections 6 and 7, hybrid methods with their strengths and challenges are presented. Conclusion and the way forward for mobile robot path planning and obstacle avoidance are given in Section 8.

2. Conventional Path Planning Methods

The conventional path planning methods (CPPM) are the traditional methods used over the years by researchers for mobile robot path planning. Most of these methods rely on distance information from objects to the robots, force of attraction and repulsion, statistical methods, clustering, or graphical map calculations to determine the path planning of the robot. They are comparatively computationally expensive. Notable among them are artificial potential field (APF), distance control algorithm, bumper event approach, wall-following, sliding mode control (SMC), dijkstra, A∗, D∗, Stereo block matching, voronoi diagram (VD), simultaneous localization and mapping (SLAM), vector field histogram (VFH), rapidly exploring random tree (RRT), curvature velocity, lane curvature, dynamic window, and tangent graph. Accounts of the use of some of these approaches are given in the next subsections.

2.1. Artificial Potential Field Path Planning Method. One of the popular methods used by researchers over the years is APF. APF is a mathematical method which causes a robot to be attracted to the goal but repelled by obstacles in the environment [6, 34, 36, 37]. There are notable research works carried out using APF with sensors to plan the path of mobile robots to provide autonomous navigation, taking obstacle avoidance into consideration [12, 37–47]. The idea of applying APF to path planning originated from Khatib [34]. APF approach was modified by some researchers to make it more efficient for path planning and obstacle avoidance [41, 43, 48, 49]. The algorithm of APF comprises the attractive force, repulsive potential force, and the total potential force. APF algorithm using Gaussian function as in [50] can be expressed as shown in the following:

$$AF(p) = f_{att}\left[1 - EXP\left(-c_{att} * d_g^2\right)\right] \tag{1}$$

$$RF(p) = \sum \begin{cases} f_{repi}\left[EXP\left(-c_{rep} * d_{obsi}^2\right)\right] & d_{obsi}(p) \le d_0 \\ 0 & else \end{cases} \tag{2}$$

$$F_{total}(p) = AF(p) + RF(p) \tag{3}$$

where AF(p) is total attraction force; f_{att} is maximum value of attraction force at any instance; c_{att} is attractive constant; d_g is

Euclidean distance between the robot and the goal; RF(p) is total repulsive force; f_{rep} is maximum value of repulsive force at any instance; c_{rep} is repulsive constant; d_{obs} is Euclidean distance between the robot and the obstacle.

One of the recent usages of APF in mobile robot path planning is found in [40]. They extended the APF method by considering household animals' motion attributes with the purpose of improving human-robot interaction. The attractive and the repulsive forces of APF were all extended. A real robot experiment was conducted using holonomic robot with six cameras to test their algorithm and to confirm the simulation results. Also, to help solve the point mass problem of APF for car-like robots, potential field window which is an extension of APF method was proposed in [51] to provide safe navigation. With this method, the potential field computations were done differently compared to the conventional APF method. APF has also been considered in multiple-robot path planning in [52].

2.2. Vision-Based Path Planning Method. Although most approaches use information from cameras and sensors to determine the execution of their algorithms, some methods rely basically on image processing of information from cameras. These methods were described as vision-based methods [33, 53–57]. A recent visual-based approach was presented in [56] to deal with path planning and obstacle avoidance problem for small unmanned vehicles by adopting the region interest extraction method used in [57]. Local blind deconvolution was utilized to classify the regions of the images collected relative to each other to help produce a feature map composed of localized structural formation of the processed images. The feature maps realized were then used as the basis for the detection and obstacle avoidance. This approach was first introduced in [58]. With this approach, before the robot decides to move to a given direction, images are captured and downsized to 320-pixel column width to aid faster computations. The feature map is then extracted. The parts of the map are then used to determine if there are obstacles before the robot makes a move to the appropriate direction. Comparative results demonstrated that this method resulted in less frequent obstacle hits of 4-5% compared to histogram-based contrast (HC), region-based contrast (RC), and spectra residual (SR) which registered between 11 and 14% hits.

A visual-based navigation algorithm for mobile robot using obstacle flow extracted from captured images to determine the estimated depth and the time to collision using the control law called balanced strategy was also presented in [53]. A combination of very low-resolution images and a sonar sensor was also used to develop a vision-based obstacle detection algorithm in unstructured indoor environment in [33]. While a digital camera was used to take images for segmentation, the sonar sensor was used to extract depth information from the image. Kim and Do [54] on the other hand presented a visual-based dynamic obstacle detection and avoidance approach with block-based motion estimation (BBME) using a single camera. A single camera sensor based on relative focus maps method was also presented in [58]. Extracted region of images was taken and divided into 3x3 regions to determine the intensity of the regions. The region with high intensity specifies an obstacle and those with low intensity determine the way to go during navigation.

Instead of using cameras as others did, Lenser and Veloso [55] demonstrated visual sonar method to detect known and unknown objects by considering the occlusions of the floor of known colors to aid detection and obstacle avoidance. With the assumption that, target velocity is known remaining that of the speeds of dynamic obstacles, a sensor-based online method termed as directive circle (DC) for motion planning and obstacle avoidance proposal was made in [59]. Both static and moving obstacles were considered in a simulation using the approach.

Recently, a visual navigation approach based on transfer learning was presented in [60] to enhance the environmental perception capacity in semantic navigation of autonomous mobile robots. The method is made up of three-layer models including place recognition, rotation region, and side recognition. Results from real experiments indicate good performance of the method in semantic navigation. The robot was able to recognize its initial state and poses and performed pose correction in real time. Improvement is required for its implementation in complex and outdoor environment.

2.3. Wall-Following Path Planning Method. Wall-following (WF) method was also considered by researchers. WF method considers the wall around the robot to guide it movement from one location to another by the help of range sensors. Gavrilut et al. [61] demonstrated a wall-following approach to mobile robot obstacle avoidance by considering the faster response time and easy integration of IR sensors into microcontrollers. Robby RP5 robot equipped with two IR proximity sensors was used to test their algorithm. Though the completion time during the experiment was low, errors were generated because of interference of emitted signals from the two sensors used. In addition, relatively small obstacles could not be detected.

2.4. Sliding Mode Control Path Planning Method. In a different development, sliding mode control (SMC) was considered by some researchers [62–64]. SMC is a nonlinear control method that brings about changes in the dynamics of a nonlinear system by applying discontinuous control signal that force the system to slide towards a cross section of the normal behavior of the system. Matveev et al. [62] presented a sliding mode strategy which is mathematically expensive for border patrolling and obstacle avoidance involving obstacles in motion. Obstacles were considered to have different shapes randomly at different periods. The velocities of the moving objects were considered to aid in the avoidance of obstacles in motion. Simulation and experiment with unicycle-like robot were performed to evaluate the approach.

2.5. A Reactive Dynamic Path Planning Method. A reactive dynamic approach employs sensor-based or vision-based approach by reacting to unforeseen obstacles and situations during navigation with appropriate decisions. A reactive dynamic approach to path planning and obstacle avoidance has been implemented in autonomous mobile robots over the years [12, 65–67]. One of such works is seen in Tang and

Ang [66]. Based on situated-activity paradigm and divide-and-conquer strategy, they adopted a reactive-navigation approach to propose a method called virtual semicircles (VSC) which put together division, evaluation, decision, and motion generation modules to enable mobile robots to avoid obstacles in complex environments. Simulation was used in the implementation of this approach. Matveev, Hoy, and Savkin [5] proposed a computational expensive approach termed as a reactive strategy for navigation of mobile robots in dynamic environment unknown to the robots with cluttered moving and deformed obstacles. This approach was implemented using simulation. A reactive-navigation approach using integrated environment representation was proposed for obstacle avoidance in dynamic environment with variety of obstacles including stationary and moving objects in [68]. The approach was experimented on a Pioneer P3-DX mobile wheeled robot in an indoor environment. A method using reactive elliptic trajectories with reactive obstacle avoidance algorithm embedded in a multicontroller architecture to aid obstacle avoidance was also presented in [69].

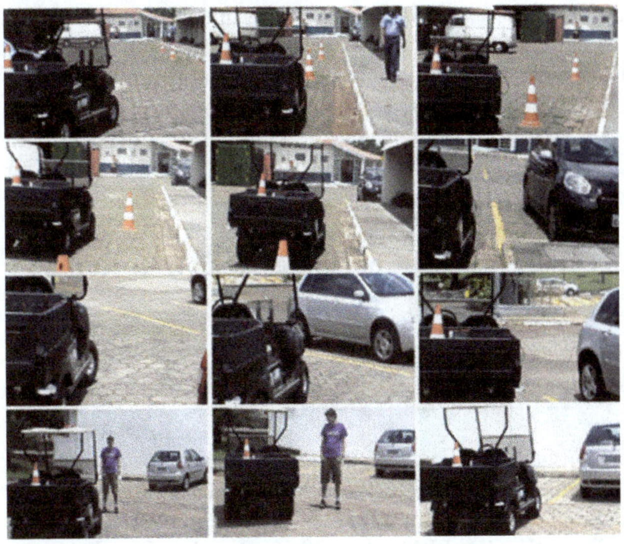

FIGURE 1: Images showing the vehicle moving towards goal while avoiding obstacles using SGBM (source [83]).

2.6. Dynamic Window Path Planning Approach. Some researchers also considered the dynamic window approach (DWA) to provide optimal obstacle avoidance [70–74]. Recently, an improved dynamic window approach was proposed in [75] for mobile robot obstacle avoidance. Considering the drawbacks of local minima of this approach causing the robot to be trapped in a U-shaped obstacle, a laser range finder was used as the sensor to ensure optima path decision making of the robot. The size of the robot was considered in this approach. Simulation was used to evaluate the approach and compared with other DWA results in MATLAB and the former performs better according to the authors.

2.7. Other Conventional Path Planning Methods. Accounts of other conventional path planning methods used by researchers other than those described above are summarized below.

A heuristic A-Star (A∗) algorithm and dynamic steering algorithm were employed in [10] to propose an approach to obtain the shortest path and the ability to avoid obstacles that are known to the robot in a predefined grid-based environment map based on information derived from sonar sensors.

A research on using robot motion strategies to enable a robot to track a target in motion in a dynamic environment was done in [76]. Four ultrasonic range sensors with two control algorithms where one controlled the stopping of the vehicle upon sensing an obstacle and the other deciding the direction of movement were presented in [77]. This method was implemented on a three-wheeled mobile robot in indoor environment.

Distance control algorithm for mobile robot navigation using range sensors was also considered by researchers for path planning. Notable among them was the research by Ullah et al. [78, 79]. They employed distance control algorithm to develop a remote-controlled robot to predict and avoid collisions between vehicles by maintaining a given safe distance between the robot and the obstacle

In another development, a bumper event approach was employed to develop an algorithm for obstacle avoidance of Turtlebot in [80]. The approach however did not provide collision free navigation.

Kunwar et al. [81] conducted a research into using rendezvous guidance approach to track moving targets of robots in a dynamic environment made up of stationary and moving obstacles. To enable efficient noise filtration of data collected from the environment, Chih-Hung, Wei-Zhun, and Shing-Tai [82] recently evaluated the performance of Extended Kalman Filtering (EKF) and Kalman Filtering (KF) for obstacle avoidance for mobile robots. Implementation was carried out using a two-wheeled mobile robot with three sonar sensors. EKF approach was described to have performed better than KF in the experiment.

Semi-global stereo method (SGBM) with local block matching algorithm (BM) where obstacles were identified using a method that checks the relative slopes and height differences of objects was used in [83]. A disparity map obtained from the processed pair of images using stereo cameras was taken through further processing and finally to the obstacle avoidance algorithm to determine the movement of the vehicle to the defined GPS point while avoiding obstacles. Real-time experiment was performed using an electric vehicle as shown in Figure 1.

Other methods including velocity space [84], SLAM, VD [85, 86], VFH [87], RRT [88, 89], and others have also gained popularity in the field of mobile robot path planning. Recently, a visual SLAM system-based method was proposed in [90] for a mobile robot equipped with depth sensor or camera to map and operate in unknown 3D structure. The purpose of the approach was to improve the performance of mapping in a 3D structure with unknown environment where positioning systems are lacking. Simulation was done using Gazebo simulator. Real experiment was performed using Husky robot equipped with Kinect v2 depth sensor and laser range scanner. Results confirmed improvement against classical frontier-based exploration algorithms in a house structure. Improvement may be required to address how to differentiate similar structures and to deal with errors in determining waypoints.

3. Strengths and Challenges of Conventional Path Planning Methods

The conventional path planning methods have their strengths and weaknesses. These are discussed in the next subsections.

3.1. Strengths of Conventional Path Planning Methods. Even though conventional methods of path planning are computational expensive, they are easy to implement. Methods like APF, DWA, A∗, PRM and other conventional methods are simple to implement. The implementation of SMC strategy for instance is easy and fast with respect to response time. It also performs well in a condition with uncertain and unconducive external factors [64]. Also, these methods can combine well with other methods and they perform better when blended with other methods. Popular among them is APF which many researchers combined with other methods. A∗ for instance is very good at obtaining the shortest path from the initial state to the goal. When combined with other methods, the hybrid method is able to generate optimal path.

3.2. Challenges of Conventional Path Planning Methods. Notwithstanding the strengths of conventional methods for path planning, there exist challenges affecting the achievement of intelligent autonomous ground vehicles. The following are some of the challenges with these methods.

To begin with, most of the approaches including conventional methods discussed used cameras and sensors to collect data from the environment to determine the execution of the respective path planning algorithms [55, 91–93]. Unfortunately, readings from cameras and sensors are interfered with noise due to changes in pressure, lightening system, temperature and others. This makes the collected data uncertain to enable control algorithms to achieve safe and optimal path planning [92, 94]. The dynamics of the vehicles also cause noise including electric noise of the mobile robots [5]. These conditions affect the accuracy, efficiency, and reliability of the acquired real-time environmental data collected to enable the robot to take a decision [95]. Research work has been done to address the noise problem, but it is still a challenge. This has effect on the practical implementation of proposed approaches.

Typical with vision-based obstacle avoidance strategies, factors including distance of objects from the robot, color, and reflection from objects affects the performance of detecting obstacles especially moving objects [54]. The use of stereo vision approaches [96, 97] is very limited and can work only within the coverage of the stereo cameras and could not work in regions that are texture-less and reflective [98]. There is a problem identifying pairs of matched points such that each of these points demonstrate the projection of the same 3D point [99]. This results in ambiguity of information between points of the images obtained which may lead to inconsistent interpretation of the scene [100].

Furthermore, some of the approaches rely on the environmental map to enable the robot take decisions in navigation. There is a problem with unnecessary stopping of the mobile robot during navigation to update its environmental map. This affects the efficiency and the real applications of mobile robots to provide safe and smooth navigation. This challenge was considered in [101] and was demonstrated through simulation. However, it could not achieve the safe optimal path to the target and was not also demonstrated in a real experiment to determine its efficiency in support of simulation results.

DWA also has the drawback of resulting in local minima problems and nonoptimal motion decision due to constraints of the mobile robot the approach could not manage [75, 102–104].

In addition, despite the popularity of the use of APF approach in path planning and obstacle avoidance of mobile robots, it has serious challenges. Navigation control of robots using APF considers the attraction forces from target and the repulsion forces from the obstacles. When performing this attraction and repulsion task, environmental information is compared to a virtual force. Computations lead to loss of important information regarding the obstacles and cause local minima problem [36, 40, 41, 105, 106]. APF can also result in unstable state of the robot where the robot finds itself in a very small space [107]. It can cause the robot to be trapped at a position rather than the goal [51]. Other problems of APF include inability of the robot to pass between closely spaced obstacles, oscillations at the point of obstacles, and oscillations when the passage is very small for the robot to navigate and goals nonreachable with obstacle nearby (GNRON). It also performed poorly in environment with different shapes of obstacles [108–110]. Limited sensing ability, bounded curvature, limited steering angle, and velocity or finite angular and linear acceleration of the mobile robot hardware have great effect on the performance of APF methods [51]. It may also be far from the optimum when planning is local and reactive [111].

Though VFH, unlike APF strategy, does well in narrow spaces, if the size and kinematics of the robot are not taken into consideration it can fail especially in narrow passages [112]. Incorrect target positioning may also result in VFH algorithm not being reliable [113].

Besides, SMC methods are fast with respect to response time and are also good transient robust when it comes to uncertain systems and other external factors that are not conducive [64]. But, it does not perform well if the longitudinal velocity of the robot is fast. It also has the problem of chattering which may result in low control accuracy [114].

Moreover, some methods like VD, RRT, and others do well in simulation environment but relatively difficult to implement in real robot platform experiments. VD, RRT, and PRM are good at generating global roadmap but a local planner is required to generate the path. They are not good at performing in dynamic environments if not combined with other methods. A∗ method on the other hand has the problem with generating smooth path and it requires path smoothening algorithms to achieve smooth path.

4. Nature-Inspired Computation-Based Methods

According to Siddique and Adeli [115], nature-inspired computing is made up of metaheuristic algorithm that try to imitate or base on some phenomena of nature given by natural sciences. A considerable number of researchers tried addressing the problem of mobile robotics path planning and obstacle avoidance using stochastic optimization algorithm techniques that imitate the behavior of some living things including bees, fish, birds, ants, flies, and cats [116–122]. These algorithms are referred to as nature-inspired paradigms and has been applied in engineering to solve research problems [123, 124]. Nature-inspired computation-based methods use ideas obtained from observing how nature behaves in different ways to solve complex problems that are characterized with imprecision, uncertainty and partial truth to achieve practical and robust solution at a lower cost [125]. Notable among nature-inspired methods used in path planning and obstacle avoidance research include artificial neural networks (ANN), genetic algorithms (GA), simulated annealing (SA), ant colony optimization (ACO), particle swarm optimization (PSO), fuzzy logic (FL), artificial bee colony (ABC), and human navigation dynamics. Nature-inspired computation-based methods were claimed to be better navigation approaches compared to conventional ones such as APF methods [126]. Most of these methods consider reinforcement learning to ensure mobile robots perform well in unknown and unstructured environments. Accounts of the use of some of these approaches are discussed in the next subsections.

4.1. Fuzzy Logic Path Planning Method. FL is one of the most significant methods used over the years for mobile robot path planning. Though FL had been studied since 1920 [127], the term was introduced in 1965 by Lotfi Zadeh when he proposed fuzzy set theory [128]. FL is a form of many-valued logic with truth values of variables of real number between 0 and 1 used in handling problems of partial truth. It is believed that Fuzzy controllers based on FL possess the ability to make inferences even in uncertain scenarios [129]. FL can extract heuristic knowledge of human expertise and imitate the control logic of human. It has the "*if-then*" human-like rules of thinking. This characteristic has made FL and other derived approaches based on FL become the most used approach by many researchers in mobile robot path planning [4, 17, 18, 28, 130–151].

Recently, fuzzy logic was employed in [152] to propose mobile robot path planning in environment composed of static and dynamic obstacles. Eight sensors were used to collect data from the environment to aid detecting the obstacles and determining trajectory planning. Implementation was done in simulation and real-time experiments in a partially known environment and the performance of the proposed approach was described to be good. Multivalued logic framework was used to propose a preference based fuzzy behavior system to control the navigation of mobile robots in cluttered environment [147]. Experiment was done using a Pioneer 2 robot with laser range finder and localization sensors in indoor modelled forest environment. The possibility of this approach performing in the real outdoor environment with considerable number of constraints and moving obstacles is yet to be determined.

In another development, FL based path planning algorithm using fuzzy logic was presented by Li et al. [139]. The positions of the obstacles and the angle between the robot and the goal positions were considered as the input variables processed, using fuzzy control system to determine the appropriate movement of the robot to avoid colliding with obstacles. The approach was implemented in simulation.

Fuzzy logic approach was employed in [153] to develop a controller with 256 fuzzy logic rules with environmental data collected using IR sensors. Webot Pro and MATLAB were used for the software development and simulation, respectively. The performance in obstacle avoidance using the method in simulation was described to be good. However, since no real experiment is implemented, we could hardly judge if the approach could work in the real environment where constraints including noise and the limitation of hardware components of mobile robots are common. An approach involving fuzzy logic to formulate the map of an input to output to determine a decision described as fuzzy inference system (FIS) was employed to propose a method to control autonomous ground vehicles in unstructured environment using sensor-based navigation technique [28, 154]. During robot navigation, environmental mapping is done in most cases to enable the robot to decide its movement [147]. At the course of navigation, the robot stops to update the environmental map to determine the next move [14]. However, Baldoni, Yang and Kim [101] are of the view that periodic stopping to update the environmental map compromises the efficiency and practical applications of these robots. Hence, they proposed a model using a graphical and mathematical modelling tool called Fuzzy Petri net with a memory module and ultrasonic sensors to control a mobile robot to provide dynamic and continuous obstacle avoidance. The approach was demonstrated using simulation which achieved 8% longer than the ideal path as against 17% longer than ideal path with the approach without memory.

Apart from the discussion above, some researchers also employed FL with the help of ultrasonic and infrared sensors to develop path planning algorithms which were successfully implemented in both simulation and real robots platform [17, 25, 135, 144–151].

4.2. Artificial Neural Network Path Planning Method. Research involving the use of ANN as intelligent strategy in solving navigation and obstacle avoidance problems in mobile robotics has also been explored extensively [102, 155–166]. The application of ANN tries to imitate the human brain to provide intelligent path planning.

Chi and Lee [157] proposed neural network control strategy with backpropagation model to control mobile robot to navigate through obstacles without collision. P3DX mobile robot was used to evaluate the developed algorithm for

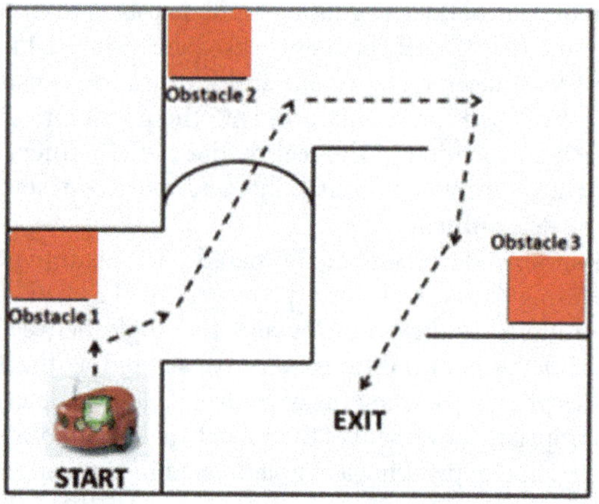

FIGURE 2: Neural Network with backpropagation approach to control P3DX navigation from the start and exit of maze (source [157]).

navigation from the start to the exit of maze as shown in Figure 2.

ANN is believed to be very good at resolving nonlinear problems that require mapping input and output relation without necessarily having knowledge of the system and its environment. Based on this strength, Akkaya, Aydogdu and Canan [167], proposed global positioning system (GPS) and dead reckoning (DR) sensor fusion approach by adopting ANN nonlinear autoregressive with external input (NARX) model to control mobile robot to avoid hitting obstacles. The mobile robot kinematics was modelled using ANN. Based on the performance of the approach through simulation and experiment, the approach was presented by the authors as an alternative to conventional navigation methods that use Kalman filter.

To achieve intelligent mobile robot control in unknown environment, Panigrahi et al. [168] proposed a redial basis function (RBF) NN approach for mobile robot path planning. Farooq et al. [169] also contributed by designing an intelligent autonomous vehicle capable of navigating noisy and unknown environment without collision using ANN. Multilayer feed forward NN controllers with back error propagation was adopted for robot safe path planning to reach goal. During the evaluation of the approach, it was identified that the efficiency of the neural controller deteriorates as the number of layers increase due to the error term that is determined using approximated function.

In another development, Medina-Satiago et al. [170] considered path planning as a problem of pattern classification. They developed a path planning algorithm using multilayer perceptron (MLP) and back propagation (BP) of ANN which was experimented on a differential drive mobile robot. Alternatively, Syed et al. [171] contributed by proposing guided autowave pulse coupled neural network (GAPCNN) which is an improvement of pulse coupled neural network (PCNN) by Qu et al. [172] for mobile robot path planning and obstacle avoidance. Dynamic thresholding technique was applied in their method. Results from simulation and experiments described the approach to be time efficient and good for static and dynamic environments compared to modified PCNN.

In [173], a data-driven end-to-end motion planning method based on CNN was introduced to control a differential drive. The method aimed at providing the ability to learn navigation strategies. Simulation and real experiment indicated that the model can be trained in simulation and transferred to a robotic platform to perform the required navigation task in the real environment. Local minima and oscillation problems of the method need to be addressed.

4.3. Genetic Algorithm Path Planning Methods. Another popular method used by researchers over the years for mobile robot path planning is GA. GA is a metaheuristic inspired by the process of natural selection that belongs to the larger class of evolutionary algorithms (EA). GA are normally considered when there is the need to generate high-quality solutions to help optimization and search problems by using bio-inspired operators including mutation, crossover, and selection [174]. GA is a category of search algorithms and optimization methods that make use of Darwin's evolutionary theory of the survival of the fittest [175]. Research on the application of GA to robotic path planning and obstacle avoidance was done in the past [38, 176–184].

Tuncer and Yildirim [178] considered GA to present a new mutation operator for mobile robot path planning in dynamic environment. According to the authors, comparing the results to other methods showed better performance of their technique in terms of path optimality. With the motivation to provide optimal and collision free mobile robot navigation, a combinatorial optimization technique based on GA approach was used in [184] to propose bacterial memetic algorithm to make a mobile robot navigate to a goal while avoiding obstacles at a minimized path length.

4.4. Memetic Algorithm Path Planning Method. Memetic algorithm is a category of evolutionary [185] method which combines evolutionary and local search methods. Memetic algorithm has been used to attempt solving optimization problems in [186]. There is remarkable research work using memetic algorithm in mobile robot path planning [187, 188].

One of such works is by Botzheim, Toda and Kubota [189]. They adopted the bacterial memetic algorithm in [190] by considering bacterial mutation and gene transfer operation as the operators. The approach was applied to mobile robot navigation and obstacle avoidance in static environment (see Figure 3).

4.5. Particle Swarm Optimization Path Planning Method. PSO technique emerged from Kennedy and Eberhart [191]. The technique was inspired by the social behavior of birds and fish in groups by considering the best positions of each bird or fish in search of food using fitness function parameter. PSO is simple to implement and requires less computing time and performs well in various optimization problems [192–195]. PSO has been adopted by a considerable number of

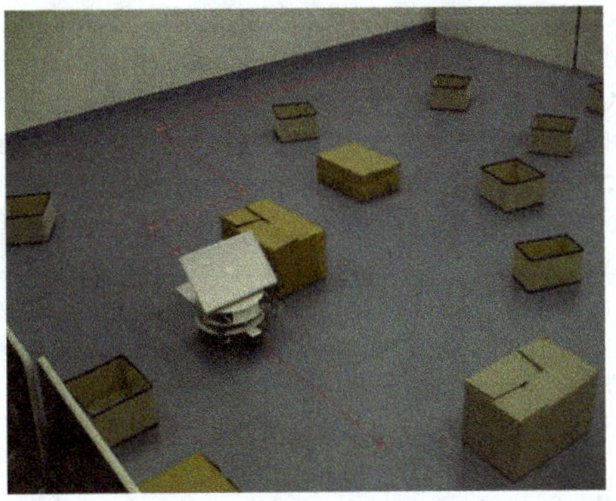

FIGURE 3: Controlling mobile robot navigation using bacteria memetic algorithm (source, Botzheim, Toda and Kubota [189]).

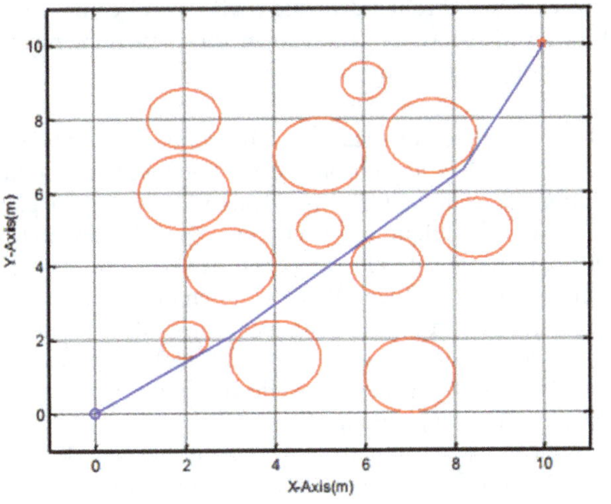

FIGURE 4: Best path results using DABC (source [122]).

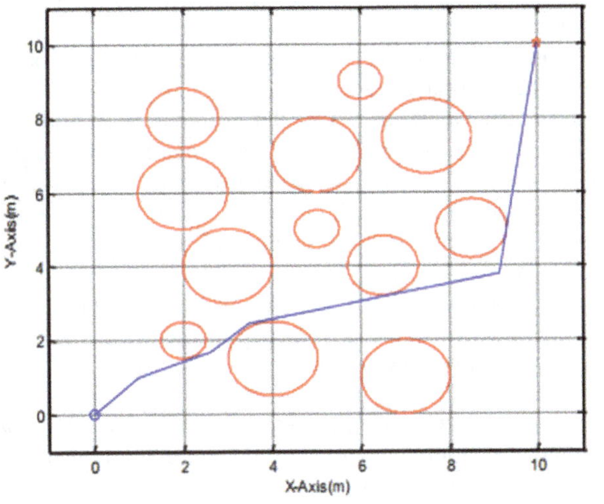

FIGURE 5: Best path results using ABC (source [122]).

researchers in mobile robot path planning task [192, 196–208]. The mathematical equations describing PSO [196] are shown in the following:

$$V_i(k+1) = w * V_i(k) + c_1 * r_1(Y_{Pbest} - y_i) + c_2 * r_2(Y_{Pbest} - y_i) \quad (4)$$

$$y_i(k+1) = y_i(k) + V_i(k+1) \quad (5)$$

where i is particle number, V_i is velocity of the particle i, y_i is position of the particle i, k is iteration counter, w is inertial weight (decreasing function), c_1 and c_2 are acceleration coefficients known as cognitive and social parameters, r_1 and r_2 are random values in $[0, 1]$, Y_{Pbest} is best position of particle, and Y_{Gbest} is global best position of particle

Recently, Rath and Deepak [196] used PSO technique to develop a motion planner for stationary and movable objects. The developed algorithm was implemented in simulation and it is yet to be implemented in robotic platform to confirm the results achieved in simulation.

PSO approach applied in [209, 210] was adopted from [205, 206, 210, 211] for motion planning and obstacle avoidance in dynamic environment. The approach was implemented in simulation environment.

4.6. Artificial Bee Colony Path Planning Method. ABC algorithm was also considered by few researchers [212]. Based on swarm intelligence and chaotic dynamic of learning, Lin and Huang [119] presented ABC algorithm for path planning for mobile robots. Though the paper stated the method was more efficient than GA and PSO, it was not tested in real robot platform to ascertain that fact. Optimal path planning method based on ABC was also introduced in [116] to provide collision free navigation of mobile robot with the aim of achieving optimal path. Simulation was used to test the algorithm which was described to be successful. However, the method was not demonstrated in experiment with real robot.

Abbass and Ali [122] on their part presented what they described as directed artificial bee colony algorithm (DABC) based on ABC algorithm for autonomous mobile robot path planning. The DABC algorithm was used to direct the bees to the direction of the best bee to help obtain optimal path while avoiding obstacles. Simulation results compared to conventional ABC algorithm was described to have performed better as demonstrated in Figures 4 and 5.

Recently, bees algorithm was used in [213] to present a real-time path planning method in an indoor dynamic environment. The bees algorithm was used to generate the path in static environment which is later updated online using modified bees algorithm to enable preventing hitting obstacles in dynamic environment. Neighborhood shrinking was used to optimize the performance of the algorithm. Simulation and experiment using AmigoBot were used to evaluate the method and it was described to have performed well with optimal path in a real time.

4.7. Simulated Annealing Algorithm Path Planning Method. SA is a probabilistic technique used for estimating the global optimum of a function. It was used in past research for mobile robot obstacle avoidance task [214–216]. SA algorithm was used in [217] to propose an algorithm to enable mobile robots to reach a goal through dynamic environment composed of obstacles with optimal path. This method was implemented successfully in simulation.

4.8. Ant Colony Optimization Algorithm Path Planning Method. ACO is a probabilistic method used for finding solution for computational problems including path planning. ACO technique has been considered by researchers for mobile robot path planning over the years [29, 218–222]. Using ACO, Guan-Zheng et al. [220] demonstrated through algorithm and simulation a path planning method for mobile robots. Results were compared to GA method and it was indicated that the method was effective and can be used in the real-time path planning of mobile robots.

Vien et al. [221] used Ant-Q reinforcement learning based on Ant Colony approach algorithm as a technique to address mobile robot path planning and obstacle avoidance problem. Results from simulation were compared with results from other heuristic based on GA. Comparatively, Ant-Q reinforcement learning approach was described to be better in terms of convergence rate.

Considering the complex obstacle avoidance problem involving multiple mobile robots, Ioannidis et al. [222] proposed a path planner using Cellular Automata (CA) and ACO techniques to provide collision free navigation for multiple robots operating in the same environment while keeping their formation immutable. Implementation was done in a simulated environment consisting of cooperative team of robots and an obstacle.

Real robot platform experimental implementation may be required to prove the efficiency of these approaches to confirm the simulation results.

4.9. Human Navigation Dynamics Path Planning Methods. Human navigation dynamics (HND) research was also given attention by researchers [68, 223–230]. However, very few considerations were given to the application of HND strategies to path planning and obstacle avoidance of mobile robots. HND strategy termed as all-or-nothing strategy was used by Frank et al. [231] to propose a minimalistic navigation model based on complex number calculus to solve mobile robots' navigation and obstacle avoidance problem. This approach did not, however, demonstrate any implementation through experiments. Much work may be required in this area to explore and experiment the use of human navigation strategies to robotics motion planning and obstacle avoidance.

5. Strengths and Challenges of Nature-Inspired Path Planning Methods

5.1. Strengths of Nature-Inspired Path Planning Methods. Nature-inspired path planning methods possess the ability to imitate the behavior of some living things that have natural intelligence. Methods like ANN and FL are good at providing learning and generalization [232]. FL for instance possess the ability to make inferences in uncertain scenarios. When combined with FL method, ANN can tune the rule base of FL to produce better results. The learning component of these methods aids in effective performance in unknown and dynamic environment of the autonomous vehicle. GA and other nature-inspired methods have good optimization ability and are good at solving problems that are difficult dealing with using conventional approaches. ACO method and memetic algorithms are noted for their fast convergence characteristics and good at obtaining optimal result [233–235]. Moreover, nature-inspired methods can combine well with other optimization algorithms [236, 237] to provide efficient path planning in the real environment.

5.2. Challenges of Nature-Inspired Path Planning Methods. Notwithstanding the strengths discussed above, there are some weaknesses of nature-inspired path planning methods, some of which include trapping in local minima, slow convergence speed, premature convergence, high computing power requirement, oscillation, difficulty in choosing initial positions, and the requirement of large data set of the environment which is difficult to obtain.

Despite the strength of ANN stated in the previous section, they require large set of data of the environment during training to achieve the best result which is difficult to obtain especially with supervised learning [232]. The use of backpropagation technique to provide efficient algorithm also have its challenges. BP method easily converges to local minima if the situation action mapping is not convex with the parameters [238]. The algorithm stops at local minima if it is above its global minimum. Moreover, it is also described to have very slow convergence speed which may result in some collisions before the robot gets to its defined goal [232]. FL systems can provide knowledge-based heuristic situation action mapping; however, their rule bases are hard to build in unstructured environment [232]. This makes it difficult using FL method to address path planning problems in unstructured environments without combining it with other methods. Despite the strength of good optimization capacity of GA, it is difficult for GA to scale well with complex scenarios and it is characterized with convenience at local minima and oscillation problems [239, 240]. Also, due to its complex principle, it may be difficult to deal with dynamic data sets to achieve good results [233]. Though PSO is described to be simple with less computing time requirement and effective in implementing with varied optimization problems with good results [192–195], it is difficult to deal with trapping into local minima problems under complex map [194, 240]. Although ACO is noted for fast convergence with optimal results [233, 234], it requires a lot of computing time and it is difficult to determine the parameters which affects obtaining quick convergence [233, 240]. Compared to conventional algorithms, memetic algorithm is described to possess the ability to produce faster convergence and good result; however, it can result in premature convergence [235]. ABC is simple with fewer control parameters with less computational time, but low convergence is a drawback of ABC algorithm [241]. It is believed that SA performs well in approximating the global optimum, but the algorithm is slow, and it is difficult choosing the initial position for SA [242].

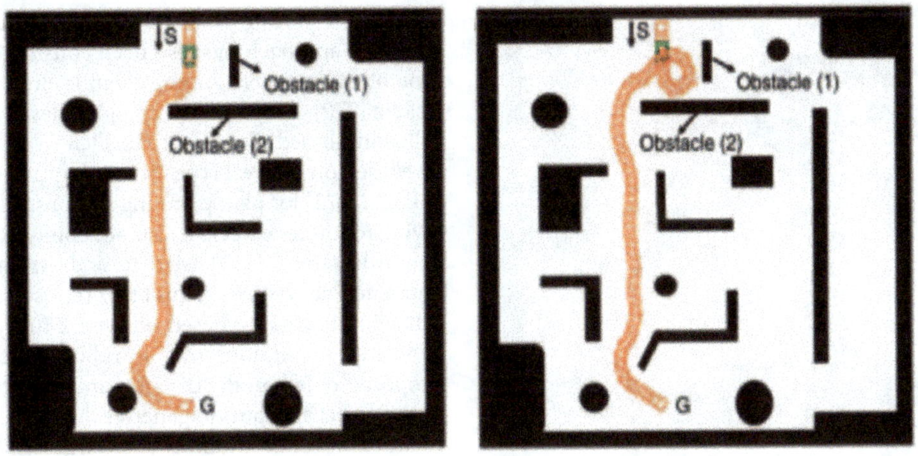

(a) (b)

FIGURE 6: Mobile robot obstacle avoidance using DRNFS with short memory (a) compared to conventional fuzzy rule-based (b) navigation system from a given start (S) and goal (G) positions (source [261]).

6. Hybrid Methods

To take advantage of the strengths of some methods while reducing the effects of their drawbacks, some researchers have combined two or more methods in their investigations to provide efficient hybrid path planning method to control autonomous ground vehicles. Some of these approaches include neuro-fuzzy, wall-following-based fuzzy logic, fuzzy logic combined with Kalman Filter, APF combined with GA, and APF combined with PSO. Some of the hybrid approaches are discussed in the next subsections.

6.1. Neuro-Fuzzy Path Planning Method. Popular among the hybrid approaches is neuro-fuzzy also termed as fuzzy neural network (FNN). It is a combination of ANN and FL. This approach considers the human-like reasoning style of fuzzy systems using fuzzy sets and linguistic model that are composed of *if-then* fuzzy rules as well as ANN [243]. The use of neuro-fuzzy system approach in obstacle avoidance for mobile robots' research has been considered by many researchers [101, 244–260].

A weightless neural network (WNN) approach using embedded interval type-2 neuro-fuzzy controller with range sensor to detect and avoid obstacles was presented in [256, 257]. This approach worked but its performance was described to be interfered by noise. Unified strategies of path planning using type-1 fuzzy neural network (T1FNN) was proposed in [16] to achieve obstacle avoidance. It is however identified in [101] that the use of (T1FNN) have challenges including the unsatisfactory control performance and the difficulty of reducing the effect of uncertainties in achieving task and the oscillation behavior that is characterized by the approach during obstacle avoidance. Kim and Chwa [14] took advantage of this limitation and modified the TIFNN to propose an obstacle avoidance method using interval type-2 fuzzy neural network (IT2FNN). Evaluation of this strategy was carried out using simulation and experiment and compared with the T1FNN approach. The IT2FNN was described to have produced better results.

AbuBaker [258] presented a neuro-fuzzy strategy termed as neuro-fuzzy optimization controller (NFOC) to control mobile robot to avoid colliding with obstacles while moving to goal. NFOC approach was evaluated in simulation and was compared with fuzzy logic controller (FLC40) and FLC625 and the performance was described to be better in terms of response time.

Chen and Richardson [261] on the other hand tried to look at path planning and obstacle avoidance of mobile robot in relation to the human driver point of view. They adopted a new fuzzy strategy to propose a dynamic recurrent neuro-fuzzy system (DRNFS) with short memory. Their method was simulated using three ultrasonic sensors to collect data from the environment, 18 obstacle avoidance trajectories and 186 training data pairs to train the DRNFS and compared to conventional fuzzy rule-based navigation system and the approach was described to be better in performance. Figure 6 demonstrates the performance of the DRNFS compared to the conventional fuzzy rule-based navigation system.

Another demonstration of neuro-fuzzy approach was presented in [259] with backpropagation algorithm to drive mobile robot to avoid obstacles in a challenging environment. Simulation and experimental results showed partial solution of reduction of inaccuracies in steering angle and reaching goal with optimal path. A proposal using FL control combined with artificial neural network was presented by Jeffril and Sariff [260] to control the navigation of a mobile robot to avoid obstacles.

To capitalize on the strengths of neural network and FL, Algabri, Mathkour and Ramdane [262] proposed adaptive neuro-fuzzy inference system (ANFIS) approach for mobile robot navigation and obstacle avoidance. Simulation was carried out using Khepera simulator (KiKs) in MATLAB as shown in Figure 7. Practical implementation was also done using Khepera robot in an environment scattered with few obstacles.

A combination of FL and spiking neural networks (SNN) approach was proposed in [263] and simulation results was

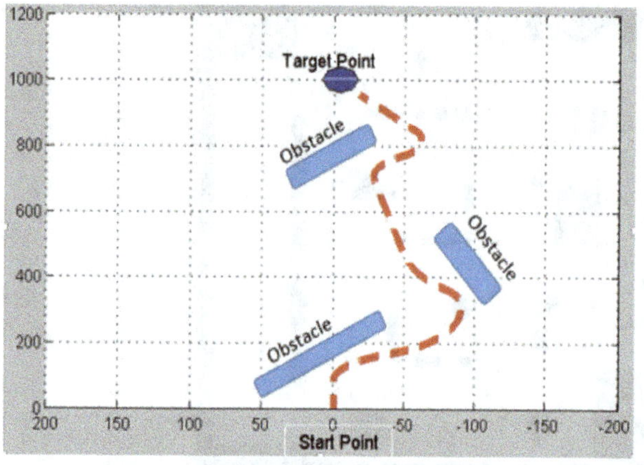

FIGURE 7: Khepera robot navigation in scattered environment with obstacles using ANFIS approach (source [262]).

compared to that of FL. The new approach was described to have performed better. Alternatively, ANFIS controller was developed using neural network approach to control multiple mobile robot to reach their target while avoiding obstacles in a dynamic indoor environment [18, 264–266]. The authors developed ROBNAV software to handle the control of mobile robot navigation using the ANFIS controller developed.

Pothal and Parhi [267] developed Adaptive Neuron Fuzzy Inference System (ANFIS) controller for mobile robot path planning. Implementation was done using single robot as well as multiple robots. The average percentage error of time taken for a single robot to reach its target comparing simulation and experiment results was 10.47%.

To deal with the effect of noise generation during information collection using sensors, a Hybrid Intelligent System (HIS) was proposed by Alves and Lopes [268]. They adopted FL and ANN to control the navigation of the robot. While, in neuro-fuzzy system, training needs to be done from scratch, this method deviated a little bit where only the fuzzy module is calibrated and trained. Simulation results showed that the method was successful.

Realizing the challenge of training and updating parameters of ANFIS in path planning which usually leads to local minimum problem, teaching-learning-based optimization (TLBO) was combined with ANFIS to present a path planning method in [269]. The aim was to exploit the benefits of the two methods to obtain the shortest path and time to goal in strange environment. The TLBO was used to train the ANFIS structure parameters without seeking to adjust parameters. PSO, invasive weed optimization (IWO), and biogeography-based optimization (BBO) algorithms were also used to train the ANFIS parameters to aid comparison with the proposed method. Simulation results indicated that TLBO-based ANFIS emerged with better path length and time. Improvement may be required to consider other path planning tasks. It may also require comparison with other path planning methods to determine its efficiency. Real experiment should be considered to prove the effectiveness of the method in the real environment.

6.2. Other Hybrid Methods That Include FL. Wall-following-based FL approach has also been considered by researchers. One of the pioneers among them is Braunstingl, Anz-J,and Ezkera [270]. They used fuzzy logic rules to implement a wall-following-based obstacle controller. Wall-following control law based on interval type-2 fuzzy logic was also proposed in [131, 271, 272] for path planning and obstacle avoidance while trying to improve noise resistance ability. Recently, Al-Mutib and Adessemed [273] tried to address the oscillation and data uncertainties problems and proposed a fuzzy logic wall-following technique based on type-2 fuzzy sets for obstacle avoidance for indoor mobile robots. The fuzzy controller proposed relied on the distance and orientation of the robot to the object as inputs to generate the needed wheel velocities to move the robot smoothly to its target while switching to the obstacle avoidance algorithm designed to avoid obstacles.

Despite the problem of sensor data uncertainties, Hank and Haddad [274] took advantage of the strengths of FL and combined trajectory-tracking and reactive-navigation mode strategy with fuzzy controller used in [275] for mobile robot path planning. Simulation and experiments were performed to test the approach which was described to have produced good results.

In [276], an approach called dynamic self-generated fuzzy Q-learning (DSGFQL) and enhanced dynamic self-generated fuzzy Q-learning (EDSGFQL) were proposed for obstacle avoidance task. The approach was compared to fuzzy Q-learning (FQL) [277], dynamic fuzzy Q-learning (DFQL) of Er and Deng [135], and clustering and Q-value based GA learning scheme for fuzzy system design (CQGAF) of Juang [278] and it was proven to perform better through simulation. Considering a learning technique in a different direction, a method termed as reinforcement ant optimized fuzzy controller (RAOFC) was designed by Juang and Hsu [279] for wheeled mobile robot wall-following under reinforcement learning environments. This approach used an online aligned interval type-2 fuzzy clustering (AIT2FC) method to generate fuzzy rules for the control automatically. This makes it possible not to initialize the fuzzy rules. Q-value aided ant colony optimization (QACO) technique in solving computational problems was employed in designing the fuzzy rules. The approach was implemented considering the wall as the only obstacle to be avoided by the robot in an indoor environment.

Recently, a path planning approach for goal seeking in unstructured environment was proposed using least mean p-norm extreme learning machine (LMP-ELM) and Q-learning by Yang et al. [232]. LMP-ELM and Q-learning are neural network learning algorithm. To control errors that may affect the performance of the robots, least mean P-power (LMP) error criterion was introduced to sequentially update the output. Simulation results indicated that the approach is good for path planning and obstacle avoidance in unknown environment. However, no real experiment was conducted to evaluate the method.

Considering the limitation of conventional APF method with the inability of mobile robots to pass between closed space, Park et al. [280] combined fuzzy logic and APF approaches as advanced fuzzy potential field method

procedure NAVIGATION(q_o, q_f, O_N, η, e, M, N_p)
  i ⟵ 0
  navigation ⟵ *True*
  $\mathcal{P}$ ⟵ BPF(q_o, q_f, O_N, η, e, M, N_p) $\mathcal{P}$ is an array
  while *navigation* **do**
    Perform verification of the environment
    if environment has changed **then**
      q_o ⟵ $\mathcal{P}$(i)
      $\mathcal{P}$⟵ BPF(q_o, q_f, O_N, η, e, M, N_p)
      i ⟵ 0
    end if
    i ⟵ i+ 1
    Perform *trajectory* planning to navigate from $\mathcal{P}$(i-1) to $\mathcal{P}$(i)
    if i >= length of $\mathcal{P}$ **then**
      navigation ⟵ *False*
      Display *Goal has been achieved*
    end if
  end while
end procedure

ALGORITHM 1: BPF algorithm.

(AFPFM) to address this limitation. The position and orientation of the robot were considered in controlling the robot. The approach was evaluated using simulation.

To address mobile robot control in unknown environment, Lee et al. [281] proposed sonar behavior-based fuzzy controller (BFC) to control mobile robot wall-following. The sub-fuzzy controllers with nine fuzzy rules were used for the control. The Pioneer 3-DX mobile robot was used to evaluate the proposed method. Although the performance was good in the environment with regular shaped obstacles, collisions were recorded in the environment with irregular obstacles. Moreover, the inability to ensure consistent distance between the robot and the wall is a challenge.

A detection algorithm for vision systems that combines fuzzy image processing algorithm, bacterial algorithm, GA and A∗ was presented in [282] to address path planning problem. The fuzzy image processing algorithm was used to detect the edges of the images of the robot's environment. The bacterial algorithm was used to optimize the output of the processed image. The optimal path and trajectory were determined using GA and A∗ algorithms. The results of the method indicate good performance in detecting edges and reducing the noise in the images. This method requires optimization to control performance in lighting environment to deal with reflection from images.

6.3. Other Hybrid Path Planning Methods. There are many other hybrid approaches used for path planning that did not include FL. Some of these include APF combined with SA [216], PSO combined with gravitational search (GS) algorithm [283], VFH algorithm combined with Kalman Filter [113], integrating game theory and geometry [284], combination of differential global positioning system (DGPS), APF, FL and A∗ path planning algorithm [41], A∗ search and discrete optimization methods [92], a combination of differential equation and SMC [243], virtual obstacle concept was combined with APF [106], and recently the use of LIDAR sensor accompanied with curve fitting. Few of such methods are discussed in this section.

Recently, Marin-Plaza et al. [285] proposed and implemented a path planning method based on Ackermann model using time elastic bands. Dijkstra method was employed to obtain the optimal path for the navigation. A good result was achieved in a real experiment conducted using iCab for the validation of the method.

Considering the strengths and weaknesses of APF in goal seeking and obstacle avoidance, Montiel et al. [6] combined APF and bacterial evolutionary algorithm (BEA) methods to present Bacterial Potential Field (BPF) to provide safe and optimal mobile robot navigation in complex environment. The purpose of the strategy was to eliminate the trapping in local minima drawbacks of the traditional APF method. Simulation results was described to have showed better results compared to genetic potential field (GPF) and pseudo-bacterial potential field (PBPF) methods. Unknown obstacles were detected and avoided during the simulation (See Figure 8). The BPF algorithm in [6] is given in Algorithm 1.

Experiment demonstrated that BPF approach compared to other APF methods used a lower computational time of a factor of 1.59 to find the optimal path.

APF method was optimized using PSO algorithm in [36] to develop a control system to control the local minima problem of APF. Implementation of this approach was done using simulation as demonstrated in Figure 9. Real robot platform implementation may be required to confirm the effectiveness of the technique demonstrated in simulation.

A visual-based approach using a camera and a range sensor was presented in [286] by combining APF, redundancy and visual servoing to deal with path planning and obstacle avoidance problem of mobile robots. This approach was improved upon in [287] by including visual path following

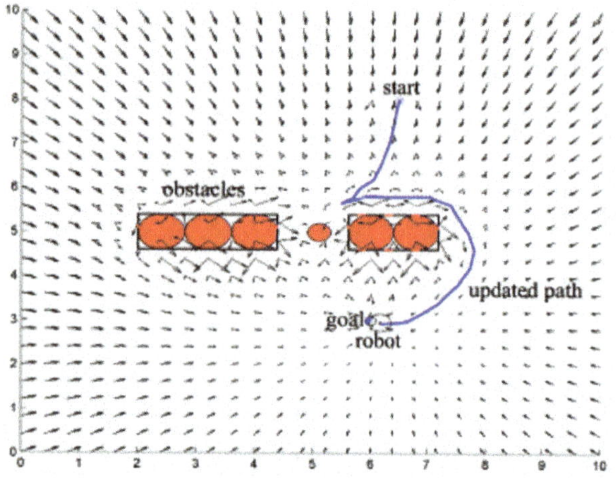

FIGURE 8: Unknown obstacle detected during navigation using BPF approach (source [6]).

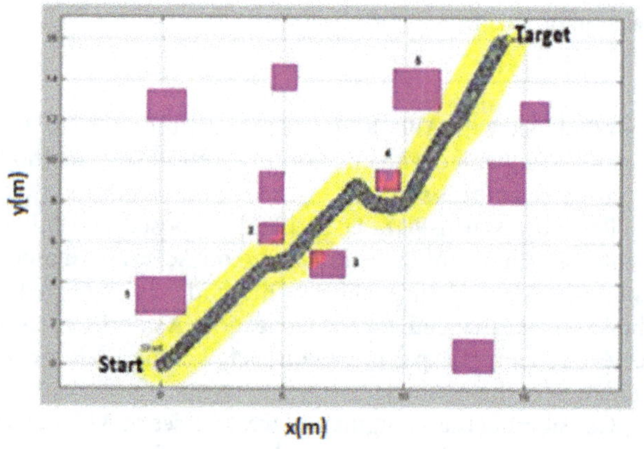

FIGURE 9: Controlling the local minima problem of APF in path planning and obstacle avoidance by optimizing APF using PSO algorithm (source [36]).

obstacle bypassing and collision avoidance with deceleration approach [288, 289]. The method was designed to enable robots progress along a path while avoiding obstacles and maintaining closeness to the 3D path. Experiment was done in outdoor environment using a cycab robot with a 70% field of view. Also, based on extended Kalman filters, a classical motion stereo vision approach was demonstrated in [290] for visual obstacle detection which could not be detected by laser range finders. Visual SLAM algorithm was also proposed in [291] to build a map of the environment in real time with the help of Field Programmable Gate Arrays (FPGA) to provide autonomous navigation of mobile robots in unknown environment. Since navigation involves obstacle avoidance, APF method was presented for the obstacle avoidance. The algorithm was tested in indoor environment composed of static and dynamic obstacles using a differential mobile robot with cameras. Different from the normal Kalman filters, computed depth estimates derived from traditional stereo method were used to initialize the filters instead of using simple heuristics to choose the initial estimates. Target-driven visual navigation method was presented in [292] to address the impractical application problem of deep reinforcement learning in real-world situations. The paper attributed this problem to lack of generalization capacity to new goals and data inefficiency when using deep reinforcement learning. Actor-critic and AI2-THOR models were proposed to address the generalization and data efficiency problems respectively. SCITOS robot was used to validate the method. Though results showed good performance, a test in environment composed of obstacles may be required to determine its efficiency in the real-world settings.

A biological navigation algorithm known as equiangular navigation guidance (ENG) law approach accompanied with the use of local obstacle avoidance techniques using sensors was employed in [13] to plan the motion of a robot to avoid obstacles in complex environment with many obstacles. The choice of the shortest path to goal was however not considered.

Savage et al. [293] also presented a strategy combining GA with recurrent neural network (RNN) to address path planning problem. GA was used to find the obstacle avoidance behavior and then implemented in RNN to control the robot. To address path planning problem in unstructured environment with relation to unmanned ground vehicles, Michel et al. [94] presented a learning algorithm approach by combining reinforcement learning (RL), computer graphics, and computer vision. Supervised learning was first used to approximate the depths from a single monocular image. The proposed algorithm then learns monocular vision cues to estimate the relative depths of the obstacles after which reinforcement learning was used to render the appropriate synthetic scenes to determine the steering direction of the robot. Instead of using the real images and laser scan data, synthetic images were used. Results showed that combining the synthetic and the real data performs better than training the algorithm on either synthetic or real data only.

To provide effective path planning and obstacle avoidance for a mobile robot responsible for transporting components from a warehouse to production lines, Zaki, Arafa and Amer [294] proposed an ultrasonic ranger slave microcontroller system using hybrid navigation system approach that combines perception and dead reckoning based navigation. As many as 24 ultrasonic sensors were used in this approach.

In [295], the authors demonstrated the effectiveness of GA and PRM path planning methods in simple and complex maps. Simulation results indicated that both methods were able to find the path from the initial state to the goal. However, the path produced by GA was smoother than that of PRM. Again, comparing GA and PRM, PRM performed path computation in lesser time which makes it more effective in reactive path planning applications. GA and PRM could be combined to exploit the benefits of the two methods to provide smooth and less time path computation. Real experiment is required to evaluate the effectiveness of the methods in a real environment.

Hassani et al. [296] aimed at obtaining safe path in complex environment for mobile robot and proposed an algorithm combining free segments and turning points algorithms. To provide stability during navigation, sliding mode control was adopted. Simulation in Khepera IV platform was used to evaluate the method using maps of known environment composed of seven static obstacles. Results indicated that safe and optimal path was generated from the initial position to the target. This method needs to be compared to other methods to determine its effectiveness. Moreover, a test is required in a more complex environment and in an environment with dynamic obstacles. Issues of replanning should be considered when random obstacles are encountered during navigation.

Path planning approach to optimize the task of mobile robot path planning using differential evolution (DE) and BSpline was given in [297]. The path planning task was done using DE which is an improved form of GA. To ensure easy navigation path, the BSpline was used to smoothen the path. Aria P3-DX mobile robot was used to test the method in a real experiment. Compared to PSO method, the results showed better optimality for the proposed method. Another hybrid method that includes BSpline is presented in [298]. The authors combined a global path planner with BSpline to achieve free and time-optimal motion of mobile robots in complex environments. The global path planner was used to obtain the waypoints in the environment while the BSpline was used as the local planner to address the optimal control problem. Simulation results showed the efficiency of the method. The KUKA youBot robot was used for real experiment to validate the approach but was carried out in simple environment. A test in a complex environment may be required to determine the efficiency of the method.

Recently, nonlinear kinodynamic constraints in path planning was considered in [299] with the aim of achieving near-optimal motion planning of nonlinear dynamic vehicles in clustered environment. NN and RRT were combined to propose NoD-RRT method. RRT was modified to perform reconstruction to address the kinodynamic constraint problem. The prediction of the cost function was done using NN. The proposed method was evaluated in simulation and real experiment using Pioneer 3-DX robot with sonar sensors to detect obstacles. Compared to RRT and RRT∗, it was described to have performed better.

7. Strengths and Challenges of Hybrid Path Planning Methods

Because hybrid methods are combination of multiple methods, they possess comparatively higher merits by taking advantage of the strengths of the integrated methods while minimizing the drawbacks of these methods. For instance, integration of appropriate methods can help improve noise resistance ability and deal better with oscillation and data uncertainties and controlling local minima problem associated with APF [36, 131, 171, 272, 273].

Though hybrid methods seem to possess the strengths of the methods integrated while reducing their drawbacks, there are still challenges using them. Compatibility of some of these methods is questionable. The integration of incompatible methods would produce worse results than using a single method. Other weaknesses not noted with the individual methods combined may emerge when the methods are integrated and implemented. New weaknesses that emerge because of the combination may also lead to unsatisfactory control performance and difficulty of reducing the effects of uncertainty and oscillations. Noise from sensors and cameras in addition to other hardware constraints, including the limitation of motor speed, imbalance mass of the robot, unequal wheel diameter, encoder sampling errors, misalignment of wheels, disproportionate power supply to the motors, uneven or slippery floors and many other systematic and nonsystematic errors affects the practical performance of these approaches.

8. Conclusion and the Way Forward

In conclusion, achieving intelligent autonomous ground vehicles is the main goal in mobile robotics. Some outstanding contributions have been made over the years in the area to address the path planning and obstacle avoidance problems. However, path planning and obstacle avoidance problem still affects the achievement of intelligent autonomous ground vehicles [77, 145]. In this paper, we analyzed varied approaches from some outstanding publications in addressing mobile robot path planning and obstacle avoidance problems. The strengths and challenges of these approaches were identified and discussed. Notable among them is the noise generated from the cameras, sensors, and the constraints of the mobile robots which affect the reliability and efficiency of the approaches to perform well in real implementations. Local minima, oscillation, and unnecessary stopping of the robot to update its environmental maps, slow convergence speed, difficulty in navigating in cluttered environment without collision, and difficulty reaching target with safe optimum path were among the challenges identified. It was also identified that most of the approaches were evaluated only in simulation environment. The challenge is how successful or efficient these approaches would be when implemented in the real environment where real conditions exist [266]. A summary of the strengths and weaknesses of the approaches considered are shown in Tables 1, 2 and 3.

The following general suggestions are given in this paper when proposing new methods for path planning for autonomous vehicles. Critical consideration should be given to the kinematic dynamics of the vehicles. The systematic and nonsystematic errors that may affect navigation should be looked at. Also, the limitation of the environmental data collection devices like sensors and cameras needs to be considered since their limitation affects the information to be processed to control the robots by the proposed algorithm. There is therefore the need to provide a method that can control noise generated from these devices to aid best estimation of data to control and achieve safe optimal path planning. To enable the autonomous vehicle to take quick decision to avoid collision into obstacles, the computation complexity of algorithms should be looked at to reduce the execution time. Moreover, the strengths and limitations of an approach or a method should be looked at before considering its implementation. Possibly, hybrid approaches that possess the ability to take advantage of the benefits and reduce the limitations of each of the methods involved can be considered to obtain

TABLE 1: Summary of Strengths and Challenges of Nature-inspired computation based mobile robot path planning and obstacle avoidance methods.

Approach	Major Implementation	Strengths	Challenges
Neural Network	Simulation and real Experiment	(i) Good at providing learning and generalization [232] (ii) Can help tune the rule base of fuzzy logic [232]	(i) Efficiency of neural controllers deteriorates as the number of layers increase [232] (ii) Difficult to acquire its required large dataset of the environment during training to achieve best results. (iii) Using BP easily results in local minima problems [238] (iv) It has a slow convergence speed [232]
Genetic Algorithm	Simulation	(i) Good at solving problems that are difficult to deal with using conventional algorithms and it can combine well with other algorithms (ii) It has good optimization ability	(i) Difficult to scale well with complex situations (ii) Can lead to convergence at local minima and oscillations [239, 240] (iii) Difficult to work on dynamic data sets and difficult to achieve results due to its complex principle [233]
Ant Colony	Simulation	(i) Good at obtaining optimal result [233] (ii) Fast convergence [234]	(i) Difficult to determine its parameters which affects obtaining quick convergence [240] (ii) Require a lot of computing time [233]
Particle Swarm Optimization	Simulation	(i) Faster than fuzzy logic in terms of convergence [132] (ii) Simple and does not require much computing time hence, effective to implement with optimization problems [192–195] (iii) Performs well on varied application problems [193]	(i) Difficult to deal with trapping into local minima under complex map [194, 240] (ii) Difficult to generalize its performance since undertaken experiments relied on objects in polygon form [240]
Fuzzy Logic	Simulation and Experiments with real robot	(i) Ability to make inferences in uncertain scenarios [129] (ii) Ability to imitate the control logic of human [129] (ii) It does well when combine with other algorithms	(i) Difficult to build rule base to deal with unstructured environment [145, 232]
Artificial Bee Colony	Simulation	(i) It does not require much computational time and it produces good results [241] (ii) It has simple algorithm and easy to implement to solve optimization problems [236, 241] (iii) Can combine well with other optimization algorithms [236, 237] (iv) It uses fewer control parameters [236]	(i) It has low convergence performance [241]
Memetic and Bacterial Memetic algorithm	Simulation	(i) Compared to conventional algorithms, it produces faster convergence and good solution with the benefit from different search methods it blends [235]	(i) It can result in premature convergence [235]
Others (*Simulated Annealing and Human Navigation strategy*)	Simulation	(i) Simulated Annealing is good at approximating the global optimum [242] (ii) Takes advantage of human navigation that does not depend on explicit planning but occur online [223]	(i) SA algorithm is slow [242] (ii) Difficult in choosing the initial position for SA [242]

better techniques for path planning and obstacle avoidance task for autonomous ground vehicles. Again, efforts should be made to implement proposed algorithms in real robot platform where real conditions are found. Implementation should be aimed at both indoor and outdoor environments to meet real conditions required of intelligent autonomous ground vehicles.

To achieve safe and efficient path planning of autonomous vehicles, we specifically recommend a hybrid approach that combines visual-based method, morphological dilation

Table 2: Summary of strengths and challenges of conventional based mobile robot path planning and obstacle avoidance methods.

Approach	Major Implementation	Strengths	Challenges
Artificial Potential Field	Simulation and real Experiment	(i) Easy to implement by considering attraction to goal and repulsion to obstacles	(i) Results in local minima problem causing the robot to be trapped at a position instead of the goal [36, 40, 41, 105, 106] (ii) Inability of the robot to pass between closely spaced obstacles and results in oscillations [107, 108, 110] (iii) Difficulty to perform in environment with obstacles of different shapes [109] (iv) Constraints of the hardware of the mobile robots affect the performance of APF methods [51]
Visual Based and Reactive approaches	Simulation and real Experiment	(i) Easy to implement by relying on the information from sensors and cameras to take decision on the movement of the robot	(i) Noise from sensors and cameras due to distance from objects, color, temperature and reflection affects performance (ii) Hardware constraints of the robots affect performance (iii) Computational expensive
Robot Motion and Sliding Mode Strategy	Simulation	(i) Is Fast with respect to response time (ii) Works well with uncertain systems and other unconducive external factors [64]	(i) Performs poorly when the longitudinal velocity of the robot is fast (ii) Computational expensive (iii) Chattering problem leading to low control accuracy [114]
Dynamic Window	Simulation	(i) Easy to implement	(i) Local minima problem (ii) Difficult to manage the effects of mobile robot constraints [75, 102–104]
Others (KF, SGBM, A∗, CSS, GS, Curvature velocity method, Bumper Event approach, Wall following.)	Simulation and Experiments with real robot	(i) Easy to implement	(i) Noise from sensors affects performance (ii) Computational expensive

Table 3: Summary of strengths and challenges of hybrid mobile robot path planning and obstacle avoidance methods.

Approach	Major Implementation	Strengths	Challenges
Neuro Fuzzy	Simulation and real Experiment	Takes advantage of the strengths of NN and Fuzzy logic while reducing the drawbacks of both methods. Example: NN can help tune the rule base of fuzzy logic which is difficult using fuzzy logic alone [145, 232]	Unsatisfactory control performance and difficulty of reducing the effect of uncertainty and oscillation of T1FNN [101]
Others (Kalman Filter and Fuzzy logic, Visual Based and APF, Wall following and Fuzzy Logic, Fuzzy logic and Q-Learning, GA and NN, APF and SA, APF and Visual Servoing, APF and Ant Colony, Fuzzy and A∗, Reinforcement and computer graphics and computer vision, PSO and Gravitational search, etc)	Simulation and real Experiments	The drawbacks of each approach in the combination is reduced. Example: Improves noise resistance ability, deal better with oscillation and data uncertainties [131, 271–273] and controlling local minima problem associated with APF [36]	Noise from sensor and cameras and the hardware constraints including the limitation of motor speed, imbalance mass of the robot, power supply to the motors and many others affects the practical performance of these approaches.

(MD), VD, neuro-fuzzy, A∗, and path smoothing algorithms. The visual-based method is to help in collecting and processing visual data from the environment to obtain the map of the environment for further processing. To reduce uncertainties of data collection tools, multiple cameras and distance sensors are required to collect the data simultaneously which should be taken through computations to obtain the best output. The MD is required to inflate the obstacles in the environment before generating the path to ensure safety of the vehicles and avoid trapping in narrow passages. The radius for obstacle inflation using MD should be based on the size of the vehicle and other space requirements to take care of uncertainties during navigation. The VD algorithm is to help in constructing the roadmap to obtain feasible way points. VD algorithm which finds perpendicular bisector of equal distance among obstacle points would add to providing safety of the vehicle and once a path exists, it would find it. The neuro-fuzzy method may be required to provide learning ability to deal with dynamic environment. To obtain the shortest path, A∗ may be used and finally a path smoothening algorithm to get a smooth navigable path. The visual-based method should also help to provide reactive path planning in dynamic environment. While implementing the hybrid method, the kinematic dynamics of the vehicle should be taken into consideration. This is a task we are currently investigating.

This study is necessary in achieving safe and optimal navigation of autonomous vehicles when the outlined recommendations are applied in developing path planning algorithms.

Conflicts of Interest

The authors declare that there are no conflicts of interest regarding the publication of this paper.

References

[1] S. G. Tzafestas, "Mobile Robot Path, Motion and Task Planning," in *Introduction of Mobile Robot Control, Elsevier Inc*, pp. 429–478, Elsevier Inc, 2014.

[2] Y. C. Kim, S. B. Cho, and S. R. Oh, "Map-building of a real mobile robot with GA-fuzzy controller," vol. 4, pp. 696–703, 2002.

[3] S. M. Homayouni, T. Sai Hong, and N. Ismail, "Development of genetic fuzzy logic controllers for complex production systems," *Computers & Industrial Engineering*, vol. 57, no. 4, pp. 1247–1257, 2009.

[4] O. R. Motlagh, T. S. Hong, and N. Ismail, "Development of a new minimum avoidance system for a behavior-based mobile robot," *Fuzzy Sets and Systems*, vol. 160, no. 13, pp. 1929–1946, 2009.

[5] A. S. Matveev, M. C. Hoy, and A. V. Savkin, "A globally converging algorithm for reactive robot navigation among moving and deforming obstacles," *Automatica*, vol. 54, pp. 292–304, 2015.

[6] O. Montiel, U. Orozco-Rosas, and R. Sepúlveda, "Path planning for mobile robots using Bacterial Potential Field for avoiding static and dynamic obstacles," *Expert Systems with Applications*, vol. 42, no. 12, pp. 5177–5191, 2015.

[7] O. Motlagh, S. H. Tang, N. Ismail, and A. R. Ramli, "A review on positioning techniques and technologies: A novel AI approach," *Journal of Applied Sciences*, vol. 9, no. 9, pp. 1601–1614, 2009.

[8] L. Yiqing, Y. Xigang, and L. Yongjian, "An improved PSO algorithm for solving non-convex NLP/MINLP problems with equality constraints," *Computers & Chemical Engineering*, vol. 31, no. 3, pp. 153–162, 2007.

[9] B. B. V. L. Deepak, D. R. Parhi, and B. M. V. A. Raju, "Advance Particle Swarm Optimization-Based Navigational Controller For Mobile Robot," *Arabian Journal for Science and Engineering*, vol. 39, no. 8, pp. 6477–6487, 2014.

[10] S. S. Parate and J. L. Minaise, "Development of an obstacle avoidance algorithm and path planning algorithm for an autonomous mobile robot," *nternational Engineering Research Journal (IERJ)*, vol. 2, pp. 94–99, 2015.

[11] S. M. LaValle, *Planning Algorithms*, Cambridge University Press, 2006.

[12] W. H. Huang, B. R. Fajen, J. R. Fink, and W. H. Warren, "Visual navigation and obstacle avoidance using a steering potential function," *Robotics and Autonomous Systems*, vol. 54, no. 4, pp. 288–299, 2006.

[13] H. Teimoori and A. V. Savkin, "A biologically inspired method for robot navigation in a cluttered environment," *Robotica*, vol. 28, no. 5, pp. 637–648, 2010.

[14] C.-J. Kim and D. Chwa, "Obstacle avoidance method for wheeled mobile robots using interval type-2 fuzzy neural network," *IEEE Transactions on Fuzzy Systems*, vol. 23, no. 3, pp. 677–687, 2015.

[15] D.-H. Kim and J.-H. Kim, "A real-time limit-cycle navigation method for fast mobile robots and its application to robot soccer," *Robotics and Autonomous Systems*, vol. 42, no. 1, pp. 17–30, 2003.

[16] C. J. Kim, M.-S. Park, A. V. Topalov, D. Chwa, and S. K. Hong, "Unifying strategies of obstacle avoidance and shooting for soccer robot systems," in *Proceedings of the 2007 International Conference on Control, Automation and Systems*, pp. 207–211, Seoul, South Korea, October 2007.

[17] C. G. Rusu and I. T. Birou, "Obstacle Avoidance Fuzzy System for Mobile Robot with IR," in *Proceedings of the Sensors*, vol. no. 10, pp. 25–29, Suceava, Romannia, 2010.

[18] S. K. Pradhan, D. R. Parhi, and A. K. Panda, "Fuzzy logic techniques for navigation of several mobile robots," *Applied Soft Computing*, vol. 9, no. 1, pp. 290–304, 2009.

[19] H.-J. Yeo and M.-H. Sung, "Fuzzy Control for the Obstacle Avoidance of Remote Control Mobile Robot," *Journal of the Institute of Electronics Engineers of Korea SC*, vol. 48, no. 1, pp. 47–54, 2011.

[20] M. Wang, J. Luo, and U. Walter, "A non-linear model predictive controller with obstacle avoidance for a space robot," *Advances in Space Research*, vol. 57, no. 8, pp. 1737–1746, 2016.

[21] Y. Chen and J. Sun, "Distributed optimal control for multi-agent systems with obstacle avoidance," *Neurocomputing*, vol. 173, pp. 2014–2021, 2016.

[22] T. Jin, "Obstacle Avoidance of Mobile Robot Based on Behavior Hierarchy by Fuzzy Logic," *International Journal of Fuzzy Logic and Intelligent Systems*, vol. 12, no. 3, pp. 245–249, 2012.

[23] C. Giannelli, D. Mugnaini, and A. Sestini, "Path planning with obstacle avoidance by G1 PH quintic splines," *Computer-Aided Design*, vol. 75-76, pp. 47–60, 2016.

[24] A. S. Matveev, A. V. Savkin, M. Hoy, and C. Wang, "Safe Robot Navigation Among Moving and Steady Obstacles," *Safe Robot Navigation Among Moving and Steady Obstacles*, pp. 1–344, 2015.

[25] S. Jin and B. Choi, "Fuzzy Logic System Based Obstacle Avoidance for a Mobile Robot," in *Control and Automation, and Energy System Engineering*, vol. 256 of *Communications in Computer and Information Science*, pp. 1-6, Springer Berlin Heidelberg, Berlin, Heidelberg, 2011.

[26] D. Xiaodong, G. A. O. Hongxia, L. I. U. Xiangdong, and Z. Xuedong, "Adaptive particle swarm optimization algorithm based on population entropy," *Journal of Electrical and Computer Engineering*, vol. 33, no. 18, pp. 222–248, 2007.

[27] B. Huang and G. Cao, "The Path Planning Research for Mobile Robot Based on the Artificial Potential Field," *Computer Engineering and Applications*, vol. 27, pp. 26–28, 2006.

[28] K. Jung, J. Kim, and T. Jeon, "Collision Avoidance of Multiple Path-planning using Fuzzy Inference System," in *Proceedings of KIIS Spring Conference*, vol. 19, pp. 278–288, 2009.

[29] Q. ZHU, "Ant Algorithm for Navigation of Multi-Robot Movement in Unknown Environment," *Journal of Software* , vol. 17, no. 9, p. 1890, 2006.

[30] J. Borenstein and Y. Koren, "The vector field histogram—fast obstacle avoidance for mobile robots," *IEEE Transactions on Robotics and Automation*, vol. 7, no. 3, pp. 278–288, 1991.

[31] D. Fox, W. Burgard, and S. Thrun, "The dynamic window approach to collision avoidance," *IEEE Robotics and Automation Magazine*, vol. 4, no. 1, pp. 23–33, 1997.

[32] S. G. Tzafestas, M. P. Tzamtzi, and G. G. Rigatos, "Robust motion planning and control of mobile robots for collision avoidance in terrains with moving objects," *Mathematics and Computers in Simulation*, vol. 59, no. 4, pp. 279–292, 2002.

[33] "Vision-based obstacle avoidance for a small, low-cost robot," in *Proceedings of the 4th International Conference on Informatics in Control, Automation and Robotics*, pp. 275–279, Angers, France, May 2007.

[34] O. Khatib, "Real-time obstacle avoida nce for manipulators and mobile robots," *International Journal of Robotics Research*, vol. 5, no. 1, pp. 90–98, 1986.

[35] M. Tounsi and J. F. Le Corre, "Trajectory generation for mobile robots," *Mathematics and Computers in Simulation*, vol. 41, no. 3-4, pp. 367–376, 1996.

[36] A. A.Ahmed, T. Y. Abdalla, and A. A. Abed, "Path Planning of Mobile Robot by using Modified Optimized Potential Field Method," *International Journal of Computer Applications*, vol. 113, no. 4, pp. 6–10, 2015.

[37] D. N. Nia, H. S. Tang, B. Karasfi, O. R. Motlagh, and A. C. Kit, "Virtual force field algorithm for a behaviour-based autonomous robot in unknown environments," *Proceedings of the Institution of Mechanical Engineers, Part I: Journal of Systems and Control Engineering*, vol. 225, no. 1, pp. 51–62, 2011.

[38] Y. Wang, D. Mulvaney, and I. Sillitoe, "Genetic-based mobile robot path planning using vertex heuristics," in *Proceedings of the Proc. Conf. on Cybernetics Intelligent Systems*, p. 1, 2006.

[39] J. Silvestre-Blanes, "Real-time obstacle avoidance using potential field for a nonholonomic vehicle," Factory Automation, InTech, 2010.

[40] B. Kovács, G. Szayer, F. Tajti, M. Burdelis, and P. Korondi, "A novel potential field method for path planning of mobile robots by adapting animal motion attributes," *Robotics and Autonomous Systems*, vol. 82, pp. 24–34, 2016.

[41] H. Bing, L. Gang, G. Jiang, W. Hong, N. Nan, and L. Yan, "A route planning method based on improved artificial potential field algorithm," in *Proceedings of the 2011 IEEE 3rd International Conference on Communication Software and Networks (ICCSN)*, pp. 550–554, Xi'an, China, May 2011.

[42] P. Vadakkepat, . Kay Chen Tan, and . Wang Ming-Liang, "Evolutionary artificial potential fields and their application in real time robot path planning," in *Proceedings of the 2000 Congress on Evolutionary Computation*, pp. 256–263, La Jolla, CA, USA.

[43] Kim. Min-Ho, Heo. Jung-Hun, Wei. Yuanlong, and Lee. Min-Cheol, "A path planning algorithm using artificial potential field based on probability map," in *Proceedings of the 2011 8th International Conference on Ubiquitous Robots and Ambient Intelligence (URAI 2011)*, pp. 41–43, Incheon, November 2011.

[44] L. Barnes, M. Fields, and K. Valavanis, "Unmanned ground vehicle swarm formation control using potential fields," in *Proceedings of the Proc of the 15th IEEE Mediterranean Conf. on Control Automation*, p. 1, Athens, Greece, 2007.

[45] D. Huang, L. Heutte, and M. Loog, *Advanced Intelligent Computing Theories and Applications. With Aspects of Contemporary Intelligent Computing Techniques*, vol. 2, Springer Berlin Heidelberg, Berlin, Heidelberg, 2007.

[46] S. S. Ge and Y. J. Cui, "Dynamic motion planning for mobile robots using potential field method," *Autonomous Robots*, vol. 13, no. 3, pp. 207–222, 2002.

[47] J.-Y. Zhang, Z.-P. Zhao, and D. Liu, "Path planning method for mobile robot based on artificial potential field," *Harbin Gongye Daxue Xuebao/Journal of Harbin Institute of Technology*, vol. 38, no. 8, pp. 1306–1309, 2006.

[48] F. Chen, P. Di, J. Huang, H. Sasaki, and T. Fukuda, "Evolutionary artificial potential field method based manipulator path planning for safe robotic assembly," in *Proceedings of the 20th Anniversary MHS 2009 and Micro-Nano Global COE - 2009 International Symposium on Micro-NanoMechatronics and Human Science*, pp. 92–97, Japan, November 2009.

[49] W. Su, R. Meng, and C. Yu, "A study on soccer robot path planning with fuzzy artificial potential field," in *Proceedings of the 1st International Conference on Computing Control and Industrial Engineering, CCIE 2010*, pp. 386–390, China, June 2010.

[50] T. Weerakoon, K. Ishii, and A. A. Nassiraei, "Dead-lock free mobile robot navigation using modified artificial potential field," in *Proceedings of the 2014 Joint 7th International Conference on Soft Computing and Intelligent Systems (SCIS) and 15th International Symposium on Advanced Intelligent Systems (ISIS)*, pp. 259–264, Kita-Kyushu, Japan, December 2014.

[51] Z. Xu, R. Hess, and K. Schilling, "Constraints of Potential Field for Obstacle Avoidance on Car-like Mobile Robots," *IFAC Proceedings Volumes*, vol. 45, no. 4, pp. 169–175, 2012.

[52] T. P. Nascimento, A. G. Conceicao, and A. P. Moreira, "Multi-Robot Systems Formation Control with Obstacle Avoidance," *IFAC Proceedings Volumes*, vol. 47, no. 3, pp. 5703–5708, 2014.

[53] K. Souhila and A. Karim, "Optical flow based robot obstacle avoidance," *International Journal of Advanced Robotic Systems*, vol. 4, no. 1, pp. 13–16, 2007.

[54] J. Kim and Y. Do, "Moving Obstacle Avoidance of a Mobile Robot Using a Single Camera," *Procedia Engineering*, vol. 41, pp. 911–916, 2012.

[55] S. Lenseir and M. Veloso, "Visual sonar: fast obstacle avoidance using monocular vision," in *Proceedings of the 2003 IEEE/RSJ International Conference on Intelligent Robots and Systems (IROS 2003) (Cat. No.03CH37453)*, pp. 886–891, Las Vegas, Nevada, USA.

[56] L. Kovacs, "Visual Monocular Obstacle Avoidance for Small Unmanned Vehicles," in *Proceedings of the 29th IEEE Conference on Computer Vision and Pattern Recognition Workshops, CVPRW 2016*, pp. 877–884, USA, July 2016.

[57] L. Kovács and T. Szirányi, "Focus area extraction by blind deconvolution for defining regions of interest," *IEEE Transactions on Pattern Analysis and Machine Intelligence*, vol. 29, no. 6, pp. 1080–1085, 2007.

[58] L. Kovács, "Single image visual obstacle avoidance for low power mobile sensing," in *Advanced Concepts for Intelligent Vision Systems*, vol. 9386 of *Lecture Notes in Comput. Sci.*, pp. 261–272, Springer, Cham, 2015.

[59] E. Masehian and Y. Katebi, "Sensor-based motion planning of wheeled mobile robots in unknown dynamic environments," *Journal of Intelligent & Robotic Systems*, vol. 74, no. 3-4, pp. 893–914, 2014.

[60] Li Wang, Lijun Zhao, Guanglei Huo et al., "Visual Semantic Navigation Based on Deep Learning for Indoor Mobile Robots," *Complexity*, vol. 2018, pp. 1-12, 2018.

[61] V. Gavrilut, A. Tiponut, A. Gacsadi, and L. Tepelea, "Wall-following Method for an Autonomous Mobile Robot using Two IR Sensors," in *Proceedings of the 12th WSEAS International Conference on Systems*, pp. 22–24, Heraklion, Greece, 2008.

[62] A. S. Matveev, C. Wang, and A. V. Savkin, "Real-time navigation of mobile robots in problems of border patrolling and avoiding collisions with moving and deforming obstacles," *Robotics and Autonomous Systems*, vol. 60, no. 6, pp. 769–788, 2012.

[63] R. Solea and D. Cernega, "Sliding Mode Control for Trajectory Tracking Problem - Performance Evaluation," in *Artificial Neural Networks - ICANN 2009*, vol. 5769 of *Lecture Notes in Computer Science*, pp. 865–874, Springer Berlin Heidelberg, Berlin, Heidelberg, 2009.

[64] R. Solea and D. C. Cernega, "Obstacle Avoidance for Trajectory Tracking Control of Wheeled Mobile Robots," *IFAC Proceedings Volumes*, vol. 45, no. 6, pp. 906–911, 2012.

[65] J. Lwowski, L. Sun, R. Mexquitic-Saavedra, R. Sharma, and D. Pack, "A Reactive Bearing Angle Only Obstacle Avoidance Technique for Unmanned Ground Vehicles," *Journal of Automation and Control Research*, 2014.

[66] S. H. Tang and C. K. Ang, "A Reactive Collision Avoidance Approach to Mobile Robot in Dynamic Environment," *Journal of Automation and Control Engineering*, vol. 1, pp. 16–20, 2013.

[67] T. Bandyophadyay, L. Sarcione, and F. S. Hover, "A simple reactive obstacle avoidance algorithm and its application in Singapore harbor," *Springer Tracts in Advanced Robotics*, vol. 62, pp. 455–465, 2010.

[68] A. V. Savkin and C. Wang, "Seeking a path through the crowd: Robot navigation in unknown dynamic environments with moving obstacles based on an integrated environment representation," *Robotics and Autonomous Systems*, vol. 62, no. 10, pp. 1568–1580, 2014.

[69] L. Adouane, A. Benzerrouk, and P. Martinet, "Mobile robot navigation in cluttered environment using reactive elliptic trajectories," in *Proceedings of the 18th IFAC World Congress*, pp. 13801–13806, Milano, Italy, September 2011.

[70] P. Ögren and N. E. Leonard, "A convergent dynamic window approach to obstacle avoidance," *IEEE Transactions on Robotics*, vol. 21, no. 2, pp. 188–195, 2005.

[71] P. Ögren and N. E. Leonard, "A tractable convergent dynamic window approach to obstacle avoidance," in *Proceedings of the 2002 IEEE/RSJ International Conference on Intelligent Robots and Systems*, pp. 595–600, Switzerland, October 2002.

[72] H. Zhang, L. Dou, H. Fang, and J. Chen, "Autonomous indoor exploration of mobile robots based on door-guidance and improved dynamic window approach," in *Proceedings of the 2009 IEEE International Conference on Robotics and Biomimetics, ROBIO 2009*, pp. 408–413, China, December 2009.

[73] F. P. Vista, A. M. Singh, D.-J. Lee, and K. T. Chong, "Design convergent Dynamic Window Approach for quadrotor navigation," *International Journal of Precision Engineering and Manufacturing*, vol. 15, no. 10, pp. 2177–2184, 2014.

[74] H. Berti, A. D. Sappa, and O. E. Agamennoni, "Improved dynamic window approach by using lyapunov stability criteria," *Latin American Applied Research*, vol. 38, no. 4, pp. 289–298, 2008.

[75] X. Li, F. Liu, J. Liu, and S. Liang, "Obstacle avoidance for mobile robot based on improved dynamic window approach," *Turkish Journal of Electrical Engineering & Computer Sciences*, vol. 25, no. 2, pp. 666–676, 2017.

[76] S. M. LaValle, H. H. Gonzalez-Banos, C. Becker, and J.-C. Latombe, "Motion strategies for maintaining visibility of a moving target," in *Proceedings of the 1997 IEEE International Conference on Robotics and Automation, ICRA. Part 3 (of 4)*, pp. 731–736, April 1997.

[77] M. Adbellatif and O. A. Montasser, "Using Ultrasonic Range Sensors to Control a Mobile Robot in Obstacle Avoidance Behavior," in *Proceedings of the In Proceeding of The SCI2001 World Conference on Systemics, Cybernetics and Informatics*, pp. 78–83, 2001.

[78] I. Ullah, F. Ullah, Q. Ullah, and S. Shin, "Sensor-Based Robotic Model for Vehicle Accident Avoidance," *Journal of Computational Intelligence and Electronic Systems*, vol. 1, no. 1, pp. 57–62, 2012.

[79] I. Ullah, F. Ullah, and Q. Ullah, "A sensor based robotic model for vehicle collision reduction," in *Proceedings of the 2011 International Conference on Computer Networks and Information Technology (ICCNIT)*, pp. 251–255, Abbottabad, Pakistan, July 2011.

[80] N. Kumar, Z. Vamossy, and Z. M. Szabo-Resch, "Robot obstacle avoidance using bumper event," in *Proceedings of the 2016 IEEE 11th International Symposium on Applied Computational Intelligence and Informatics (SACI)*, pp. 485–490, Timisoara, Romania, May 2016.

[81] F. Kunwar, F. Wong, R. B. Mrad, and B. Benhabib, "Guidance-based on-line robot motion planning for the interception of mobile targets in dynamic environments," *Journal of Intelligent & Robotic Systems*, vol. 47, no. 4, pp. 341–360, 2006.

[82] W. Chih-Hung, H. Wei-Zhon, and P. Shing-Tai, "Performance Evaluation of Extended Kalman Filtering for Obstacle Avoidance of Mobile Robots," in *Proceedings of the International Multi Conference of Engineers and Computer Scientist*, vol. 1, 2015.

[83] C. C. Mendes and D. F. Wolf, "Stereo-Based Autonomous Navigation and Obstacle Avoidance*," *IFAC Proceedings Volumes*, vol. 46, no. 10, pp. 211–216, 2013.

[84] E. Owen and L. Montano, "A robocentric motion planner for dynamic environments using the velocity space," in *Proceedings of the 2006 IEEE/RSJ International Conference on Intelligent Robots and Systems, IROS 2006*, pp. 4368–4374, China, October 2006.

[85] H. Dong, W. Li, J. Zhu, and S. Duan, "The Path Planning for Mobile Robot Based on Voronoi Diagram," in *Proceedings of the in 3rd Intl Conf on Intelligent Networks Intelligent Syst*, Shenyang, 2010.

[86] Y. Ho and J. Liu, "Collision-free curvature-bounded smooth path planning using composite Bezier curve based on Voronoi diagram," in *Proceedings of the 2009 IEEE International Symposium on Computational Intelligence in Robotics and Automation - (CIRA 2009)*, pp. 463–468, Daejeon, Korea (South), December 2009.

[87] P. Qu, J. Xue, L. Ma, and C. Ma, "A constrained VFH algorithm for motion planning of autonomous vehicles," in *Proceedings of the IEEE Intelligent Vehicles Symposium, IV 2015*, pp. 700–705, Republic of Korea, July 2015.

[88] K. Lee, J. C. Koo, H. R. Choi, and H. Moon, "An RRT* path planning for kinematically constrained hyper-redundant inpipe robot," in *Proceedings of the 12th International Conference on Ubiquitous Robots and Ambient Intelligence, URAI 2015*, pp. 121–128, Republic of Korea, October 2015.

[89] S. Kamarry, L. Molina, E. A. N. Carvalho, and E. O. Freire, "Compact RRT: A New Approach for Guided Sampling Applied to Environment Representation and Path Planning in Mobile Robotics," in *Proceedings of the 12th LARS Latin American Robotics Symposium and 3rd SBR Brazilian Robotics Symposium, LARS-SBR 2015*, pp. 259–264, Brazil, October 2015.

[90] M. Srinivasan Ramanagopal, A. P. Nguyen, and J. Le Ny, "A Motion Planning Strategy for the Active Vision-Based Mapping of Ground-Level Structures," *IEEE Transactions on Automation Science and Engineering*, vol. 15, no. 1, pp. 356–368, 2018.

[91] C. Fulgenzi, A. Spalanzani, and C. Laugier, "Dynamic obstacle avoidance in uncertain environment combining PVOs and occupancy grid," in *Proceedings of the IEEE International Conference on Robotics and Automation (ICRA '07)*, pp. 1610–1616, April 2007.

[92] Y. Wang, A. Goila, R. Shetty, M. Heydari, A. Desai, and H. Yang, "Obstacle Avoidance Strategy and Implementation for Unmanned Ground Vehicle Using LIDAR," *SAE International Journal of Commercial Vehicles*, vol. 10, no. 1, pp. 50–55, 2017.

[93] R. Lagisetty, N. K. Philip, R. Padhi, and M. S. Bhat, "Object detection and obstacle avoidance for mobile robot using stereo camera," in *Proceedings of the IEEE International Conference on Control Applications*, pp. 605–610, Hyderabad, India, August 2013.

[94] J. Michels, A. Saxena, and A. Y. Ng, "High speed obstacle avoidance using monocular vision and reinforcement learning," in *Proceedings of the ICML 2005: 22nd International Conference on Machine Learning*, pp. 593–600, Germany, August 2005.

[95] H. Seraji and A. Howard, "Behavior-based robot navigation on challenging terrain: A fuzzy logic approach," *IEEE Transactions on Robotics and Automation*, vol. 18, no. 3, pp. 308–321, 2002.

[96] S. Hrabar, "3D path planning and stereo-based obstacle avoidance for rotorcraft UAVs," in *Proceedings of the 2008 IEEE/RSJ International Conference on Intelligent Robots and Systems, IROS*, pp. 807–814, France, September 2008.

[97] S. H. Han. Na and H. Jeong, "Stereo-based road obstacle detection and tracking," in *Proceedings of the In 13th Int. Conf. on ICACT, IEEE*, pp. 1181–1184, 2011.

[98] D. N. Bhat and S. K. Nayar, "Stereo and Specular Reflection," *International Journal of Computer Vision*, vol. 26, no. 2, pp. 91–106, 1998.

[99] P. S. Sharma and D. N. G. Chitaliya, "Obstacle Avoidance Using Stereo Vision: A Survey," *International Journal of Innovative Research in Computer and Communication Engineering*, vol. 03, no. 01, pp. 24–29, 2015.

[100] A. Typiak, "Use of laser rangefinder to detecting in surroundings of mobile robot the obstacles," in *Proceedings of the The 25th International Symposium on Automation and Robotics in Construction*, pp. 246–251, Vilnius, Lithuania, June 2008.

[101] P. D. Baldoni, Y. Yang, and S. Kim, "Development of Efficient Obstacle Avoidance for a Mobile Robot Using Fuzzy Petri Nets," in *Proceedings of the 2016 IEEE 17th International Conference on Information Reuse and Integration (IRI)*, pp. 265–269, Pittsburgh, PA, USA, July 2016.

[102] D. Janglová, "Neural Networks in Mobile Robot Motion," *International Journal of Advanced Robotic Systems*, vol. 1, no. 1, p. 2, 2004.

[103] D. Kiss and G. Tevesz, "Advanced dynamic window based navigation approach using model predictive control," in *Proceedings of the 2012 17th International Conference on Methods & Models in Automation & Robotics (MMAR)*, pp. 148–153, Miedzyzdroje, Poland, August 2012.

[104] P. Saranrittichai, N. Niparnan, and A. Sudsang, "Robust local obstacle avoidance for mobile robot based on Dynamic Window approach," in *Proceedings of the 2013 10th International Conference on Electrical Engineering/Electronics, Computer, Telecommunications and Information Technology (ECTI-CON 2013)*, pp. 1–4, Krabi, Thailand, May 2013.

[105] C. H. Hommes, "Heterogeneous agent models in economics and finance," in *Handbook of Computational Economics*, vol. 2, pp. 1109–1186, Elsevier, 2006.

[106] L. Chengqing, M. H Ang, H. Krishnan, and L. S. Yong, "Virtual Obstacle Concept for Local-minimum-recovery in Potential-field Based Navigation," in *Proceedings of the IEEE Int. Conf. on Robotics Automation*, pp. 983–988, San Francisco, 2000.

[107] Y. Koren and J. Borenstein, "Potential field methods and their inherent limitations for mobile robot navigation," in *Proceedings of the IEEE International Conference on Robotics and Automation*, pp. 1398–1404, April 1991.

[108] S. S. Ge and Y. J. Cui, "New potential functions for mobile robot path planning," *IEEE Transactions on Robotics and Automation*, vol. 16, no. 5, pp. 615–620, 2000.

[109] Q. Jia and X. Wang, "An improved potential field method for path planning," in *Proceedings of the Chinese Control and Decision Conference (CCDC '10)*, pp. 2265–2270, Xuzhou, China, May 2010.

[110] J. Canny and M. Lin, "An opportunistic global path planner," in *Proceedings of the IEEE International Conference on Robotics and Automation*, pp. 1554–1559, Cincinnati, OH, USA.

[111] K.-Y. Chen, P. A. Lindsay, P. J. Robinson, and H. A. Abbass, "A hierarchical conflict resolution method for multi-agent path planning," in *Proceedings of the 2009 IEEE Congress on Evolutionary Computation, CEC 2009*, pp. 1169–1176, Norway, May 2009.

[112] Y. Morales, A. Carballo, E. Takeuchi, A. Aburadani, and T. Tsubouchi, "Autonomous robot navigation in outdoor cluttered pedestrian walkways," *Journal of Field Robotics*, vol. 26, no. 8, pp. 609–635, 2009.

[113] D. Kim, J. Kim, J. Bae, and Y. Soh, "Development of an Enhanced Obstacle Avoidance Algorithm for a Network-Based Autonomous Mobile Robot," in *Proceedings of the 2010 International Conference on Intelligent Computation Technology and Automation (ICICTA)*, pp. 102–105, Changsha, China, May 2010.

[114] V. I. Utkin, "Sliding mode control," in *Variable structure systems: from principles to implementation*, vol. 66 of *IEE Control Eng. Ser.*, pp. 3–17, IEE, London, 2004.

[115] N. Siddique and H. Adeli, "Nature inspired computing: an overview and some future directions," *Cognitive Computation*, vol. 7, no. 6, pp. 706–714, 2015.

[116] M. H. Saffari and M. J. Mahjoob, "Bee colony algorithm for real-time optimal path planning of mobile robots," in *Proceedings of the 5th International Conference on Soft Computing, Computing with Words and Perceptions in System Analysis, Decision and Control (ICSCCW '09)*, pp. 1–4, IEEE, September 2009.

[117] L. Na and F. Zuren, "Numerical potential field and ant colony optimization based path planning in dynamic environment," in *Proceedings of the 6th World Congress on Intelligent Control and Automation, WCICA 2006*, pp. 8966–8970, China, June 2006.

[118] C. A. Floudas, *Deterministic global optimization*, vol. 37 of *Nonconvex Optimization and its Applications*, Kluwer Academic Publishers, Dordrecht, 2000.

[119] J.-H. Lin and L.-R. Huang, "Chaotic Bee Swarm Optimization Algorithm for Path Planning of Mobile Robots," in *Proceedings of the 10th WSEAS International Conference on Evolutionary Computing*, Prague, Czech Republic, 2009.

[120] L. S. Sierakowski and C. A. Coelho, "Bacteria Colony Approaches with Variable Velocity Applied to Path Optimization of Mobile Robots," in *Proceedings of the 18th International Congress of Mechanical Engineering, Ouro Preto*, MG, Brazil, 2005.

[121] C. A. Sierakowski and L. D. S. Coelho, "Study of Two Swarm Intelligence Techniques for Path Planning of Mobile Robots, 16thIFAC World Congress," Prague, 2005.

[122] N. HadiAbbas and F. Mahdi Ali, "Path Planning of an Autonomous Mobile Robot using Directed Artificial Bee Colony Algorithm," *International Journal of Computer Applications*, vol. 96, no. 11, pp. 11–16, 2014.

[123] H. Chen, Y. Zhu, and K. Hu, "Adaptive bacterial foraging optimization," *Abstract and Applied Analysis*, Art. ID 108269, 27 pages, 2011.

[124] X. Yang, *Nature-Inspired Metaheuristic Algorithms*, Luniver Press, United Kingdom, 2 edition, 2010.

[125] D. K. Chaturvedi, "Soft Computing Techniques and Their Applications," in *Mathematical Models, Methods and Applications*, Industrial and Applied Mathematics, pp. 31–40, Springer Singapore, Singapore, 2015.

[126] N. B. Hui and D. K. Pratihar, "A comparative study on some navigation schemes of a real robot tackling moving obstacles," *Robotics and Computer-Integrated Manufacturing*, vol. 25, no. 4-5, pp. 810–828, 2009.

[127] F. J. Pelletier, "Review of Mathematics of Fuzzy Logics," *The Bulleting ofn Symbolic Logic*, vol. 6, no. 3, pp. 342–346, 2000.

[128] L. A. Zadeh, "Fuzzy sets," *Information and Computation*, vol. 8, pp. 338–353, 1965.

[129] R. Ramanathan, "Service-Driven Computing," in *Handbook of Research on Architectural Trends in Service-Driven Computing*, Advances in Systems Analysis, Software Engineering, and High Performance Computing, pp. 1–25, IGI Global, 2014.

[130] F. Cupertino, V. Giordano, D. Naso, and L. Delfine, "Fuzzy control of a mobile robot," *IEEE Robotics and Automation Magazine*, vol. 13, no. 4, pp. 74–81, 2006.

[131] H. A. Hagras, "A hierarchical type-2 fuzzy logic control architecture for autonomous mobile robots," *IEEE Transactions on Fuzzy Systems*, vol. 12, no. 4, pp. 524–539, 2004.

[132] K. P. Valavanis, L. Doitsidis, M. Long, and R. R. Murphy, "A case study of fuzzy-logic-based robot navigation," *IEEE Robotics and Automation Magazine*, vol. 13, no. 3, pp. 93–107, 2006.

[133] Rui Wang, Ming Wang, Yong Guan, and Xiaojuan Li, "Modeling and Analysis of the Obstacle-Avoidance Strategies for a Mobile Robot in a Dynamic Environment," *Mathematical Problems in Engineering*, vol. 2015, pp. 1–11, 2015.

[134] J. Vaščák and M. Palá, "Adaptation of fuzzy cognitive maps for navigation purposes by migration algorithms," *International Journal of Artificial Intelligence*, vol. 8, pp. 20–37, 2012.

[135] M. J. Er and C. Deng, "Online tuning of fuzzy inference systems using dynamic fuzzy Q-learning," *IEEE Transactions on Systems, Man, and Cybernetics, Part B: Cybernetics*, vol. 34, no. 3, pp. 1478–1489, 2004.

[136] X. Yang, M. Moallem, and R. V. Patel, "A layered goal-oriented fuzzy motion planning strategy for mobile robot navigation," *IEEE Transactions on Systems, Man, and Cybernetics, Part B: Cybernetics*, vol. 35, no. 6, pp. 1214–1224, 2005.

[137] F. Abdessemed, K. Benmahammed, and E. Monacelli, "A fuzzy-based reactive controller for a non-holonomic mobile robot," *Robotics and Autonomous Systems*, vol. 47, no. 1, pp. 31–46, 2004.

[138] M. Shayestegan and M. H. Marhaban, "Mobile robot safe navigation in unknown environment," in *Proceedings of the 2012 IEEE International Conference on Control System, Computing and Engineering, ICCSCE 2012*, pp. 44–49, Malaysia, November 2012.

[139] X. Li and B.-J. Choi, "Design of obstacle avoidance system for mobile robot using fuzzy logic systems," *International Journal of Smart Home*, vol. 7, no. 3, pp. 321–328, 2013.

[140] Y. Chang, R. Huang, and Y. Chang, "A Simple Fuzzy Motion Planning Strategy for Autonomous Mobile Robots," in *Proceedings of the IECON 2007 - 33rd Annual Conference of the IEEE Industrial Electronics Society*, pp. 477–482, Taipei, Taiwan, November 2007.

[141] D. R. Parhi and M. K. Singh, "Intelligent fuzzy interface technique for the control of an autonomous mobile robot," *Proceedings of the Institution of Mechanical Engineers, Part C: Journal of Mechanical Engineering Science*, vol. 222, no. 11, pp. 2281–2292, 2008.

[142] Y. Qin, F. Da-Wei, H. Wang, and W. Li, "Fuzzy Control Obstacle Avoidance for Mobile Robot," *Journal of Hebei University of Technology*, 2007.

[143] H.-H. Lin and C.-C. Tsai, "Laser pose estimation and tracking using fuzzy extended information filtering for an autonomous mobile robot," *Journal of Intelligent & Robotic Systems*, vol. 53, no. 2, pp. 119–143, 2008.

[144] S. Parasuraman, "Sensor fusion for mobile robot navigation: Fuzzy Associative Memory," in *Proceedings of the 2nd International Symposium on Robotics and Intelligent Sensors 2012, IRIS 2012*, pp. 251–256, Malaysia, September 2012.

[145] M. Dupre and S. X. Yang, "Two-stage fuzzy logic-based controller for mobile robot navigation," in *Proceedings of the 2006 IEEE International Conference on Mechatronics and Automation, ICMA 2006*, pp. 745–750, China, June 2006.

[146] S. Nurmaini and S. Z. M. Hashim, "An embedded fuzzy type-2 controller based sensor behavior for mobile robot," in *Proceedings of the 8th International Conference on Intelligent Systems Design and Applications, ISDA 2008*, pp. 29–34, Taiwan, November 2008.

[147] M. F. Selekwa, D. D. Dunlap, D. Shi, and E. G. Collins Jr., "Robot navigation in very cluttered environments by preference-based fuzzy behaviors," *Robotics and Autonomous Systems*, vol. 56, no. 3, pp. 231–246, 2008.

[148] U. Farooq, K. M. Hasan, A. Raza, M. Amar, S. Khan, and S. Javaid, "A low cost microcontroller implementation of fuzzy logic based hurdle avoidance controller for a mobile robot," in *Proceedings of the 2010 3rd IEEE International Conference on Computer Science and Information Technology (ICCSIT 2010)*, pp. 480–485, Chengdu, China, July 2010.

[149] Fahmizal and C.-H. Kuo, "Trajectory and heading tracking of a mecanum wheeled robot using fuzzy logic control," in *Proceedings of the 2016 International Conference on Instrumentation, Control, and Automation, ICA 2016*, pp. 54–59, Indonesia, August 2016.

[150] S. H. A. Mohammad, M. A. Jeffril, and N. Sariff, "Mobile robot obstacle avoidance by using Fuzzy Logic technique," in *Proceedings of the 2013 IEEE 3rd International Conference on System Engineering and Technology, ICSET 2013*, pp. 331–335, Malaysia, August 2013.

[151] J. Berisha and A. Shala, "Application of Fuzzy Logic Controller for Obstacle," in *Proceedings of the 5th Mediterranean Conf. on Embedded Comp*, pp. 200–205, Bar, Montenegro, 2016.

[152] M. M. Almasri, K. M. Elleithy, and A. M. Alajlan, "Development of efficient obstacle avoidance and line following mobile robot with the integration of fuzzy logic system in static and dynamic environments," in *Proceedings of the IEEE Long Island Systems, Applications and Technology Conference, LISAT 2016*, USA.

[153] M. I. Ibrahim, N. Sariff, J. Johari, and N. Buniyamin, "Mobile robot obstacle avoidance in various type of static environments using fuzzy logic approach," in *Proceedings of the 2014 International Conference on Electrical, Electronics and System Engineering (ICEESE)*, pp. 83–88, Kuala Lumpur, December 2014.

[154] A. Al-Mayyahi and W. Wang, "Fuzzy inference approach for autonomous ground vehicle navigation in dynamic environment," in *Proceedings of the 2014 IEEE International Conference on Control System, Computing and Engineering (ICCSCE)*, pp. 29–34, Penang, Malaysia, November 2014.

[155] U. Farooq, M. Amar, E. Ul Haq, M. U. Asad, and H. M. Atiq, "Microcontroller based neural network controlled low cost autonomous vehicle," in *Proceedings of the 2010 The 2nd International Conference on Machine Learning and Computing, ICMLC 2010*, pp. 96–100, India, February 2010.

[156] U. Farooq, M. Amar, K. M. Hasan, M. K. Akhtar, M. U. Asad, and A. Iqbal, "A low cost microcontroller implementation of neural network based hurdle avoidance controller for a car-like robot," in *Proceedings of the 2nd International Conference on Computer and Automation Engineering, ICCAE 2010*, pp. 592–597, Singapore, February 2010.

[157] K.-H. Chi and M.-F. R. Lee, "Obstacle avoidance in mobile robot using neural network," in *Proceedings of the 2011 International Conference on Consumer Electronics, Communications and Networks, CECNet '11*, pp. 5082–5085, April 2011.

[158] V. Ganapathy, S. C. Yun, and H. K. Joe, "Neural Q-Iearning controller for mobile robot," in *Proceedings of the IEEEIASME Int. Conf. on Advanced Intelligent Mechatronics*, pp. 863–868, 2009.

[159] S. J. Yoo, "Adaptive neural tracking and obstacle avoidance of uncertain mobile robots with unknown skidding and slipping," *Information Sciences*, vol. 238, pp. 176–189, 2013.

[160] J. Qiao, Z. Hou, and X. Ruan, "Application of reinforcement learning based on neural network to dynamic obstacle avoidance," in *Proceedings of the 2008 IEEE International Conference on Information and Automation, ICIA 2008*, pp. 784–788, China, June 2008.

[161] A. Chohra, C. Benmehrez, and A. Farah, "Neural Navigation Approach for Intelligent Autonomous Vehicles (IAV) in Partially Structured Environments," *Applied Intelligence*, vol. 8, no. 3, pp. 219–233, 1998.

[162] R. Glasius, A. Komoda, and S. C. A. M. Gielen, "Neural network dynamics for path planning and obstacle avoidance," *Neural Networks*, vol. 8, no. 1, pp. 125–133, 1995.

[163] V. Ganapathy, C. Y. Soh, and J. Ng, "Fuzzy and neural controllers for acute obstacle avoidance in mobile robot navigation," in *Proceedings of the IEEE/ASME International Conference on Advanced Intelligent Mechatronics (AIM '09)*, pp. 1236–1241, July 2009.

[164] . Guo-Sheng Yang, . Er-Kui Chen, and . Cheng-Wan An, "Mobile robot navigation using neural Q-learning," in *Proceedings of the 2004 International Conference on Machine Learning and Cybernetics*, pp. 48–52, Shanghai, China.

[165] C. Li, J. Zhang, and Y. Li, "Application of Artificial Neural Network Based on Q-learning for Mobile Robot Path Planning," in *Proceedings of the 2006 IEEE International Conference on Information Acquisition*, pp. 978–982, Veihai, China, August 2006.

[166] K. Maček, I. Petrović, and N. Perić, "A reinforcement learning approach to obstacle avoidance of mobile robots," in *Proceedings of the 7th International Workshop on Advanced Motion Control (AMC '02)*, pp. 462–466, IEEE, Maribor, Slovenia, July 2002.

[167] R. Akkaya, O. Aydogdu, and S. Canan, "An ANN Based NARX GPS/DR System for Mobile Robot Positioning and Obstacle Avoidance," *Journal of Automation and Control*, vol. 1, no. 1, p. 13, 2013.

[168] P. K. Panigrahi, S. Ghosh, and D. R. Parhi, "A novel intelligent mobile robot navigation technique for avoiding obstacles using RBF neural network," in *Proceedings of the 2014 International Conference on Control, Instrumentation, Energy and Communication (CIEC)*, pp. 1–6, Calcutta, India, January 2014.

[169] U. Farooq, M. Amar, M. U. Asad, A. Hanif, and S. O. SaleH, "Design and Implementation of Neural Network Based," vol. 6, pp. 83–89, 2014.

[170] A. Medina-Santiago, J. L. Camas-Anzueto, J. A. Vazquez-Feijoo, H. R. Hernández-De León, and R. Mota-Grajales, "Neural control system in obstacle avoidance in mobile robots using ultrasonic sensors," *Journal of Applied Research and Technology*, vol. 12, no. 1, pp. 104–110, 2014.

[171] U. A. Syed, F. Kunwar, and M. Iqbal, "Guided autowave pulse coupled neural network (GAPCNN) based real time path planning and an obstacle avoidance scheme for mobile robots," *Robotics and Autonomous Systems*, vol. 62, no. 4, pp. 474–486, 2014.

[172] H. Qu, S. X. Yang, A. R. Willms, and Z. Yi, "Real-time robot path planning based on a modified pulse-coupled neural network model," *IEEE Transactions on Neural Networks and Learning Systems*, vol. 20, no. 11, pp. 1724–1739, 2009.

[173] M. Pfeiffer, M. Schaeuble, J. Nieto, R. Siegwart, and C. Cadena, "From perception to decision: A data-driven approach to end-to-end motion planning for autonomous ground robots," in *Proceedings of the 2017 IEEE International Conference on Robotics and Automation, ICRA 2017*, pp. 1527–1533, Singapore, June 2017.

[174] D. FOGEL, "An introduction to genetic algorithms Melanie Mitchell. MIT Press, Cambridge MA, 1996. %0.00 (cloth), 270 pp," *Bulletin of Mathematical Biology*, vol. 59, no. 1, pp. 199–204, 1997.

[175] Tu. Jianping and S. Yang, "Genetic algorithm based path planning for a mobile robot," in *Proceedings of the IEEE International Conference on Robotics and Automation. IEEE ICRA 2003*, pp. 1221–1226, Taipei, Taiwan.

[176] Y. Zhou, L. Zheng, and Y. Li, "An improved genetic algorithm for mobile robotic path planning," in *Proceedings of the 2012 24th Chinese Control and Decision Conference, CCDC 2012*, pp. 3255–3260, China, May 2012.

[177] K. H. Sedighi, K. Ashenayi, T. W. Manikas, R. L. Wainwright, and H.-M. Tai, "Autonomous local path planning for a mobile robot using a genetic algorithm," in *Proceedings of the 2004 Congress on Evolutionary Computation, CEC2004*, pp. 1338–1345, USA, June 2004.

[178] A. Tuncer and M. Yildirim, "Dynamic path planning of mobile robots with improved genetic algorithm," *Computers and Electrical Engineering*, vol. 38, no. 6, pp. 1564–1572, 2012.

[179] I. Chaari, A. Koubaa, H. Bennaceur, S. Trigui, and K. Al-Shalfan, "smartPATH: A hybrid ACO-GA algorithm for robot path planning," in *Proceedings of the 2012 IEEE Congress on Evolutionary Computation (CEC)*, pp. 1–8, Brisbane, Australia, June 2012.

[180] R. S. Lim, H. M. La, and W. Sheng, "A robotic crack inspection and mapping system for bridge deck maintenance," *IEEE Transactions on Automation Science and Engineering*, vol. 11, no. 2, pp. 367–378, 2014.

[181] T. Geisler and T. W. Monikas, "Autonomous robot navigation system using a novel value encoded genetic algorithm," *Midwest Symposium on Circuits and Systems*, vol. 3, pp. III45–III48, 2002.

[182] T. Geisler and T. Manikas, "Autonomous robot navigation system using a novel value encoded genetic algorithm," in *Proceedings of the Midwest Symposium on Circuits and Systems*, pp. III-45–III-48, Tulsa, OK, USA.

[183] P. Calistri, A. Giovannini, and Z. Hubalek, "Epidemiology of West Nile in Europe and in the Mediterranean basin," *The Open Virology Journal*, vol. 4, pp. 29–37, 2010.

[184] Hu. Yanrong, S. Yang, Xu. Li-Zhong, and M. Meng, "A Knowledge Based Genetic Algorithm for Path Planning in Unstructured Mobile Robot Environments," in *Proceedings of the 2004 IEEE International Conference on Robotics and Biomimetics*, pp. 767–772, Shenyang, China.

[185] P. Moscato, *On evolution, search, optimization, genetic algorithms and martial arts: towards memetic algorithms*, Caltech Concurrent Computation Program, California Institute of Technology, Pasadena, 1989.

[186] A. Caponio, G. L. Cascella, F. Neri, N. Salvatore, and M. Sumner, "A fast adaptive memetic algorithm for online and offline control design of PMSM drives," *IEEE Transactions on Systems, Man, and Cybernetics, Part B: Cybernetics*, vol. 37, no. 1, pp. 28–41, 2007.

[187] F. Neri and E. Mininno, "Memetic compact differential evolution for cartesian robot control," *IEEE Computational Intelligence Magazine*, vol. 5, no. 2, pp. 54–65, 2010.

[188] N. Shahidi, H. Esmaeilzadeh, M. Abdollahi, and C. Lucas, "Memetic algorithm based path planning for a mobile robot," in *Proceedings of the int. conf*, pp. 56–59, 2004.

[189] J. Botzheim, Y. Toda, and N. Kubota, "Bacterial memetic algorithm for offline path planning of mobile robots," *Memetic Computing*, vol. 4, no. 1, pp. 73–86, 2012.

[190] J. Botzheim, C. Cabrita, L. T. Kóczy, and A. E. Ruano, "Fuzzy rule extraction by bacterial memetic algorithms," *International Journal of Intelligent Systems*, vol. 24, no. 3, pp. 312–339, 2009.

[191] J. Kennedy and R. Eberhart, "Particle swarm optimization," in *Proceedings of the IEEE International Conference on Neural Networks (ICNN '95)*, vol. 4, pp. 1942–1948, Perth, Western Australia, November-December 1995.

[192] A.-M. Batiha and B. Batiha, "Differential transformation method for a reliable treatment of the nonlinear biochemical reaction model," *Advanced Studies in Biology*, vol. 3, pp. 355–360, 2011.

[193] Y. Tang, Z. Wang, and J.-A. Fang, "Feedback learning particle swarm optimization," *Applied Soft Computing*, vol. 11, no. 8, pp. 4713–4725, 2011.

[194] Y. Tang, Z. Wang, and J. Fang, "Parameters identification of unknown delayed genetic regulatory networks by a switching particle swarm optimization algorithm," *Expert Systems with Applications*, vol. 38, no. 3, pp. 2523–2535, 2011.

[195] Yong Ma, M. Zamirian, Yadong Yang, Yanmin Xu, and Jing Zhang, "Path Planning for Mobile Objects in Four-Dimension Based on Particle Swarm Optimization Method with Penalty Function," *Mathematical Problems in Engineering*, vol. 2013, pp. 1–9, 2013.

[196] M. K. Rath and B. B. V. L. Deepak, "PSO based system architecture for path planning of mobile robot in dynamic environment," in *Proceedings of the Global Conference on Communication Technologies, GCCT 2015*, pp. 797–801, India, April 2015.

[197] Y. Ma, H. Wang, Y. Xie, and M. Guo, "Path planning for multiple mobile robots under double-warehouse," *Information Sciences*, vol. 278, pp. 357–379, 2014.

[198] E. Masehian and D. Sedighizadeh, "Multi-objective PSO- and NPSO-based algorithms for robot path planning," *Advances in Electrical and Computer Engineering*, vol. 10, no. 4, pp. 69–76, 2010.

[199] Y. Zhang, D.-W. Gong, and J.-H. Zhang, "Robot path planning in uncertain environment using multi-objective particle swarm optimization," *Neurocomputing*, vol. 103, pp. 172–185, 2013.

[200] Q. Li, C. Zhang, Y. Xu, and Y. Yin, "Path planning of mobile robots based on specialized genetic algorithm and improved particle swarm optimization," in *Proceedings of the 31st Chinese Control Conference, CCC 2012*, pp. 7204–7209, China, July 2012.

[201] Q. Zhang and S. Li, "A global path planning approach based on particle swarm optimization for a mobile robot," in *Proceedings of the 7th WSEAS Int. Conf. on Robotics, Control and Manufacturing Tech*, pp. 111–222, Hangzhou, China, 2007.

[202] D.-W. Gong, J.-H. Zhang, and Y. Zhang, "Multi-objective particle swarm optimization for robot path planning in environment with danger sources," *Journal of Computers*, vol. 6, no. 8, pp. 1554–1561, 2011.

[203] Y. Wang and X. Yu, "Research for the Robot Path Planning Control Strategy Based on the Immune Particle Swarm Optimization Algorithm," in *Proceedings of the 2012 Second International Conference on Intelligent System Design and Engineering Application (ISDEA)*, pp. 724–727, Sanya, China, January 2012.

[204] S. Doctor, G. K. Venayagamoorthy, and V. G. Gudise, "Optimal PSO for collective robotic search applications," in *Proceedings of the 2004 Congress on Evolutionary Computation, CEC2004*, pp. 1390–1395, USA, June 2004.

[205] L. Smith, G. K. Venayagamoorthy, and P. G. Holloway, "Obstacle avoidance in collective robotic search using particle swarm optimization," in *Proceedings of the in IEEE Swarm Intelligence Symposium*, Indianapolis, USA, 2006.

[206] L. Lu and D. Gong, "Robot path planning in unknown environments using particle swarm optimization," in *Proceedings of the 4th International Conference on Natural Computation (ICNC '08)*, pp. 422–426, IEEE, Jinan, China, October 2008.

[207] R. Bibi, B. S. Chowdhry, and R. A. Shah, "PSO based localization of multiple mobile robots emplying LEGO EV3," in *Proceedings*

of the 2018 International Conference on Computing, Mathematics and Engineering Technologies (iCoMET), pp. 1–5, Sukkur, March 2018.

[208] A. Z. Nasrollahy and H. H. S. Javadi, "Using particle swarm optimization for robot path planning in dynamic environments with moving obstacles and target," in *Proceedings of the UKSim 3rd European Modelling Symposium on Computer Modelling and Simulation, EMS 2009*, pp. 60–65, Greece, November 2009.

[209] . Hua-Qing Min, . Jin-Hui Zhu, and . Xi-Jing Zheng, "Obstacle avoidance with multi-objective optimization by PSO in dynamic environment," in *Proceedings of 2005 International Conference on Machine Learning and Cybernetics*, pp. 2950–2956 Vol. 5, Guangzhou, China, August 2005.

[210] . Yuan-Qing Qin, . De-Bao Sun, . Ning Li, and . Yi-Gang Cen, "Path planning for mobile robot using the particle swarm optimization with mutation operator," in *Proceedings of the 2004 International Conference on Machine Learning and Cybernetics*, pp. 2473–2478, Shanghai, China.

[211] B. Deepak and D. Parhi, "PSO based path planner of an autonomous mobile robot," *Open Computer Science*, vol. 2, no. 2, 2012.

[212] P. Curkovic and B. Jerbic, "Honey-bees optimization algorithm applied to path planning problem," *International Journal of Simulation Modelling*, vol. 6, no. 3, pp. 154–164, 2007.

[213] A. Haj Darwish, A. Joukhadar, M. Kashkash, and J. Lam, "Using the Bees Algorithm for wheeled mobile robot path planning in an indoor dynamic environment," *Cogent Engineering*, vol. 5, no. 1, 2018.

[214] M. Park, J. Jeon, and M. Lee, "Obstacle avoidance for mobile robots using artificial potential field approach with simulated annealing," in *Proceedings of the IEEE International Symposium on Industrial Electronics Proceedings (ISIE '01)*, pp. 1530–1535, June 2001.

[215] H. Martínez-Alfaro and S. Gómez-García, "Mobile robot path planning and tracking using simulated annealing and fuzzy logic control," *Expert Systems with Applications*, vol. 15, no. 3-4, pp. 421–429, 1998.

[216] Q. Zhu, Y. Yan, and Z. Xing, "Robot Path Planning Based on Artificial Potential Field Approach with Simulated Annealing," in *Proceedings of the Sixth International Conference on Intelligent Systems Design and Applications*, pp. 622–627, Jian, China, October 2006.

[217] H. Miao and Y. Tian, "Robot path planning in dynamic environments using a simulated annealing based approach," in *Proceedings of the 2008 10th International Conference on Control, Automation, Robotics and Vision (ICARCV)*, pp. 1253–1258, Hanoi, Vietnam, December 2008.

[218] O. Montiel-Ross, R. Sepúlveda, O. Castillo, and P. Melin, "Ant colony test center for planning autonomous mobile robot navigation," *Computer Applications in Engineering Education*, vol. 21, no. 2, pp. 214–229, 2013.

[219] C.-F. Juang, C.-M. Lu, C. Lo, and C.-Y. Wang, "Ant colony optimization algorithm for fuzzy controller design and its FPGA implementation," *IEEE Transactions on Industrial Electronics*, vol. 55, no. 3, pp. 1453–1462, 2008.

[220] G.-Z. Tan, H. He, and A. Sloman, "Ant colony system algorithm for real-time globally optimal path planning of mobile robots," *Acta Automatica Sinica*, vol. 33, no. 3, pp. 279–285, 2007.

[221] N. A. Vien, N. H. Viet, S. Lee, and T. Chung, "Obstacle Avoidance Path Planning for Mobile Robot Based on Ant-Q Reinforcement Learning Algorithm," in *Advances in Neural Networks – ISNN 2007*, vol. 4491 of *Lecture Notes in Computer Science*, pp. 704–713, Springer Berlin Heidelberg, Berlin, Heidelberg, 2007.

[222] K. Ioannidis, G. C. Sirakoulis, and I. Andreadis, "Cellular ants: A method to create collision free trajectories for a cooperative robot team," *Robotics and Autonomous Systems*, vol. 59, no. 2, pp. 113–127, 2011.

[223] B. R. Fajen and W. H. Warren, "Behavioral dynamics of steering, obstacle avoidance, and route selection," *J. Exp. Psychol. Hum. Percept. Perform*, vol. 29, pp. 343–362, 2003.

[224] P. J. Beek, C. E. Peper, and D. F. Stegeman, "Dynamical models of movement coordination," *Human Movement Science*, vol. 14, no. 4-5, pp. 573–608, 1995.

[225] J. A. S. Kelso, *Coordination Dynamics: Issues and Trends*, Springer-Verlag, Berlin, 2004.

[226] R. Fajen, W. H. Warren, S. Temizer, and L. P. Kaelbling, "A dynamic model of visually-guided steering, obstacle avoidance, and route selection," *Int. J. Comput. Vision*, vol. 54, pp. 13–34, 2003.

[227] K. Patanarapeelert, T. D. Frank, R. Friedrich, P. J. Beek, and I. M. Tang, "Theoretical analysis of destabilization resonances in time-delayed stochastic second-order dynamical systems and some implications for human motor control," *Physical Review E: Statistical, Nonlinear, and Soft Matter Physics*, vol. 73, no. 2, Article ID 021901, 2006.

[228] G. Schöner, M. Dose, and C. Engels, "Dynamics of behavior: theory and applications for autonomous robot architectures," *Robotics & Autonomous Systems*, vol. 16, no. 2-4, pp. 213–245, 1995.

[229] T. D. Frank, M. J. Richardson, S. M. Lopresti-Goodman, and M. T. Turvey, "Order parameter dynamics of body-scaled hysteresis and mode transitions in grasping behavior," *Journal of Biological Physics*, vol. 35, no. 2, pp. 127–147, 2009.

[230] H. Haken, *Synergetic Cmputers and Cognition: A Top-Down Approach to Neural Nets*, Springer, Berlin, Germany, 1991.

[231] T. D. Frank, T. D. Gifford, and S. Chiangga, "Minimalistic model for navigation of mobile robots around obstacles based on complex-number calculus and inspired by human navigation behavior," *Mathematics and Computers in Simulation*, vol. 97, pp. 108–122, 2014.

[232] J. Yang, P. Chen, H.-J. Rong, and B. Chen, "Least mean p-power extreme learning machine for obstacle avoidance of a mobile robot," in *Proceedings of the 2016 International Joint Conference on Neural Networks, IJCNN 2016*, pp. 1968–1976, Canada, July 2016.

[233] Q. Zheng and F. Li, "A survey of scheduling optimization algorithms studies on automated storage and retrieval systems (AS/RSs)," in *Proceedings of the 2010 3rd International Conference on Advanced Computer Theory and Engineering (ICACTE 2010)*, pp. V1-369–V1-372, Chengdu, China, August 2010.

[234] M. A. P. Garcia, O. Montiel, O. Castillo, R. Sepúlveda, and P. Melin, "Path planning for autonomous mobile robot navigation with ant colony optimization and fuzzy cost function evaluation," *Applied Soft Computing*, vol. 9, no. 3, pp. 1102–1110, 2009.

[235] F. Neri and C. Cotta, "Memetic algorithms and memetic computing optimization: a literature review," *Swarm and Evolutionary Computation*, vol. 2, pp. 1–14, 2012.

[236] J. M. Hall, M. K. Lee, B. Newman et al., "Linkage of early-onset familial breast cancer to chromosome 17q21," *Science*, vol. 250, no. 4988, pp. 1684–1689, 1990.

[237] A. L. Bolaji, A. T. Khader, M. A. Al-Betar, and M. A. Awadallah, "Artificial bee colony algorithm, its variants and applications: a survey," *Journal of Theoretical and Applied Information Technology*, vol. 47, no. 2, pp. 434–459, 2013.

[238] S. Haykin, *Neural Networks: A Comprehensive Foundation*, Prentice Hall, New Jersey, 1999.

[239] H. Burchardt and R. Salomon, "Implementation of path planning using genetic algorithms on mobile robots," in *Proceedings of the IEEE Congress on Evolutionary Computation (CEC '06)*, pp. 1831–1836, July 2006.

[240] K. Su, Y. Wang, and X. Hu, "Robot Path Planning Based on Random Coding Particle Swarm Optimization," *International Journal of Advanced Computer Science and Applications*, vol. 6, no. 4, 2015.

[241] D. Karaboga, B. Gorkemli, C. Ozturk, and N. Karaboga, "A comprehensive survey: artificial bee colony (ABC) algorithm and applications," *Artificial Intelligence Review*, vol. 42, pp. 21–57, 2014.

[242] L. A. Barroso and U. Hölzle, "The case for energy-proportional computing," *The Computer Journal*, vol. 40, no. 12, pp. 33–37, 2007.

[243] A. Abraham, "Adaptation of Fuzzy Inference System Using Neural Learning," in *Fuzzy Systems Engineering*, vol. 181 of *Studies in Fuzziness and Soft Computing*, pp. 53–83, Springer Berlin Heidelberg, Berlin, Heidelberg, 2005.

[244] O. Obe and I. Dumitrache, "Adaptive neuro-fuzzy controler with genetic training for mobile robot control," *International Journal of Computers, Communications & Control*, vol. 7, no. 1, pp. 135–146, 2012.

[245] A. Zhu and S. X. Yang, "Neurofuzzy-Based Approach to Mobile Robot Navigation in Unknown Environments," *IEEE Transactions on Systems, Man and Cybernetics, Part C (Applications and Reviews)*, vol. 37, no. 4, pp. 610–621, 2007.

[246] C.-F. Juang and Y.-W. Tsao, "A type-2 self-organizing neural fuzzy system and its FPGA implementation," *IEEE Transactions on Systems, Man, and Cybernetics, Part B: Cybernetics*, vol. 38, no. 6, pp. 1537–1548, 2008.

[247] S. Dutta, "Obstacle avoidance of mobile robot using PSO-based neuro fuzzy technique," vol. 2, pp. 301–304, 2010.

[248] A. Garcia-Cerezo, A. Mandow, and M. Lopez-Baldan, "Fuzzy modelling operator navigation behaviors," in *Proceedings of the 6th International Fuzzy Systems Conference*, pp. 1339–1345, Barcelona, Spain.

[249] M. Maeda, Y. Maeda, and S. Murakami, "Fuzzy drive control of an autonomous mobile robot," *Fuzzy Sets and Systems*, vol. 39, no. 2, pp. 195–204, 1991.

[250] C.-S. Chiu and K.-Y. Lian, "Hybrid fuzzy model-based control of nonholonomic systems: A unified viewpoint," *IEEE Transactions on Fuzzy Systems*, vol. 16, no. 1, pp. 85–96, 2008.

[251] C.-L. Hwang and L.-J. Chang, "Internet-based smart-space navigation of a car-like wheeled robot using fuzzy-neural adaptive control," *IEEE Transactions on Fuzzy Systems*, vol. 16, no. 5, pp. 1271–1284, 2008.

[252] P. Rusu, E. M. Petriu, T. E. Whalen, A. Cornell, and H. J. W. Spoelder, "Behavior-based neuro-fuzzy controller for mobile robot navigation," *IEEE Transactions on Instrumentation and Measurement*, vol. 52, no. 4, pp. 1335–1340, 2003.

[253] A. Chatterjee, K. Pulasinghe, K. Watanabe, and K. Izumi, "A particle-swarm-optimized fuzzy-neural network for voice-controlled robot systems," *IEEE Transactions on Industrial Electronics*, vol. 52, no. 6, pp. 1478–1489, 2005.

[254] C.-F. Juang and Y.-C. Chang, "Evolutionary-group-based particle-swarm-optimized fuzzy controller with application to mobile-robot navigation in unknown environments," *IEEE Transactions on Fuzzy Systems*, vol. 19, no. 2, pp. 379–392, 2011.

[255] K. W. Schmidt and Y. S. Boutalis, "Fuzzy discrete event systems for multiobjective control: Framework and application to mobile robot navigation," *IEEE Transactions on Fuzzy Systems*, vol. 20, no. 5, pp. 910–922, 2012.

[256] S. Nurmaini, S. Zaiton, and D. Norhayati, "An embedded interval type-2 neuro-fuzzy controller for mobile robot navigation," in *Proceedings of the 2009 IEEE International Conference on Systems, Man and Cybernetics, SMC 2009*, pp. 4315–4321, USA, October 2009.

[257] S. Nurmaini and S. Z. Mohd. Hashim, "Motion planning in unknown environment using an interval fuzzy type-2 and neural network classifier," in *Proceedings of the 2009 IEEE International Conference on Computational Intelligence for Measurement Systems and Applications, CIMSA 2009*, pp. 50–55, China, May 2009.

[258] A. AbuBaker, "A novel mobile robot navigation system using neuro-fuzzy rule-based optimization technique," *Research Journal of Applied Sciences, Engineering & Technology*, vol. 4, no. 15, pp. 2577–2583, 2012.

[259] S. Kundu, R. Parhi, and B. B. V. L. Deepak, "Fuzzy-Neuro based Navigational Strategy for Mobile Robot," vol. 3, p. 1, 2012.

[260] M. A. Jeffril and N. Sariff, "The integration of fuzzy logic and artificial neural network methods for mobile robot obstacle avoidance in a static environment," in *Proceedings of the 2013 IEEE 3rd International Conference on System Engineering and Technology, ICSET 2013*, pp. 325–330, Malaysia, August 2013.

[261] C. Chen and P. Richardson, "Mobile robot obstacle avoidance using short memory: A dynamic recurrent neuro-fuzzy approach," *Transactions of the Institute of Measurement and Control*, vol. 34, no. 2-3, pp. 148–164, 2012.

[262] M. Algabri, H. Mathkour, and H. Ramdane, "Mobile robot navigation and obstacle-avoidance using ANFIS in unknown environment," *International Journal of Computer Applications*, vol. 91, no. 14, pp. 36–41, 2014.

[263] Z. Laouici, M. A. Mami, and M. F. Khelfi, "Hybrid Method for the Navigation of Mobile Robot Using Fuzzy Logic and Spiking Neural Networks," *International Journal of Intelligent Systems and Applications*, vol. 6, no. 12, pp. 1–9, 2014.

[264] S. M. Kumar, D. R. Parhi, and P. J. Kumar, "ANFIS approach for navigation of mobile robots," in *Proceedings of the ARTCom 2009 - International Conference on Advances in Recent Technologies in Communication and Computing*, pp. 727–731, India, October 2009.

[265] A. Pandey, "Multiple Mobile Robots Navigation and Obstacle Avoidance Using Minimum Rule Based ANFIS Network Controller in the Cluttered Environment," *International Journal of Advanced Robotics and Automation*, vol. 1, no. 1, pp. 1–11, 2016.

[266] R. Zhao and H.-K. Lee, "Fuzzy-based path planning for multiple mobile robots in unknown dynamic environment," *Journal of Electrical Engineering & Technology*, vol. 12, no. 2, pp. 918–925, 2017.

[267] J. K. Pothal and D. R. Parhi, "Navigation of multiple mobile robots in a highly clutter terrains using adaptive neuro-fuzzy inference system," *Robotics and Autonomous Systems*, vol. 72, pp. 48–58, 2015.

[268] R. M. F. Alves and C. R. Lopes, "Obstacle avoidance for mobile robots: A Hybrid Intelligent System based on Fuzzy Logic and

Artificial Neural Network," in *Proceedings of the 2016 IEEE International Conference on Fuzzy Systems, FUZZ-IEEE 2016*, pp. 1038–1043, Canada, July 2016.

[269] Awatef Aouf, Lotfi Boussaid, and Anis Sakly, "TLBO-Based Adaptive Neurofuzzy Controller for Mobile Robot Navigation in a Strange Environment," *Computational Intelligence and Neuroscience*, vol. 2018, pp. 1–8, 2018.

[270] R. Braunstingl, P. Sanz, and J. M. Ezkerra, "Fuzzy logic wall following of a Mobile Robot based on the Concept of General Perception," in *Proceedings of the in 7th Int. Conf. on Advanced Robotics, Sant Feliu De Guixols*, Spain, 1995.

[271] C.-H. Hsu and C.-F. Juang, "Evolutionary robot wall-following control using type-2 fuzzy controller with species-DE-activated continuous ACO," *IEEE Transactions on Fuzzy Systems*, vol. 21, no. 1, pp. 100–112, 2013.

[272] N. Baklouti, R. John, and A. M. Alimi, "Interval Type-2 Fuzzy Logic Control of Mobile Robots," *Journal of Intelligent Learning Systems and Applications*, vol. 04, no. 04, pp. 291–302, 2012.

[273] K. Al-Mutib, M. Faisal, M. Alsulaiman, F. Abdessemed, H. Ramdane, and M. Bencherif, "Obstacle avoidance using wall-following strategy for indoor mobile robots," in *Proceedings of the 2nd IEEE International Symposium on Robotics and Manufacturing Automation, ROMA 2016*, Malaysia, September 2016.

[274] M. Hank and M. Haddad, "A hybrid approach for autonomous navigation of mobile robots in partially-known environments," *Robotics and Autonomous Systems*, vol. 86, pp. 113–127, 2016.

[275] T. Takagi and M. Sugeno, "Fuzzy identification of systems and its applications tomodeling and control," *IEEE Transactions on Systems, Man, and Cybernetics, Part B: Cybernetics*, vol. SMC-15, no. 2, pp. 116–132, 1985.

[276] Y. Zhou and M. Er, "Self-Learning in Obstacle Avoidance of a Mobile Robot via Dynamic Self-Generated Fuzzy Q-Learning," in *Proceedings of the 2006 International Conference on Computational Inteligence for Modelling Control and Automation and International Conference on Intelligent Agents Web Technologies and International Commerce (CIMCA'06)*, pp. 116-116, Sydney, Australia, November 2006.

[277] L. Jouffe, "Fuzzy inference system learning by reinforcement methods," *IEEE Transactions on Systems, Man, and Cybernetics, Part C: Applications and Reviews*, vol. 28, no. 3, pp. 338–355, 1998.

[278] C.-F. Juang, "Combination of online clustering and Q-Value based GA for reinforcement fuzzy system design," *IEEE Transactions on Fuzzy Systems*, vol. 13, no. 3, pp. 289–302, 2005.

[279] C.-F. Juang and C.-H. Hsu, "Reinforcement ant optimized fuzzy controller for mobile-robot wall-following control," *IEEE Transactions on Industrial Electronics*, vol. 56, no. 10, pp. 3931–3940, 2009.

[280] Jong-Wook Park, Hwan-Joo Kwak, Young-Chang Kang, and D. W. Kim, "Advanced Fuzzy Potential Field Method for Mobile Robot Obstacle Avoidance," *Computational Intelligence and Neuroscience*, vol. 2016, Article ID 6047906, pp. 1–13, 2016.

[281] C.-L. Lee, C.-J. Lin, and H.-Y. Lin, "Smart robot wall-following control using a sonar behavior-based fuzzy controller in unknown environments," *Smart Science*, vol. 5, no. 3, pp. 160–166, 2017.

[282] Rami Al-Jarrah, Mohammad Al-Jarrah, and Hubert Roth, "A Novel Edge Detection Algorithm for Mobile Robot Path Planning," *Journal of Robotics*, vol. 2018, pp. 1–12, 2018.

[283] C. Purcaru, R. Precup, D. Iercan, L. Fedorovici, and R. David, "Hybrid PSO-GSA robot path planning algorithm in static environments with danger zones," in *Proceedings of the 2013 17th International Conference on System Theory, Control and Computing (ICSTCC)*, pp. 434–439, Sinaia, Romania, October 2013.

[284] P. Fabiani, H.-H. González-Baños, J.-C. Latombe, and D. Lin, "Tracking an unpredictable target among occluding obstacles under localization uncertainties," *Robotics and Autonomous Systems*, vol. 38, no. 1, pp. 31–48, 2002.

[285] P. Marin-Plaza, A. Hussein, D. Martin, and A. de la Escalera, "Global and Local Path Planning Study in a ROS-Based Research Platform for Autonomous Vehicles," *Journal of Advanced Transportation*, vol. 2018, pp. 1–10, 2018.

[286] A. Cherubini and F. Chaumette, "A redundancy-based approach for obstacle avoidance in mobile robot navigation," in *Proceedings of the 2010 IEEE/RSJ International Conference on Intelligent Robots and Systems (IROS 2010)*, pp. 5700–5705, Taipei, October 2010.

[287] A. Cherubini and F. Chaumette, "Visual navigation with obstacle avoidance," in *Proceedings of the 2011 IEEE/RSJ International Conference on Intelligent Robots and Systems (IROS 2011)*, pp. 1593–1598, San Francisco, CA, September 2011.

[288] S. Šegvić, A. Remazeilles, A. Diosi, and F. Chaumette, "A mapping and localization framework for scalable appearance-based navigation," *Computer Vision and Image Understanding*, vol. 113, no. 2, pp. 172–187, 2009.

[289] M. Cherubini, G. Colafrancesco, L. Oriolo, L. Freda, and F. Chaumette, *Comparing appearance-based controllers for nonholonomic navigation from a visual memory, ICRA Workshop on safe navigation in open and dynamic environments: application to autonomous vehicles*, ICRA Workshop on safe navigation in open and dynamic environments, application to autonomous vehicles, 2009.

[290] C. Einhorn, H. J. Schr, Böhme., H. M. Gross, and H. J. Schröter, "A Hybrid Kalman Filter Based Algorithm for Real-Time Visual Obstacle Detection," in *Proceedings of the in Proceedings of the 3rd European Conference on Mobile Robots (ECMR)*, Freiburg, 2007.

[291] M. A. Moreno-Armendariz and H. Calvo, "Visual SLAM and Obstacle Avoidance in Real Time for Mobile Robots Navigation," in *Proceedings of the 2014 International Conference on Mechatronics, Electronics and Automotive Engineering (ICMEAE)*, pp. 44–49, Cuernavaca, Mexico, November 2014.

[292] Y. Zhu, R. Mottaghi, E. Kolve et al., "Target-driven visual navigation in indoor scenes using deep reinforcement learning," in *Proceedings of the 2017 IEEE International Conference on Robotics and Automation, ICRA 2017*, pp. 3357–3364, Singapore, June 2017.

[293] J. Savage, S. Muñoz, M. Matamoros, and R. Osorio, "Obstacle Avoidance Behaviors for Mobile Robots Using Genetic Algorithms and Recurrent Neural Networks," *IFAC Proceedings Volumes*, vol. 46, no. 24, pp. 141–146, 2013.

[294] A. M. Zaki, O. Arafa, and S. I. Amer, "Microcontroller-based mobile robot positioning and obstacle avoidance," *Journal of Electrical Systems and Information Technology*, vol. 1, no. 1, pp. 58–71, 2014.

[295] R. M. Santiago, A. L. De Ocampo, A. T. Ubando, A. A. Bandala, and E. P. Dadios, "Path planning for mobile robots using genetic algorithm and probabilistic roadmap," in *Proceedings of the 2017 IEEE 9th International Conference on Humanoid, Nanotechnology, Information Technology, Communication and Control, Environment, and Management (HNICEM)*, pp. 1–5, Manila, December 2017.

[296] Imen Hassani, Imen Maalej, and Chokri Rekik, "Robot Path Planning with Avoiding Obstacles in Known Environment Using Free Segments and Turning Points Algorithm," *Mathematical Problems in Engineering*, vol. 2018, pp. 1-13, 2018.

[297] C. H. Do and H. Lin, "Differential evolution for optimizing motion planning of mobile robot," in *Proceedings of the 2017 IEEE/SICE International Symposium on System Integration (SII)*, pp. 399–404, Taipei, December 2017.

[298] T. Mercy, E. Hostens, and G. Pipeleers, "Online motion planning for autonomous vehicles in vast environments," in *Proceedings of the 2018 15th International Workshop on Advanced Motion Control (AMC)*, pp. 114–119, Tokyo, March 2018.

[299] Y. Li, R. Cui, Z. Li, and D. Xu, "Neural Network Approximation Based Near-Optimal Motion Planning With Kinodynamic Constraints Using RRT," *IEEE Transactions on Industrial Electronics*, vol. 65, no. 11, pp. 8718–8729, 2018.

18

Automatic Estimation Method for Intersection Saturation Flow Rate based on Video Detector Data

Linhong Wang,[1] Yunhao Wang,[2] and Yiming Bie[1]

[1]*School of Transportation, Jilin University, Changchun 130022, China*
[2]*School of Economics, Northeast Normal University, Changchun 130117, China*

Correspondence should be addressed to Yiming Bie; yimingbie@126.com

Academic Editor: David Z. W. Wang

Saturation flow rate (SFR) is a fundamental parameter to the level of service evaluation, lane capacity calculation, and signal timing plan optimization at signalized intersections. It is affected by a variety of factors including weather conditions, lane width, and the type of the driver. How to accurately estimate the SFR remains one of the most important tasks in traffic engineering. Existing studies generally rely on the field measurement method which requires a large number of people collecting data at the intersection. As a result, the method incurs a high economic cost and cannot adapt to the dynamic change of SFR. In recent years, video detectors have been widely installed at intersections which are capable of recording the time each vehicle passes the stop line, the number plate of each vehicle, and the vehicle type. This paper therefore aims to propose an automatic estimation method for the SFR based on video detector data in order to overcome the limitation of the field measurement method. A prerequisite for estimating the SFR is to recognize the saturation headway. We consider the actual vehicle headway as time series and build an auxiliary regression equation whose parameters are estimated through the ordinary least squares method. We employ the Dickey-Fuller test to verify whether the headways in the time series are saturation headways. An iterative method using quantiles is proposed to filter out abnormal data. The SFR is finally calculated using the average value of saturation headways. To demonstrate the proposed method, we conduct a case study using data from an intersection with three entrance lanes in Qujing city, Yunnan Province, China. The overall estimation process is displayed and the impacts of quantile selection and data duration on the estimation accuracy are analyzed.

1. Introduction

In traffic engineering domain, the saturation flow rate (SFR) of an approaching lane at the intersection is a fundamental parameter for signal timing optimization, capacity calculation, level of service evaluation, and intersection design. As defined by the Highway Capacity Manual (HCM), the SFR is the flow in vehicles per hour that can be accommodated by the lane group assuming that the green phase was displayed 100 percent of the time [1]. The SFR cannot be measured directly and hence needs to be estimated by traffic engineers. While different intersections may have different SFRs, the same intersection may also exhibit a changing SFR because it is affected by a variety of factors including weather conditions, lane width and slope, the type of the driver, and the ratio of heavy vehicles [2, 3].

Given the importance of the SFR, a number of studies have been devoted to its estimation methods. Two methods are provided by HCM. The first method is called the adjustment method. Computations begin with the selection of a base SFR, usually 1900 passenger cars per hour per lane (pcu/h/ln). This value is adjusted for a variety of factors, such as number of lanes in lane group, lane width, heavy vehicles, approach grade, parking, bus blockage, type of area, lane utilization, left turns, right turns, pedestrian, and pedestrian-bicycle. The adjustment formula for each factor is given. However, adjusting the SFR considering such a variety of factors would result in a huge amount of work [4]. HCM therefore provides the second method which is called the field measurement method. Field-measured values of SFR will produce more accurate results than the estimation procedure described in the adjustment method and can be used directly without further adjustment. SFR is the maximum discharge rate during the green time. It is usually achieved after about 10 to 14 s of green, which corresponds to the front axle of the fourth to sixth passenger car crossing the stop line after the beginning of green. The period of saturation flow begins when the front axle of the fourth vehicle in the queue crosses the stop line or reference point and ends when the front axle of the last queued vehicle crosses the stop line. The last queued vehicle may be a vehicle that joined the queue during the green time. The times when the fourth to the last queued vehicle cross the stop line are recorded and used to calculate the average headway per vehicle. The SFR is 3,600 divided by this value. In order to obtain a statistically significant value, a minimum of 15 signal cycles with more than eight vehicles in the initial queue is typically required. An average of the SFR values in individual cycles represents the prevailing local SFR for the surveyed lane [5, 6].

Shanteau [7] proposed a method to estimate SFR based on cumulative curve method. In the cumulative curve figure, the horizontal axis represented time and the vertical axis represented the number of the vehicle crossing the stop line. A diagram was presented to show the average number of vehicles to enter the intersection versus time. Then a straight line was selected to best fit the data. The straight line portion of the plot could be determined by eye (the easiest and usually sufficient way) or by some formalized method such as linear regression. Then the SFR could be determined according to the slope of the straight line. Hossain [8] developed a microsimulation model to depict the traffic flows in developing cities, where the traffic comprised both motorized and nonmotorized vehicles. The saturation flows at signalized intersections were investigated in an aggregate form of vehicles per hour. It was found that the saturation flows were dependent on the approach width, turning proportion, and composition of the traffic mix. A regression model was proposed to estimate the saturation flow from the influencing variables. Shao and Liu [9] found that using the average value of observed queue discharge headways to estimate the saturation headway might lead to underestimate SFR. Based on the field data, normal and log-normal distributions were used to fit the headways. The median value of discharge headways was suggested to estimate the saturation headway. Zhao et al. [10] studied the headway distribution function at signalized intersections using field collected data. The relationship between headway and vehicle position were modeled and the 78th percentile of headway at each position was selected to calibrate the SFR.

The SFR is equal to 3,600 divided by the saturation headway. When the saturation headway is obtained, the SFR is automatically determined. Therefore, a few methods for estimating the saturation headway have been proposed. Tong and Hung [11] proposed a neural network approach to simulate the queued vehicle discharge headway. A computer-based three-layered neural network model was developed for the estimation of discharge headway. Hung et al. [12] found that discharge headway of individual vehicles at different queue position follows the Type I Extreme Value Distribution. They proposed a method to estimate site-specific parameters for this distribution. Murat and Gedizlioglu [13] investigated the vehicle time headways at signalized intersection using the data in Turkey. Seven distribution models were taken into account to fit the headway data and log-normal, Weibull, gamma, or Pearson distribution models performed better than other models. Jin et al. [14] found that distributions of the departure headways at each position (except the first position) in a queue were revealed to approximately follow a certain log-normal distribution. Then they proposed a new car-following model from the driver's perspective to explain why the departure headways follow such a log-normal distribution. Tan et al. [15] compared the departure headway collected from intersections in Beijing and generated by VISSIM and TransModeler. Results showed that slight difference in car-following models and parameters would result in significant difference of the distribution. Yin et al. [16] compared the departure headways in Beijing and Atlanta. Urban traffic flows in Beijing, China, were often hindered by pedestrians at intersections and queuing vehicles often cannot be totally discharged during the green time. It was found that the departure headways of the first and second vehicles and after the 5th vehicle in Beijing were significantly different from those in Atlanta. Tan et al. [17] developed two distribution models to fit startup lost time and effective flow rate, respectively. The dependence of departure headways in three types of simulation software (VISSIM, PARAMICS, and TransModeler) was also examined. Hao and Ma [18] tried to validate the assumption that the mean headway of saturation flow was saturated headway. They investigated the distribution of departure headways based on the field data in Shanghai, China. It was found that shifted log-normal was more suitable to capture the distribution of steady-flow headways, and the mode of steady-flow headway distribution could be considered as theoretical saturation headway and the mean could be considered as practical saturation headway.

While quite some research effort has been made to estimate the intersection SFR, all the aforementioned methods are essentially the field measurement method provided by HCM. These methods require traffic engineers to conduct filed surveys at intersections. The number of vehicles queuing behind the stop line is recorded manually and the times when the fourth to the last queued vehicle cross the stop line are extracted to calculate the saturation headways. Although the field measurement method can estimate the saturation headway and hence the SFR, it has the following two limitations:

(i) High labor and economic costs: as suggested by HCM, two surveyors are needed to estimate the SFR for one entrance lane [1]. While the field measurement method may be feasible when the number of entrance lanes is small, it can hardly be applied in a city where there are hundreds of intersections and hence thousands of entrance lanes. Therefore, to be able to estimate the SFR, a large number of surveyors are required to collect data for each lane resulting in a high economic cost.

(ii) Inability to adapt to the dynamic change in SFR: traffic data collected by the field measurement method and the resulting estimations of the SFR only represent a single or a few typical traffic scenarios. Since the SFR is affected by a number of factors, there can be hundreds of traffic scenarios and, hence, the method can hardly be applied to estimate the SFR under all possible scenarios. Moreover, the real-time SFR is an input parameter for adaptive signal control at intersections. When one of the influencing factors changes (e.g., weather or the ratio of heavy vehicles), the SFR changes accordingly. While it may take at least 30 minutes to send traffic engineers to the site and obtain the adjusted SFR, this obviously cannot fulfill the real-time requirement of adaptive signal control.

In recent years, video detectors have been widely installed at signalized intersections in many cities. They can provide fruitful traffic information to advanced traffic management system, such as traffic volumes of different types of vehicles, volume of pedestrian, vehicle trajectory, and traffic OD (origin-destination) matrix. For example, with the help of video detectors, adaptive traffic control system can be developed to adjust signal timing plans in real time and estimate the queue length behind the stop line. In addition, video detectors can also provide time headways by recording in real time the number plate and time of each vehicle crossing the stop line and the type of the vehicle. The advantages of such an automatic estimation method for the SFR using video detector data include the: (i) vehicle discharge headways during the green phase can be automatically recorded without the need for manual surveys, which reduces the economic cost; (ii) vehicle operating data under different traffic scenarios can be recorded in real time, which fulfills the real-time requirement of traffic control; (iii) the method can be applied to the entire urban road network to automatically estimate the SFR for each entrance lane under each traffic scenario.

However, vehicle discharge headways obtained from video detector data cannot be used directly to estimate the SFR. This is because, as shown in HCM, the SFR can only be estimated using the saturation headways from the queuing vehicles. While the field measurement method allows traffic engineers to identify the queuing vehicles on-site, video detectors only record the time each vehicle crosses the stop line rather than automatically identifying the queuing vehicles. Therefore, further analysis on video detector data is needed.

Different video detectors can record different traffic flow parameters. The video detectors used in this study cover the area surrounding the stop line and record the following parameters: the number plate, speed, direction, and time of each vehicle crossing the stop line, the type of the vehicle, and the number of the entrance lane. Such a technique has already been widely applied at intersections in a number of cities in China. Therefore, the objective of this study is to propose an automatic estimation method for the intersection SFR based on video detector data that can overcome the limitations of the field measurement method (i.e., high labor cost and inability to fulfill the real-time requirement of traffic control). The rest of the paper is organized as follows. Section "Automatic Estimation Method for Saturation Flow Rate" proposes the automatic estimation methodology, the core of which is the automatic recognition of the saturation flow. Section "Case Study" conducts a case study on a field intersection and analyzes the effects of different parameters on the SFR. The last section concludes the paper.

2. Automatic Estimation Method for Saturation Flow Rate

2.1. Recognition of Saturation Flow. As described in the HCM, the period of saturation flow begins when the front axle of the fourth vehicle in the queue crosses the stop line or reference point and ends when the front axle of the last queued vehicle crosses the stop line. The first step of estimating the SFR is to recognize the saturation flow during the green phase, i.e., to recognize the fourth to the last queued vehicles. Since video detectors record the time each vehicle crosses the stop line, it is rather easy to obtain the time the fourth vehicle crosses the stop line. However, it is impossible to tell which vehicle is the last in the queue because video detectors do not record the real-time queue length. Therefore, a new method is needed to recognize the saturation flow.

During the green phase, the queuing vehicles cross the stop line successively. If the traffic flow is saturated, vehicle headways should fluctuate around the saturation headway. Therefore, from a statistics perspective, vehicle headways should fluctuate around a certain constant and the fluctuations are irregular. In this sense, the error series of the vehicle headway series should be independently and identically distributed (IID) with zero mean. Let X_m denote the time series of all sampled saturation headways during the peak period, with observations denoted by $X_m(1), X_m(2), \ldots, X_m(n), \ldots, X_m(N_m)$, where $1 \le n \le N_m$ and N_m is the number of samples. The n-th time headway $X_m(n)$ is therefore expressed as follows:

$$X_m(n) = \mu + e(n) \tag{1a}$$

where μ is the true saturation headway and $e(n)$ is the random error term of the n-th headway. We assume that e follows a normal distribution with zero mean and variance σ_e^2 and that any two of the random errors are IID. Since μ is unknown and to be estimated, the fluctuations of the observed headways around μ are unknown. Thus, σ_e^2 is also unknown.

Since e follows a normal distribution and μ is a constant, the expectation and variance of X_m can be obtained based on the following two equations:

$$E[X_m(n)] = \mu \tag{1b}$$

$$\operatorname{var}[X_m(n)] = \sigma_e^2 \tag{1c}$$

That is, the saturation headway series X_m follows a normal distribution with mean equal to μ and variance equal to σ_e^2.

According to (1a), the (n-1)-th time headway in X_m can be expressed as follows:

$$X_m(n-1) = \mu + e(n-1) \tag{1d}$$

Both (1a) and (1d) contain the real saturation flow rate μ; thus we can subtract (1d) from (1a) to eliminate μ and then obtain the following relation:

$$X_m(n) - X_m(n-1) = e(n) - e(n-1) \tag{2a}$$

TABLE 1: Critical values of the *DF* distribution.

Number of samples	Quantile α 0.01	0.025	0.05	0.10	0.90	0.95	0.975	0.99
25	-2.66	-2.26	-1.95	-1.60	0.92	1.33	1.70	2.16
50	-2.62	-2.25	-1.95	-1.61	0.91	1.31	1.66	2.08
100	-2.60	-2.24	-1.95	-1.62	0.90	1.29	1.64	2.03
250	-2.58	-2.23	-1.95	-1.62	0.89	1.29	1.63	2.01
500	-2.58	-2.23	-1.95	-1.62	0.89	1.28	1.62	2.00
∞	-2.58	-2.23	-1.95	-1.62	0.89	1.28	1.62	2.00

By setting $\Delta e(n) = e(n) - e(n-1)$ and $\Delta X_m(n) = X_m(n) - X_m(n-1)$, (2a) is further simplified:

$$\Delta X_m(n) = \Delta e(n) \tag{2b}$$

Let Δe and ΔX_m denote the time series comprised $\Delta e(n)$ and ΔX_m, respectively. Since any two of the random errors are IID, the expectation and variance of $\Delta e(n)$ are expressed as follows:

$$\begin{aligned} E[\Delta e(n)] &= E[e(n) - e(n-1)] \\ &= E[e(n)] - E[e(n-1)] = 0 \end{aligned} \tag{3a}$$

$$\begin{aligned} \operatorname{var}[\Delta e(n)] &= \operatorname{var}[e(n) - e(n-1)] \\ &= \operatorname{var}[e(n)] + \operatorname{var}[e(n-1)] = 2\sigma_e^2 \end{aligned} \tag{3b}$$

Equations (3a) and (3b) show that Δe follows a normal distribution with zero mean and $2\sigma_e^2$ variance, which provides a first method to test whether the vehicle headways in the time series are saturation headways. Specifically, we first obtain ΔX_m and Δe through the observed X_m and calculate the sample mean and variance. We then perform a hypothesis test of the population mean (i.e., t-test with unknown population variance) by checking whether there is a significant difference between the sample mean and zero. If there is no significant difference, the expectation of Δe is considered zero. While the sample variance of Δe can be calculated, σ_e^2 cannot be obtained through (1a). As a result, we cannot check whether the sample variance of Δe is equal to $2\sigma_e^2$ and the method cannot work properly.

To better test whether the observed time headways are saturation headways, we set $k(n) = \Delta e(n)$ and $\sigma_k^2 = 2\sigma_e^2$ and propose the following auxiliary regression equation:

$$X_m(n) = \rho X_m(n-1) + k(n) \tag{4a}$$

where ρ is a parameter to be estimated.

After (4a) has been developed, the next step is to perform a hypothesis test with the null hypothesis of $\rho = 1$. If the null hypothesis cannot be rejected, the observed headways can be considered as saturation headways. In such condition, $X_m(n) = X_m(n-1) + k(n)$. Because $k(n) = \Delta e(n)$ and $E[\Delta e(n)] = E[k(n)] = 0$ (please refer to (3a)), then we can get $E[X_m(n)] = E[X_m(n-1)]$. Thus, it can be concluded that the headways in X_m are saturation headways.

The parameters ρ and $k(n)$ in (4a) are estimated using the ordinary least squares (OLS) method. The resulting estimations are denoted by $\hat{\rho}$ and $\hat{k}(n)$ as follows:

$$\hat{\rho} = \frac{\sum_{n=2}^{N_m} X_m(n-1) X_m(n)}{\sum_{n=2}^{N_m} [X_m(n-1)]^2} \tag{4b}$$

$$\hat{k}(n) = X_m(n) - \hat{\rho} X_m(n-1) \tag{4c}$$

According to classical regression theory, $(\hat{\rho} - \rho)/SE(\hat{\rho})$ should follow a t distribution with a $(N_m - 2)$ degree of freedom. If $\rho = 1$ is satisfied, $(\hat{\rho} - 1)/SE(\hat{\rho})$ should follow a t distribution with a $(N_m - 2)$ degree of freedom where $SE(\hat{\rho})$ is the standard error of $\hat{\rho}$. A smaller standard error indicates more closeness between the sample statistics and the population parameters. The samples are therefore a better representative of the population which results in more reliable estimations of the population parameters.

$$SE(\hat{\rho}) = \frac{\hat{\sigma}_k^2}{\sum_{n=2}^{N_m} \left[X_m(n-1) - \overline{X}_m\right]^2} \tag{4d}$$

$$\overline{X}_m = \frac{1}{N_m - 1} \sum_{n=2}^{N_m} X_m(n-1) \tag{4e}$$

$$\hat{\sigma}_k^2 = \frac{1}{N_m - 2} \sum_{n=2}^{N_m} \left[\hat{k}(n)\right]^2 \tag{4f}$$

Since (4a) involves both the n-th and (n-1)-th vehicle headways, the characteristics of the constructed statistic for the hypothesis test of ρ may change. It is proved by Dickey and Fuller [19] that when a hypothesis test is conducted on ρ in (4a), the constructed "t" statistic based on the null hypothesis of $\rho = 1$ no longer follows the t distribution. Instead, it follows the *DF* distribution. To distinguish between the two statistics, we construct the following *DF* statistic:

$$DF = \frac{(\hat{\rho} - 1)}{SE(\hat{\rho})} \tag{4g}$$

The constructed *DF* statistic in (4g) follows the *DF* distribution whose critical values are needed during the hypothesis test. Table 1 shows the critical values of the *DF* distribution corresponding to different quantiles.

Table 1 shows that when we perform a two-sided test with a significance level of 5%, the acceptance region lies between -2.25 and 1.66 given that the number of samples is larger than 50. That is, when the *DF* statistic lies within [-2.25, 1.66], the null hypothesis of $\rho = 1$ cannot be rejected. The observed headways can be considered as saturation headways. In statistics, the time series X_m satisfying $X_m(n) = X_m(n-1) + k(n)$ is termed a random walk.

2.2. Exclusion of Abnormal Data. While section "Recognition of Saturation Flow" proposes a method to test whether the observed headways are saturation headways, this section aims to propose a method to filter out abnormal data.

Estimating the SFR usually needs traffic data collected during a peak period. Let t_{start} and t_{end} denote the beginning and ending times of the peak period, respectively. Based on the outputs of video detectors (i.e., the time each vehicle crosses the stop line), we can obtain the time series of the observed headways within $[t_{start}, t_{end}]$, denoted by X_0. The number of samples is N_0 and the n-th sample is $X_0(n)$ where $1 \le n \le N_0$. When we construct X_0, various factors that may affect the headway are not taken into account. As a result, the headways in X_0 may not necessarily be at saturation.

There are three main types of abnormal headways in X_0: (a) due to the red light, the headway may be large between the first vehicle crossing the stop line in current cycle and the last vehicle crossing the stop line in the previous cycle; (b) at the beginning of the green phase, the headways between the first few vehicles crossing the stop line may be large because they are still accelerating; (c) during the green phase, there may be a large headway as a result of the inattention of drivers or the random arrivals of vehicles.

When the calculated headway is larger than or equal to the red time of the signal phase, the (a) type of headways occurs and is excluded from X_0. The time series after the exclusion is denoted by X_1 and the number of samples is N_1. Since X_1 still includes the (b) and (c) types of headways, the next step is to find a threshold θ whereby the headways smaller than θ are considered as saturation headways (i.e., through the *DF* test shown in section "Recognition of Saturation Flow").

Since X_1 is calculated using traffic data collected during a peak period, the number of abnormal samples of types (b) and (c) only accounts for a small proportion of all the samples, and the majority of observed headways are fluctuating around the saturation headway. The mode of the sampled data in X_1 remains around the saturation headway rather than becoming larger due to abnormal data. If the mode exists for X_1 containing abnormal data, the mean value should be larger than the mode; i.e., X_1 is a right-skewed distribution. As a result, using the mean value of X_1 as the threshold θ is inappropriate. Using the mode as the threshold is, however, also inappropriate because, under some circumstances, the mode does not exist for X_1. Moreover, the mode of some headway samples may be smaller than the saturation headway. Therefore, this paper uses sample quantiles to set the threshold θ.

When the number of samples is large, we can use the 60%-80% quantile to set θ which excludes more data from X_1. When the number of samples is small, we can use the 80%-95% quantile as the threshold which excludes fewer data from X_1. The detailed steps of filtering out abnormal data are as follows.

Step 1. Initialize and set $i = 1$.

Step 2. Calculate the headway corresponding to the quantile β in X_i and set as the threshold θ.

Step 3. Exclude data larger than θ from X_i and generate the new time series X_{i+1} (the number of samples becomes N_{i+1}). Conduct a *DF* test on X_{i+1}. If the test passes, terminate the iterative algorithm; otherwise go to Step 4.

Step 4. Set $i = i + 1$ and go to Step 2.

The time series generated at the last iteration is denoted by X_m consisting of N_m samples. The algorithmic steps show that each iteration reduces the number of samples by a factor of β ($\beta < 1$) and that the headway corresponding to the quantile β changes accordingly. The number of samples in X_m, i.e., N_m, is equal to $\beta^{m-1} \times N_1$.

2.3. Calculation of the Saturation Flow Rate. From section "Exclusion of Abnormal Data", we obtain the time series X_m where the headways are denoted by $X_m(1), X_m(2), \ldots, X_m(N_m)$, respectively. Since X_m passes the *DF* test, it contains saturation headways and should follow a normal distribution as proved by (1a), (1b), (1c), and (1d). In this sense, the mean value, median, and mode of X_m are nearly equal.

We use the sample mean value to estimate the true saturation headway μ. The estimation $\hat{\mu}$ is expressed as follows:

$$\hat{\mu} = \frac{1}{N_m} \sum_{n=1}^{N_m} X_m(n) \tag{5a}$$

By taking the expectation on both sides of (5a) and plugging it into (1b), we obtain $E[\hat{\mu}] = \mu$. Therefore, $\hat{\mu}$ is an unbiased estimation of μ and the estimation accuracy can be expressed by var$[\hat{\mu}]$ where var$[\hat{\mu}] = \sigma_{\hat{\mu}}^2 / N_m$. $\sigma_{\hat{\mu}}^2$ is the unknown variance of the true saturation headway μ. We can use the variance of X_m, i.e., $S_{\hat{\mu}}^2$, as the estimation of $\sigma_{\hat{\mu}}^2$:

$$S_{\hat{\mu}}^2 = \frac{1}{N_m - 1} \sum_{n=1}^{N_m} \left[X_m(n) - \hat{\mu}\right]^2 \tag{5b}$$

The interval estimation of μ with a significance level of α is as follows:

$$\begin{aligned} \hat{\mu} - t_{\alpha/2}(N_m - 1)\frac{S_{\hat{\mu}}^2}{\sqrt{N_m}} &\le \mu \\ &\le \hat{\mu} + t_{\alpha/2}(N_m - 1)\frac{S_{\hat{\mu}}^2}{\sqrt{N_m}} \end{aligned} \tag{6}$$

where $t_{\alpha/2}(N_m - 1)$ is the critical value of the t distribution with a $(N_m - 1)$ degree of freedom and a significance level of α.

The limit error of the estimated μ, i.e., δ, is expressed as follows:

$$\delta = t_{\alpha/2}(N_m - 1)\frac{S_{\hat{\mu}}^2}{\sqrt{N_m}} \tag{7a}$$

TABLE 2: Statistics of the raw time headways.

Statistical parameters	Lane 1	Lane 2	Lane 3
Number of samples	810	1247	986
Mean	11.029	8.634s	9.526s
Median	3.063s	2.521s	2.874
Mode	2.867s	2.400s	2.704s
Standard deviation	15.292	12.619s	18.835s
Minimum	1.597s	1.259s	1.332s
Maximum	163.006s	145.572s	145.384s

Assuming that the minimum number of samples to ensure that δ is smaller than ε is N_ω, we obtain the following relation:

$$t_{\alpha/2}\left(N_\omega - 1\right)\frac{S_{\hat{\mu}}^2}{\sqrt{N_\omega}} \leq \varepsilon \quad (7b)$$

We can use (7b) to solve the interval for N_ω. The minimum value of N_ω is achieved when the error precision ε is reached. When the number of samples N_m is much larger than N_ω, the estimated SFR using the mean value of the samples can achieve a very high precision. If N_m is smaller than N_ω, we can use headway data in peak hours of multiple days (e.g., a week or a month) to increase N_m.

3. Case Study

We apply the proposed method to a signalized intersection in Qujing, China, and present the overall estimation process of the SFR. Qujing is the second largest city in Yunan province, with the main urban area of about 130 square kilometers and a population of 1.0 million. The intersection is made up of Qilin South Road and Wenchang Street and the layout is shown in Figure 1. There are three approaching lanes and three exiting lanes on each leg. To illustrate the estimation of SFR, the three approaching lanes on the north leg are selected as examples: lane 1 dedicated to left turn vehicles, lane 2 for through vehicles, and lane 3 shared by through and right-turn vehicles. The lane width is 3.5 meters. Video detectors are installed for each lane recording data such as the number plate, type, and time of each vehicle crossing the stop line. The cycle length of the signal timing during the morning peak is 180 seconds. The green time, red time, and amber time for lane 1 are 19 seconds, 158 seconds, and 3 seconds, respectively. The green time, red time, and amber time for lanes 2 and 3 are the same because they are controlled by the same signal phase, which are 36 seconds, 141 seconds, and 3 seconds, respectively.

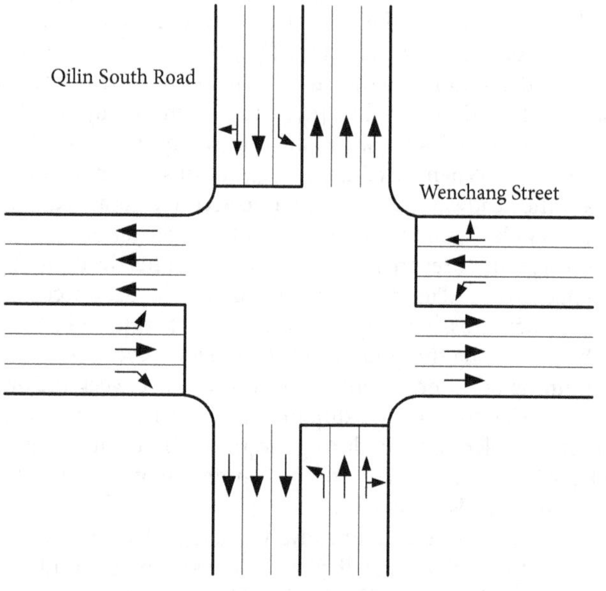

FIGURE 1: The layout of the survey intersection.

3.1. Data Description. We use data collected during the morning peak (7:00 – 10:00) on October 30, 2017, to estimate the SFRs for the three lanes. The weather was fine without rains. The numbers of raw samples of headways for the three lanes are 810, 1,247, and 986, respectively. Table 2 shows the resulting statistics. It is obvious that the raw samples contain quite a few abnormal data. For example, the maximum headway of lane 2 reaches 145.572 seconds, which actually represents the headway between the first vehicle crossing the stop line in the current cycle and the last vehicle crossing the stop line in the previous cycle.

3.2. Saturation Flow Estimation Using the 80% Quantile. We first use the red time for each lane as the threshold and exclude the (a) type of abnormal data shown in the section "Exclusion of Abnormal Data". That is, we exclude headways larger than 158 seconds and 141 seconds for lane 1 and lanes 2 and 3, respectively. The numbers of samples for the three lanes after the exclusion become 743, 1,183, and 922, respectively. Table 3 shows the resulting statistics. By comparing Table 2 with Table 3, we observe that the mean values of the samples reduce after the (a) type of abnormal data is excluded. However, the differences between the mean values, medians, and modes are still large. The maximum values of the samples for the three lanes reach 10.256 s, 8.884 s, and 9.857 s, respectively.

To exclude the (b) and (c) types of abnormal data, we apply the algorithmic steps proposed in Section "Exclusion of Abnormal Data" with $\beta = 80\%$. We take lane 2 as an example and show the process in Table 4 where H_0 represents the null hypothesis. The samples pass the *DF* test after six

TABLE 3: Statistics of the samples after excluding the (a) type of abnormal data.

Statistical parameters	Lane 1	Lane 2	Lane 3
Number of samples	743	1184	922
Mean	4.643s	3.648s	4.257s
Median	2.937s	2.441s	2.661s
Mode	2.867s	2.400s	2.704s
Standard deviation	7.421s	5.037s	6.633s
Maximum	1.597s	1.259s	1.332s
Minimum	10.256s	8.884s	9.857s

TABLE 4: Applying the iterative method using samples from lane 2.

Number of iteration	Number of samples	*DF* statistic	Reject or not $H_0 : \rho = 1$	80% quantile (θ)
1	1184	-20.77	Reject	3.73s
2	947	-5.23	Reject	2.88s
3	759	-3.63	Reject	2.49s
4	607	-2.82	Reject	2.28s
5	486	-2.27	Reject	2.16s
6	392	-1.83	Do not reject	/

TABLE 5: Saturation headway estimations based on the 80% quantile for the three lanes.

Data	Lane 1	Lane 2	Lane 3
T.N.I.	3	6	5
N.S.	380	392	302
Final *DF* statistic	-1.91	-1.83	-1.87
Mean value of SH(s)	2.421	1.790	2.232
Median of SH (s)	2.404	1.801	2.229
Mode of SH (s)	2.419	1.796	2.224
S.D. of SH (s)	0.544	0.251	0.362
Limit error of SH (s)	0.055	0.025	0.041
I.E. of SH (s)	[2.366, 2.476]	[1.765, 1.815]	[2.191, 2.273]
Estimated SFR (pcu/h)	1487	2011	1613
I.E. of SFR (pcu/h)	[1454, 1522]	[1983, 2039]	[1583, 1643]

* T.N.I. represents total number of iterations; N.S. denotes number of the samples; S.D. represents standard deviation; I.E. denotes interval estimation; SH is the abbreviation of saturation headways.

iterations and the number of samples reduces to 392. The mean value, median, and mode of these samples are 1.790 s, 1.801 s, and 1.796 s, respectively, which are nearly equal. This implies that the 392 samples obtained after six iterations follow a normal distribution and that the value of each sample fluctuates around a certain constant (i.e., the true saturation headway).

We therefore use the mean value of the samples to estimate the saturation headway for lane 2; i.e., $\widehat{\mu}_2 = 1.790$, where the subscript 2 represents the number of the lane. Hence, the SFR of lane 2 is estimated to be $\widehat{\lambda}_2 = 2011$ pcu/h. The standard deviation of the samples is 0.251 s. Given $\alpha = 0.05$ and $t_{\alpha/2}(391) = 1.97$, the limit error of the saturation headway of lane 2 is $\delta_2 = 1.97 \cdot 0.251/\sqrt{392} = 0.025$ s. We are therefore 95% confident that the saturation headway of lane 2 lies within $[\widehat{\mu}_2 - \delta_2, \widehat{\mu}_2 + \delta_2]$, i.e., [1.765 s, 1.815 s]. The corresponding interval estimation of the SFR is [1983 pcu/h, 2039 pcu/h].The method is also applied to lanes 1 and 3 and the results are shown in Table 5. The results of lane 2 are included for the comparison.

3.3. Effect of Quantile Selection on the Estimated Saturation Flow Rate. When filtering out abnormal data in section "Saturation Flow Estimation using the 80% Quantile", we assume $\beta = 80\%$ indicating that the number of samples reduces by 20% during each iteration. In this section, we aim to analyze the effect of different quantiles on the iterative process and the estimated saturation headway.

We choose β to be 60%, 70%, and 90%, respectively, and estimate the corresponding SFRs based on the collected headway samples of lane 2. The results are compared with what we have obtained in section "Saturation Flow Estimation

TABLE 6: SFR estimations based on the 60%, 70%, 80%, and 90% quantiles for lane 2.

Data	$\beta = 60\%$	$\beta = 70\%$	$\beta = 80\%$	$\beta = 90\%$
T.N.I.	3	4	6	12
N.S.	258	407	392	383
Final *DF* statistic	-1.34	-1.89	-1.83	-1.79
Mean value of SH(s)	1.661s	1.804s	1.790s	1.782s
Median of SH (s)	1.720s	1.810s	1.801s	1.800s
Mode of SH (s)	1.920s	1.805s	1.805s	1.805s
S.D. of SH (s)	0.210s	0.256s	0.251s	0.247s
Limit error of SH (s)	0.026s	0.025s	0.025s	0.025s
I.E. of SH (s)	[1.635, 1.687]	[1.779, 1.829]	[1.765, 1.815]	[1.757, 1.807]
Estimated SFR (pcu/h)	2167	1995	2011	2020
I.E. of SFR (pcu/h)	[2133, 2201]	[1968, 2023]	[1983, 2039]	[1992, 2049]

TABLE 7: SFR estimations of lane 2 under different data durations.

Data duration (min)	30	60	90	120	150	180
N.R.S.	166	399	644	855	1054	1247
T.N.I.	3	4	4	5	6	6
N.S.	99	197	315	335	338	392
Final *DF* statistic	-1.47	-1.52	-1.92	-1.82	-1.75	-1.83
Mean value of SH(s)	2.276s	1.977s	1.924s	1.838s	1.788s	1.790s
Median of SH (s)	2.285s	2.040s	1.921s	1.880s	1.800s	1.801s
Mode of SH (s)	2.400s	2.400s	1.796s	1.796s	1.796s	1.796s
S.D. of SH (s)	0.459s	0.354s	0.324s	0.280	0.246	0.251
Limit error of SH (s)	0.091	0.049	0.036	0.030	0.027	0.025
I.E. of SH (s)	[2.185, 2.367]	[1.928, 2.026]	[1.888, 1.960]	[1.808, 1.868]	[1.761, 1.815]	[1.765, 1.815]

*N.R.S. denotes number of raw samples.

using the 80% Quantile", which are further summarized in Table 6.

Table 6 shows that when β is equal to 70%, 80%, or 90%, the mean value, median, and mode of the saturation headways are nearly equal. However, there is a big difference when β is equal to 60% which implies that the saturation headways do not follow a normal distribution. This is because when a small quantile is selected, a large number of samples are excluded by the iterative method which, however, may not all be abnormal data. As a result, the remaining headways tend to be smaller. When excluding the (a) type of abnormal data for lane 2 in Table 3, we are left with 1,184 samples. In Table 6, the numbers of iterations corresponding to β equal to 60%, 70%, 80%, and 90% are 3, 4, 6, and 12, respectively, and the numbers of the remaining samples are 258, 407, 392, and 383, respectively. It is obvious that when β is equal to 60%, a large number of samples are excluded and, hence, a small β is not recommended.

A larger β, however, may result in a larger number of iterations which reduces the real-time capability of the estimation method without increasing its accuracy. Therefore, we suggest β be set at 80%.

3.4. Effect of Data Duration on the Estimated Saturation Flow Rate. While video detectors are capable of continuously recording traffic data for estimating the SFR, excessive data do not necessarily result in higher accuracy. However, the accuracy may reduce if the number of samples is insufficient. This section therefore aims to analyze the effect of data duration on the estimated SFR.

The data used in the two previous subsections were collected during the 7:00-10:00 morning peak on October 30, 2017. We take lane 2 as an example and estimate the SFRs corresponding to 30 min, 60 min, 90 min, 120 min, 150 min, and 180 min of data duration. The estimated SFRs are summarized in Table 7 with β equal to 80%.

Table 7 shows that when the data duration is small, the mean value, median and mode of the saturation headways exhibit a large difference which violates the assumption of a normal distribution. When the data duration increases to 150 min and 180 min, all the three values are nearly equal. Therefore, we suggest using the data duration no less than 150 min to ensure the estimation accuracy and robustness.

4. Conclusion

By using video detector data, this study proposes an automatic estimation method for the intersection SFR consisting of the following three sections: automatic recognition of the saturation flow, exclusion of abnormal data, and calculation of the SFR based on the saturation headway. The following conclusions are made.

(i) Recognition of saturation flow is the key to the estimation of the SFR. In this study, saturation flow is recognized by constructing the *DF* statistic and conducting the *DF* test. We find that the time series of the saturation headways is a random walk in statistics.

(ii) While headways can be obtained through video detector data, there exist quite a few abnormal data that need to be filtered out to ensure the estimation accuracy. The iterative method based on quantiles proposed in this study is able to exclude these abnormal data and maintain a high level of flexibility.

(iii) Both quantile selection and data duration have an impact on the estimation accuracy of the SFR. We suggest using the quantile of 80% and the data duration of no less than 150 min.

Conflicts of Interest

The authors declare that there are no conflicts of interest regarding the publication of this paper.

Acknowledgments

This study is supported by the National Natural Science Foundation of China (nos. 71771062, 71771050, and 71501038), the Postdoctoral Startup Foundation of Heilongjiang Province (LBH-Q15050), and the Opening Foundation of the Intelligent Traffic Engineering Technology Research Center of Yunan Province Universities.

References

[1] Transportation Research Board, "Highway Capacity Manual," *National Research Council*, 2000.

[2] S. Wang and X. Qu, "Rural bus route design problem: Model development and case studies," *KSCE Journal of Civil Engineering*, vol. 19, no. 6, pp. 1892–1896, 2015.

[3] Y. Bie, X. Gong, and Z. Liu, "Time of day intervals partition for bus schedule using GPS data," *Transportation Research Part C: Emerging Technologies*, vol. 60, pp. 443–456, 2015.

[4] Z. Liu, S. Wang, W. Chen, and Y. Zheng, "Willingness to board: a novel concept for modeling queuing up passengers," *Transportation Research Part B: Methodological*, vol. 90, pp. 70–82, 2016.

[5] Z. Liu, S. Wang, B. Zhou, and Q. Cheng, "Robust optimization of distance-based tolls in a network considering stochastic day to day dynamics," *Transportation Research Part C: Emerging Technologies*, vol. 79, pp. 58–72, 2017.

[6] Z. Liu, Y. Yan, X. Qu, and Y. Zhang, "Bus stop-skipping scheme with random travel time," *Transportation Research Part C: Emerging Technologies*, vol. 35, pp. 46–56, 2013.

[7] J. Bared, G. L. Giering, and D. L. Warren, "Safety evaluation of acceleration and deceleration lane lengths," *ITE Journal*, vol. 69, no. 5, pp. 50–54, 1999.

[8] M. Hossain, "Estimation of saturation flow at signalised intersections of developing cities: a micro-simulation modelling approach," *Transportation Research Part A: Policy and Practice*, vol. 35, no. 2, pp. 123–141, 2001.

[9] C. Shao and X. Liu, "Estimation of Saturation Flow Rates at Signalized Intersections," *Discrete Dynamics in Nature and Society*, vol. 2012, pp. 1–9, 2012.

[10] Y. Zhao, W. Zhang, J. Lu, W. Zhang, and Y. Ma, "Estimation of Saturation Flow Rate and Start-Up Lost Time for Signal Timing Based on Headway Distribution," *Discrete Dynamics in Nature and Society*, vol. 2015, pp. 1–7, 2015.

[11] H. Y. Tong and W. T. Hung, "Neural network modeling of vehicle discharge headway at signalized intersection: Model descriptions and results," *Transportation Research Part A: Policy and Practice*, vol. 36, no. 1, pp. 17–40, 2002.

[12] W. T. Hung, F. Tian, and H. Y. Tong, "Discharge headway at signalized intersections in Hong Kong," *Journal of Advanced Transportation*, vol. 37, no. 1, pp. 105–117, 2003.

[13] Y. S. Murat and E. Gedizlioglu, "Investigation of vehicle time headways in Turkey," *Proceedings of the Institution of Civil Engineers: Transport*, vol. 160, no. 2, pp. 73–78, 2007.

[14] X. Jin, Y. Zhang, F. Wang et al., "Departure headways at signalized intersections: a log-normal distribution model approach," *Transportation Research Part C: Emerging Technologies*, vol. 17, no. 3, pp. 318–327, 2009.

[15] J. Tan, L. Li, Z. Li, and Y. Zhang, "A comparison of departure headway at signalized intersection produced by VISSIM and TransModeler," in *Proceedings of the Paper presented at the IEEE International Conference on Vehicular Electronics and Safety*, Beijing, China, July 2011.

[16] S. Yin, Y. Su, C. Wang, D. Yao, L. Li, and Y. Zhang, "Comparison of vehicle departure headways in Beijing and Atlanta," *Tsinghua Science and Technology*, vol. 16, no. 3, pp. 332–336, 2011.

[17] J. Tan, L. Li, Z. Li, and Y. Zhang, "Distribution models for start-up lost time and effective departure flow rate," *Transportation Research Part A: Policy and Practice*, vol. 51, pp. 1–11, 2013.

[18] H. Hao and W. Ma, "Revisiting distribution model of departure headways at signalised intersections," *Transportmetrica B: Transport Dynamics*, vol. 5, no. 1, pp. 1–16, 2017.

[19] D. A. Dickey and W. A. Fuller, "Distribution of the estimators for autoregressive time series with a unit root," *Journal of the American Statistical Association*, vol. 74, no. 366, part 1, pp. 427–431, 1979.

19

Heterogeneous Wireless Sensor Network for Transportation System Applications

K. Selvarajah,[1] C. Shooter,[2] L. Liotti,[3] and A. Tully[1]

[1] *School of Computing Science, Newcastle University, Newcastle NE1 7RU, UK*
[2] *TRW Conekt, Stratford Road, Solihull B90 4AX, UK*
[3] *Process Research, Centro Ricerche FIAT, Strada Torino, 50, 10043 Orbassano (TO), Italy*

Correspondence should be addressed to K. Selvarajah, k.selvarajah@ncl.ac.uk

Academic Editor: Syed R. Rizvi

The important innovations in wireless and digital electronics will support many applications in the areas of safety, environmental and emissions control, driving assistance, diagnostics, and maintenance in the transport domain. The last few years have seen the emergence of many new technologies that can potentially have major impacts on transportation systems. One of these technologies is Wireless Sensor Networks. A wireless sensor device is typically composed of a processing unit, memory, and a radio chip which allows it to communicate wirelessly with other devices within range. The Embedded Middleware in Mobility Applications (EMMA) project delivers a middleware that aims to facilitate the interaction between sensing technologies in transportation systems. This paper outlines our experience in the EMMA project and provides an illustration of the important role that wireless sensor technology can play in future transportation system. The paper discusses our experience of using heterogeneous sensors to develop transportation system applications in the EMMA project and focuses on how cooperation between vehicle and infrastructure can be addressed. It also presents encouraging results obtained from the experiments in investigating the feasibility of utilising wireless sensor in vehicle and vehicle-to-infrastructure communication in real transportation applications.

1. Introduction

A recent study by the UK Government's Office of Science and Innovation, which examined how future intelligent infrastructure would evolve to support transportation over the next 50 years, looked at a range of new technologies, systems, and services that may emerge over that period [1, 2]. One key class of technology that was identified as having a significant role in delivering future intelligence to the transport sector is Wireless Sensor Networks (WSN) and in particular the fusion of fixed and mobile networks to help deliver a safe, sustainable, and robust future transport system based on the better collection of data, its processing and dissemination, and the intelligent use of the data in a fully connected environment. As future intelligent infrastructure will bring together and connect individuals, vehicles and infrastructure through wireless communications, it is critical that robust communication protocols are developed.

Mobile wireless ad-hoc networks (MANETs) are self-organising networks where nodes exchange data without the need for an underlying infrastructure [3]. MANETs have attracted extraordinary attention from the research community in recent years including in real transport applications. In the road transport domain, schemes which are fully infrastructureless and those which use a combination of fixed (infrastructure) devices and mobile devices fitted to vehicles and other moving objects are of significant interest to the transport community as they have the potential to deliver a "connected environment" where individuals, vehicles, and infrastructure can coexist and cooperate, thus delivering more knowledge about the transport environment, the state of the network, and who indeed is travelling or wishes to travel [4]. This may offer benefits in terms of real-time management, optimisation of transport systems, intelligent design, and the use of such systems for innovative road charging and possibly carbon trading schemes as well as through the CVHS (Cooperative Vehicle and Highway Systems) for safety and control applications. Within the vehicle, the devices may provide wireless connection to various Information and Communications Technologies (ICT)

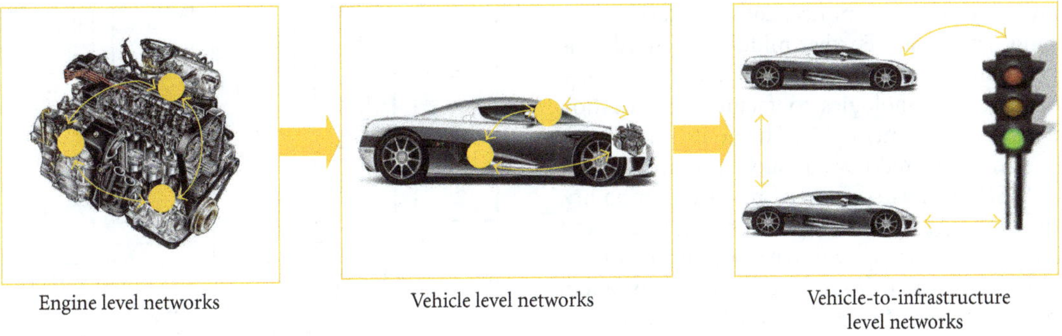

FIGURE 1: EMMA hierarchical network.

components in the vehicle and connect with sensors and other devices within the engine management system [5]. Advances in wireless sensor networking techniques which offer tiny, low power and MEMS (Micro-Electro Mechanical Systems) integrated devices for sensing and networking will exploit the possibility of vehicle-to-vehicle and vehicle-to-infrastructure communications [6].

In this paper, wireless sensor network applications in the transport systems and using middleware to integrate heterogeneous Wireless Cooperative Objects (WICOs) are discussed. Section 2 describes the EMMA project and its hierarchical approach and communication technologies. An overview of wireless sensor networks middleware and components of the EMMA middleware is presented in Section 3. Section 4 describes the different hardware platforms which are used as wireless cooperating objects in the prototype applications. Three applications for each hierarchical level and an inter-hierarchical level application are given in Section 5. Section 5 also presents encouraging results obtained from the experiments in investigating the feasibility of utilising EMMA middleware in real transport system applications. Conclusions are then presented in Section 6.

2. The EMMA Project

The EMMA project (Embedded Middleware in Mobility Applications project) is partly funded by the European Commission under the Information Society Technologies (IST) Priority of the 6th Framework Programme. The EMMA project was committed to deliver a middleware platform and a development environment which facilitates the design and implementation of embedded software for cooperative sensing objects [7, 8]. The EMMA network architecture (Figure 1) can be considered at three levels: within an engine level, at a vehicle level, and at the supra-vehicle level. Recently, many wireless sensor network applications have been developed for a variety of applications including transport monitoring and control. However, there are still numerous challenges to be overcome if wireless sensor devices are to communicate with each other in an intelligent, cost-effective, and reliable way.

EMMA communication networks at the supra-vehicle level can be considered as mobile wireless sensor networks while vehicle level networks and engine level networks can be considered as static wireless sensor networks. The current wireless sensor networks employ conventional technologies to interact with other devices in the network, and many companies and organizations are developing various wireless communication interface and protocols for sensors. Bluetooth is currently the most widely used automotive wireless technology for in-vehicle communication and Wi-Fi is used for vehicle to vehicle communication by several pilot research projects such as the Car2Car consortium [9]. ZigBee technology is able to provide the interconnection of low power wireless sensors within vehicles and vehicle to infrastructure. The ZigBee standard has evolved since its original release in 2004 and it is a low cost low power wireless networking standard for sensors and control devices. ZigBee provides network speed of up to 250 kbps and is expected to be largely used in typical wireless sensor network applications where high data rates are not required [10, 11].

The EMMA project needed to discover which communication technologies are more suitable and how the networks are formed by WICOs from different levels. ZigBee, Bluetooth and Wi-Fi have been designed for short-range wireless applications with low power solutions and could be used at the EMMA infrastructure level. ZigBee can accommodate larger numbers of devices than Bluetooth. On the other hand, Bluetooth offers high bandwidth with relatively high throughput. EMMA network applications do not require high data rate communication technology as it is based on data exchange. ZigBee provides 250 kbps data rate and is expected to be enough for the EMMA sensor network applications. Notably, ZigBee uses low overhead data transmission and requires low system resources which are vitally important factors for embedded wireless sensor networks. Also mesh networking features in ZigBee technology allow devices to extend coverage and optimize radio resources. These features show that ZigBee is a suitable communication technology for the EMMA project applications.

3. Middleware for Wireless Sensor Networks

The term middleware refers to the software layer between the operating system and the applications. A middleware layer seeks primarily to hide the underlying network environment complexity by insulating applications from explicit protocol handling, disjoint memories, data replication, network faults, and parallelism. Further middleware masks the heterogeneity of computer architectures, operating systems, and communication technologies to facilitate application programming and management [12]. The design and development of a successful middleware should address many challenges in WSN such as scarcity of resources, mobility, heterogeneity, data aggregation, quality of services, and security. Several middleware systems have been proposed for WSN but each addresses a different part of the problem space. Notable middleware for sensor networks are Impala, Mate, TinyDB, TinyCubus, TinyLime, and MiLAN [13]. Most of these middleware are built on top of TinyOS [14] which is an open source operating system mainly designed for wireless sensor networks. The middleware can be classified as service-centric middleware and data-centric middleware. Service centric middleware is driven by commands while data centric middleware is driven by data.

Service-centric middleware is described as a well-defined and self-contained function that does not depend on the context or the state of other services. Such a service is executed by explicitly calling it. After the completion of the service, a response is returned. This type of middleware is the principally used paradigm in traditional distributed systems, either with a procedural abstraction or based on object-orientation. Data-centric middleware is mostly concerned with the communication of data and provides a small general purpose API to send and receive data. There is no client-server relationship but there is a distinction between data providers and data consumers. The data-centric approach is mainly followed in the area of sensor networks where the naming and type of data play a more important role than the specific device responsible for its processing.

The EMMA Embedded Middleware Platform (EM2P) is designed to support a range of applications running on different WICOs. EM2P is designed in a modular fashion. The communication adapters form the interface between the communication module and the actual hardware drivers. The communication module has a generic part and two specialised parts for message and data-centric communication. The security add-on is configurable via the middleware API, but is actually used in the communication module. The same applies for the data connector. Installation and the configuration and monitoring module use message communication and are, therefore, built on top of the message communication. A general communication abstraction is used by the synchronisation module. The main components of the EM2P is shown in Figure 2.

The middleware abstracts from the underlying communication technology by providing a high-level addressing mechanism and the communication functions do not imply a specific communication technology. The middleware converts the local representation of a data value to its network representation and vice versa when sending or receiving data. Messages can be sent directly to a specific WICO by knowing only its EMMA WICO address. An application can register for the reception of messages. A call back function is called when a message is received. The content of a message is completely controlled by the application and no data conversion is done by the middleware. Therefore, the application has to assure that the receiver understands the message contents. EM2P uses publish/subscribe communication, request/response communication, and data connector functionalities.

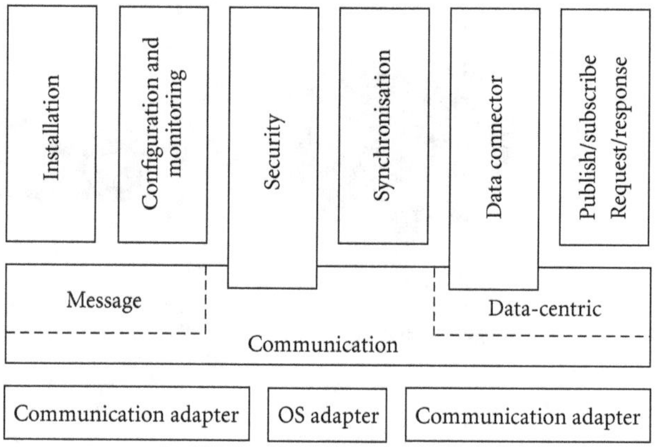

FIGURE 2: Components of EMMA Middleware.

4. EMMA Wireless Cooperative Objects

The following sections explain three different platforms which are used in EMMA project prototype applications: Commercially available Crossbow MicaZ and TelosB and an off-the-shelf Xilinx ML403 FPGA board. C-based multi-threaded NanoQplus [15] operating system is used in MicaZ and TelosB motes while a Linux-based Qplus [15] operating system is used in Xilinx ML403 FPGA. These devices with sensors, actuators, and related software are called WIreless Cooperating Objects (WICOs) which may be heterogeneous, but nevertheless able to cooperate together to achieve specific goals [8].

4.1. Supra-Vehicle Level WICO. Smartdust (or mote) is a new concept for wireless sensor networks which offers tiny, low-power, and MEMS integrated devices for sensing and networking. It is interesting for the low-power sensing technologies and also the low-power communication and networking capability which it has demonstrated. Fundamentally, it provides a convenient and economic means of gathering and disseminating environmental and other useful information in the transport domain. The existence of a ZigBee-based networking capability between the motes and other devices will benefit many applications in the transportation domain. The motes have sensors attached to them to monitor the physical environment in some way. These sensors can be built directly onto the mote or can come as daughter boards which can be connected in to the motes main mother board. Initial studies suggest environmental monitoring, vehicle-to-vehicle, vehicle-to-infrastructure, and infrastructure to infrastructure applications may exist for motes in the transportation domain. The vital application of the devices is beginning to be tested in the road vehicle environment. Even though a range of mote platforms are

FIGURE 3: MicaZ WICO.

FIGURE 4: TelosB WICO.

available in the market, Crossbow MicaZ [16] family motes had been chosen for EMMA project as it features sensing and networking capabilities with low-power consumption using ZigBee as communication protocol. Figure 3 shows a Crossbow MicaZ mote.

The MicaZ is a family of the Crossbow Mica motes where the radio transceiver uses the Chipcon CC2420 IEEE 802.15.4 (ZigBee) compliant chipset. This allows the MicaZ to communicate with other ZigBee compliant equipment. The software stack includes a MicaZ mote-specific layer with ZigBee support and platform device drivers, as well as a network layer for topology establishment and single/multihop routing features. It is mainly used for research and development of low power wireless sensor network applications. The MicaZ mote platform is built around the Atmel AtMega128L processor which is capable of running at 7.37 MHz. The MicaZ motes have 128 Kbytes of program memory, 512 Kbytes of flash data logger memory, and 4 Kbytes of SRAM. Power is provided by two AA batteries, and the devices have a battery life of roughly one year depending on the application (very low duty cycle assumed). Sensor boards can be attached through a surface mount 51 pin connector, Inter-IC (I2C), Digital Input Output (DIO), Universal Asynchronous Receiver Transmitter (UART), and a multiplexed address/data bus.

4.2. Engine Level WICO. In the Engine level application, the Crossbow TelosB [16] mote platform was chosen for the EMMA project prototype applications. Compared to the MicaZ mote, the TelosB mote has higher processing power which is required to implement with a multitasking approach applications: to run several threads of the middleware, the acquisition task, and keeping low latencies in the communication among the Engine WICOs. The high ADC resolution is necessary to satisfy one of the needs of the engine application that is the acquisition of an accurate analogue signal. Less important but good features are the presence of a USB connection, the on board sensors, and LEDs for demonstration and development purposes.

The Crossbow TelosB (Figure 4) mote is a commercially available mote platform with Chipcon CC2420 IEEE 802.15.4 (ZigBee) compliant chipset with integrated onboard antenna. This allows the TelosB mote to communicate with other ZigBee compliant equipment such as MicaZ. It is also mainly used for research and development of low-power wireless sensor network applications. The TelosB mote platform is built around the TI MSP430 (a 16-bit microcontroller) which is capable of running at 8 MHz. The TelosB mote has 48 Kbytes of program memory, 1024 Kbytes of flash data logger memory, and 10 Kbytes of SRAM. Power is provided by two AA batteries, and the devices have a battery life of roughly 1 year depending on the application (very low duty cycle assumed). The TelosB mote has 12 bit ADC resolution, while the MicaZ mote has 10 bit resolution. Sensor boards can be attached through Inter-IC (I2C), Digital Input Output (DIO), Universal Asynchronous Receiver Transmitter (UART), and SPI.

4.3. Vehicle Level WICO. At the vehicle level, there is a need for introducing wireless communication to existing sensing technologies. In addition, it may be necessary to increase the processing power and memory space in order to ensure the EMMA middleware runs seamlessly without compromising the performance of each sensor. It is important to keep the low-power communications available in the other hardware alternatives used but it may be necessary for the units themselves to be capable of running much more complex algorithms. The hardware chosen for EMMA project for this level of WICO, therefore, reflects that extra processing power needed. The vehicle level WICO consists of two elements. Firstly, an off-the-shelf Xilinx ML403 FPGA [17] board is used as the foundation of the system. This FPGA contains a powerful Power PC microprocessor. Secondly, the functionality in this board is extended using a custom, built daughter board. This daughter board contains a number of different hardware devices required by the project including a 12 V automotive power supply, a CAN [18] port for interfacing to automotive ECUs and 2 further RS232 ports which are used to send and receive data over ZigBee and from the other devices as appropriate (e.g., GPS). Figure 5 shows Xilinx ML403 FPGA with TRW Conekt interface board.

5. Prototype Applications

This section provides four different prototype applications, their implementation, and related experiments. The applications are developed here not only to evaluate the overall

Figure 5: Xilinx processor board with TRW Conekt interface board mounted on top.

results of the project, but also to demonstrate the validity of the EMMA approach and its potential applications of heterogeneous wireless networks in transport systems.

5.1. Engine Level Application. The application proposed in the EMMA project is a solution for new engine control architecture (Figure 6), characterised by the integration of new sensors (in-cylinder pressure, oil pressure, and valve lift, not available on current engines) without redesigning the ECU engine. The engine network (Figure 7), wirelessly connected with the ZigBee technology, is so composed:

(i) 4 Cylinder WICOs, sampling the two sensors (a pressure sensor and a valve lift sensor) on each cylinder,

(ii) oil pressure WICO, sampling by a sensor for the measurement of the oil pressure in the oil delivery head,

(iii) ECU WICO, composed by a wireless node connected to the ECU (Electronic Control Unit).

The role of the four Cylinder WICOs is to sample the two analogue channels (connected to the valve lift and the in-cylinder pressure sensors) and calculate the maximum value for the first and the integral over a whole engine cycle for the second. The oil pressure sensor is responsible for sampling the oil pressure sensor. Upon a request from the ECU (simulated by a LabVIEW software on a Laptop), the ECU WICO queries the other engine WICOs, collects the received messages, calculates their latency and validity, and returns the data to the ECU.

5.1.1. Implementation. The Engine WICO consists of two main components: a TelosB mote and hardware adaptation module, necessary to properly interface the connection available on the board to the required conditioning electronics required by the engine sensors. The application has been evaluated by an ad-hoc test bench, where the values acquired from the real sensors are reproduced on the analogue outputs of an acquisition board, reading them from a set of measurement previously acquired from a real Multijet engine. The ECU is implemented by the software running on a Laptop.

All the WICO applications have been programmed by the EM2P (EMMA middleware) functionalities. Several set of experiments have been carried out for data centric paradigm (request/response).

5.1.2. Experimental Results. This engine application has been validated in three different scenarios: the whole test bench has been tested in the laboratory environment and in the environmental chamber, while the engine nodes (without the acquisition board) have been tested directly on a real engine. A set of tests have been performed for the three different environments based on EMMA request/response mechanism (*data-centric*), and for several values of RPM (from 1000 rpm to 6000 rpm). For each test, a set of log files (registering latencies, packet loss, and other WSN-related data) have been collected for offline analysis of the application performance.

Laboratory Environment. The laboratory tests were carried out on a test bench using an engine simulator for the six different rpm values (1000 to 6000 rpm, step 1000). Table 1 summarises the results of each log file generated from each test run. The average and standard deviation of each cylinder's latency has been calculated using only complete packets, that is, where all cylinder WICOs returned a packet flag of 0.

Figure 8 shows that the average latency of all Cylinder WICOs increases slightly in-line with the rpm, whilst the standard deviations are consistent throughout all rpm. These results suggest that it is also possible to make a *data-centric* measurement of the engine sensors for a single engine cycle by simply performing a request with the necessary advance. Packet loss is under 2% for all rpm values, which demonstrates good communication stability for the *data-centric* paradigm.

Engine Environment. The engine environment tests were conducted as before, using a petrol engine to measure the influence of electromagnetic noise and the presence of metal objects in close proximity to the WICOs, in order to model the effects of "real world" conditions. The tests were carried out using an engine simulator for four different rpm values (1000, 2000, 3000, and 6000 rpm), all with the engine switched on. Table 2 summarises the results of each log file generated from each test run. The average and standard deviation of each cylinder's latency has been calculated using only complete packets, that is, where all cylinder WICOs returned a packet flag of 0.

Figure 9 shows that the average WICO latency increases in-line with the rpm for Cylinder 1 WICO, Cylinder 3 WICO, and Cylinder 4 WICO. Although the average latency of Cylinder 2 WICO actually decreases slightly at rpm value 2000 it increases for higher rpm values. The standard deviations of all Cylinder WICO latencies are consistent at all rpm. This further supports the notion that for *data-centric* communication, it is possible to make a measurement of the engine sensors for a single engine cycle by simply performing a request with the necessary advance. Packet loss

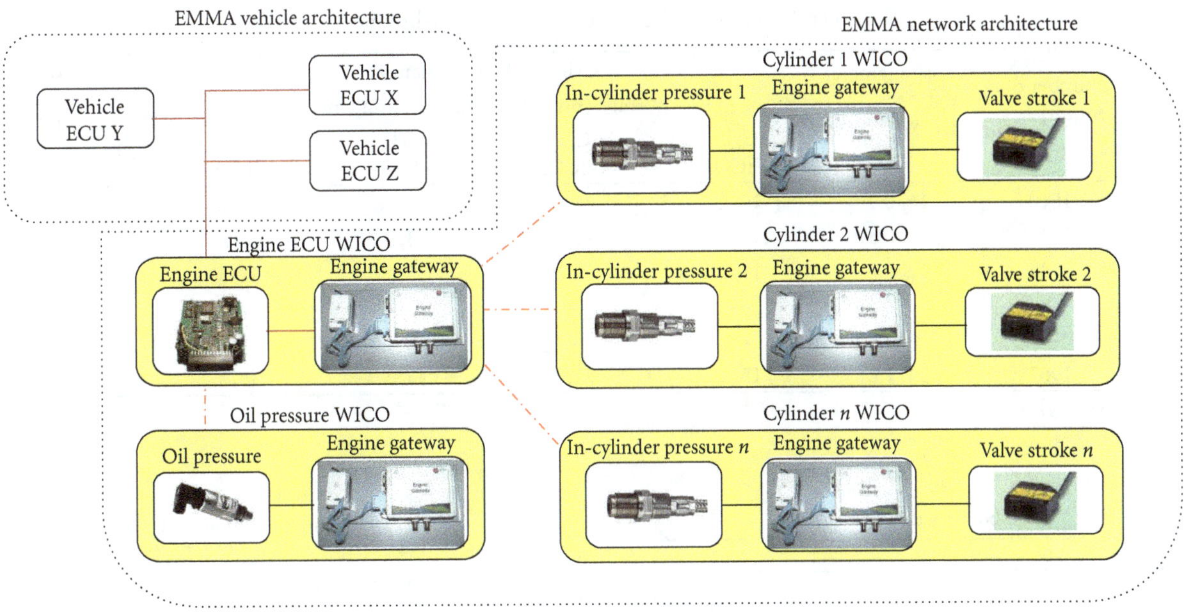

Figure 6: Engine level application diagram.

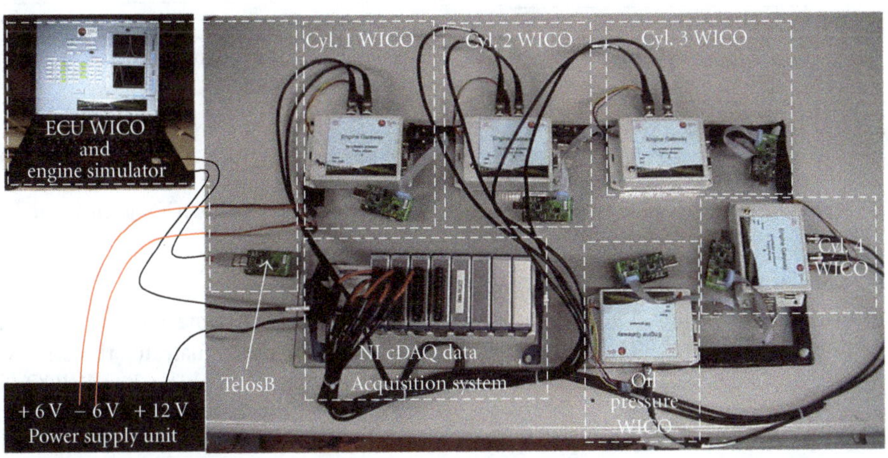

Figure 7: Engine WICOs experimental setup: acquisition board, 6 WICOs and ECU implemented on a Laptop.

Table 1: Summary of test results for laboratory environment.

rpm	Number of enquiries	Lost engine data* (n, %)	Cylinder WICO latency average (ms), standard deviation (ms) Cyl. WICO1	Cyl. WICO2	Cyl. WICO3	Cyl. WICO4	Serial latency average (ms), standard deviation (ms)
1000	5717	88, 1.5	27, 3	4, 2	20, 3	16, 3	119, 7
2000	2603	12, 0.5	33, 3	11, 2	27, 3	23, 4	125, 6
3000	2952	36, 1.2	35, 2	13, 2	29, 3	25, 3	128, 3
4000	7445	72, 1.0	37, 2	15, 2	30, 2	26, 3	128, 6
5000	2631	24, 0.9	37, 2	15, 2	31, 2	27, 3	128, 3
6000	2024	12, 0.6	38, 3	16, 2	31, 3	27, 3	130, 4

*Excluding lost packets from Oil WICO.

TABLE 2: Summary of test results for engine environment.

rpm	Number of enquiries	Lost engine data* (n, %)	Cylinder WICO latency average (ms), standard deviation (ms) Cyl. WICO1	Cyl. WICO2	Cyl. WICO3	Cyl. WICO4	Serial latency average (ms), standard deviation (ms)
1000	1810	16, 0.9	10, 2	26, 3	3, 2	19, 2	124, 5
2000	3731	30, 0.8	11, 2	19, 3	33, 3	28, 3	132, 6
3000	1644	12, 0.7	13, 2	21, 2	35, 2	31, 3	132, 5
4000	—	—	—	—	—	—	—
5000	—	—	—	—	—	—	—
6000	1967	24, 1.2	15, 3	23, 2	37, 3	33, 3	133, 3

*Excluding lost packets from Oil WICO.

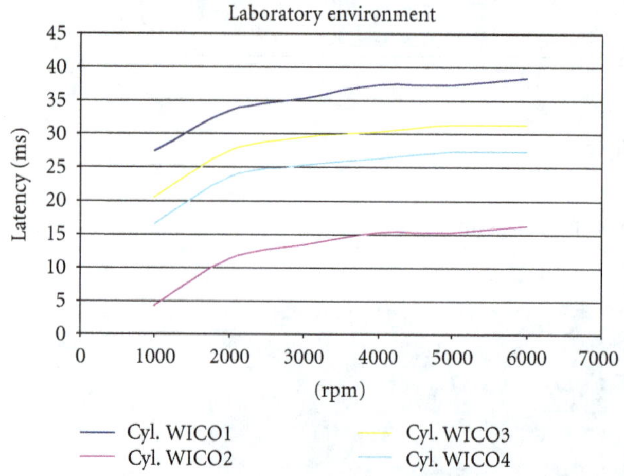

FIGURE 8: Cylinder WICOs latencies in laboratory environment.

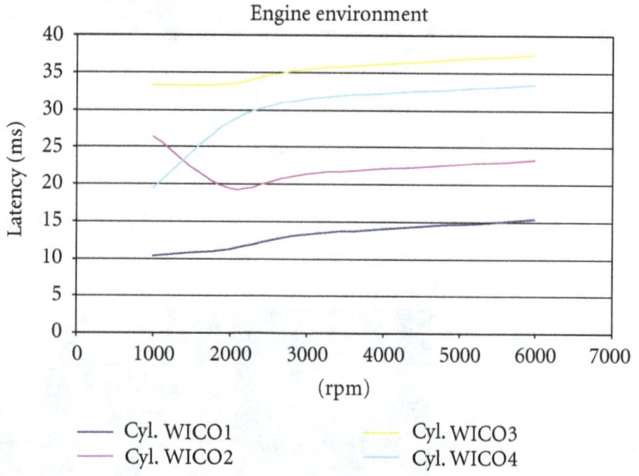

FIGURE 9: Cylinder WICOs latencies in engine environment.

in the engine environment is low, around 1% for all rpm, demonstrating good communication stability for the *data-centric* paradigm.

Environmental Chamber. Tests were carried out using an environmental chamber to control temperature and humidity conditions. The Cylinder 1 and Cylinder 2 WICOs were placed inside the chamber and the internal temperature varied for each test, whilst the remaining WICOs were placed outside the chamber during each test. Test runs were conducted using constant temperatures of −10, 10, 30, 50, and 80°C but unlike the laboratory and engine tests, these tests were only undertaken at two different rpm, 1000 and 3000, so the main variable investigated here was the effect on the WICO performance of the temperature inside the chamber. Tables 3 and 4 summarise these tests.

The results of the *data-centric* environmental chamber tests indicate that there is an effect of higher temperatures on the performance of the WICOs. To confirm this finding, the constant temperature tests were followed by a further variable temperature test. This applied a positive temperature ramp to the WICOs in the environmental chamber from -30 to 90°C using an rpm of 1000.

Table 5 presents the results of the variable temperature test which further illustrates the effects of temperature on the WICO latency values. The WICOs inside the chamber (Cylinder 1 and Cylinder 2) have higher standard deviations than those outside and overall packet loss is also slightly higher than in the other tests, at 3%. Figure 10 illustrates how the latencies of each WICO changed as the temperature ramp was applied, and clearly demonstrates the effect of increasing temperature on the WICO latency.

For the laboratory and engine environments, the latency standard deviations of all Cylinder WICOs were remarkably consistent, which demonstrates robust data communication stability across both environments for the *data-centric* paradigms. For the Environmental Chamber tests, an increase in the average latency of the Cylinder WICOs within the chamber was observed along with an increase in the standard deviations, which indicates a slight loss in communication stability for the *data-centric* paradigms.

The results of the environmental chamber tests clearly illustrate the impact of higher temperatures on WICO latency performance and the packet loss rate. The impact on latency performance is particularly noticeable in the Environmental Chamber tests where the differences in individual

TABLE 3: Summary of test results for environmental chamber at 1000 rpm.

Temp °C	Number of enquiries	Lost engine data* (n, %)	Cylinder WICO latency average(ms), standard deviation (ms)				Serial latency average (ms), standard deviation (ms)
			Cyl. WICO1	Cyl. WICO2	Cyl. WICO3	Cyl. WICO4	
−10	2295	44, 1.9	28, 3	5, 2	22, 3	17, 3	123, 6
10	1529	8, 0.5	9, 3	16, 3	26, 2	5, 2	119, 6
30	1868	16, 0.9	19, 2	30, 5	27, 3	6, 5	122, 7
50	1359	8, 0.6	2, 4	1, 7	8, 6	17, 5	92, 5
80	5652	68, 1.2	45, 4	39, 4	13, 5	6, 5	123, 6

*Excluding lost packets from Oil WICO.

TABLE 4: Summary of test results for environmental chamber at 3000 rpm.

Temp °C	Number of enquiries	Lost engine data* (n, %)	Cylinder WICO latency average (ms), standard deviation (ms)				Serial latency average (ms), standard deviation (ms)
			Cyl. WICO1	Cyl. WICO2	Cyl. WICO3	Cyl. WICO4	
−10	2354	48, 2.0	36, 2	14, 2	30, 2	25, 3	128, 3
10	1608	28, 1.7	19, 2	26, 3	35, 2	14, 2	127, 2
30	2147	24, 1.1	23, 2	32, 4	35, 3	17, 3	127, 2
50	1350	8, 0.6	35, 3	34, 7	26, 6	17, 5	122, 6
80	2236	23, 1.0	28, 3	40, 4	36, 3	14, 3	127, 4

*Excluding lost packets from Oil WICO.

TABLE 5: Summary of test results for temperature ramp within the environmental chamber.

Ramp	Number of enquiries	Lost engine data* (n, %)	WICO latency average (ms), standard deviation (ms)				Serial latency average (ms), standard deviation (ms)
			Cyl. WICO1	Cyl. WICO2	Cyl. WICO3	Cyl. WICO4	
+ve	2225	67, 3.0	34, 15	32, 13	7, 5	14, 5	122, 8

*Excluding lost packets from Oil WICO.

WICO latencies are primarily due to the experimental setup of the tests performed.

During these tests, two Cylinder WICOs were placed in the Environmental Chamber (Cylinder 1 and 2 WICOs) while the others (ECU WICO and Cylinder 3 and 4 WICOs) were placed outside. As the temperature increases, the time to perform the channel sampling and the calculation increases for the TelosB in the Environmental Chamber. For this reason, both the operations are completed first for the Cylinder WICOs 3 and 4, and then for the Cylinder WICOs 1 and 2 at higher temperatures, not in the serial order as expected.

As an example, where the bit rate of the ZigBee channel is around 40 kbps, if the answer length of each Cylinder WICO is around 800 bits, each answer will keep the ZigBee channel busy for at least 800 b/40000 bps (=20 ms). With the overhead of the ZigBee and the overhead introduced by the EMMA middleware, a difference between the latency values of two contiguous Cylinder WICOs of about 40 ms is understandable. This explains why the latency increases for the Cylinder WICOs 1 and 2 in the Environmental Chamber, and therefore the reason why they are the last in the answer message queue.

The reasons for the increase in the loss rate can be attributed to the fact that all electronic devices have an operating point at which they can operate normally. As the temperature increases, I/V (current-voltage) characteristic of a device changes and the behaviour of the device can be different from what would be expected under "normal" conditions. For the WICOs, this change in I/V characteristics means that there is a possibility that a ZigBee transmitter chip behaves erroneously, which in turn produces corrupted data at the chip level. If an error does occur, repeated retransmission occurs on a chip as well as at the MAC level.

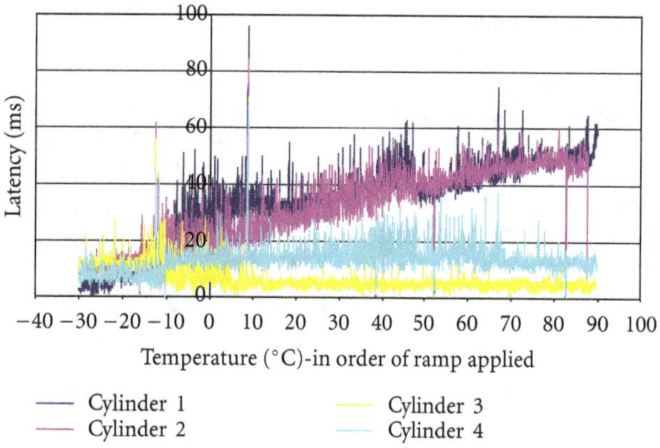

FIGURE 10: Cylinder WICO Latencies, positive temperature ramp applied.

The experimental results of all tests performed on the three scenarios have highlighted some key findings and issues.

(1) The latencies are quite stable with the RPM, but they increase when the temperature reaches 50–60 degrees: this suggests that the TelosB mote requires further hardware design development for an automotive engine level application.

(2) For the message centric version of the application, a loss packet rate lower than the data centric version has been observed: this suggest that in the two paradigms there is a different call-back implementation, or in the thread management in the OS.

(3) For high values of RPM, ZigBee was unable to manage synchronously the connection between the ECU WICO and the 4 Cylinder WICOs: in fact the amount of data transmitted by each Cylinder WICO keep the RF channel busy for a number of millisecond comparable with the engine period, and this does not allow the ECU WICO to collect the data from the Cylinder WICO at the same time.

5.2. Vehicle Level Application. Figure 11 shows the example application that was developed to test the EMMA system. The overall purpose of the vehicle WICO was to provide the absolute position of targets being tracked by the radar. This was achieved by combining the absolute position of the vehicle (based on GPS & vehicle dynamics data) and the relative position of the detected target (using automotive ACC radar).

NMEA 0183 format data was used to transmit the GPS and vehicle data to the Radar WICO within the system. This standardised format was selected to ensure maximum interoperability of the individual WICOs in future setups.

5.2.1. Implementation. All implementations on the Xilinx ML403 FPGA board follow the generic architecture layout described below in Figure 12. Creation of the applications was carried out utilising the Qplus [15] operating system, which is based on Linux and the EMMA middleware.

TABLE 6: Summary of latency results for vehicle-level request/response.

Metric (in ms)	Run 1	Run 2	Run 3
Minimum	66	58	54
Maximum	475	436	452
Average	267	267	273
Standard deviation	400	483	566

GPS WICO. The interface to the Radar WICO was implemented using the publish/subscribe EM2P functionality to send the required GPS NMEA sentences to the main Radar WICO when they have been correctly received from the GPS unit. During development of the host vehicle tracking algorithm, it was discovered that only the GPGGA NMEA sentence was required for the application so all other sentences were filtered out by the GPS WICO.

Vehicle Dynamics WICO. The interface to the Radar WICO was implemented using the request/response EM2P functionality to send the latest vehicle dynamics data to the main Radar WICO when requested. The WICO buffers the received data from the individual sensors and sends the latest full update when requested. The NMEA sentence received from the digital compass was modified to filter out all data except the required heading data. The integrity of all of the data received was checked before it is passed on to the Radar WICO.

Radar WICO. The interfaces to the other two WICOs are defined above and attempt to test as many of the EM2P interface options as possible. The overall scheduling of the application was implemented so that the tracking algorithm runs after a full update of radar data has been received from the ACC radar. The request to the vehicle dynamics data was sent after the algorithm has run so that the next run of the algorithm has the latest vehicle dynamics and radar data. The GPS data was published from the GPS WICO totally asynchronous to the rest of the application.

The host vehicle's GPS position was calculated in the Radar WICO using a Kalman filter algorithm that fuses GPS and Vehicle Dynamics data in order to update position at higher rate and overcome synchronism issues between the sensor WICOs. All of the targets reported from the ACC radar were in relative coordinates relative to the centre of the radar. These coordinates are then converted to the full GPS coordinate system and referenced to the tracked host vehicle position. This data was then available for fusing with other on board sensor data using the GPS co-ordinate system or for passing to the infrastructure for use in traffic management.

5.2.2. Experimental Results. To validate the WICOs in a "real-world" environment for this application it was decided that a test bench environment (Figure 13) would be used to playback a variety of recorded scenarios using data from the Radar, GPS, and Vehicle Dynamics devices to ensure consistency. Three runs of the same data were undertaken,

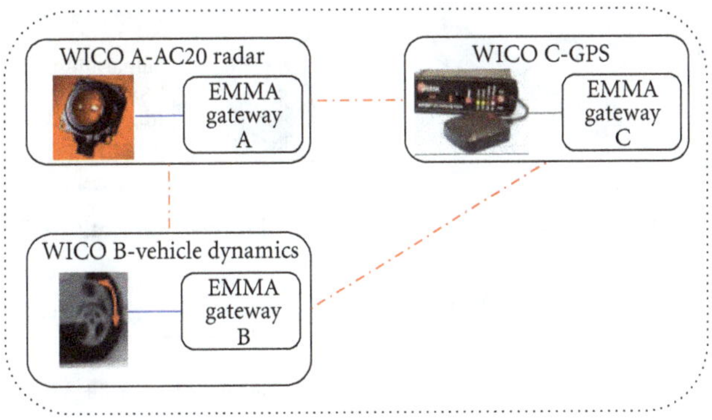

FIGURE 11: Vehicle level application diagram.

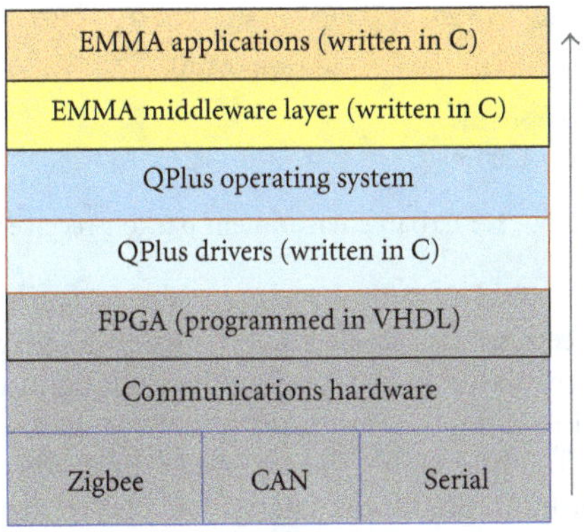

FIGURE 12: Application architecture on Xilinx board.

FIGURE 13: Three WICOs as tested in lab and in vehicle demonstration.

which would allow for the relevant metrics (message latency and lost messages) to be evaluated.

(a) Message Latency

Publish/Subscribe. It was only possible to check the timing of the publish/subscribe mechanism using a timestamp recorded in a log file. This was only set up to record to a one-second level of precision, and the GPS messages were only updated every second anyway. There was very little evidence of latency except in the third run where there was evidence some messages were delayed by one second.

TABLE 7: Summary of lost message results for car-level communications.

WICO-WICO interaction	Missing messages		
	Run 1 (n, %)	Run 2 (n, %)	Run 3 (n, %)
GPS-Radar	15, 12.6	36, 28.1	16, 13.6
VehDyn-Radar	7, 1.2	9, 1.4	6, 1.0

Request/Response. The request/response time was measured internally in microseconds and reported in the log file. The results as shown in Table 6.

(b) Lost WICO to WICO Messages. The results showed in Table 7 a small message loss (around the 1% level for all runs) between the Vehicle Dynamics WICO and the Radar WICO, whereas there was a much higher message loss between the GPS WICO and the Radar, as high as 28.1% during Run 2.

There was minimal message loss between the Vehicle Dynamics WICO and the Radar WICO. It is supposed that these missed messages could have been caused by the Radar WICO simultaneously receiving a successful GPS message. However, comparison of log files proved inconclusive as all records of a missing Vehicle Dynamic message in the Radar logs coincided with a missing GPS message, which suggest a temporary total loss of communications between all WICOs. The only exception to this could be found in 4 records from all the Radar logs which had a missing Vehicle Dynamics message followed one second later by a successful GPS message, but these occurrences were not cyclical in the log files.

For the missing GPS messages, the causes for the higher loss rate were again not clear. A small number of messages in the GPS log files had a fault code which indicates that ZigBee communications were not allowed due to the system not acknowledging that the previous message had successfully

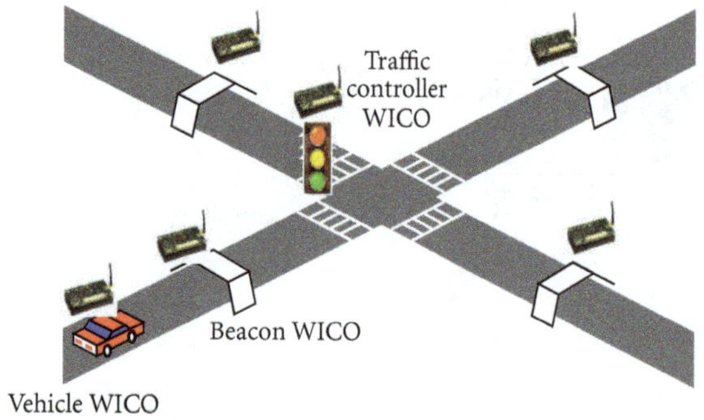

FIGURE 14: Giving priorities for emergency vehicle.

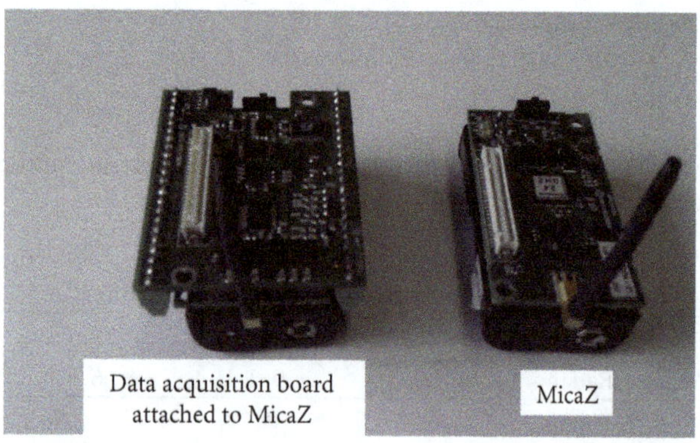

FIGURE 15: Micaz and MDA300.

been sent. Further investigation of these results is required to improve message lost between the Radar and GPS WICOs.

5.3. Supra-Vehicle Level Application. An application was developed in order to demonstrate the benefits of the middleware in priority to emergency vehicles. For that purpose, an emergency vehicle (ambulance, fire engine or police car, etc) would be equipped with a MicaZ WICO which would broadcast a beacon message if it was on an emergency mission (Figure 14). For example, in a busy intersection controlled by traffic lights, emergency vehicles are detected and given priority by regulating the state of the traffic lights.

5.3.1. Implementation. The implementation consists of two elements. The first element consists of a MicaZ WICO. The second element consists of a CITY traffic controller, manufactured by ETRA I+D, Spain. The CITY traffic controller is a well-proven controller that implements advanced capabilities for traffic management and control. A Cross-bow commercial data acquisition board MDA 300 (Figure 15) is used to provide as an interface between MicaZ WICO and CITY traffic controller with a small electric signal adaption stage.

In the demonstration, a MicaZ WICO was connected to a CITY traffic controller (Figure 16) that acted directly by providing information to the regulator about an emergency situation. Another MicaZ WICO was placed in the infrastructure to relay this message to the traffic light regulator.

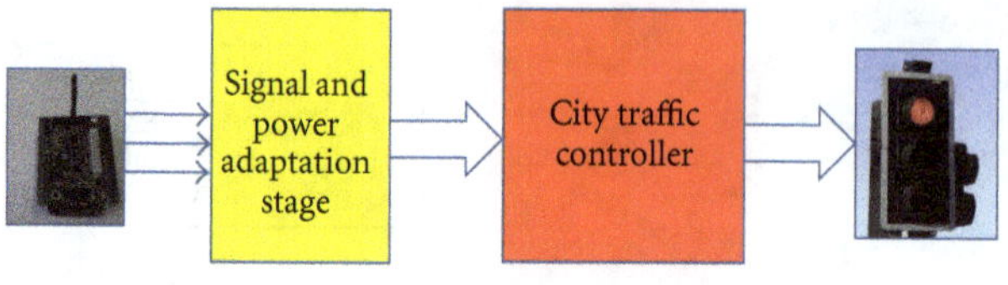

FIGURE 16: MicaZ to CITY Traffic Controller Interface.

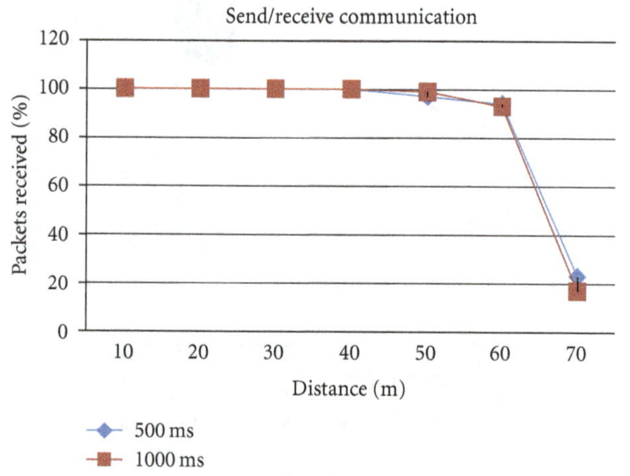

FIGURE 17: Urban environment packets received.

It was placed so far as needed in order to give time to the traffic regulator to change its status taking into account the time lost due to communication mechanisms (publish/subscribe) and other time periods the traffic regulator needs in order to guarantee safety first. The traffic regulator has been programmed to attend to the trigger signal provided by the mote activating an emergency control sequence. The demonstration was successfully carried out in a real road environment in Valencia, Spain.

5.3.2. Experimental Results. Several sets of experiments were carried out with EMMA middleware to evaluate possible use of the MicaZ WICO in the supra-vehicle level application. These experiments evaluated the use of MicaZ WICO with EMMA middleware for the application scenario at the supra-vehicle level. Two MicaZ WICOs (EMMA middleware running on them) were used for data-centric (request/response) and message-centric (send/receive) communication both in urban environment and mobile environment.

(a) Send/Receive Communication

Urban Environment Experiment. This experiment was carried out on Claremont Road, a busy road near Newcastle University. In each scenario, 100 packets were sent for every 500 ms, 1000 ms and those packets were received with another WICO which was connected to a Laptop via MIB 520 programming board. Both WICOs were placed at 1m above the ground. The MicaZ WICO's power level was set to default (NanoQplus power level 31). Each scenario was repeated three times, and calculations were performed offline to determine how many messages were lost at each distance and average values reported in Figure 17.

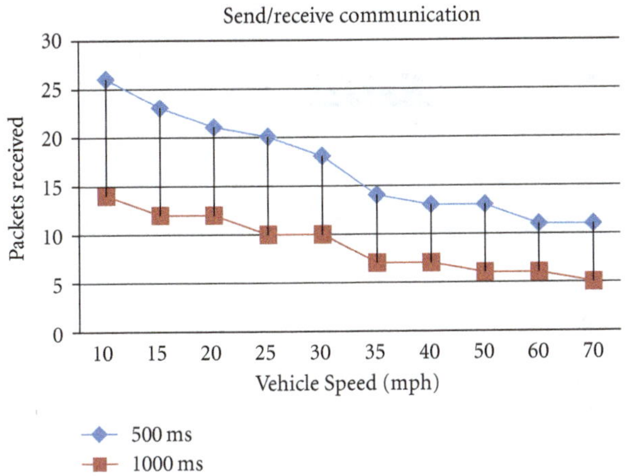

FIGURE 18: Mobile environment packets received.

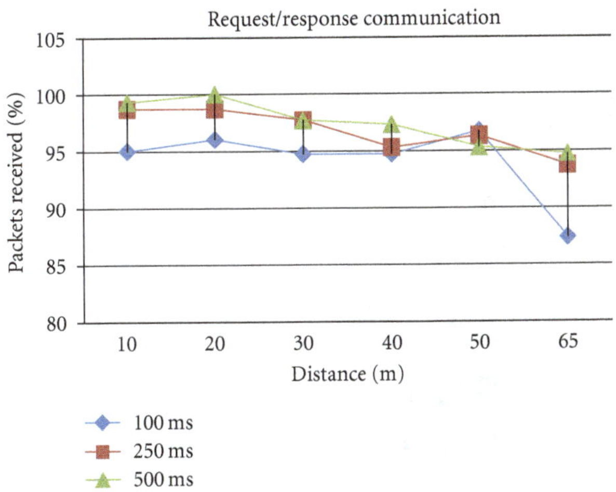

FIGURE 19: Urban environment packets received.

Mobile Environment Experiment. This experiment was carried out on Claremont Road up to 40 mph and on a Motorway near to the Newcastle Airport for higher speeds. The first MicaZ WICO was placed on a roadside stand 1 m from the ground, and the second MicaZ WICO was placed on the middle of the dashboard of a vehicle and connected to a Laptop via MIB 520 programming board. The MicaZ WICO at the road side sent messages periodically (500 ms, 1000 ms) which were received by the MicaZ WICO in the vehicle. Each scenario was repeated three times, and calculations were performed offline to determine how many messages were lost at each distance and average values reported in Figure 18.

In the mobile environment experiment, the received packets decreased with an increase in speed as the WICO is in range for a shorter period of time. This means that communication time window decreased with the increase in vehicle speed. At 70 mph speed, the WICO in the roadside received 5 and 11 packets for sending intervals 1000 ms, 500 ms, respectively. There were no packets lost between the first packet and the last packet received. This Experiment demonstrated that

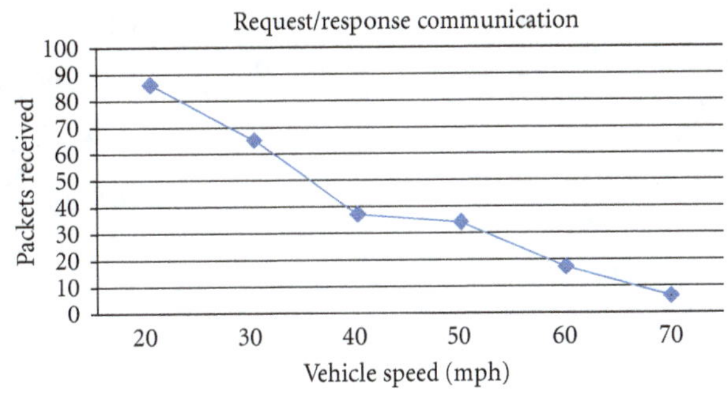

FIGURE 20: Mobile environment packets received.

MicaZ WICO can be used with EMMA middleware communication methods between a fixed infrastructure WICO and also fast moving vehicle-based WICO application. This is an important finding which proves that MicaZ WICOs do not suffer from any Doppler effects at normal motorway (70 mph) speeds.

(b) Request/Response Communication

Urban Environment Experiment. This experiment was carried out on Claremont Road, a busy road near to Newcastle University. Two WICOs were used for request/response communication with a request message transmitted every 100 ms, 250 ms, and 500 ms for different WICO-WICO separations from 10 m to 65 m. In each scenario, response packets were received and recorded. Both WICOs were placed at 1m above the ground. The MicaZ WICOs power level was set to default (NanoQplus power level 31). Each scenario was repeated three times, and calculations were performed offline to determine how many messages were lost at each distance and average values reported in Figure 19.

Mobile Environment Experiment. This experiment was carried out on Claremont Road up to 40 mph and on a Motorway near to Newcastle airport for higher speeds. The first MicaZ WICO was placed on a road side stand 1m from the ground and the second MicaZ WICO was placed on the middle of the dashboard of a vehicle and connected to a Laptop via MIB520 programming board. The MicaZ WICO at the roadside sent messages periodically which were received by the MicaZ WICO in the vehicle. Due to limited access to public roads and for safety reasons, the experiment was conducted only at a packet transmission interval of 100 ms. The experiment was repeated three times, and calculations were performed offline to determine how many messages were received at each distance and average values reported in Figure 20.

The urban environment experiment shows that packets can be received without any packet lost up to 45 m distance. The percentage of packets lost increases above 45 m distance in both cases. In the mobile environment experiment, the received packets decrease with the speed increases as the WICO is in range for a shorter period of time. This means that communication time window is decreasing with the

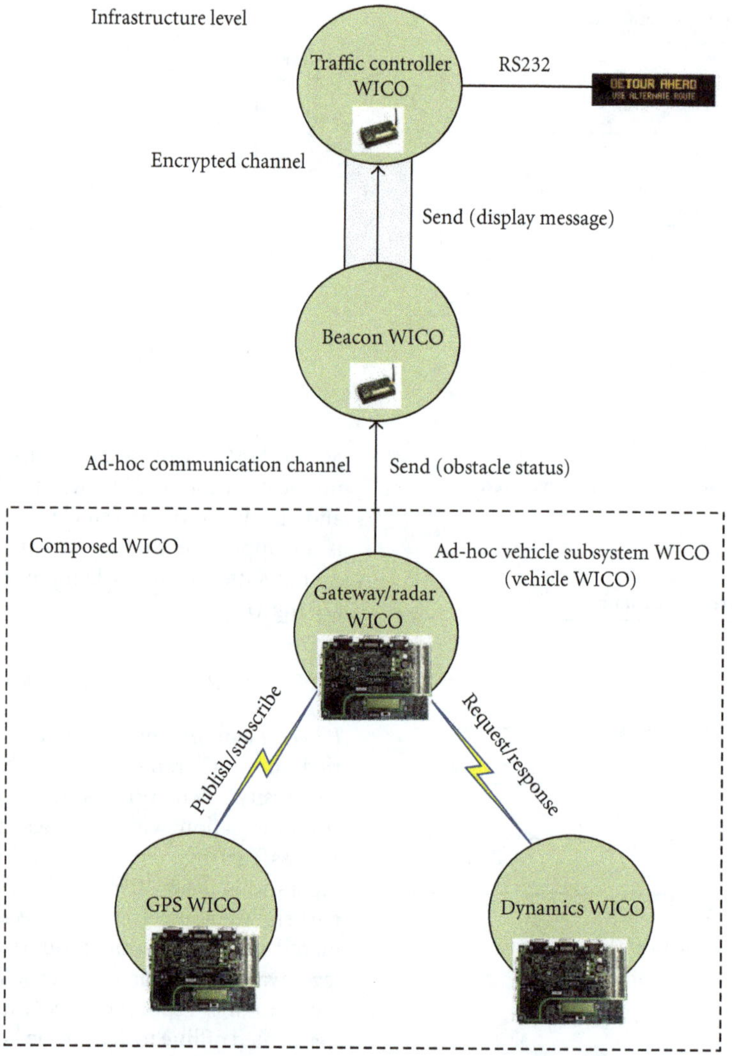

FIGURE 21: Deployment of WICOs in the inter-hierarchical demonstrator.

vehicle speed. At the 70 mph speed, The WICO in the roadside received 5 and 11 packets for sending intervals 1000 ms, 500 ms, respectively. And interestingly, there were no packets lost between the first packet and the last packet received. This experiment demonstrated that MicaZ WICO can be used with EMMA middleware communication methods between a fixed infrastructure WICO and also fast moving vehicle-based WICO applications. This is an important finding which proves that the MicaZ WICOs do not suffer from any Doppler effects at normal motorway (70 mph) speeds.

5.4. Inter-Hierarchical Level Application. One of the main objectives of the project was to achieve a middleware able to abstract complex subsystems formed by different kinds of WICOs into simpler elements (composed WICOs) that behave in the upper level system as a single unit. This way, complex applications could be built with a hierarchical shape, each group of WICOs working together on the same functionality appearing a unique element providing certain types of data to the remaining. In addition, the possibility to form ad-hoc WICOs (i.e., to discover previously unknown elements on the system), and to propagate published data through the different abstraction layers, allowing their transformation and combination as it crosses certain points of the hierarchies, does really enhance the achievable possibilities of applications built on EMMA Middleware.

An inter-hierarchical application, integrating different hierarchical levels developed in the project: the car level and the supra-vehicle level, demonstrated how heterogeneity issues could be solved by developing middleware such as EMMA. In order to demonstrate the inter-hierarchical collaboration of the WICOs developed on the project, the application consisted of transforming the information provided by the vehicles at both automotive and vehicle subsystem levels into specific traffic control actions at infrastructure (i.e., supra-vehicle) level.

The inter-hierarchical demonstration made use of the WICOs at the vehicle level and supra-vehicle level. This demonstration aimed to provide advanced warning to a vehicle behind that there is an obstacle ahead. As can be seen in Figure 21 the inter-hierarchical demonstrator made use of all communications mechanisms in the EM2P middleware and exercised most of the functionality of the middleware. In this demonstrator, inter-hierarchical collaboration of the WICOs developed on the project consisted of transforming the information provided by the vehicles at vehicle level (GPS, Vehicle Dynamics and Radar sensors based on Xilinx ML 403 platform with TRW Conekt daughter board) into specific traffic control actions at infrastructure (to MicaZ) level. In the infrastructure level, two MicaZ WICOs were used. First MicaZ WICO was used as beacon WICO to relay any message received by ad-hoc vehicle subsystem WICO to the second MicaZ WICO which was connected to a portable VMS panel to display the information sent by the vehicle level WICO. This application was successfully demonstrated for the EMMA project final review in a real road environment in London.

6. Conclusions

It is clear that the next generation of vehicles will be required to have increased safety, lower emissions, and more entertainment with higher performance than those of today. The innovations in wireless sensor devices will enable novel automotive applications which will become very common in future transportation applications. The challenges such as integrating heterogeneous wireless devices for specific transportation application can be met by developing middleware technologies such as in EMMA. This paper has presented the EMMA project that has been undertaken to investigate the suitability of using heterogeneous wireless sensors in transportation system applications. The validation of the prototype applications shows that wireless sensor networking technologies can be used at the engine level, vehicle level, and supra-vehicle level. The ability to communicate between vehicle and roadside illustrates that wireless sensor networks will enable efficient and discrete communications between vehicle and roadside.

Acknowledgments

Authors would like to thank Alberto Zambrano Galbis (ETRA I+D) for his contribution to the supra-vehicle level and inter-hierarchical level application. The work presented here was sponsored by EC under FP6 programme (IST-2005-5-034097) and authors also would like to thank all the project partners for their contributions.

References

[1] The Eddington Transport Study, UK Department of Transport, May 2011, http://webarchive.nationalarchives.gov.uk/+/http:/www.dft.gov.uk/about/strategy/transportstrategy/eddingtonstudy/.

[2] A. Tully, "Pervasive tagging, sensors and data collection: a science and technology review for the foresight project on intelligent infrastructure systems," *IEE Proceedings Intelligent Transport Systems*, vol. 153, no. 2, pp. 129–146, 2006.

[3] F. Li and Y. Wang, "Routing in vehicular ad hoc networks: a survey," *IEEE Vehicular Technology Magazine*, vol. 2, no. 2, pp. 12–22, 2007.

[4] N. Navet, Y. Song, F. Simonot-Lion, and C. Wilwert, "Trends in automotive communication systems," *Proceedings of the IEEE*, vol. 93, no. 6, pp. 1204–1223, 2005.

[5] X. Li, W. Shu, M. Li, H. Y. Huang, P. E. Luo, and M. Y. Wu, "Performance evaluation of vehicle-based mobile sensor networks for traffic monitoring," *IEEE Transactions on Vehicular Technology*, vol. 58, no. 4, pp. 1647–1653, 2009.

[6] H. Cai and Y. Lin, "A roadside ITS data bus prototype for intelligent highways," *IEEE Transactions on Intelligent Transportation Systems*, vol. 9, no. 2, pp. 344–348, 2008.

[7] K. Selvarajah, A. Tully, and P. T. Blythe, "Integrating Smartdust into the Embedded Middleware in Mobility Applications (EMMA) Project," in *4th IET International Conference on Intelligent Environments, IE08*, Seattle USA, 21-22 July 2008.

[8] T. Kosch, I. Kulp, M. Bechler, M. Strassberger, B. Weyl, and R. Lasowski, "Communication architecture for cooperative systems in Europe," *IEEE Communications Magazine*, vol. 47, no. 5, pp. 116–125, 2009.

[9] CAR 2 CAR Communication Consortium, July 2010, http://www.car-to-car.org/.

[10] Zigbee Alliance, July 2010, http://zigbee.org/.

[11] G. Cena, A. Valenzano, and S. Vitturi, "Advances in automotive digital communications," *Computer Standards and Interfaces*, vol. 27, no. 6, pp. 665–678, 2005.

[12] K. Geihs, "Middleware challenges ahead," *Computer*, vol. 34, no. 6, pp. 24–31, 2001.

[13] S. Hadim and N. Mohamed, "Middleware: middleware challenges and approaches for wireless sensor networks," *IEEE Distributed Systems Online*, vol. 7, no. 3, pp. 1–23, 2006.

[14] TinyOS, July 2010, http://www.tinyos.net.

[15] NanoQplus and Qplus, July 2010, http://www.qplus.or.kr.

[16] Crossbow Technology: "MPR/MIB User's Manual", May 2011, http://www.memsic.com/products/wireless-sensor-networks/wireless-modules.html.

[17] Xilinx, December 2010, http://www.xilinx.com/.

[18] M. Peller and N. Berwanger, "Road vehicle Controller area Network (CAN)—part 4: time-triggered communication," ISO IS 11898.

Permissions

All chapters in this book were first published in JAT, by Hindawi Publishing Corporation; hereby published with permission under the Creative Commons Attribution License or equivalent. Every chapter published in this book has been scrutinized by our experts. Their significance has been extensively debated. The topics covered herein carry significant findings which will fuel the growth of the discipline. They may even be implemented as practical applications or may be referred to as a beginning point for another development.

The contributors of this book come from diverse backgrounds, making this book a truly international effort. This book will bring forth new frontiers with its revolutionizing research information and detailed analysis of the nascent developments around the world.

We would like to thank all the contributing authors for lending their expertise to make the book truly unique. They have played a crucial role in the development of this book. Without their invaluable contributions this book wouldn't have been possible. They have made vital efforts to compile up to date information on the varied aspects of this subject to make this book a valuable addition to the collection of many professionals and students.

This book was conceptualized with the vision of imparting up-to-date information and advanced data in this field. To ensure the same, a matchless editorial board was set up. Every individual on the board went through rigorous rounds of assessment to prove their worth. After which they invested a large part of their time researching and compiling the most relevant data for our readers.

The editorial board has been involved in producing this book since its inception. They have spent rigorous hours researching and exploring the diverse topics which have resulted in the successful publishing of this book. They have passed on their knowledge of decades through this book. To expedite this challenging task, the publisher supported the team at every step. A small team of assistant editors was also appointed to further simplify the editing procedure and attain best results for the readers.

Apart from the editorial board, the designing team has also invested a significant amount of their time in understanding the subject and creating the most relevant covers. They scrutinized every image to scout for the most suitable representation of the subject and create an appropriate cover for the book.

The publishing team has been an ardent support to the editorial, designing and production team. Their endless efforts to recruit the best for this project, has resulted in the accomplishment of this book. They are a veteran in the field of academics and their pool of knowledge is as vast as their experience in printing. Their expertise and guidance has proved useful at every step. Their uncompromising quality standards have made this book an exceptional effort. Their encouragement from time to time has been an inspiration for everyone.

The publisher and the editorial board hope that this book will prove to be a valuable piece of knowledge for researchers, students, practitioners and scholars across the globe.

List of Contributors

Zongxin Tang and Jianchuan Cheng
School of Transportation, Southeast University, 2 Sipailou, Nanjing, Jiangsu 210096, China

Sikai Chen, Seyed Ali Ghahari and Samuel Labi
Lyles School of Civil Engineering, Purdue University, 550 Stadium Mall Dr.,West Lafayette, IN 47907, USA

Pablo Marin-Plaza, Ahmed Hussein, David Martin and Arturo de la Escalera
Universidad Carlos III de Madrid, Leganes, Spain

Lixian Fan and Sijie Zhang
School of Management, Shanghai University, 599 Shangda Rd., Shanghai 200444, China

Jingbo Yin
State Key Laboratory of Ocean Engineering, School of NavalArchitecture, Ocean and Civil Engineering, Shanghai Jiao Tong University, 800 Dongchuan Rd., Shanghai 200240, China

Jang-Ho Cho, Dong-Sung Pae and Myo-Taeg Lim
School of Electrical Engineering, Korea University, Seoul, Republic of Korea

Tae-Koo Kang
Department of Information and Telecommunication Engineering, Sangmyung University, Cheonan, Republic of Korea

Velaphi Msomi and Shandene E. A. Solomons
Mechanical Engineering Department, Engineering Faculty, Cape Peninsula University of Technology, Bellville 7535, South Africa

Przemysław Sekuła, Zachary Vander Laan, Kaveh Farokhi Sadabadi and Mirosław J. Skibniewski
Department of Civil and Environmental Engineering, University of Maryland, College Park, MD, USA

Przemysław Sekuła
Faculty of Informatics and Communication, University of Economics in Katowice, Katowice, Poland

Mirosław J. Skibniewski
Institute for Theoretical and Applied Informatics, Polish Academy of Sciences, Gliwice, Poland
Chaoyang University of Technology, Taichung, Taiwan

G. Birgillito, C. Rindone and A. Vitetta
Dipartimento di ingegneria dell'Informazione, delle Infrastrutture e dell'Energia Sostenibile, Università Mediterranea di Reggio Calabria, Reggio Calabria, Italy

Chao Qian and Shuguang Li
School of Electronic & Control Engineering, Chang'an University, Xi'an 710064, China

Jianxun Chen and Yanbin Luo
School of Highway, Chang'an University, Xi'an 710064, China

Karam Eliker, Guoqing Zhang and Weidong Zhang
Department of Automation, Shanghai Jiao Tong University, Shanghai 200240, China

Said Grouni
Faculty of Sciences Engineering, University of Boumerdes, 35000 Boumerdes, Algeria

Weidong Zhang
School of Computer Engineering and Science, Shanghai University, Shanghai 200444, China

Mina Ghanbarikarekani
School of Civil and Environmental Engineering, University of Technology Sydney, Sydney, NSW2007, Australia

Xiaobo Qu
Department of Architecture and Civil Engineering, Chalmers University of Technology, Gothenburg, Sweden

Michelle Zeibots
Institute for Sustainable Futures, University of Technology Sydney, Sydney, NSW2007, Australia

Weiwei Qi
School of Civil Engineering and Transportation, South China University of Technology, Guangzhou 510641, China

Tan Zhao, Jincai Huang, Jianmai Shi and Chao Chen
Science and Technology on Information Systems Engineering Laboratory, National University of Defense Technology, Changsha, Hunan 410073, China

Jianmai Shi
School of Traffic and Transportation Engineering, Central South University, Changsha 410075, China

D. C. Duives, W. Daamen and S. P. Hoogendoorn
Civil Engineering and Geosciences, Department of Transport & Planning, Delft University of Technology, Stevinweg 1, 2628 CN, Delft, Netherlands

Yi Hong and Hua Xu
Information Engineering College, Beijing Institute of Petrochemical Technology, Beijing 102617, China

Deying Li
School of Information, Renmin University of China, Beijing 100872, China

Qiang Wu
National Engineering Research Center of Coal Mine Water Hazard Controlling, China University of Mining and Technology, Beijing, Beijing 100083, China

Siliang Luan, Qingfang Yang, Wei Wang, Zhongtai Jiang, Ruru Xing and Ruijuan Chu
School of Transportation, Jilin University, Changchun 130022, Jilin, China
Jilin Research Center for Intelligent Transportation System, Jilin University, Changchun 130022, Jilin, China
Jilin Province Key Laboratory of Road Traffic, Jilin University, Changchun 130022, Jilin, China

Xiaoli Ma, Yingying Xing and Jian Lu
The Key Laboratory of Road and Traffic Engineering, Ministry of Education, Tongji University, 4800 Cao'an Road, Shanghai 201804, China

António Lobo, Marco Amorim, Carlos Rodrigues and António Couto
Research Centre for Territory, Transports and Environment, Faculty of Engineering, University of Porto, Rua Dr. Roberto Frias, s/n, 4200-465 Porto, Portugal

Ben Beklisi Kwame Ayawli, Albert Yaw Appiah and Frimpong Kyeremeh
College of Electrical Engineering and Control Science, Nanjing Tech. University, China

Ben Beklisi Kwame Ayawli
Computer Science Department, Sunyani Technical University, Sunyani, Ghana

Ryad Chellali
Nanjing Forestry University, Nanjing, China
Kita Technologies, Nanjing, China

Albert Yaw Appiah
Electrical and Electronic Engineering Department, Sunyani Technical University, Sunyani, Ghana

Linhong Wang and Yiming Bie
School of Transportation, Jilin University, Changchun 130022, China

Yunhao Wang
School of Economics, Northeast Normal University, Changchun 130117, China

Index